geoENV IV – GEOSTATISTICS FOR ENVIRONMENTAL APPLICATIONS

Quantitative Geology and Geostatistics

VOLUME 13

The titles published in this series are listed at the end of this volume.

geoENV IV – GEOSTATISTICS FOR ENVIRONMENTAL APPLICATIONS

Proceedings of the Fourth European Conference on Geostatistics for Environmental Applications held in Barcelona, Spain, November 27–29, 2002

Edited by

XAVIER SANCHEZ-VILA

Universitat Politècnica de Catalunya,
Barcelona, Spain

JESUS CARRERA

Universitat Politècnica de Catalunya,
Barcelona, Spain

and

JOSÉ JAIME GÓMEZ-HERNÁNDEZ

Universidad Politecnica de Valencia,
Valencia, Spain

KLUWER ACADEMIC PUBLISHERS
DORDRECHT / BOSTON / LONDON

A C.I.P. Catalogue record for this book is available from the Library of Congress.

ISBN 1-4020-2114-3 (PB)
ISBN 1-4020-2007-4 (HB)
ISBN 1-4020-2115-1 (e-book)

Published by Kluwer Academic Publishers,
P.O. Box 17, 3300 AA Dordrecht, The Netherlands.

Sold and distributed in North, Central and South America
by Kluwer Academic Publishers,
101 Philip Drive, Norwell, MA 02061, U.S.A.

In all other countries, sold and distributed
by Kluwer Academic Publishers,
P.O. Box 322, 3300 AH Dordrecht, The Netherlands.

Printed on acid-free paper

Contents

Foreword

The fourth edition of the European Conference on Geostatistics for Environmental Applications (geoENV IV) took place in Barcelona, November 27-29, 2002. As a proof that there is an increasing interest in environmental issues in the geostatistical community, the conference attracted over 100 participants, mostly Europeans (up to 10 European countries were represented), but also from other countries in the world. Only 46 contributions, selected out of around 100 submitted papers, were invited to be presented orally during the conference. Additionally 30 authors were invited to present their work in poster format during a special session.

All oral and poster contributors were invited to submit their work to be considered for publication in this Kluwer series. All papers underwent a reviewing process, which consisted on two reviewers for oral presentations and one reviewer for posters. The book opens with one keynote paper by Philippe Naveau. It is followed by 40 papers that correspond to those presented orally during the conference and accepted by the reviewers. These papers are classified according to their main topic. The list of topics show the diversity of the contributions and the fields of application. At the end of the book, summaries of up to 19 poster presentations are added.

The geoENV conferences stress two issues, namely geostatistics and environmental applications. Thus, papers can be classified into two groups. The reader will find a number of papers dedicated to the most recent methodological developments, with examples predominantly in environmental sciences. The remaining ones provide a good indication of the wide variety of environmental applications in which geostatistics plays its role.

The fourth volume in the geoENV conference series proves how dynamic the geostatistical community is, and confirms the relevance of geostatistics as a tool to be included as a standard procedure in environmental sciences. We now look forward to geoENV 2004 for new applications and new methodological advances.

Barcelona, November 2002

The editors
Jesus Carrera
J. Jaime Gómez-Hernández
Xavier Sánchez-Vila

Organizing Committee

SÁNCHEZ–VILA, Xavier, *Universitat Politècnica de Catalunya, Chairman*
CARRERA, Jesús, *Universitat Politècnica de Catalunya, Secretary*
ALLARD, Denis, *Unité de Biométrie, INRA*
FROIDEVAUX, Roland, *FSS International*
GÓMEZ–HERNÁNDEZ, J. Jaime, *Universidad Politécnica de Valencia*
MONESTIEZ, Pascal, *Unité de Biométrie, INRA*
SOARES, Amilcar, *Instituto Superior Tecnico, CMRP*

The editors are grateful to the following persons for their work as referees:

ALCOLEA, Andrés, *Universitat Politècnica de Catalunya*
ALLARD, Denis, *Unité de Biométrie, INRA*
ALMEIDA, José Antonio, *Instituto Superior Tecnico*
ATKINSON, Peter M., *University of Southampton*
AXNESS, Carl, *Universidad Politécnica de Valencia*
BALBUENA, Camino, *Universitat Politècnica de Catalunya*
BELLIN, Alberto, *Università di Trento*
BIO, Ana, *Instituto Superior Tecnico, CMRP*
BOGAERT, Patrick, *Université Catholique de Louvain*
CAPILLA, José, *Universidad Politécnica de Valencia*
CARRERA, Jesús, *Universitat Politècnica de Catalunya*
CASSIRAGA, Eduardo, *Universidad Politécnica de Valencia*
CHICA–OLMO, Mario, *Universidad de Granada*
CHÌLES, Jean–Paul, *BRGM*
CHRISTAKOS, G., *University of North Carolina at Chapel Hill*
CHRISTENSEN, Ole Fredslund, *Lancaster University*
CRESSIE, Noel, *Ohio State University*
D'OR, Dimitri, *Université Catholique deLouvain*
DE FOUQUET, *Chantal, Ecole des Mines de Paris*
DELGADO, Jorge, *Universidad de Jaén*
EGOZCUE, Juan José, *Universitat Politècnica de Catalunya*
FROIDEVAUX, Roland, *FSS International*
GILI, Josep, *Universitat Politècnica de Catalunya*
GÓMEZ-HERNÀNDEZ, J. Jaime, *Universidad Politécnica de Valencia*
GOOVAERTS, Pierre, *University of Michigan*
GUADAGNINI, Alberto, *Politecnico de Milano*
GUADAGNINI, Laura, *Politecnico de Milano*
HENDRICKS–FRANSSEN, Harri–Jan, *ETH Zurich*
HERVADA, Carme, *Universitat Politècnica de Catalunya*

JARAUTA–BRAGULAT, Eusebi, *Universitat Politècnica de Catalunya*
JIMÉNEZ–ESPINOSA, Rosario, *Universidad de Jaén*
KNUDBY, Christen, *Universitat Politècnica de Catalunya*
LOPHAVEN, Søren, *Danmarks Tekniske Universitet*
MCSORLEY, Claire, *Joint Nature Conservation Committe*
MILITINO, Ana F., *Universidad Pública de Navarra*
MONESTIEZ, Pascal, Unité de Biométrie, *INRA*
NUNES, Carla, *Universidade de Évora*
OLIVELLA, Sebastià, *Universitat Politècnica de Catalunya*
PAZ, Antonio, *Universidade da Coruña*
PEREIRA, Maria João, *Instituto Superior Técnico*
PTAK, Thomas, *University of Tübingen*
RIVA, Mónica, *Politecnico Milano*
SAHUQUILLO, Andrés, *Universidad Politécnica de Valencia*
SÁNCHEZ–VILA, Xavier, *Universitat Politècnica de Catalunya*
SERRE, Marc, *University of North Carolina–Chapel Hill*
SOARES, Amilcar, Instituto Superior Tecnico, *CMRP*
STEIN, Alfred, *Wageningen University*
WACKERNAGEL, Hans, *Ecole des Mines de Paris*

Sponsors of the Conference

ESCOLA TÉCNICA SUPERIOR D'ENGINYERS DE CAMINS CANALS I PORTS DE BARCELONA

HYDROGEOLOGY GROUP–UPC–CSIC

CENTRE INTERNACIONAL DE MÈTODES NUMÈRICS EN ENGINYERIA

UNIVERSITAT POLITÈCNICA DE CATALUNYA

AGENCIA DE GESTIÓ D'AJUTS UNIVERSITARIS I DE RECERCA

CENTRO DE INVESTIGACIÓN CIENTÍFICA Y TECNOLÓGICA

TWO STATISTICAL METHODS FOR IMPROVING THE ANALYSIS OF LARGE CLIMATIC DATA SETS: GENERAL SKEWED KALMAN FILTERS AND DISTRIBUTIONS OF DISTRIBUTIONS

P. Naveau[1], M. Vrac[2,3], M.G. Genton[4], A. Chédin[2] and E. Diday[3]

[1]*Dept. of Applied Mathematics, University of Colorado, Boulder, USA.*
[2]*Institut Pierre Simon Laplace, Ecole Polytechnique, France.*
[3] *Université Paris IX Dauphine, France.*
[4]*Dept. of Statistics, North Carolina State University, USA.*

Abstract: This research focuses on two original statistical methods for analyzing large data sets in the context of climate studies. First, we propose a new way to introduce skewness to state-space models without losing the computational advantages of the Kalman filter operations. The motivation stems from the popularity of state-space models and statistical data assimilation techniques in geophysics, specially for forecasting purposes in real time. The added skewness comes from the extension of the multivariate normal distribution to the *general multivariate skew-normal distribution*. A new specific state-space model for which the Kalman Filtering operations are carefully described is derived. The second part of this work is dedicated to the extension of clustering methods into the *distributions of distributions*} framework. This concept allows us to cluster distributions, instead of simple observations. To illustrate the applicability of such a method, we analyze the distributions of 16200 temperature and humidity vertical profiles. Different levels of dependencies between these distributions are modeled by copulas. The distributions of distributions are decomposed as mixtures and the algorithm to estimate the parameters of such mixtures is presented. Besides providing realistic climatic classes, this clustering method allows atmospheric scientists to explore large climate data sets into a more meaningful and global framework.

X. Sanchez-Vila et al. (eds.), geoENV IV – Geostatistics for Environmental Applications, 1-14.

1. INTRODUCTION

In geophysical studies, the dimension of data sets from most oceanic, atmospheric numerical models and satellites is extremely large. There exists a variety of recent techniques to deal with such an issue in the special context of climate studies. For example, Bayesian methods (e.g. Wikle et al., 2002), data mining, imaging and statistical visualization procedures have provided interesting and innovative ways to analyze large climatic data sets. In addition to the computational problem, the distribution of climatic random vectors is often supposed to be Gaussian or a mixture of Gaussian distributions, although this assumption is not always satisfied for a wide range of atmospheric variables. For example, the distribution of daily precipitation amounts is by nature skewed. In this paper, we attend to address these two problems, large size and skewness, with two different approaches. Because the scope of these problems is very large, we will focus our attention on two specific statistical methods used in climate studies. In Section 2, we will present a simple way to incorporate skewness in Kalman filtering techniques without losing the computational advantages associated with the normal distribution (Naveau and Genton, 2002). In Section 3, the concept of distributions of distributions (Diday et al., 1985; Vrac 2002; Vrac et al., 2001) will be used in order to improve classical clustering methods for large climatic data sets. This application is closely linked to the algorithm of inversion of the equation of radiative transfer (Chédin et al., 1985).

2. GENERAL SKEWED KALMAN FILTERS

Before presenting the details of our research on Kalman filters, we want to clarify some climatic terms to the statistician who may not be familiar with atmospheric sciences. In particular, we would like to recall the meaning of numerical models and data assimilation in the context of this work. For the former, a numerical computer model solves the governing physical, thermodynamics and micro-physical processes at different scales of interest and over a specific region (depending on the scientific problem under study). It provides deterministic outputs of different atmospheric variables (temperature, humidity, winds, etc) according to certain forcings (inputs). It is worthwhile to note that the evaluation of such computer simulations has generated an interdisciplinary effort between scientists and statisticians in recent years. The interested reader can look at Berk and collaborators' work (Berk et al., 2002) on the statistical assessment of such models. Data assimilation can be seen as a way of incorporating observations into a numerical model as it runs. From a statistical point of view, the objective of data assimilation is to use both sources of data, observations and model outputs, to provide a better statistical analysis, in particular to give better forecasts. In the context of numerical weather prediction, updates and forecasts have be performed routinely and in real time. This compounds with the large size of data sets and implies that very efficient but slow methods have to be disregarded. The data-assimilation or update step is closely related to Kalman filter which is the best known filtering algorithm in the context of Gaussian distributions and linear system dynamics. Before presenting the details of our method, we would like to underline there exists a very large body of literature dedicated to Kalman filters and its extensions. For example, ensemble Kalman filter (Bengstton et al, 2002; Anderson 2001), space-time Kalman filters (Wikle and Cressie, 1999), partially Non-Gaussian state-space models (Shepard, 1994) and particle filters (Doucet et al., 2001) have been recently used for many different applications. In particular, the approximation of non-Gaussian distributions by a mixture of Gaussian distributions has already been implemented via Monte-Carlo methods. In the same way, a mixture of general skew-normal distributions could be defined and extends the range of applications of our method. But because of the limited space available, we will restrict our exposition to the simplest form of Kalman filter, the linear one, in this paper. Future work will present the extension to more complex Kalman filter models.

The overwhelming assumption of normality in the Kalman filter literature can be understood for many reasons. A major one is that the multivariate distribution is completely characterized by its first two moments. In addition, the stability of multivariate normal distribution under summation and conditioning offers tractability and simplicity. Therefore, the Kalman filter operations can be performed rapidly and efficiently whenever the normality assumption holds. However, this assumption is not satisfied for a large number of applications. For example, some distributions used in a state-space model can be skewed. In this work, we propose a novel extension of the Kalman filter by working with a larger class of distributions than the normal distribution. This class is called *general multivariate skew-normal distributions*. Besides introducing skewness to the normal distribution, it has the advantages of being closed under marginalization and conditioning. This class has been introduced by Domínguez-Molina et al. (2001) and is an extension of the multivariate skew-normal distribution first proposed by Azzalini and his coworkers (1996, 1999). These distributions are particular types of generalized skew-elliptical distributions recently introduced by Genton and Loperfido (2002), i.e. they are defined as the product of a multivariate elliptical density with a skewing function.

2.1 The general multivariate skew-normal distribution

The general multivariate skew-normal distribution is a family of distributions including the normal one, but with extra parameters to regulate skewness. It allows for a continuous variation from normality to non-normality, which is useful in many situations (Azzalini and Capitanio, 1999) who emphasized statistical applications for the skew-normal distribution. An n-dimensional random vector X is said to have a general multivariate skew-normal distribution (Domínguez-Molina et al., (2001)), denoted by $GMSN_{n,m}(\mu,\Sigma,D,\nu,\Delta)$ if it has a density function of the form:

$$\frac{1}{\Phi_m\left(D\mu;\nu,\Delta+D\Sigma D^T\right)}\phi_n\left(x;\mu,\Sigma\right)\Phi_m\left(Dx;\nu,\Delta\right), \qquad x\in\mathbb{R}^n, \tag{1}$$

where $\mu \in \mathbb{R}^n$, $\nu \in \mathbb{R}^m$, $\Sigma \in \mathbb{R}^{n\times n}$ and $\Delta \in \mathbb{R}^{m\times m}$ are both covariance matrices, $D \in \mathbb{R}^{m\times n}$, $\phi_n(x;\mu,\Sigma)$ and $\Phi_n(x;\mu,\Sigma)$ are the n-dimensional normal pdf and cdf with mean μ and covariance matrix Σ. When $D = 0$, the density (1) reduces to the multivariate normal one, whereas it reduces to Azzalini and Capitanio's (1999) density when $m = 1$ and $\nu = D\mu$. The matrix parameter D is referred to as a "shape parameter". The moment generating function $M(t)$ for a GMSN distribution is given by:

$$M(t)=\frac{\Phi_m\left(D\left(\mu+\Sigma t\right);\nu,\Delta+D\Sigma D^T\right)}{\Phi_m\left(D\mu;\nu,\Delta+D\Sigma D^T\right)}\exp\left\{\mu^T t+\frac{1}{2}\left(t^T\Sigma t\right)\right\}, t\in\mathbb{R}^n. \tag{2}$$

The simulation of random vectors from the GMSN distribution is rather simple. Indeed, Domí}nguez-Molina et al. (2001) showed that if $X \in \mathbb{R}^n$ and $Y \in \mathbb{R}^m$ are two random vectors with joint distribution given by:

$$\begin{pmatrix} X \\ Y \end{pmatrix} \sim N_{n+m}\left(\begin{pmatrix} \mu \\ \nu \end{pmatrix},\begin{pmatrix} \Sigma & -\Sigma D^T \\ -D\Sigma & \Delta+D\Sigma D^T \end{pmatrix}\right), \tag{3}$$

then the conditional distribution of X given $Y \leq D\mu$ is a general multivariate skew-normal distribution $GMSN_{n,m}(\mu,\Sigma,D,\nu,\Delta)$.

The three basic tools when implementing the Kalman filter are the closure under linear transformation, under summation and conditioning. In section 2.3, we will present how the general skew-normal distribution behaves under such constraints.

2.2 The state-space model and the Kalman filter

The *State Space Model* has been widely studied (e.g. Cressie and Wilke, 2002; Shepard, 1994; Shumway and Stoffer, 1991; Harrison and Stevens, 1976). This model has become a powerful tool for modeling and forecasting dynamical systems and it has been used in a wide range of disciplines such as biology, economics, engineerings and geophysics (Naveau et al. 2002; Guo et al., 1999; Kitagawa and Gersch, 1984). The basic idea of the state-space model is that the d-dimensional vector of observation Y_t at time t is generated by two equations, the *observational* and the *system* equations. The first equation describes how the observations vary in function of the unobserved state vector X_t of length h: $Y_t = F_t X_t + \varepsilon_t$ where ε_t represent an added noise and F_t is a $d \times h$ matrix of scalars. The essential difference between the state-space model and the conventional linear model is that the state vector X_t is not assumed to be constant but may change in time. The temporal dynamical structure is incorporated via the system equation: $X_t = G_t X_{t-1} + \eta_t$ where η_t represents an added noise and G_t is an $h \times h$ matrix of scalars. There exists a long literature about the estimation of the parameters for such models. In particular, the Kalman filter provides an optimal way to estimate the model parameters if the assumption of gaussianity holds. Following the definition by Meinhold and Singpurwalla (1983). The term "Kalman filter" used in this work refers to a recursive procedure for inference about the state vector. To simplify the exposition, we assume that the observation errors ε_t are independent of the state errors η_t and that the sampling is equally spaced, $t = 1,..., n$. The results shown in this paper could be easily extended without such constraints. But, the loss of clarity in the notations would make this work more difficult to read without bringing any new important concepts.

2.3 Kalman filtering and general skew-normal distributions

From Equation (2), it is straightforward to see that the sum of two independent general multivariate skew-normal distributions is not necessary a general multivariate skew-normal distribution. In order to obtain the closure under summation needed for the Kalman Filtering, we extend the linear state-space model to a wider state-space model for which the stability under summation is better preserved. In order to pursue this goal, we need the following lemma. Its proof can be found in Domínguez-Molina et al. (2001).

Lemma 1 Suppose $Y = GMSN_{n,m}(\mu,\Sigma,D,\nu,\Delta)$ and A is a $r \times n$ matrix. Then, we have $X = AY \sim GMSN_{r,m}(A\mu,A\Sigma A^T,DA^{\leftarrow},\nu,\Delta)$ where $A^{\leftarrow}$ is the left inverse of A and $A^{\leftarrow}=A^{-1}$ when A is an $n \times n$ nonsingular matrix. If Y is partitioned into two components, Y_1 and Y_2, of dimensions h and $n\text{-}h$ respectively and with a corresponding partition for μ, Σ, D and ν. Then the conditional distribution of Y_2 given $Y_1 = y_1$ is:

$$GMSN_{n-h,m}\left(\mu_2 + \Sigma_{21}\Sigma_{11}^{-1}\left(y_1 - \mu_1\right), \Sigma_{22} - \Sigma_{21}\Sigma_{11}^{-1}\Sigma_{12}, D_2, \nu - D_1 y_1, \Delta\right). \quad (4)$$

The converse is also true, i.e. if (4) is the conditional distribution of Y_2 given $Y_1 = y_1$ and $Y_1 \sim GMSN_{h,m}(\mu_1,\Sigma_{11},D_1,\nu_1,\Delta)$, then the joint distribution of Y_1 and Y_2 is $GMSN_{n,m}(\mu,\Sigma,D,\nu,\Delta)$.

The proof is the same as for the multivariate Gaussian distribution.

2.4 Extension of the linear state-space model

Our strategy to derive a model with a more flexible skewness is to directly incorporate a skewness term, say S_t, into the observation equation

$$\begin{aligned} Y_t &= F_t X_t + \varepsilon_t \\ &= P_t U_t + Q_t S_t + \varepsilon_t, \quad \text{with} \quad F_t = \left(P_t, Q_t\right) \quad \text{and} \quad X_t = \left(U_t^T, S_t^T\right)^T \end{aligned} \quad (5)$$

where the random vector U_t of length k and the $d \times k$ matrix of scalar P_t represent the linear part of the observation equation. In comparison, the random vector S_t of length l and the $d \times l$ matrix of scalar Q_t correspond to the additional skewness. The most difficult task in this construction is to propose a simple dynamical structure of the skewness vector S_t and the "linear" vector U_t while keeping the independence between these two vectors (the last condition is not theoretically necessary but it is useful when interpreting the parameters). To reach this goal, we suppose that the bi-variate random vector $(U^T_t, V^T_t)^T$ is generated from a linear system:

$$\begin{cases} U_t &= K_t U_{t-1} + \eta_t^* \\ V_t &= -L_t V_{t-1} + \eta_t^+ \end{cases} \tag{6}$$

where the Gaussian noise $\eta^*_t \sim N(\mu^*_\eta, \Sigma^*_\eta)$ is independent of $\eta^+_t \sim N(\mu^+_\eta, \Sigma^+_\eta)$ and where K_t, respectively L_t represents a $k \times k$ matrix of scalars, respectively a $l \times l$ matrix of scalars. The multivariate normal distribution of the vector $(U^T_t, V^T_t)^T$ is denoted by

$$\begin{pmatrix} U_t \\ V_t \end{pmatrix} \sim N_{k+1}\left(\begin{pmatrix} \psi_t^* \\ \psi_t^+ \end{pmatrix}, \begin{pmatrix} \Omega_t^* & 0 \\ 0 & \Omega_t^+ \end{pmatrix}\right). \tag{7}$$

The parameters of such vectors can be sequentially derived from any initial vector $(U^T_0, V^T_0)^T$ with a normal distribution. From (3), we define the skewness part S_t of the state vector $X_t = (U^T_t, S^T_t)^T$ as the following conditional variable $S_t = [V_{t-1} \mid V_t \le L_t \psi^+_{t-1}]$. It follows a general multivariate skew-normal distribution $S_t \sim GMSN_{l,l}(\psi^+_{t-1}, \Omega^+_{t-1}, L_t, \psi^+_t, \Sigma^+_v)$. Consequently the state vector has also a general multivariate skew-normal distribution

$$X_t = \begin{pmatrix} U_t \\ S_t \end{pmatrix} \sim GMSN_{k+1,k+l}\left(\psi_t, \Omega_t, D_t, v_t, \Delta_t\right), \quad with \quad \psi_t = \begin{pmatrix} \psi_t^* \\ \psi_{t-1}^+ \end{pmatrix}, \tag{8}$$

$$\Omega_t = \begin{pmatrix} \Omega_t^* & 0 \\ 0 & \Omega_{t-1}^+ \end{pmatrix}, \quad D_t = \begin{pmatrix} 0 & 0 \\ 0 & L_t \end{pmatrix}, \quad v_t = \begin{pmatrix} 0 \\ \psi_t^+ \end{pmatrix}, \quad and \quad \Delta_t = \begin{pmatrix} I & 0 \\ 0 & \Sigma_v^+ \end{pmatrix}.$$

The price for this gain in skewness flexibility is that this state vector does not have anymore a linear structure like the one defined by the system equation. If $P_t = 0$ or $L_t = 0$ then the classical state-space model is obtained.

Proposition 1 Suppose that the initial vector $(U^T_0, V^T_0)^T$ of the linear system defined by (6) follows the normal distribution defined by

$$\begin{pmatrix} U_0 \\ V_0 \end{pmatrix} \sim N_{k+1}\left(\begin{pmatrix} \psi_0^* \\ \psi_0^+ \end{pmatrix}, \begin{pmatrix} \Omega_0^* & 0 \\ 0 & \Omega_0^+ \end{pmatrix}\right). \tag{9}$$

Then both the state vector $X_t = (U^T_t, S^T_t)^T$ and the observation vector Y_t follow general multivariate skew-normal distributions, $X_t \sim GMSN_{h,m}(\psi_t, \Omega_t, D_t, \psi_t, \Delta_t)$ and $Y_t \sim GMSN_{d,m}(\mu_t, \Gamma_t, E_t, v_t, \Delta_t)$ for $t \ge 1$. The parameters of these distributions satisfy

$$\psi_t^* = K_t \psi_{t-1}^* + \mu_\eta^*, \quad \psi_t^+ = -L_t \psi_{t-1}^+ + \mu_\eta^+ \quad and \quad \mu_t = F_t \psi_t + \mu_\varepsilon,$$

and

$$\Omega_t^* = K_t \Omega_{t-1}^* K_t^T + \Sigma_\eta^*, \quad \Omega_t^+ = L_t \Omega_{t-1}^+ L_t^T + \Sigma_\eta^+ \quad and \quad \Gamma_t = F_t \Omega_t F_t^T + \Sigma_\varepsilon$$

$$E_t = D_t F_t^{\leftarrow}, D_t = D_{t-1} G_t^{\leftarrow} \quad and \quad v_t = \left(0^T, \psi_t^{+T}\right)^T.$$

The proofs of our propositions about the Skewed Kalman filter can be found in Naveau and Genton (2002).

2.5 Sequential estimation procedure: Kalman Filtering

To extend the Kalman filter to general skewed normal distributions, we follow the work of Meinfold and Singpurwalla (1983) who derived a Bayesian formulation to derive the different steps of the Kalman filtering. The key notion is that given the data $\mathbf{Y}_t = (Y_1,..., Y_t)$, inference about the state vector values can be carried out through a direct application of Bayes' theorem. In the Kalman literature, the conditional distribution of $(X_t\text{-}1 \mid \mathbf{Y}_{t-1})$ is usually assumed to follow a Gaussian distribution at time t-1. In our case, this assumption at time t-1 is expressed in function of the general multivariate skew-normal distribution:

$$\left(X_{t-1} \middle| Y_{t-1}\right) = GMSN_{n,m}\left(\hat{\psi}_{t-1}, \hat{\Omega}_{t-1}, \hat{D}_{t-1}, \hat{v}_{t-1}, \hat{\Delta}_{t-1}\right), \tag{10}$$

where $\hat{.}$ represents the location, scale, shape, and skewness parameters of $(X_t\text{-}1 \mid \mathbf{Y}_{t-1})$. Then, we look forward in time t, but in two stages: prior to observing Y_t, and after observing Y_t. To implement these two steps, Lemma 1 is used to determine the conditional distribution of a general multivariate skew-normal distribution.

Proposition 2 Suppose that the initial vector $(U^T_0, V^T_0)^{\mathrm{T}}$ follows the normal distribution defined by (9), that the posterior distribution of X_t follows (10) at time t-1 and that we have for U_t and V_t introduced in (5)

$$\begin{pmatrix} U_{t-1} \\ V_{t-1} \end{pmatrix} \Big| Y_{t-1} \sim N_{k+1}\left(\begin{pmatrix} \hat{\psi}^{*}_{t-1} \\ \hat{\psi}^{+}_{t-1} \end{pmatrix}, \begin{pmatrix} \hat{\Omega}^{*}_{t-1} & \hat{\Omega}^{*+}_{t-1} \\ \hat{\Omega}^{*+}_{t-1} & \hat{\Omega}^{+}_{t-1} \end{pmatrix}\right), \tag{11}$$

where $\hat{.}$ represents the posterior mean and covariance. We define the following quantities: $R^+_t = L_t\, \hat{\Omega}^+_{t-1} L_t + \Sigma^+_{\eta\text{-}1}$, $R^*_t = \hat{\Omega}^+_{t-1} K_t + \Sigma^*_{\eta}$,

$$\Sigma_t = Q_t R^*_t Q^T_t + P_t R^+_t P^T_t + \Sigma_\varepsilon \quad and \quad \hat{\Omega}_t = L_t\left(\hat{\Omega}^+_{t-1} + C_t P^T_t \Sigma^{-1}_t P_t C_t\right) L^T_t$$

and $e_t = Y_t - Q_t\,[K_t\, \hat{\psi}^*_{t-1} + \mu^*_\eta] - P_t\,[E(S_t|Y_{t-1})] - \mu_\in$, where $E(S_t|Y_{t-1})$ is the conditional expectation of S_t given $\mathbf{Y}_{t-1}$ and C_t is the conditional covariance $C_t = cov(V_{t-1}, S_t \mid \mathbf{Y}_{t-1})$. The parameters of the posterior distributions are computed through the next cycle by the following sequential procedure:

$$\left(X_t \middle| \mathbf{Y}_t\right) \sim GMSN_{k+l,k+l}\left(\hat{\psi}_t, \hat{\Omega}_t, \hat{D}_t, \hat{v}_t, \hat{\Delta}_t\right), \; with\; \hat{\psi}_t = \begin{pmatrix} \hat{\psi}^*_t \\ \hat{\psi}^*_{t-1} \end{pmatrix},$$

and where

$$\hat{\Omega}_t = \begin{pmatrix} \hat{\Omega}^*_t & 0 \\ 0 & \bar{\Omega}_t \end{pmatrix}, \hat{D}_t = \begin{pmatrix} 0 & 0 \\ 0 & L_t \end{pmatrix}, \hat{v}_t = \begin{pmatrix} 0 \\ \hat{\psi}^+_t \end{pmatrix}, \quad \text{and} \quad \hat{\Delta}_t = \begin{pmatrix} I & 0 \\ 0 & \hat{\Sigma}^+_v \end{pmatrix}$$

and with

$$\begin{pmatrix} \hat{\psi}_t^* \\ \hat{\psi}_t^+ \end{pmatrix} = \begin{pmatrix} K_t \hat{\psi}_{t-1}^* + \mu_\eta^* + R_t^* Q_t^T \Sigma_t^{-1} e_t \\ -L_t \hat{\psi}_{t-1}^+ + \mu_\eta^+ - L_t C_t P_t^T \Sigma_t^{-1} e_t \end{pmatrix},$$

and

$$\begin{pmatrix} \hat{\Omega}_t^* & \hat{\Omega}_t^{*+} \\ \hat{\Omega}_t^{*+} & \hat{\Omega}_t^+ \end{pmatrix} = \begin{pmatrix} R_t^* - R_t^* Q_t^T \Sigma_t^{-1} Q_t R_t^* & -K_t \hat{\Omega}_{t-1}^{*+} L_t^T + R_t^* Q_t^T \Sigma_t^{-1} P_t C_t L_t^T \\ -L_t \hat{\Omega}_{t-1}^{*+} K_t^T + L_t C_t P_t^T \Sigma_t^{-1} Q_t R_t^* & R_t^+ - L_t C_t P_t^T \Sigma_t^{-1} P_t C_t L_t^T \end{pmatrix}$$

Although the notations are a little more complex, the Kalman filtering steps for the skewed extended state-space model does not present any particular computational difficulties.

3. DISTRIBUTIONS OF DISTRIBUTION WITH APPLICATION TO CLIMATOLOGY

3.1 Motivations and data

The data set under study comes from the European Center for Meteorological Forecasting (ECMWF). The temporal resolution is of 6 hour (0 a.m., 6 a.m., 12 a.m., 6 a.m.) and the data covers the period from December 1998 to December 1999. For each latitude and each longitude, the values of different atmospheric variables (pressure values, temperature, specific humidity, winds, etc) are available at 50 different vertical levels. These levels are not equally spaced and vary from one location to another. This implies that we can not choose a specific altitude (or pressure level) and simply apply classical methods at different chosen altitudes. Despite this difficulty, the atmospheric scientist would like to summarize the information contained in this multi-variate 3D grid into a 2D map, i.e. on the surface of the Earth. Being able to recognize different climatic behaviors is of particular interest. An accurate partition of these vertical profiles is essential to interpret satellite observations into atmospheric variables (inversion of equations of radiative transfer, Chédin et al., 1985). From a statistical point of view, we rephrase this scientific question as a clustering problem, classifying multi-variate vertical profile distributions into clusters with similar physical properties inside a cluster and distinct physical characteristics between clusters. Consequently, a fundamental difference with classical clustering algorithms is that a classification method has been directly applied to distributions (vertical profiles) instead of observations. As an application, 16200 multi-variate vertical profile distributions have to be decomposed as a mixture of K=7 classes. This number was chosen by atmospheric scientists and each class should correspond to a specific climatic situation. The distributions will either be of temperatures, humidities, or both. To illustrate the clustering procedure, we will focus on a particular date (the 15th of December 1998 at midnight). Before showing the results of this analysis, we need to establish a basic statistical framework.

3.2 Defining distributions of distributions

Suppose that the vector **F**=(F_1,..., F_n) represents the temperature vertical profile distributions over the entire globe. To work with such sets of distributions, the concept of *distributions of distributions* developed by Diday (2001) is needed. The details of the clustering methodology of distribution of distributions can be found in the work by Vrac (2002, 2001).

Let t be a real. A *distribution function of distributions* is defined by

$$D_t(x) = P\left(\left\{F \in \Omega_F \text{ such that } F(t) \le x\right\}\right),$$

where Ω_F is the set of all possible temperature distributions. From a more practical point of view, $D_t(x)$ could be estimated by

$$\hat{D}_t(x) = \frac{1}{n}\sum_{i=1}^{n} I[\hat{F}_i(t) \le x], \quad \text{with} \quad \hat{F}_i(t) = \frac{1}{n_i}\sum_{j=1}^{n_i} I[X_{i,j} \le t],$$

where $I[A]$ represents the indicator function, equal to 1 if A true and 0 otherwise, and $\hat{F}_i(t)$ denotes the empirical distribution of the ith profile that has n_i observations. Although this estimation strategy has the advantage of being simple, the clustering algorithm converges slowly due to the step-functions. Instead, we use the "Parzen estimation method" to model the vertical profile distributions

$$\hat{f}_i(x) = \frac{1}{n_i h_i}\sum_{j=1}^{n_i} K\left(\frac{x - X_{i,j}}{h_i}\right), \quad \text{and} \quad \hat{F}_i(t) = \int_{-\infty}^{t} \hat{f}_i(x)dx$$

where K is a kernel function and h_i the window width (Silverman, 1986). Because the density $d_t(x)=D'_t(x)$ takes its values on [0,1], we choose to model it by a Beta density

$$d_{t,\gamma_t}(x) = \frac{\Gamma(\rho_t + \nu_t)}{\Gamma(\rho_t)\Gamma(\nu_t)} x^{\rho_t - 1}(1-x)^{\nu_t - 1}, \quad \text{with} \quad \gamma_t = (\rho_t, \nu_t) > 0 \qquad (12)$$

Hence, $\hat{D}_{t,\hat{\gamma}_t}(x) = \int_0^x \hat{d}_{t,\hat{\gamma}_t}(u)$ with $\hat{\gamma}_t$ estimated from the sample $\{\hat{F}_i(t)\}$ with i=1,...,n.

For the practitioner, studying the relationship between two given temperatures, say t_1 and t_2, is of primary interest. To investigate such a link, the definition of D_t with t real is extended to the bi-vector **t**=(t_1,t_2) by setting

$$D_t(x_1, x_2) = P\left(\left\{F \in \mathbf{\Omega}_\mathrm{F} \text{ such that } F(t_1) \le x_1 \quad \text{and} \quad F(t_2) \le x_2\right\}\right).$$

The extension to higher dimensions does not present any major difficulty, but to reduce the notational complexity we restrict our exposition to the bi-variate case for the remainder of this paper.

3.3 Mixture of distribution of distributions and copulas

Our goal is to cluster the different vertical profile distributions into K=7 classes. To perform this task, we assume that the distribution D_t can be expressed as a mixture of distributions

$$D_t(x_1, x_2) = \sum_{k=1}^{K} \pi_k D_{t,k}(x_1, x_2)$$

where $\sum \pi = 1$, $0 < \pi_k < 1$ and $D_{t,k}$ represents a bi-variate distribution. We express the relationship between the distribution $D_{t,k}$ and its two marginals by directly applying Sklar's theorem (Sklar, 1959; Nelsen, 1998). This gives

$$D_t(x_1, x_2) = \sum_{k=1}^{K} \pi_k C_{t,k}\left(D_{t_1,k}(x_1), D_{t_2,k}(x_2)\right),$$

where $C_{t,k}$ is a copula function. There exists a variety of parametric forms to model this copula. In our applications, we use Frank's copula (Nelsen 1998)

$$C_{t,k}(u,v) = \frac{1}{\log \beta_{t,k}} \log\left(1 + \frac{(\beta_{t,k}^u - 1)(\beta_{t,k}^v - 1)}{\beta_{t,k} - 1}\right), \text{ with } u, v \in [0,1],$$

where the positive parameter $\beta_{t,k} \neq 1$ is a indicator of dependence, $C_{t,k}(u,v) \sim uv$ for $\beta_{t,k} \uparrow 1$, $C_{t,k}(u,v) \sim \min(u,v)$ for $\beta_{t,k} \downarrow 0$ and $C_{t,k}(u,v) \sim \max(u+v-1,0)$ for $D_{t,k}$ $\beta_{t,k} \uparrow \infty$. The first case, respectively the second case, corresponds to the independence, respectively to the total dependence.

3.4 Parameters estimation and clustering algorithm

The next step is to sequentially cluster the n=16200 vertical profile distributions and to estimate all parameters from the previous sections. The chosen method is an extension to distributions of the "Nuées Dynamiques" method (Diday et al., 1974). Given a partition $\Pi = \{\Pi_1,...,\Pi_K\}$ (the first one is randomly generated), the clustering algorithm constitutes of 3 main steps: (1) estimation of the mixture proportions $\{\pi_k\}$, (2) estimation of other mixture parameters, $(\gamma_{t1,k}, \gamma_{t2,k})$ for the Beta laws and $\{\beta_{t,k}\}$ for the copula's parameter, (3) re-allocation of all individuals ω_i into K new classes with i=1,...,n. This 3 step procedure is repeated until the desired convergence is reached. The first step is undertaken by setting π_k as the number of elements in the kth class divided by the total number of individuals. Other alternatives can be used (Celeux and Govaert, 1993). The second step is realized by maximizing the *classifier log-likelihood*

$$l(\Pi,\theta) = \sum_{k=1}^{K} \sum_{\omega_i \in \Pi_k} \log[d_{k,t}(x_1^{(i)}, x_2^{(i)}; \theta_k)], \text{ with } \theta = \{\beta_{t,k}, \gamma_{t_1,k}, \gamma_{t_2,k}\} k = 1,...K,$$

where $\omega_i = \{\text{i}: \hat{F}_i(t_1) \leq x_1,\ \hat{F}_i(t_2) \leq x\}$ and $d_{k,t}(x_1,x_2;\boldsymbol{\theta}_k)$ is the density derived from $D_{k,t}(x_1,x_2;\boldsymbol{\theta}_k)$. The last step is implemented by defining the new classes as $\Pi_k = \{\omega: \pi_k d_{k,t}(\omega;\boldsymbol{\theta}_k) \geq \max\{\pi_l d_{l,t}(\omega;\boldsymbol{\theta}_l): l=1,...,K\}\}$.

3.5 Application to the temperature profiles

Figure 1 shows a classification of the 16200 vertical temperature profiles into 7 clusters. This result was obtained after applying the clustering procedure for two iterations. Although no spatial dependence was introduced in the model, the spatial coherence obtained from the clustering procedure is a positive indicator of the quality of the algorithm. From a scientific perspective, the clusters provides realistic classes. Cluster 4 can be identified as a "tropical class". Two "polar" clusters can be linked to the winter season at the South pole (cluster 1) and to the summer season at the North pole (cluster 7). Cluster 3 makes the transition between moderate and tropical zones, cluster 6 between polar and moderate zones. The high mountains are clearly identified (Himalaya, Andes).

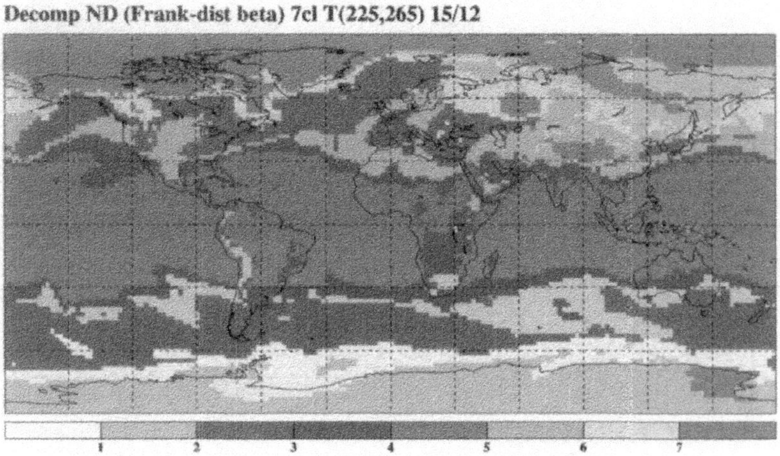

Figure 1. Clustering of the 16200 temperature vertical profiles into 7 clusters.

3.6 Extension to multi-dimensional distributions

In the previous sections, we exclusively focused on the temperature profiles but extending the procedure to multi-dimensional atmospheric vectors, e.g. the bi-variate vector of the temperature and humidity profiles, will greatly increase the range of applications of this work. The coupling method is based on the following mixture decomposition

$$D_{(r)}(x^{(r)}) = \sum_{k=1}^{K} \pi_{r,k} C_{r,t^{(r)},k} \left(D_{r,t_{1,r},k}(x_1) D_{r,t_{2,r},k}(x_2) \right), \quad \text{with} \quad x^{(r)} = \left(x_1^{(r)}, x_2^{(r)} \right),$$

where the integer r represents either the temperature (r=1) or the humidity (r=2). Then this couple of distributions can be linked by Sklar's theorem. There exists a copula function C such that

$$D(\mathbf{x}^{(1)}, \mathbf{x}^{(2)}) = C(D_{(1)}(\mathrm{x}^{(1)}), D_{(2)}(\mathrm{x}^{(2)})).$$

Although the notations become more complex, the same overall principles of the algorithm described in Section 3.4 can be applied. A main difference is that, in addition of setting two temperature levels ($t^{(1)}{}_1, t^{(1)}{}_2$), we also need to fix two humidity levels ($t^{(2)}{}_1, t^{(2)}{}_2$). Figure 2 represents the output of such a coupling procedure. Cluster 7, respectively cluster 1, corresponds to the winter season at the North pole, respectively the summer season at the South pole. This two regions were already identified in the temperature clustering, but additional variations are generated from humidity in Figure 2. Two tropical classes are identified, very humid (cluster 4) and humid (clusters 3). Cluster 4 is in better agreement with existing humid zones than the ones obtained before. The other clusters represent transition regions from tropical classes (hot and humid) to polar classes (dry and cold).

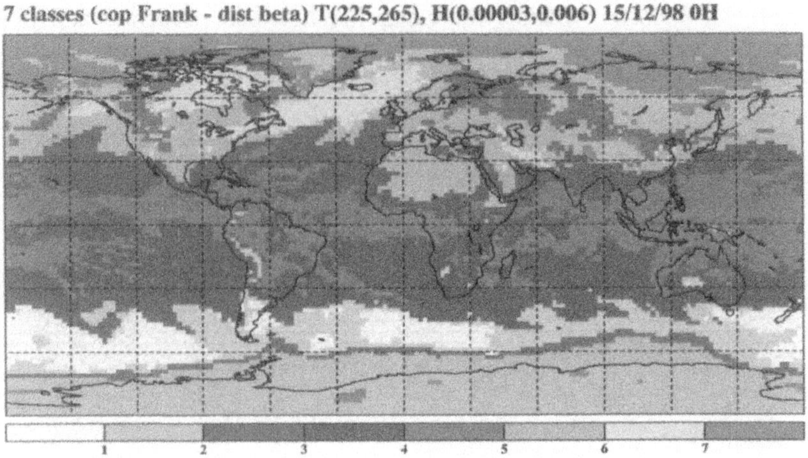

Figure 2. Clustering in 7 classes by coupling the temperature and the humidity.

4. CONCLUSIONS

In the first part of this work, we showed that extending the normal distribution to the general multivariate skew-normal distribution for state-space models did neither reduce the flexibility nor the traceability of the operations associated with Kalman filtering. To the contrary, the introduction of a few skewness parameters provides a simple source of asymmetry needed for many applications. Further research is currently conducted to illustrate the capabilities of such extended state-space models for real case studies.

By introducing a higher abstraction level in clustering methods, the concept of distributions of distributions and copulas extends the applicability of current procedures (Diday et al., 2001; Vrac, 2002). In addition, it allows to model different dependence levels for probabilistic data, internal dependencies inside a distribution of distributions (Section 3.5) and external ones, for example between the humidity and temperature vertical profile distributions. Besides providing realistic climatic classifications, these results emphasize the strong potential of this clustering method at helping the understanding of other atmospheric variables and their inter-relationships. Other algorithms have been generalized in the same way with copulas : the theoretically extensions of the algorithms EM, SEM, SAEM, and CEM was derived by Vrac (2002). Comparisons between these extended methods and "classical" algorithms of classification indicate that the procedures based on the concept of distributions of distributions perform better in the context of climatic studies (Vrac, 2002). It is worthwhile to note that the proposed method can also be applied to classical numerical observations and functional data. Finally, multi-variate versions of the algorithm exist and are based on multidimensional generalized Archimedian copulas (Vrac 2002). This extension to multi-variate cases constitutes a strong axis of current research.

ACKNOWLEDGEMENTS

The authors would like to thank the organizers of the 2002 GeoENV conference for their financial and logistic supports.

REFERENCES

1. Anderson, J. (2001). An ensemble adjustment Kalman filter for data assimilation, *Monthly Weather Review*, **129**: 2884-2903.
2. Azzalini, A., Dalla Valle A. (1996). The multivariate skew-normal distribution, *Biometrika*, **83**: 715-726.
3. Azzalini, A., Capitanio, A. (1999). Statistical applications of the multivariate skew normal distribution, *J. R. Statist. Soc. B*, **61**: 579-602.
4. Bengtsson, T., Nychka, D., Snyder, C. (2002). A frame work for data assimilation and forecasting in high-dimensional non-linear dynamical systems. Submitted to *J. R. Statist. Soc. B*.
5. Berk, R., Bickel, P., Campbell, K., Fovell, R., Keller-McNulty, S., Kelly, E., Linn, R., Park, B., Perelson, A., Rouphail, N., Sacks, J., Schoenberg, F. (2002). Workshop on statistical approaches for the evaluation of complex computer models. *Statistical Science*, **17:** 173-192.
6. Bock, H.H., Diday,E. (2000). Analysis of symbolic data. Exploratory methods for extracting statistical information from complex data, publisher Springer-Verlag, Heidelberg.
7. Chen, R., Liu, J.S. (2000). Mixture Kalman filters, *J. R. Statist. Soc. B*, **62:** 493-508.
8. Celeux, G., Govaert, G. (1993). Comparison of the mixture and the classification maximum likelihood in cluster analysis, *Journal of statist. computer*, **47**: 127-146
9. Chédin, A., Scott, N., Wahiche, C., Moulinier, P. (1985). he improved initialization inversion method: a high resolution physical method for temperature retrievals from satellites of the TIROS-N series, *J. Clim. Appl. Meteor.*, **24**: 128-143.
10. Cressie, N., Wikle, C.K. (2002). Space-time Kalman filter. *Entry in Encyclopedia of Environmetrics*, **4**, eds. A.H. El-Shaarawi and W.W. Piegorsch. Wiley, New York, pp. 2045-2049.
11. Doucet, A., Freitas, N., Gordon, N. (2001). Sequential Monte Carlo Methods in Practice. Springer.
12. Diday, E. (2001).A generalisation of the mixture decomposition problem in the symbolic data analysis framework, *Cahiers du CEREMADE*, **0112**.
13. Diday, E. (1974). The dynamic clusters method in pattern recognition. *Proceeding of IFIP*, Stockolm.
14. Domínguez-Molina, González-Farías, G., Gupta, A.K. (2001). A general multivariate skew-normal distribution. Submitted to *Math. Methods of Statistics*.
15. Genton, M.G., Loperfido, N. (2002). Generalized skew-elliptical distributions and their quadratic forms. *Scandinavian Journal of Statistics*. (revised).

16. Guo, W., Wang, Y., Brown, M. (1999). A signal Extraction Approach to Modeling Hormones Time series with Pulses and a Changing Baseline. *J. Amer. Stat. Assoc.*, vol. 94, **447**: 746-756.
17. Harrison, P.J., Stevens, C.F. (1971). A Bayesian approach to short-term forecasting. *Operational Res. Quart.* **22**: 341-362.
18. Kitagawa & Gersch (1984). A Smoothness Priors State-Space Modeling of Times Series With Trend and Seasonality. *J. Amer. Statist. Assoc.* **79**: 378-389.
19. Meinhold, R.J., Singpurwalla, N.D. (1983). Understanding the Kalman filter. *The American Statistician*. **37**: 123-127.
20. Naveau, P., Genton, M. (2002). The Multivariate General Skewed Kalman Filter. *Submitted to the J. of Multi-Variate Analysis.*
21. Naveau, P., Ammann, C.M., Oh, H.S., Guo, W. (2002). A Statistical Methodology to Extract Volcanic Signal in Climatic Times Series. *Submitted to the J. of Geophysical Research.*
22. Nelsen, R.B. (1998). An introduction to Copulas, publisher Springer Verlag, Lectures Notes in Statistics.
23. Shepard, N. (1994). Partially Non-Gaussian State-space Models. *Biometrika*, **81**: 115-131.
24. Shumway, R.H., Stoffer, D.S. (1991). Dynamic linear models with switching. *J. Amer. Statist. Assoc.* **86**: 763-769.
25. Silverman, B.W. (1986). Density Estimation for Statistics and Data Analysis, publisher Chapman and Hall, London.
26. Vrac, M., Diday, E., Chédin, A., Naveau, P. (2001). Mélange de distributions de distributions, *SFC'2001 8èmes Rencontres de la Société Francophone de Classification*, Université des Antilles et de Guyane, Guadeloupe.
27. Vrac, M. (2002). Analyse et mod\'elisation de données probabilistes par Décomposition de Mélange de Copules et Application à une base de données climatologiques, *Thèse de doctorat*, Université Paris IX Dauphine.
28. Wikle, C.K., Cressie, N. (1999). A dimension reduced approach to space-time Kalman filtering. *Biometrika*, **86**: 815-829.
29. Wikle, C.K., Milliff, R.F., Nychka, D., Berliner, L.M. (2001). Spatiotemporal hierarchical Bayesian modeling: Tropical ocean surface winds. *JASA*, **96**: 382-397.

SUPER-RESOLUTION LAND COVER CLASSIFICATION USING THE TWO-POINT HISTOGRAM

P.M. Atkinson
Department of Geography, University of Southampton, Highfield, Southampton SO17 1BJ, United Kingdom; pma@soton.ac.uk

Abstract: A geostatistical optimization algorithm is proposed for super-resolution land cover classification from remotely sensed imagery. The algorithm requires as input, a soft classification of land cover obtained from a remotely sensed image. A super-resolution (sub-pixel scale) grid is defined. The soft land cover proportions (pixel scale) are then transformed into a hard classification (sub-pixel scale) by allocating hard classes randomly to the sub-pixels. The number allocated per pixel is determined in proportion to the original land cover proportion per pixel. The algorithm optimizes the match between a target and current realization of the two-point histogram by swapping sub-pixel classes within pixels such that the original class proportions defined per pixel are maintained. The algorithm is demonstrated for two simple simulated images. The advantages of the approach are its ability to recreate any target spatial distribution and to work with features that are both large and small relative to the pixel size, in combination.

1. INTRODUCTION

Land cover is an important variable for many scientific investigations and operational applications. For example, land cover data are required to provide boundary conditions for atmospheric (e.g., global climate circulation) modelling, hydrological modelling, geomorphological modelling and so on. However, accurate land cover data at the required (coarse) spatial resolution are often not available because of the difficulties and expense of surveying large areas. Remote sensing has been invaluable for mapping, and ultimately monitoring, land cover over large areas because of the complete

X. Sanchez-Vila et al. (eds.), geoENV IV – Geostatistics for Environmental Applications, 15-28.

synoptic coverage provided. However, current state-of-the-art techniques do not make full use of the available data within remotely sensed images. In particular, techniques are limited by the spatial resolution of the original multiple waveband imagery. The objective of this paper was to demonstrate a geostatistical technique for land cover classification from remotely sensed imagery that actually maps at a spatial resolution that is finer that that of the original imagery, thus, making greater use of the available data.

Hard classification techniques, such as maximum likelihood (ML) classification, have been popular in remote sensing for many years (e.g., Thomas, 1987). In hard classification, every pixel is allocated to one class (for ML classification, the class to which it is most likely to belong). A criticism of hard classification is that many pixels actually contain a mixture of land cover classes. Such pixels are referred to as 'mixed'. Mixed pixels can arise for two main reasons: (i) more than one distinct (crisp) class is represented within a pixel and (ii) classes intergrade within a pixel (e.g., an ecotone). In (i) the mixing leads to ambiguity and in (ii) the mixing leads to vagueness demanding the definition of fuzzy sets (e.g., Bezdek et al., 1984). The concern in this paper is the unmixing of pixels that contain crisp classes.

The mixed pixel problem led to the adoption of techniques for soft classification, originally for geological remote sensing (Adams, et al., 1985). Soft classifiers (also sometimes referred to as fuzzy classifiers) map each pixel onto many classes and assign membership values to each class which predict the proportion of the pixel that each class represents. Examples include the linear mixture model (Adams, et al., 1985), fuzzy c-means (Bezdek et al., 1984), feed-forward, back-propagation neural networks (Atkinson et al., 1997) and support vector machines (Brown et al., 1999). Soft classification represents greater information in the land cover prediction at no extra cost: hard classification simply omits the land cover proportion information, presenting only the most likely class. Indeed, the only drawback of soft classification appears to be the difficulty in displaying more than three or four class proportions simultaneously in the same map.

While soft classification is preferable to hard classification, almost ubiquitously, because class proportions are predicted per pixel, no attempt is made to predict where, within each pixel, the land cover actually exists. Thus, if a soft classifier has predicted that a pixel contains 50% woodland, 30% grassland and 20% built-land, the user (e.g., decision-maker) does not know where the woodland, grassland and heathland patches are located within the pixel. The new technique demonstrated in this paper is designed to post-process a soft classified remotely sensed image to classify (in a hard sense) land cover at the sub-pixel scale. This objective is referred to as super-resolution classification.

Several researchers have attempted super-resolution mapping based on remotely sensed imagery of radiance or reflectance (e.g., Flack et al., 1994; Foody, 1998, Steinwendner et al., 1998; Schneider, 1999). Atkinson (1997) suggested super-resolution mapping based solely on the output from a soft classification. The idea proposed was to convert soft land cover proportions to hard (per-sub-pixel) land cover classes (that is, at a finer spatial resolution). The most intuitive (most visually appealing) solution was attained by maximizing the spatial statistical correlation between neighbouring sub-pixels. The basic idea was to maximize the spatial correlation between neighbouring sub-pixels under the constraint that the original pixel proportions were maintained (Atkinson, 1997). This basic idea was extended by Verhoeye et al. (2000). A fundamental limitation of the approach was that the relation between pixels and sub-pixels was modelled, thereby mixing scales of measurement (Atkinson and Tate, 2000).

A solution to the super-resolution problem may be achieved by comparing sub-pixels to sub-pixels meaning that a non-linear model should be used to achieve solution. A pixel swapping algorithm in which the goal is to maximize the spatial correlation between neighbours was demonstrated recently for simple simulated images (Atkinson, 2001). This algorithm works for both the binary (e.g., target detection) and categorical (e.g., land cover) cases.

Recently, Tatem et al. (2001a) developed a Hopfield neural network (HNN) technique (Hopfield and Tank, 1985) for super-resolution target mapping. The HNN was used essentially as an optimization tool. To solve the super-resolution mapping problem, with the pixel-level class proportions as initial conditions, the HNN architecture was arranged as a super-resolution grid of sub-pixels. The HNN was then set up to minimize an energy function that comprises a goal and constraints:

$$E = k_1 G + k_2 C + b \tag{1}$$

where G is the goal (to increase the spatial correlation between neighbouring sub-pixels), C is the constraint (that original class proportions per-pixel are maintained), b is a bias term and and are weights. The HNN was applied initially to detect targets (two-class problem) (Tatem et al., 2001a), but eventually extended to super-resolution land cover mapping (multiple class problem) (Tatem et al., 2001b).

Tatem et al. (2002) developed an extension of the HNN super-resolution mapping technique in which the spatial clustering goal was replaced by a K-class variogram-matching goal. This new goal allowed replication of spatial pattern, which was particularly useful for objects that were smaller than a pixel. In this paper, a geostatistical optimization algorithm is described which is capable of producing super-resolution maps from soft classified input images. It represents an alternative to the HNN variogram-matching algorithm.

2. THEORY

The spatial optimization algorithm used here is based on the two-point histogram as defined and used in the program ANNEAL.for, which is part of the GSLIB library of Fortran routines (Deutsch and Journel, 1998). The present optimization algorithm was coded in S-PLUS. The two-point histogram and its use in optimization are presented, followed by a description of its use in super-resolution classification.

2.1 Two-Point Histogram

This section, which describes the two-point histogram, is adapted from Deutsch and Journel (1998). Given a random variable Z that can take one of k=1, ..., K outcomes (i.e., a categorical variable) the two-point histogram for a particular lag (distance and direction of separation) $\mathbf{h}$ is the set of all bivariate transition probabilities:

$$p_{k,k'}(\mathbf{h}) = \Pr\left\{\begin{matrix} Z(\mathbf{u}) \in \text{category } k, \\ Z(\mathbf{u}+\mathbf{h}) \in \text{category } k' \end{matrix}\right\} \tag{2}$$

independent of $\mathbf{u}$, for all k, k' = 1, ..., K. The objective function corresponding to the two-point histogram control statistic is as follows:

$$O = \sum_{\mathbf{h}} \left(\sum_{k=1}^{K} \sum_{k'=1}^{K} \left[p_{k,k'}^{training}(\mathbf{h}) - p_{k,k'}^{realization}(\mathbf{h}) \right]^2 \right) \tag{3}$$

where $p_{k,k'}^{training}(\mathbf{h})$ are the target transition probabilities, for example, calculated from a training image and $p_{k,k'}^{realization}(\mathbf{h})$ are the corresponding transition probabilities of the realization image (i.e., the current image being altered).

2.2 Optimization Algorithm

While equation 2 lies at the heart of the optimization algorithm, it is insufficient alone for super-resolution mapping. First, a scheme must be devised for altering the sub-pixel values. This can either be via a change to the attribute (as for HNN, ANNEAL.for) or via a swap in sub-pixel location (as here). In either case, it is important that the optimization goal (Equation 2) is constrained so that the original pixel proportions are maintained as closely as possible. Where the attribute values are changed this constraint should be added to the goal to form a single energy function. In this way, the original pixel proportions will be maintained approximately in the solution. Where sub-pixels are swapped, the constraint can either be added to the goal, or the sub-pixels to be swapped can be constrained to the same pixel. The latter strategy, which results in the pixel proportions being maintained perfectly in the solution, is adapted here.

2.3 Initialization

The decision to swap sub-pixels within pixels means that the attribute values in the initial image must correspond to those desired in the solution. In the present case, this means that the pixels of the initial image must contain hard classified sub-pixels, with the number of sub-pixels per class determined in proportion to the class proportions. Thus, in a pixel of 10 by 10 sub-pixels (for which proportions are predicted as woodland (50%), grassland (30%) and built-land (20%)) there will be 50 sub-pixels of woodland, 30 of grassland and 20 of built-land. The image to be presented to the optimization algorithm is initialized by distributing spatially the required number of sub-pixels randomly within each pixel.

2.4 Summary of algorithm

The full algorithm is summarized below.

1. Create current image at sub-pixel scale by randomly distributing land cover proportions
2. Calculate two-point histogram for training image
3. Calculate two-point histogram for current image
4. For each iteration
5. For every pixel
6. For every sub-pixel (visited in random order within the current pixel)
7. Compare to another sub-pixel (drawn randomly from the same pixel)
8. If swap results in smaller objective function, retain swap and update two-point histogram.

Two checks were added to the algorithm to increase its efficiency. First, it was found that many pixels contained only one land cover class. Such pixels were ignored. It should be noted however, that sub-pixels within such pixels may be used in comparison with sub-pixels within adjacent pixels because the two-point histogram was computed for eight directions (at 45° to each other) and at various lags. Second, sub-pixels were compared only if their classes were different. While not fast, the current implementation in S-PLUS was sufficient to demonstrate the utility of the optimization algorithm on simulated images. It is anticipated that the algorithm will be written in C or C++ in the future for operational use.

3. SIMULATED DATA

3.1 Circles

To provide a simple test of the optimization algorithm two circles of different class were simulated on a background in an image of 35 by 35 sub-pixels (Figure 1a). The spatial resolution of the image was then coarsened by a factor of 7, to provide an image of 5 by 5 pixels. The proportions of each of the three classes in each pixel of the image are shown in Figure 1b-d. These three images are taken to represent the output of a soft classifier. That is, Figure 1b-d represents the final result of applying a soft classifier to a remotely sensed image (i.e., the current state-of-the-art solution). It also represents the sole data input to the geostatistical optimization algorithm.

3.2 Simulated remotely sensed image

A simple Boolean simulation was used to provide a more realistic image with which to test the optimization algorithm. First, n_w=7 rectangles of varying height and width $r \sim$ U ($\min_r$, $\max_r$) were drawn at locations $l \sim$ U ($\min_l$, $\max_l$) where $\min_r = 2$, $\max_r = 14$, $\min_l = 0$ and $\max_l = \frac{2}{3}(n_p . n_{sp}) = 46.7$ sub-pixels, where n_p is the number of pixels along the image edge and n_{sp} is the number of sub-pixels along a pixel edge. These n_w rectangles, which superimposed themselves naturally, were meant to simulate an area of woodland (Figure 2a). Second, n_b = 20 rectangles of varying height and width r ($\min_r$ = 2 and $\max_r$ = 3) were drawn at locations l with $\min_l$ = $\frac{1}{3}(n_p . n_{sp})$ and $\max_l$ = $n_p . n_{sp}$. These rectangles, some of which were superimposed, were meant to simulate built-land (i.e., buildings). Because buildings were simulated after woodland, the buildings appear to nestle inside the woodland, as desired.

The third class, the background, is meant to represent grassland. The effect of overlapping the positions of draws of woodland and build-land objects is to create several pixels in the centre of the image that contain all three classes.

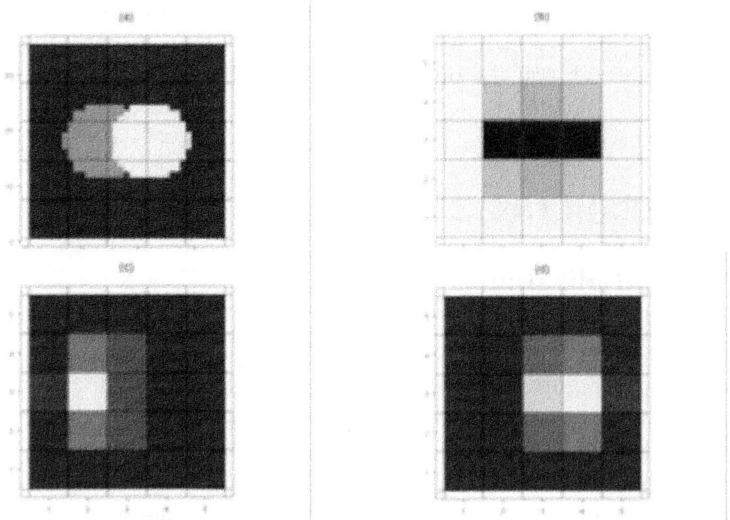

Figure 1. (a) Target image (35 by 35 sub-pixels) defined at the sub-pixel scale and (b-d) class proportions (5 by 5 pixels) defined at the pixel scale for classes (b) background, (c) left circle and (d) right circle. Note that images b-d provide the only input to the optimization algorithm.

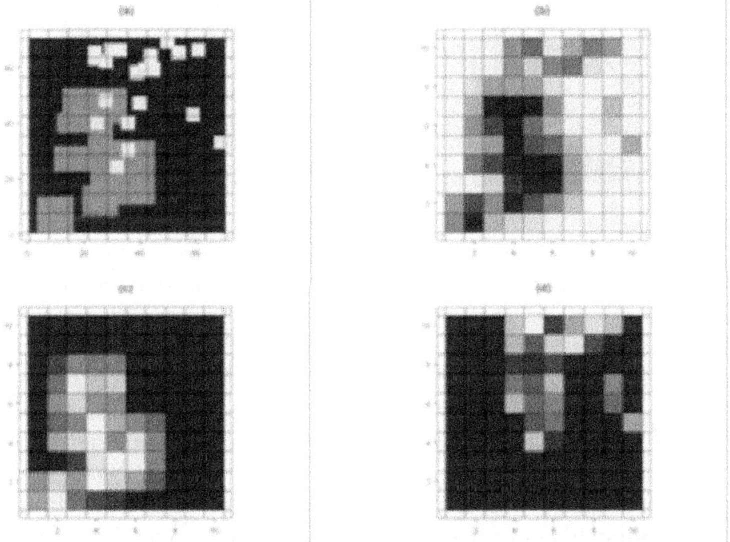

Figure 2. (a) Target image (70 by 70 sub-pixels) defined at the sub-pixel scale and (b-d) land cover class proportions (10 by 10 pixels) defined at the pixel scale for simulated classes (b) grassland, (c) woodland and (d) built-land. Note that images b-d provide the only data input to the optimization algorithm.

The spatial resolution of the image (Figure 2a) was coarsened by a factor of 7 to create an image of 10 by 10 pixels representing proportional land cover (Figure 2b-d). Figure 2b-d is taken to represent the output from a soft classifier applied to a remotely sensed image. Again, it represents the state-of-the-art solution, and the only data input to the geostatistical optimization algorithm.

4. RESULTS

4.1 Initialization

To provide an input to the optimization algorithm, the pixel proportions represented in Figures 1b-d and 2b-d were allocated to locations selected randomly, as described in section 2.3. The resulting images are shown in Figure 3. It is interesting to note that this sub-pixel allocation presents a useful method of visualizing a soft classification, particularly where the number of classes K > 3. However, that benefit is coincidental to the present goal.

(a) (b)

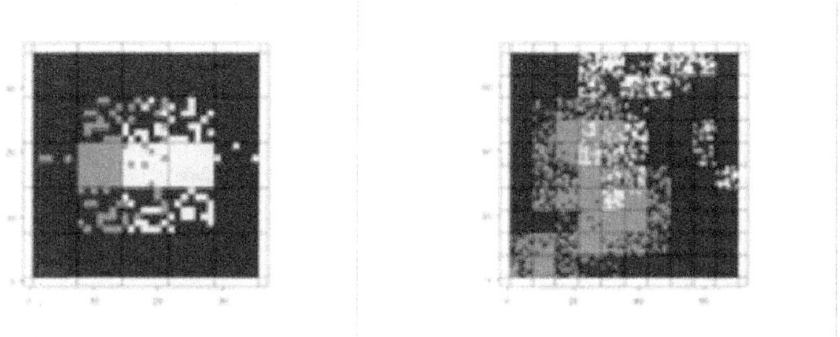

Figure 3. Initial images for (a) circles (35 by 35 sub-pixels) and (b) land cover (70 by 70 sub-pixels). For each pixel, the (soft) class proportion for each class defined at the pixel scale (Figure 1b-d) is allocated (hard classification) to sub-pixels whose spatial location is defined randomly. The number of sub-pixels for each class is determined by the pixel scale class proportion. This image is the input to the optimization algorithm.

4.2 Training

In the practical or operational situation, training (i.e., definition of the target two-point histogram for use in Equation 2) would be provided by a training image with the desired super-resolution. A practical example might be super-resolution classification of Landsat Thematic Mapper (TM) imagery (spatial resolution of 30 m by 30 m) via training with a classified IKONOS image (spatial resolution of 4 m by 4 m). This strategy is sensible because Landsat TM images cover a much larger area than IKONOS images.

Many other sensor combinations can be used to illustrate the utility of this approach.

In the absence of training data, and to test the utility of the algorithm in the ideal case, the training two-point histogram was obtained from the target image. This choice is justified because (i) the two-point histogram is calculated at a limited number of lags only ($\mathbf{h}_{\max} = n_{sp}$) such that it contains only partial information on the original image and (ii) while in the practical situation the accuracy of the predicted super-resolution classification will depend on the extent to which the training image represents the spatial character of the true target, that is not the present interest.

4.3 Optimization

The first six iterations of the optimization algorithm are shown in Figure 4 (circles) and Figure 5 (remotely sensed classification). The super-resolution classifications achieved after 100 iterations are shown in Figure 6a (circles) and Figure 6b (remotely sensed classification). Additional iteration may have decreased the energy function (Equation 2) further, but the results are sufficient to illustrate the utility of the technique.

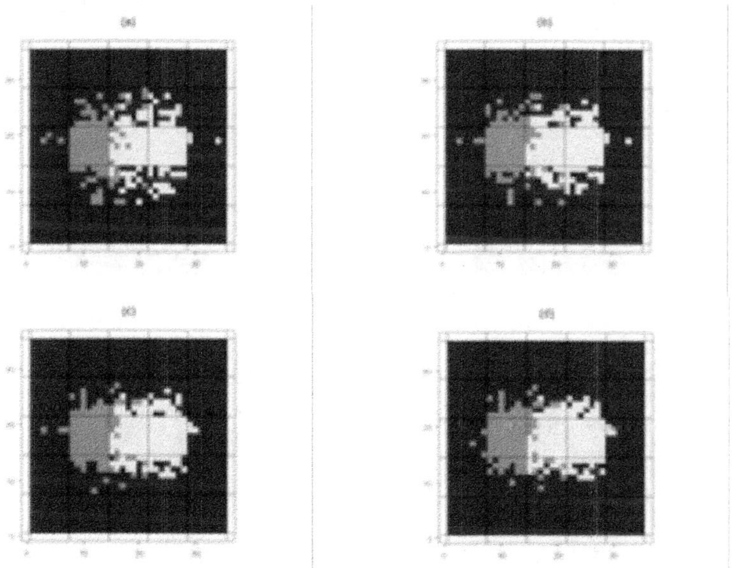

Figure 4. The first six iterations of the optimization algorithm. Each iteration involves, for every pixel, a comparison of every sub-pixel (chosen in a random sequence) with another sub-pixel chosen randomly from the same pixel.

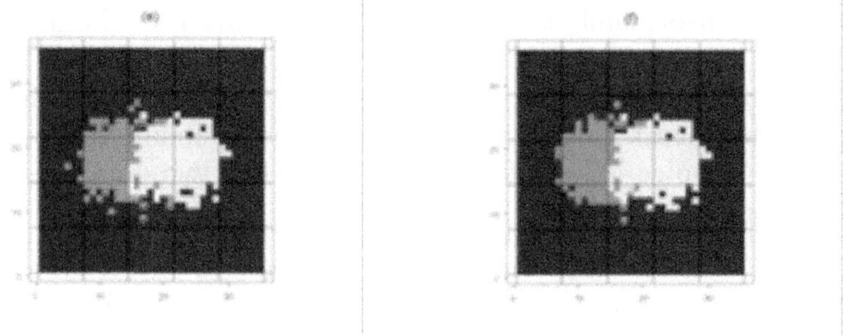

Figure 4. The first six iterations of the optimization algorithm. Each iteration involves, for every pixel, a comparison of every sub-pixel (chosen in a random sequence) with another sub-pixel chosen randomly from the same pixel.

The algorithm appears to have reproduced the circles almost perfectly, and the land cover target reasonably closely in spatial character (built-land).

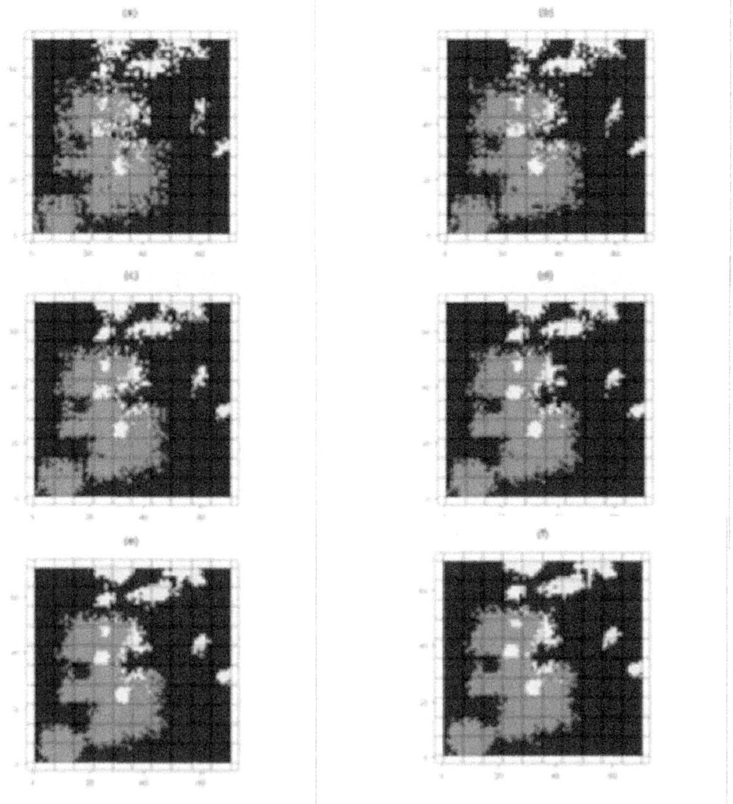

Figure 5. The first six iterations of the optimization algorithm. Each iteration involves, for every pixel, a comparison of every sub-pixel (chosen in a random sequence) with another sub-pixel chosen randomly from the same pixel.

4.4 Assessment

The circles represent Woodcock and Strahler's (1987) H-resolution case in which the spatial resolution is fine relative to the size of objects in the scene (in this case the circle diameter). In the H-resolution case, the target (Figure 1a) can be reproduced with a spatial clustering algorithm in which the objective is to *maximise* the spatial correlation between neighbours (c.f., Atkinson, 2001). The advantage of the present technique is its ability to *match* any prior target spatial distribution. The potential of this is not realized fully for the circles, although the example does illustrate the generality of the technique.

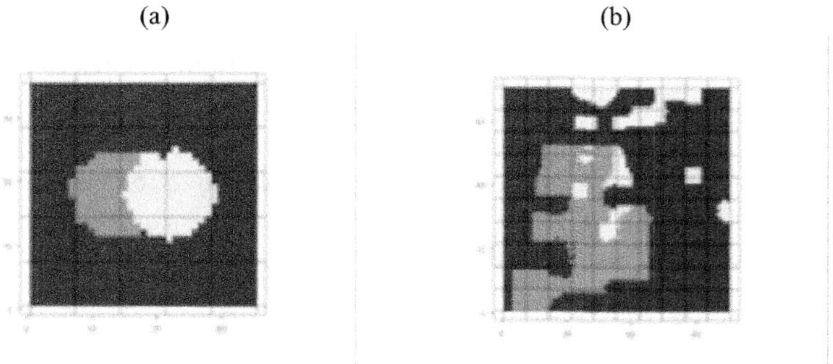

Figure 6. Super-resolution images of (a) circles (35 by 35 sub-pixels) and (b) land cover (70 by 70 sub-pixels) after 100 iterations.

While the woodland (and grassland) areas in the simulated remotely sensed scene represent the H-resolution case, this is to a lesser extent than for the circles because of the greater curvature of the feature (object) boundaries. The parcels of built-land, however, represent the L-resolution case in which the spatial resolution is coarse relative to the size of objects in the scene. In the L-resolution case, a spatial clustering algorithm would fail to provide a realistic solution, joining together patches of a given class where possible. Not only does the optimization algorithm recreate the spatial character of the target, but it also allows a realistic solution for both the H-resolution (woodland, grassland) and L-resolution (built-land) cases simultaneously.

The buildings in the solution do not match the buildings in the target on a sub-pixel-by-sub-pixel basis. Neither are they expected to. Insufficient data and constraints are provided to achieve such spatial definition. However, the spatial character of the target is recreated reasonably well. Such a map would find utility in many circumstances, but particularly as input to spatially distributed process models (e.g., as boundary conditions for flood inundation models where water flows around buildings etc.).

5. DISCUSSION

While the algorithm presented represents a useful basic tool for super-resolution classification, several possible refinements have been identified and these are described in this section.

5.1 Simulated annealing

Convergence appears to be fast (the main features are identifiable within the first eight iterations). However, it should be remembered that n_{sp} by n_{sp} comparisons are made per pixel at each iteration (where n_{sp} is, in this case, 7). The rapid rate of convergence may be a cause for concern in that it is possible for the solution to become trapped in local minima. If that is a supportable concern then the algorithm can be modified readily to include full spatial simulated annealing with an annealing schedule designed to avoid local minima (e.g., van Groenigen, 1999). A further possibility is to run the algorithm several times with different initializations and compare the solutions. However, for the present application, where the solution is constrained by the original pixel proportions, the risk of local minima is believed to be small.

5.2 Non-stationarity and regularization

It is clear from the target image that the local spatial character of variation differs from place-to-place. It might be useful then if the target two-point histogram also varied from place-to-place. The problem, of course, is that in the practical situation the small training image available at the target spatial resolution will not relate to any particular spatial location in the image being optimized. Some location-specific information is provided, however, by the initial image output from the soft classifier (i.e., Figure 1b-d and 2b-d). In particular, the local two-point histogram may be computed at the pixel-scale. The problem is that this information is provided at the pixel scale, whereas information is required at the sub-pixel scale. A solution to this problem may be possible via regularization of the *modelled* sub-pixel two-point histogram, thereby providing a link between the two scales of measurement (Journel and Huijbregts, 1978; Jupp *et al.*, 1988). This will be the subject of future research.

5.3 Error and the point-spread function

Two issues which have been deliberately overlooked are error and the point-spread function (PSF). In the simulated soft classifications (Figures

1b-d and 2b-d) zero error was assumed. That is, the (land cover) class proportions were assumed to be predicted perfectly. Research has shown that in practice the accuracy of soft classification is typically 80% (e.g., Atkinson *et al.*, 1997). This error will have a detrimental effect on super-resolution mapping. In the presence of such error it would be sensible to allow the sub-pixel values (i.e., the original class proportions) to change to an extent determined by the expectation of the error. Whether or not adequate convergence is possible in this essentially under-constrained scenario is unclear.

The PSF provides a second practical problem that has been ignored in the analysis above. The PSF is usually shaped like a two-dimensional step function (termed a square wave response) convolved with a smoothing filter. It means that the remotely sensed response for a given pixel is, in part, a function of spatial variation in neighbouring pixels. This introduces ambiguity into the class proportions predicted by the soft classifier. It means that the pixels in the class proportion image (Figure 1b-d and 2b-d) should actually overlap (in fact, should have the same shape as the PSF). In practice, therefore, it may be desirable to allow some swapping of sub-pixels between neighbouring pixels, restricted to zones of PSF overlap, and in number determined by the amount of overlap.

6. CONCLUSION

A new geostatistical optimization technique has been demonstrated for super-resolution land cover prediction from remotely sensed imagery. While no quantitative assessment of accuracy was provided, the results are encouraging. In particular, the algorithm provides acceptable solutions in both the H-resolution and L-resolution cases, and when both are combined. The super-resolution map (L-resolution case) is likely to be useful as input to spatially distributed process models. The optimization technique will be applied in future research to real remotely sensed imagery. Further, the algorithm will be extended to incorporate the suggestions made in section 5, in particular, the use of a non-stationary model.

REFERENCES

1. Adams, J.B., Smith, M.O. and Johnson, P.E. (1985) Spectral mixture modelling: a new analysis of rock and soil types at the Viking Lander 1 site, Journal of Geophysical Research, vol. 91, pp. 8098-8112.

2. Atkinson, P.M. (1997) Mapping sub-pixel boundaries from remotely sensed images, in Innovations in GIS IV (ed., Z. Kemp) (Taylor an1d Francis: London), p. 166-180.
3. Atkinson, P.M., Cutler, M.E.J. and Lewis, H. (1997) Mapping sub-pixel proportional land cover with AVHRR imagery, *International Journal of Remote Sensing,* vol.18, pp. 917-935.
4. Atkinson, P.M. and Tate, N.J. (2000) Spatial scale problems and geostatistical solutions: a review, *Professional Geographer*, vol. 52, pp. 607-623.
5. Atkinson, P.M. (2001) Super-resolution target mapping from soft classified remotely sensed imagery, *Fifth International Conference on GeoComputation*. University of Leeds: Leeds, CD-ROM.
6. Bezdek, J.C., Ehrlich, R. and Full, W. (1984) FCM: The fuzzy *c*-means clustering algorithm, *Computers and Geosciences*, vol. 10, pp. 191-203.
7. Brown, M., Gunn, S.R. and Lewis, H.G. (1999) Support vector machines for optimal classification and spectral unmixing, *Ecological Modelling*, vol.120, pp. 167-179.
8. Deutsch, C.V. and Journel, A. G. (1998) *GSLIB: Geostatistical Software and User's Guide*, Second Edition. Oxford University Press: Oxford.
9. Flack, J., Gahegan, M. and West, G. (1994) The use of sub-pixel measures to improve the classification of remotely sensed imagery of agricultural land, *Proceedings of the 7th Australasian Remote Sensing Conference,* Melbourne, pp. 531-541.
10. Foody, G.M. (1998) Sharpening fuzzy classification output to refine the representation of sub-pixel land cover distribution, *International Journal of Remote Sensing,* vol.19, pp. 2593-2599.
11. Hopfield, J. and Tank, D.W. (1985) Neural computation of decisions in optimization problems, *Biological Cybernetics*, vol.52, pp. 141-152.
12. Journel, A.G. and Huijbregts, C. J. (1978) *Mining Geostatistics*. Academic Press: London.
13. Jupp, D.L.B., Strahler, A.H. and Woodcock, C.E. (1988) Autocorrelation and regularization in digital images I. Basic theory, *IEEE Transactions on Geoscience and Remote Sensing*, vol. 26, pp. 463-473.
14. Schneider, W. (1999) Land cover mapping from optical satellite images employing subpixel segmentation and radiometric calibration, in I. Kanellopoulos, G. Wilkinson and T. Moons, *Machine Vision and Advanced Image Processing in Remote Sensing*. Springer: London.
15. Steinwendner, J., Schneider, W. and Suppan, F. (1998) Vector segmentation using spatial subpixel analysis for object extraction, *International Archives of Photogrammetry and Remote Sensing*, vol. 32, pp. 265-271.
16. Tatem, A.J., Lewis, H.G., Atkinson, P.M. and Nixon, M.S. (2001a) Super-resolution target identification from remotely sensed images using a Hopfield neural network, *IEEE Transactions on Geoscience and Remote Sensing*, vol. 39, pp. 781-796.
17. Tatem, A.J., Lewis, H.G., Atkinson, P.M. and Nixon, M.S. (2001b) Multiple class land cover mapping at the sub-pixel scale using a Hopfield neural network, *International Journal of Applied Earth Observation and Geoinformation* (in press).
18. Tatem, A.J., Lewis, H.G., Atkinson, P.M. and Nixon, M.S. (2002) Land cover simulation and estimation at the sub-pixel scale using a Hopfield neural network, *Remote Sensing of Environment*, vol. 79, pp. 1-14.
19. Thomas, I.L., Benning, V.M., and Ching, N.P. (1987) *Classification of Remotely Sensed Images*, Bristol, Adam Hilger.
20. Verhoeye *et al.* (2000) *IGARSS2001 Scanning The Present and Resolving the Future,* IEEE, Sydney, Australia, CD-ROM.
21. Van Groenigen, J.-W. (1999) *Constrained optimisation of spatial sampling. A Geostatistical Approach*, ITC Publication Series, No. 65, ITC: Enschede, the Netherlands.

NON PARAMETRIC VARIOGRAM ESTIMATOR. APPLICATION TO AIR POLLUTION DATA

L. Bel

Laboratoire de Mathématique. Université Paris-Sud. Bât 425. 91405 Orsay Cedex, France

Abstract: Environmental processes are rarely stationary and isotropic. In order to produce maps of pollutant concentration over a region where few measurements are available, classical kriging performs badly. To have more accurate maps, it is necessary from one hand to take into account external information, such as emissions and meteorological data and from the other hand to release the stationary assumption, modelling the variogram when kriging. In this paper we propose a non parametric estimator of the variogram, we study its theoretical properties and its behaviour on a simulation case. We use this estimator and a chemistry-transport model to produce maps of ozone concentration over Paris area and compare to maps obtained with classical kriging methods.

1. INTRODUCTION

Beyond forecasting the level pollutant concentration for the next day, air monitoring agencies are in charge of estimating the level of pollutant concentration over an entire area, including locations where no measurement has been made.

In that aim we have two methods in mind:

– a statistical interpolation from observations made on the monitoring network,

X. Sanchez-Vila et al. (eds.), geoENV IV – Geostatistics for Environmental Applications, 29-40.

– a simulation by means of a chemistry-transport model on a grid.

As the monitoring network is often too sparse and not well located, no interpolation method can render the complexity of the pollution phenomenon with only measurement data. Besides, deterministic physical models are very complicated and often have biases that can be very high.

It is worth combining the two approaches: the chemistry-transport model is used to catch the phenomenon structure, while measurements on the monitoring network are used to adjust the outputs on the observations.

In order to perform kriging, the variogram needs to be estimated. Usually assumptions of stationarity and isotropy are made, but it is widely recognized that real environment processes are rarely stationary and isotropic. In the case of pollutant concentration, the behaviour of two sites depends more on their typology (rural, urban) than on their distance.

The question is to know if it is better to bypass the assumptions of stationarity and isotropy and use parametric fitting of variograms which are robust and well-tried or if it is more suitable to use more sophisticated variograms, adjusting well the data instationarity but with the drawback of unstability.

Several attempts to model nonstationary covariance or variogram functions have been made, see for example Sampson and Guttorp (1992), Hall and Patil (1994), or Fuentes (2001).

Following Guillot, Monestiez and Senoussi (2000), we propose a non parametric, kernel based estimator of the variogram for nonstationary fields. Firstly we show that it is admissible, that is, it is conditionally negative definite and propose some practical improvements. Then this estimator is compared to classical parametric fitting of the variogram through a simulation study and on a dataset of ozone concentration over Paris area.

2. NON-PARAMETRIC ESTIMATOR OF THE VARIOGRAM

Let us consider a second order, non stationary random field $Z(s)$, defined on a domain D of IR^2, with a covariance function C, and a semivariogram function Γ on $D \times D$.

Let $S = \{s_1,\ldots,s_n\}$ be a set of points of D and $z(s_i,t)$, $1 \leq i \leq n$, $1 \leq t \leq T$ be a set of T i.i.d. observations of Z at sites i.

Let us denote by C_{emp} the empirical covariance matrix of Z, and Γ_{emp} the empirical semivariogram of Z, namely,

$$C_{emp}(i,j) = c_{ij} = \frac{1}{T}\sum_{t=1}^{T}(z(s_i,t)-\overline{z}(s_i))(z(s_j,t)\overline{z}(s_j))$$

$$\Gamma_{emp}(i,j) = \gamma_{ij} = \frac{1}{2T}\sum_{t=1}^{T}(z(s_i,t)-z(s_j,t)-(\overline{z}(s_i)-\overline{z}(s_j)))^2$$

Let K be a non-negative kernel defined on $D \times D$. We study the nonparametric estimator of C and Γ obtained by regularization of C_{emp} and Γ_{emp}:

$$\widehat{C}_h(u,v) = \sum_{i,j} c_{ij} \frac{K_h(u-s_i, v-s_j)}{\sum_{k,l} K_h(u-s_k, v-s_l)}$$

$$\widehat{\Gamma}_h(u,v) = \sum_{i,j} \gamma_{ij} \frac{K_h(u-s_i, v-s_j)}{\sum_{k,l} K_h(u-s_k, v-s_l)}$$

where $K_h(u,v) = K(u/h,v/h)$ for any positive real h. We suppose K is a separable kernel i.e. $K(u,v) = k(u)k(v)$ for all $(u,v) \in \mathrm{IR}^2$. This assumption is sufficient to prove properties of $\hat{C}_h$ and $\hat{\Gamma}_h$, but it is not clear that it is necessary. This estimator is quite the same as the one proposed by Hall and Patil (1994), in the stationary case.

2.1 Positive definiteness of the covariance estimator

Proposition 1 $\hat{C}_h$ *is positive definite.*

Proof. K is positive definite, i.e. for any $(u_1,\ldots,u_m)$ in D and complex numbers $(\theta_1,\ldots,\theta_m)$, $\Sigma_{i,j}\theta_i\overline{\theta}_j K_h(u_i,u_j) \geq 0$. For any integer m, complex valued vector $(\alpha_1,\ldots,\alpha_m)$ and points $u_1,\ldots,u_m$ of D, we have

$$\sum_{k,l}\alpha_k\overline{\alpha}_l \widehat{C}_h(u_k,u_l) = \sum_{i,j} c_{ij}\sum_{k,l}\alpha_k\overline{\alpha}_l \frac{K_h(u_k-s_i,u_l-s_j)}{\sum_{i,j} K_h(u_k-s_i,u_l-s_j)}$$

$$= \sum_{i,j} c_{ij}\theta_{ij}$$

The empirical covariance matrix C is positive definite, therefore, by Fejer's theorem, it is enough to prove the positive definiteness of the matrix $\Theta = (\theta_{ij})$ to prove that $\Sigma_{i,j}\, c_{ij}\theta_{ij} \geq 0$. Let $(\beta_1,\ldots,\beta_n)$ be a complex valued vector,

$$\sum_{i,j} \beta_i \bar{\beta}_j \theta_{ij} = \sum_{i,j} \sum_{k,l} \alpha_k \bar{\alpha}_l \beta_i \bar{\beta}_j \frac{K_h(u-s_i, v-s_j)}{\sum_i k_h(u_k - s_i) \sum_j k_h(u_l - s_j)}$$

$$= \sum_{i,j} \sum_{k,l} \frac{\alpha_k \beta_i}{\sum_i k_h(u_k - s_i)} \frac{\bar{\alpha}_l \bar{\beta}_j}{\sum_j k_h(u_l - s_j)} K_h(u_k - s_i, u_l - s_j)$$

$$\geq 0$$

and Θ is positive definite.

2.2 Conditional negative definiteness of the variogram estimator

It is straightforward to verify that the empirical variogram is conditionally negative definite, that is for any complex valued vector $(\alpha_1,\ldots,\alpha_m)$ such that $\Sigma_i\ \alpha_i = 0$, $\Sigma_{i,j}\ \alpha_i \bar{\alpha}_j \gamma_{i,j} \leq 0$. The point is that we can write

$$2\gamma_{i,j} = \sigma_i^2 + \sigma_j^2 - 2c_{ij}$$

where $\sigma^2_i = \frac{1}{T} \Sigma_t (z(s_i,t) - \bar{z}(s_i))^2$ is the empirical variance of $Z(s_i)$. So

$$\sum_{i,j} \alpha_i \bar{\alpha}_j \gamma_{i,j} = \sum_i \alpha_i \sigma_i^2 \sum_j \bar{\alpha}_j + \sum_j \bar{\alpha}_j \sigma_j^2 \sum_i \alpha_i - 2 \sum_{i,j} \alpha_i \bar{\alpha}_j c_{ij}$$

$$= -2 \sum_{i,j} \alpha_i \bar{\alpha}_j c_{ij}$$

$$\leq 0$$

Proposition 2 $\hat{\Gamma}_h$ *is conditionally negative definite.*

Proof For any integer m, complex valued vector $(\alpha_1,\ldots,\alpha_m)$ such that $\Sigma_i \alpha_i = 0$ and points $u_1,\ldots,u_m$ of D, we have

$$
\begin{aligned}
\sum_{k,l}\alpha_k\bar{\alpha}_l\hat{\Gamma}_h(u_k,u_l) &= \sum_i\sum_k\sigma_i^2\alpha_k\frac{k_h(u_k-s_i)}{\sum_i k_h(u_k-s_i)}\sum_l\bar{\alpha}_l \\
&+\sum_j\sum_l\sigma_j^2\bar{\alpha}_l\frac{k_h(u_l-s_j)}{\sum_j k_h(u_l-s_j)}\sum_k\alpha_k \\
&-2\sum_{i,j}c_{ij}\sum_{k,l}\alpha_k\bar{\alpha}_l\frac{K_h(u_k-s_i,u_l-s_j)}{\sum_{i,j}K_h(u_k-s_i,u_l-s_j)} \\
&= -2\sum_{i,j}\sum_{k,l}c_{i,j}\alpha_k\bar{\alpha}_l\frac{K_h(u_k-s_i,u_l-s_j)}{\sum_{i,j}K_h(u_k-s_i,u_l-s_j)} \\
&\leq 0
\end{aligned}
$$

2.3 Practical design

The covariance and variogram estimators will be used in kriging systems. In the case the process Z is not continuous, for example if there is a nugget effect or if there is a measurement error, $\hat{C}(s_i,s_i)$ is not an estimation of $Var(Z(s_i))$ and $\hat{\Gamma}(s_i,s_i)$ is not necessarily 0. Hence the diagonal of the matrix involved in the kriging system has to be replaced by the empirical variance when the covariance is used and by 0 when the variogram is used. In both cases, it remains to check that they are admissible covariance or variogram.

Case of the covariance Let $\tilde{C}_h = (\tilde{c}_{ij})$ be the matrix such that

$$
\begin{aligned}
\tilde{c}_{ij} &= \hat{c}_{ij} \quad \text{if} \quad i \neq j \\
\tilde{c}_{ii} &= \sigma_i^2
\end{aligned}
$$

Since $\tilde{c}_{ii} = \sum_{k,l} c_{k,l}\,(k_h(s_k - s_i) \,/\, (\sum_k k_h(s_k - s_i)) = Var(\sum_k \alpha_k Z_k)$, for suitable h we will have $\hat{c}_{ii} < \tilde{c}_{ii}$ C_h can be written $\tilde{C}_h = \tilde{C}_h + \Sigma$, Σ being a diagonal matrix with positive elements, hence $\tilde{C}_h$ stills being positive definite.

Case of the variogram Let $\hat{\Gamma}_h = (\tilde{\gamma}_{ij})$ be the matrix such that

$$
\begin{aligned}
\tilde{\gamma}_{ij} &= \hat{\gamma}_{ij} \quad \text{if} \quad i \neq j \\
\tilde{\gamma}_{ii} &= 0
\end{aligned}
$$

For any complex valued vector $(\alpha_1,\ldots,\alpha_n)$ such that $\sum_i = 0$ we have

$$\begin{aligned}\sum_{i,j}\alpha_i\bar{\alpha}_j\tilde{\gamma}_{ij}(s_i,s_j) &= \sum_{i,j}\alpha_i\bar{\alpha}_j\hat{\gamma}_{ij}(s_i,s_j)-\sum_{i=j}\alpha_i\bar{\alpha}_j\hat{\gamma}_{ij}\\ &= \sum_{i,j}\alpha_i\bar{\alpha}_j\hat{\gamma}_{ij}(s_i,s_j)-\sum_{i}|\alpha_i|^2\hat{\gamma}_{ii}\\ &\le 0\end{aligned}$$

since $\hat{\Gamma}_h$ is conditionnally negative definite and γ_{ii} is a weighted sum of positive terms.

It has to be noticed that if the set S is sparse and if s_0 is close to an isolated point s_{i0}, it may happen that $\hat{\Gamma}_h(s_0,s_i)$ and $\hat{C}_h(s_0,s_i)$ be very close to $\hat{\Gamma}_h(s_{i0},s_i)$ and $\hat{C}_h(s_{i0},s_i)$. When ordinary kriging is performed, this leads to kriging weights equal to 0 if $i \neq i_0$ and 1 if $i = i_0$. In such a case we have $\hat{z}(s_0) = Z(s_{i0})$ and a vanishing kriging variance.

While it is well established that the choice of the kernel K is not a crucial point, the choice of the bandwith h is an important issue. Large h lead to oversmooth the covariance or the variogram and in the kriging setting measurements at monitoring sites will be considered as almost independant. Small h lead to the empirical covariance or variogram at monitoring sites, but estimation at non-monitoring sites will lack robustness and kriging results can be quite inappropriate. In the framework of kriging the choice of parameter h will generally be driven by the minimization of a cross validation criteria.

3. SIMULATION EXAMPLE

In order to check whether non parametric estimation leads to an improvement in kriging nonstationary fields, we simulate a Gaussian random function Z on a domain D = [0;10] × [0;10], deforming the space:

$$\operatorname{cov}\left(Z(s),Z(t)\right) = \exp(-\|\phi(s)-\phi(t)\|)$$

with $\phi(s) = \phi(x,y) = ((1/\sqrt{2.5}\,||s\text{-}O||)\,(x\text{-}5.05),\ (1/\sqrt{2.5}\,||s\text{-}O||)\,(y\text{-}5.05))$. O is the point (5.05,5.05). Points of the domain which are located near the center are slightly correlated with their neighbours, points which are far from the center are highly correlated with their neighbours and this process is strongly nonstationary. We consider 100 realizations of Z at 51 × 51 sites on a regular grid of D. The simulations are carried using a turning bands algorithm, written by Lantuéjoul, 2001. 100 points are randomly sampled as observed points and we perform kriging on a grid of points with entire coordinates (these points cannot be sampled as observed points). Figure 1 shows a realization of field Z together with the sampled points.

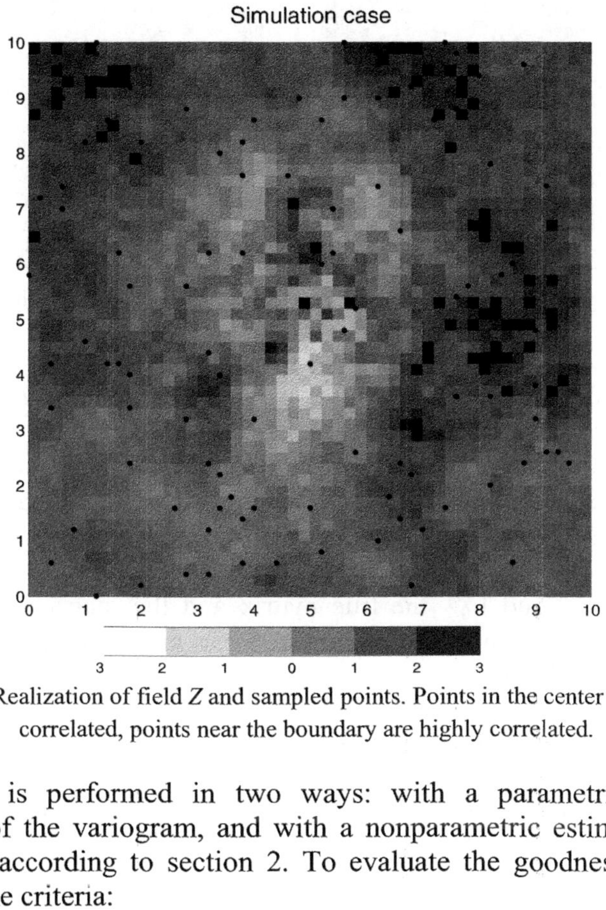

Figure 1. Realization of field Z and sampled points. Points in the center are slightly correlated, points near the boundary are highly correlated.

Kriging is performed in two ways: with a parametric isotropic estimator of the variogram, and with a nonparametric estimator of the variogram according to section 2. To evaluate the goodness of fit we calculate the criteria:

$$C = \frac{1}{100}\sum_{t=1}^{100} c(t) \qquad c^2(t) = \frac{1}{121}\sum_{i=1}^{121}(z(s_i,t) - \hat{z}(s_i,t))^2$$

in both cases.

Fitting the variogram with an exponential model gives a range a = 4.0897, a sill c = 1.0156 and the value of the criteria is 0.4116.

The nonparametric estimator of the variogram is built from the empirical variogram with Gaussian kernel and bandwith h = 0.5. The value of the criteria is 0.4326.

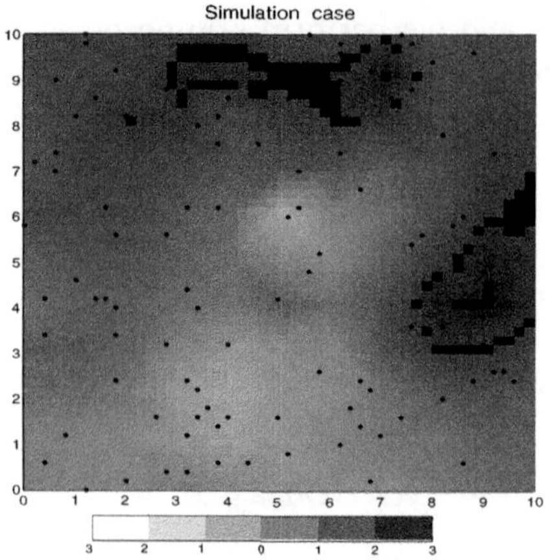

Figure 2. Kriging with parametric estimator, $c(1) = 0.3630$.

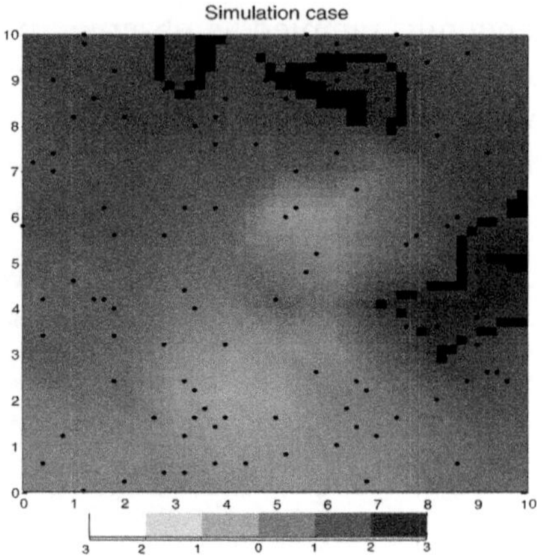

Figure 3. Kriging with nonparametric estimator, $c(1) = 0.3859$.

Figures 2 and 3 show the kriging map in both cases for $t = 1$.

With the true variogram (that is the one used to simulate the data) the criteria would be 0.3990. These results are quite surprising: if we seek the matrices Γ_0 and the right hand side of the kriging linear system to be solved, those given by the nonparametric estimator are closer to the true ones than those given by the parametric isotropic estimator. Indeed noting Γ_0, Γ_0^{p} and Γ_0^{np} the true matrix and the matrices given by the parametric and the nonparametric estimator, and γ_0, γ_0^{P} and γ_0^{np} the right hand sides we have

$$\left\|\Gamma_0 - \Gamma_0^{p}\right\| = 18.85 \qquad \left\|\Gamma_0 - \Gamma_0^{np}\right\| = 5.1275$$

$$\left\|\gamma_0 - \gamma_0^{p}\right\| = 22.69 \qquad \left\|\gamma_0 - \gamma_0^{np}\right\| = 5.8548$$

but

$$\left\|\Lambda - \Lambda^{p}\right\| = 0.5463 \qquad \left\|\Lambda - \Lambda^{np}\right\| = 0.9974$$

where Λ, Λ^{P}, and Λ^{np} are the weighting coefficients given by solving the kriging system in each case. That is, the kriging weights are closer to the "best" ones solving the system with the parametric estimator than the ones obtained, solving the system with the nonparametric estimator. This is probably due to the fact that the matrix to be inversed is better conditioned in the parametric case. This is illustrated by Figure 4 that shows the value of $\gamma_0(s_0,s_i)_{i=1,100}$ for $s_0=(5,6)$ a point in the center of the domain, together with $\gamma_0^{P}(s_0,s_i)_{i=1,100}$ and $\gamma_0^{np}(s_0,s_i)_{i=1,100}$ and $\lambda(s_0,s_i)_{i=1,100}$. $\lambda_0^{P}(s_0,s_i)_{i=1,100}$ and $\lambda_0^{np}(s_0,s_i)_{i=1,100}$. Figure 5 is for $s_0=(8,8)$ a point near the boundary.

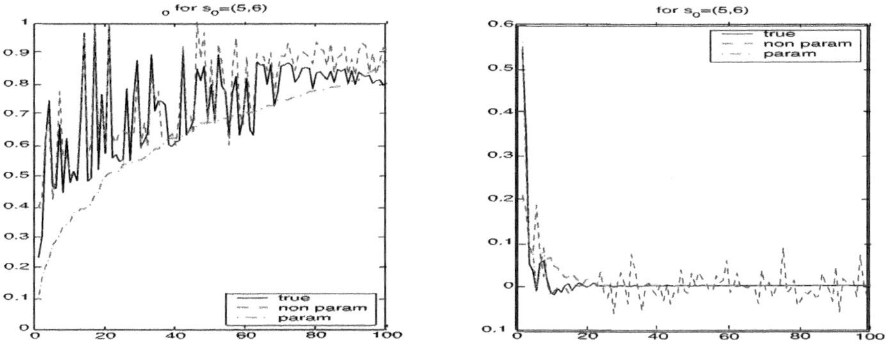

Figure 4. Right-hand side γ_0 and solution λ_0 of the kriging system, for point s=(5,6). In solid line the true one, in dashed line the nonparametric estimator and dotted line the parametric estimator.

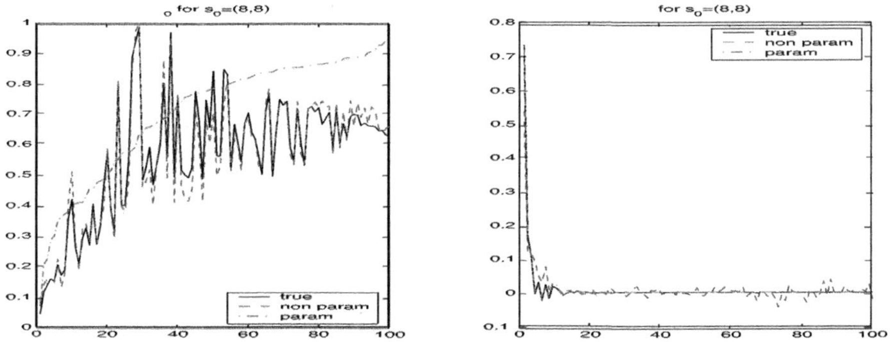

Figure 5. Right-hand side γ_0 and solution λ_0 of the kriging system, for point s=(8,8). In solid line the true one, in dashed line the nonparametric estimator and dotted line the parametric estimator.

4. AIR POLLUTION DATA

We now deal with a dataset of ozone concentration measured each day at 15h at 21 monitoring stations in the Paris area during summer 99. The aim is to estimate the pollutant concentration over an area of 150 km by 150 km. There are 6 rural stations located a few tens of kilometers away from the city center, 3 suburban stations and 12 urban stations.

The monitoring network is obviously too sparse to render the complexity of the phenomenon, just kriging the observations. Figure 6 shows the kriging map obtained for 17 July, a highly polluted day. All the North West area shows very high concentrations due to the influence of the stations located in this zone. Physically this map doesn't make sense. We have at our disposal outputs of a deterministic chemistry-transport model with resolution of 6km × 6km (Blond *et al.*, 2002). Figure 7 shows the map performed by this model together with observations at monitoring stations for 17 July.

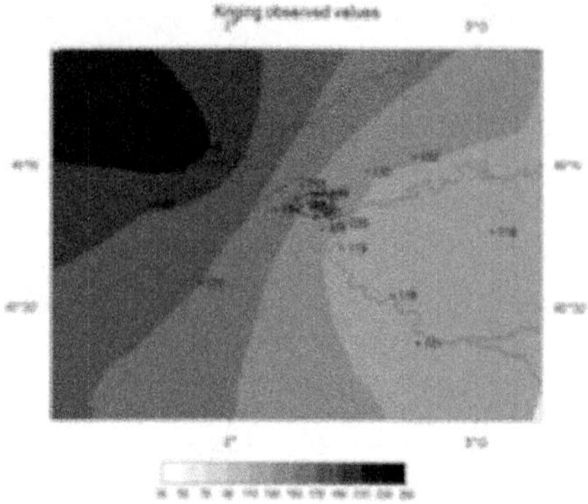

Figure 6. Map of ozone concentration over Paris (center of the map) area, kriging observed values.

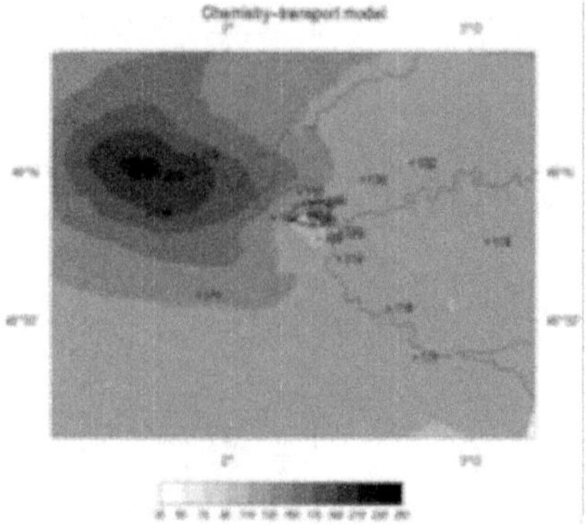

Figure 7. Map of ozone concentration over Paris area, given by the Chemistry-Transport-Model.

The shape is totally different than the one obtained when kriging the observations, and it is physically satisfactory, but at monitoring stations predicted values are quite different from observed values. Kriging differences between the values of the deterministic chemistry-transport model at monitoring stations and their observed values give an estimate of the difference between the real concentration and the model output for every point of the grid. The results are added to the model output to give an estimate of the concentration field over the area. This is shown in Figure 8 for 17 July, ordinary kriging is performed with an exponential model for the variogram.

It is widely recognized that real environment processes are rarely stationary and isotropic. In our case, we guess that most of the nonstationarity has been taken off taking into account the deterministic chemistry-transport model, but rural and urban stations still have very distinct behaviour with respect to the model. Working on the differences we compare the nonstationary, nonparametric method with the parametric stationary method. The empirical variogram is computed over the entire period (actually only on 53 days instead of 123 because of missing data), a classical exponential variogram for the entire period is fitted and a nonparametric nonstationary variogram is estimated according to section 2. Figure 9 shows the result of cross validation for each station for both methods. The nonparametric method is slightly more accurate than the parametric one, still it is not true for every station.

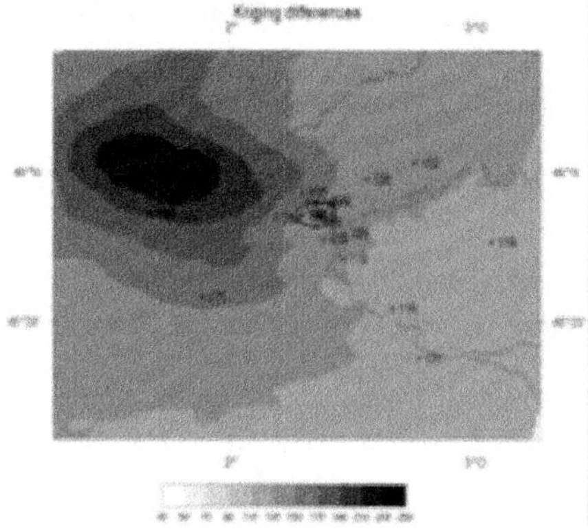

Figure 8. Map of ozone concentration over Paris (center of the map) area, kriging the differences between observed values and model outputs.

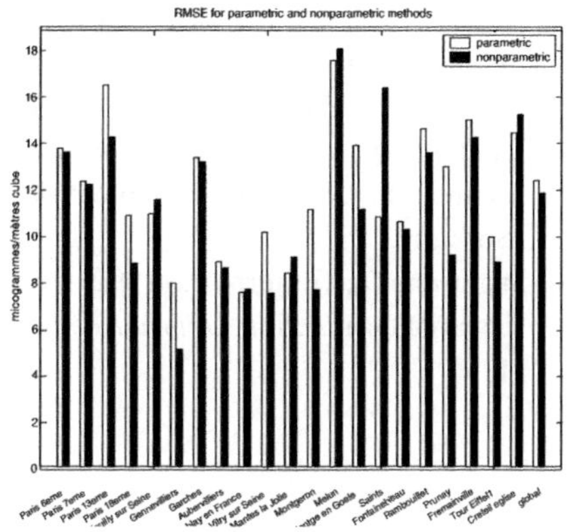

Figure 9. Values of the Root Mean Square Errors obtained by cross validation for each station with the parametric estimator (open bars) and the nonparametric estimator (filled bars).

5. CONCLUSION

The nonparametric variogram estimator gives better estimates of the variogram than the parametric one when the process is strongly nonstationary. However the simulation study shows that even in this case it is too instable to improve the kriging algorithm. Applied to a real dataset, cross validation show a little improvement with the nonparametric estimator, but observations are too rare to make the cross validation criteria really relevant.

In order to improve the skill of this estimator, some modifications can be tried: truncating to 0 the variogram function for high distances like Patil's and Hall's method, or using stable inverse matrice solving the kriging system to avoid instability effects.

REFERENCES

1. Blond, N., Bel, L., Vautard, R. (2002). Three-dimensional ozone data analysis with an air quality model over Paris area. *(submitted)*
2. Fuentes, M. (2001). A high frequency approachfor non-stationary environmental processes, *Environmetrics*, **12**:469-483.
3. Guillot, G., Senoussi, R., Monestiez, P. (2000). A positive definite estimator of the non stationary covariance of random fields. In: *GeoENV 2000: third European*

Conference on Geostatistics for Environmental Applications, P.Monestiez, D.Allard and R.Froidevaux, eds, Kluwer, Dordrecht.

4. Hall, P., Patil, P. (1994). Properties of nonparametric estimators of autocovariance for stationnary random fields. *Prob. Theory Relat. Fields*, **99**:399-424.
5. Lantuéjoul, Ch. (2001). *Geostatistical Simulation. Models and algorithms*. Berlin, Springer.
6. Sampson, D., Guttorp, P. (1992). Nonparametric estimation of nonstationnary spatial covariance function. *Journal of the American Statistical Association*, **87(147)**:108-119.

IMPROVING SATELLITE IMAGE FOREST COVER CLASSIFICATION WITH FIELD DATA USING DIRECT SEQUENTIAL CO-SIMULATION

A.M. Bio[1], J. Carvalho, P. Maio and L. Rosário[2]
[1]Environmental Group of the Centre for Modelling Petroleum Reservoirs, CMRP/IST, Av. Rovisco Pais, 1049-001 Lisbon, Portugal. E-mail: anabio@ist.utl.pt; [2]Direcção Geral de Florestas, Portuguese Ministry of Agriculture, Rua 5, Bro. Calçada dos Mestres 13, 1070-058 Lisbon, Portugal

Abstract: Land surface cover classification is assessed using Direct Sequential Co-Simulation, combining field observations with classified remote sensing data. Local co-regionalisation models are applied to account for local differences in both, field data availability and distribution, and the correlation between these hard data and the classified satellite images as soft data. The suggested methodology is based on two criteria: influence of the field observations dependent on field data availability and proportional to field data proximity; and, influence of the soft data dependent on their local correlation to the hard data. The method is applied to a study of four economically important forest tree species on the Setúbal peninsula. Local correlations between field observations (hard data) and satellite image classification results (soft data) are computed and interpolated for the whole study area. Direct Sequential Co-Simulation is performed conditioned to the local correlation estimates, yielding estimates and uncertainties for forest cover proportions. Cover-probabilities are combined into one forest cover classification map, constrained to reproducing the global proportions for the different classes. Direct Sequential Co-Simulation results show more contiguous forest covers – i.e. more spatial contiguity – than the classified satellite image. In comparison to the field data used for calibration during satellite image classification, the proposed simulation method improved forest cover estimations for species with good local correlation between hard and soft data and worsened those for species with poor local correlations.

X. Sanchez-Vila et al. (eds.), geoENV IV – Geostatistics for Environmental Applications, 41-54.

1. INTRODUCTION

To evaluate forest resources over large areas, remote sensing data are a low-cost and abundant source of information. Satellite images can be classified into cover classes, with map units (grid cells) assigned to the most likely coverage class.

Satellite images, though abundant, are evidently *soft data*, given the uncertainties in attaining and inferring information from satellite images. Comparison with forest coverage records collected in the field – i.e. reliable *hard data* – will reveal wrongly classified classes or geographic areas. Classical satellite image analysis is done on a pixel-by-pixel basis yielding an often too scattered image of land cover, as spatial continuity between neighbouring pixels is generally disregarded.

Field observations, on the other hand, are costly and therefore scarce, most of the times too scarce to adequately estimate resources for large areas.

Combination of abundant soft with scarce *hard data* in a geostatistical framework is, therefore, common practice to increase the accuracy of forest resource estimates based on satellite images (Fouquet and Mandallaz 1993, Nunes *et al.* 2000). Using a co-regionalisation model, relating field to satellite-image data allows for the combination of the scarce primary and the abundant secondary variables (Soares *et al.* 1997). A regional co-regionalisation model, e.g. based on the correlation between the two variables for the entire region (Goovaerts 2000) has its drawbacks, as the same spatial model – i.e. identical cross-variograms and cross-variances – may not be valid for the whole region. To overcome that problem, recent studies propose co-located co-kriging with local co-regionalisation models (Pereira *et al.* 2000).

Estimations based on a limited number of field observations often fail to reproduce the spatial variability of the studied variable, frequently producing estimation artefacts (like *bull eyes* surrounding sample locations), especially in the presence of a variable with a large-range variogram.

Given the nature of our data – forest species covers on distinct, bounded areas – kriging estimates would result in unrealistic, smooth surfaces. Simulation is here likely to perform more accurately.

The proposed methodology constitutes a geostatistical satellite image calibration based on the stochastic Direct Sequential Co-Simulation (DSCoS) procedure, as proposed by Soares (2001) and on local co-regionalisation models. This is an alternative approach to spatial forest species characterisation from satellite imagery and field data. It is based on two simultaneously applied criteria: the influence of the field observations is dependent on field-data availability and proportional to field-data proximity (i.e. the influence of field observations on species cover estimates is greatest at field sample locations, decreasing with increasing distance to these locations); and, the influence of the *soft data* is dependent on their local correlation to the *hard data* (i.e. it is higher in regions of high correlation than in regions of low correlation).

The procedure builds on a classical supervised satellite image classification into forest species and other covers. This constitutes a first forest cover image that will function as secondary variable for a geostatistical calibration process based on forest-cover field observations. Geostatistical simulation will reproduce a spatial pattern similar to that of the classified satellite image, but following the statistics of the 'reliable' field data. Its results hence constitute a form of calibration of the original satellite image classification.

In the present study, the probability of land surface cover with four forest tree species – Eucalypt, Umbrella Pine, Maritime Pine and Cork oak – is assessed through Direct Sequential Co-Simulation using local co-regionalisation models to account for local differences in both, field-observations' (i.e. hard data) availability and distribution, and the correlation between field data and probabilities of species cover obtained from a supervised Landsat7 satellite image classification as secondary variable or soft data. The study is part of a LIFE-Environment project supported by the European Community.

2. HARD AND SOFT DATA

The study area is the Setúbal peninsula, south of Lisbon, Portugal, covering about 154000 ha of forested and bare mountainous grounds and urban areas. The *hard data* consist of 70 field observations (Figure 1) of forest species cover with records of Eucalypt (*Eucalyptus globulus*), Umbrella Pine (*Pinus pinea*), Maritime Pine (*Pinus pinaster*), Cork oak (*Quercus suber*), among others, collected from 2000 to 2002. The *soft data* are posterior probabilities for the same cover classes and of the sum of all other cover classes obtained from an assisted Landsat7 satellite image maximum-likelihood classification (on a 30×30m grid), calibrated on a set of 214 training areas (collected in 1995), with post-classification smoothing of noise using a 3×3 grid-cell low-pass majority filter (Figures 1). Comparison of the satellite image classification with a set of 214 training areas (areas covered by a single type of coverage) of variable size, revealed better agreement for Eucalypt and Cork oak (87% and 92% of correctly classified grid cells, respectively), than for Umbrella and Maritime pine (52% and 55%, respectively). For the latter two species, misclassification consisted mainly of Umbrella pine field observations classified as Maritime pine cover

(30%) and of Maritime pine field observations classified as Eucalypt (20%) or Umbrella pine cover (10%). Classified map and training areas extend beyond the study area; the 156 training areas within the study area (Table 1) were selected to validate the forest cover classification calibration obtained through Direct Sequential Co-Simulation.

The posterior probabilities obtained from the classified satellite image for each of the original 21 cover classes were scaled to unit sum and all but the four forest species classes were aggregated into one class, hereafter denominated "other". Posterior probabilities were upscaled and averaged to the 90×90 m grid used throughout this study.

Satellite image forest cover classification, with each grid cell assigned to the most likely coverage class, is summarized in Table 1. This classification, yet upscaled to a 90×90 m grid resolution, will be used for comparison with the classification results obtained using direct sequential co-simulation (Figure 5). During upscaling each new grid cell was assigned to the most frequent class of the nine underlying 30×30 m grid cells. Apart from unavoidable errors in the presence of equally frequent competing classes, this led to a penalization of the less frequent and more scattered cover classes. Therefore, Eucalypt, Umbrella Pine, Maritime Pine and Cork oak had their cover percentages reduced to 6.2, 2.4, 5.7 and 16.1, respectively, while the "other" cover percentage raised to 69.7.

Table 1. Land cover of the study area according to the classified satellite image and number and area of the respective training data.

	Coverage		Training data areas	
Class	Area (ha)	Area (%)	Number	Area (ha)
Eucalypt	10460	6.8	14	87
Umbrella pine	4654	3.0	18	56
Maritime pine	10410	6.8	34	48
Cork oak	26741	17.4	11	64
Other	101570	66.0	79	162
Sum	153835	100.0	156	418

3. DIRECT SEQUENTIAL CO-SIMULATION

The simulation of the histogram for the field data measurements was hampered by a problem: their shape. These histograms are bi-modal with a high spike (comprising > 70% of values) at the origin (see Figure 3). The use of Sequential Indicator Simulation (SIS), however, requires the estimation of indicator variograms, unfeasible for such uneven classes. Hence we applied Direct Sequential Simulation not on the original data but on their cumulative probability – a more continuous and uniformly distributed variable.

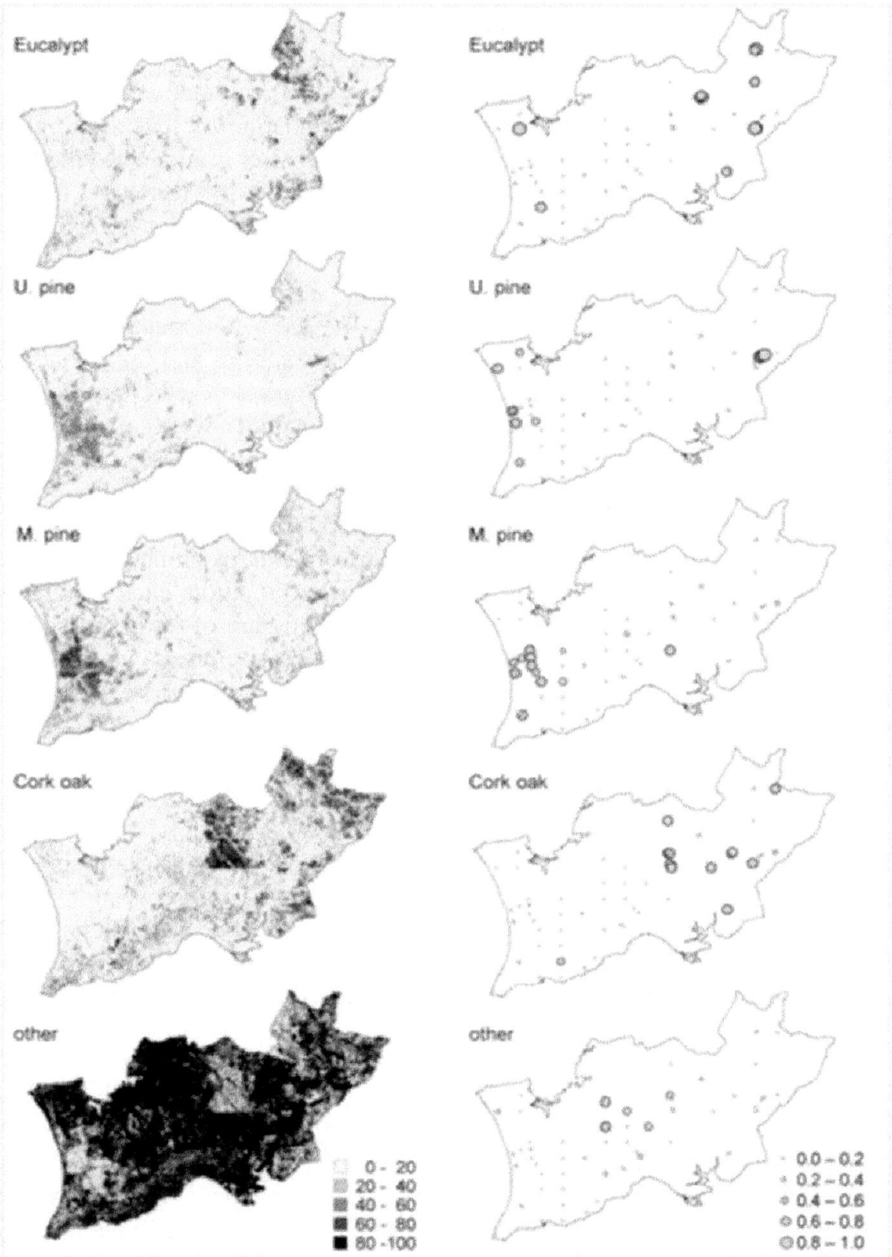

Figure 1. Proportion of land covered forest species and by any other coverage obtained through classification of satellite images (left column) and observed in the field (right column)

Direct Sequential Simulation (DSS) and Co-Simulation (DSCoS) were first proposed by Soares (2001), based on previous work of Journel (1994) and Caers (1999). DSS allows for the simulation of untransformed continuous variables, using local simple kriging estimates of the variable's mean and variance to sample from the global cumulative distribution function. Analogously, DSCoS allows joint simulation of several variables without previous transformation. In this study, DSCoS is applied to the field observations (as *hard data*) and satellite classification of forest cover (as *soft data*), based on a co-regionalisation model that reproduces local correlation between these two variables.

Prior to the co-simulation procedure, *hard* and *soft data* were transformed to a uniform distribution U(0,1) with mean = 0.5. Ten simulations were computed for each of the forest species and for the sum of other land covers. The statistics of each simulation result were compared to those of the *hard data*. Subsequently, all simulations were averaged and back-transformed to the original scale. Simulation variance was computed as a measure of uncertainty. An example is given for Cork oak (Figure 3). All simulations had approximately uniform distributions with statistics similar to those of the transformed data. Inherent to the procedure, simulations respect the *hard data* at their locations. After back-transformation, each simulation reproduces approximately the histogram and variogram of the original *hard data*, while the simulation average has statistics similar to those of the original *soft data*. Analogous results were obtained for all cover classes.

4. LOCAL CORRELATIONS

DSS is based on co-located simple co-kriging estimates, applying the knowledge of spatial covariances for the *hard data* and between *soft* and *hard data*. According to the Markov-Bayes approximation (Goovaerts 1997) one needs only to estimate the covariances of *hard data* and the correlation coefficients between *soft* and *hard data* which is the correlogram between *soft* and *hard data* at distance $h = 0$.

As field observations are scarce, co-regionalisation models are based on the indirect measure of local correlation between the scarce but reliable field observations (as primary variable or *hard data*) and the classified satellite image (secondary variable or *soft data*) available for the whole study area.

Global Pearson correlation coefficients between *hard* and *soft data* (*i.e.* scaled posterior probabilities from the classified image for the nine 30×30 m grid cells centred at *hard data* locations) are: 0.60 for Eucalypt, 0.50 for Umbrella pine, 0.57 for Maritime pine, 0.74 for Cork oak and 0.55 for other cover classes. Because, correlation is not homogeneous for the whole study area (Figure 2) and *hard data* are few and not regularly distributed, local correlations were estimated to account for local differences in both, *hard data* availability and hard-soft data correlation.

For each of the 70 *hard data* points, local correlations between *hard data* within a given radius and the corresponding *soft data* were computed. Several neighbourhood radii were tested; a radius of 15000 m was considered most appropriate considering both the number of neighbourhood samples involved and spatial correlation variability in the data. The local correlations were subsequently interpolated through ordinary kriging for the whole study area, on a 90×90m grid. This way, correlation surfaces were obtained for each of the forest species and for the sum of the remaining coverage classes (Figure 2). Local correlations range from 0 to 0.95, 0.98, 0.91, 0.96 and 0.83, and average 0.56, 0.65, 0.28, 0.78 and 0.61, for Eucalypt, Umbrella Pine, Maritime Pine, Cork oak and others, respectively. More than half of the local correlation map for Maritime pine displays low correlations due to the lack of hard data in Maritime pine-covered areas in the east of the study area.

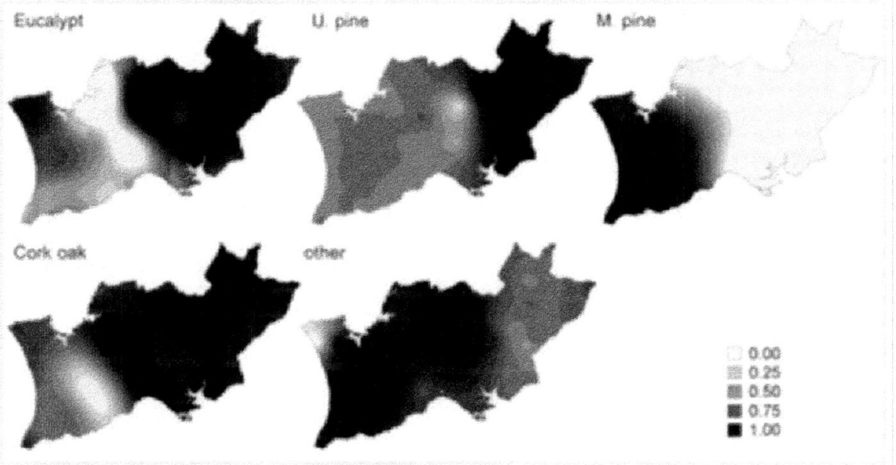

Figure 2. Correlation surfaces for each of the four studied forest species and other cover classes obtained by ordinary kriging of local correlations.

5. COVERAGE PROPORTION ESTIMATES

The coverage proportion estimates obtained through Direct Sequential Co-Simulation, and their simulation variance are presented in Figure 4. In comparison to the coverage probabilities derived from the satellite image (Figure 1, left column), which were used as *soft data* input, simulation results show smoother coverage proportion distributions. Forest species distribution does not always match; *e.g.* Umbrella pine appears in the upper western corner of the study area in the simulation, but not in the satellite image classification.

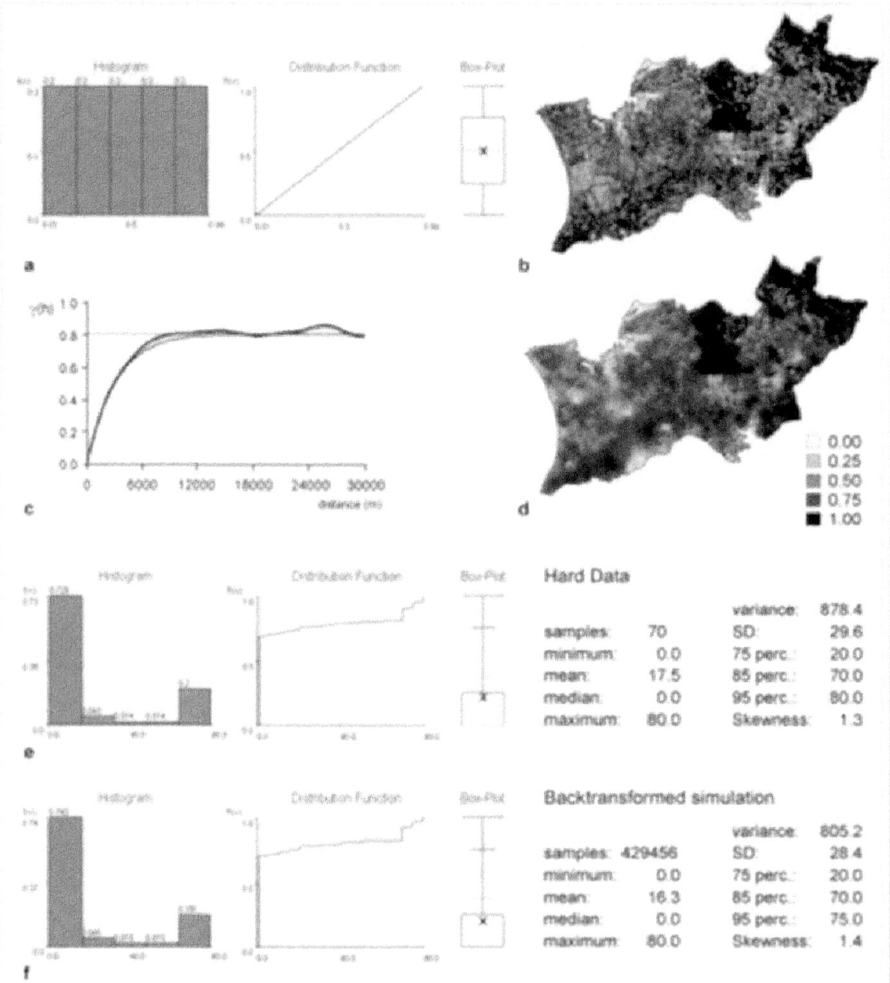

Figure 3. Example of DSCoS applied to Cork oak forest cover: a) statistics of the transformed soft data (with Cork oak coverage transformed to a uniform distribution on the x-axes); b) transformed soft data; c) variogram model for the transformed hard data (black curve) and experimental variogram for one of the simulations (thicker grey curve); d) average simulation result; e) statistics of the hard data and f) statistics of one back-transformed simulation (both with Cork oak cover percentage on the x-axes).

Spatial uncertainty measured by the variance of the simulated images, is closely related to the degree of local correlation. Comparison of the variance and local correlation surfaces (Figure 2) reveals coincidence between low correlation and high uncertainty zones, and vice versa. Furthermore, uncertainty appears to be less for areas with higher proportions of land covered by the respective class.

6. FOREST COVER CLASSIFICATION

The forest cover of the study area was classified, combining the simulation estimates in a single classification map. In order to aggregate the five resulting cover-probability maps (for the four forest species and other cover) into one grid map, each grid cell was assigned to the most likely cover class under the constraint of reproduction of global proportions for the different classes; *i.e.* the grid cells with the highest probabilities for a given class were assigned to that class until its global proportion was reached using a dynamical classification procedure (Soares 1992, Goovaerts 1997). The classes' cover proportions for the study area given by the original classified satellite image were taken as indicative of the classes' real global proportions.

Analogous to the cover-proportion maps (Figure 4) the DSCoS classification yields a smoother image than the satellite image *a priori* classification (for comparison, upscaled to the 90×90 m resolution used for simulation; Figure 5). Classifications differ most for the Umbrella pine (with only 21% overlap) and Maritime pine cover (26% overlap), with large areas originally classified as Maritime pine and "others", respectively (Table 2). Notice that the distinction between Umbrella and Maritime pine cover was particularly difficult in the SI classification (*cf.* Table 3). Comparison of the original 30´30 m satellite image classification is slightly more divergent, as it is evidently less smooth.

Table 2. Comparison of land cover percentages of study areas according to the classified satellite image and the DSCoS

		Satellite image classification				
		Eucalypt	U. pine	M. pine	Cork oak	Other
DSCoS	Eucalypt	41.6	3.3	5.2	16.3	33.5
Class	U. pine	8.2	21.1	49.4	2.4	18.9
	M. pine	2.8	6.3	26.1	4.6	60.2
	Cork oak	4.4	0.4	0.8	62.7	31.7
	Other	3.2	1.6	2.9	5.6	86.7

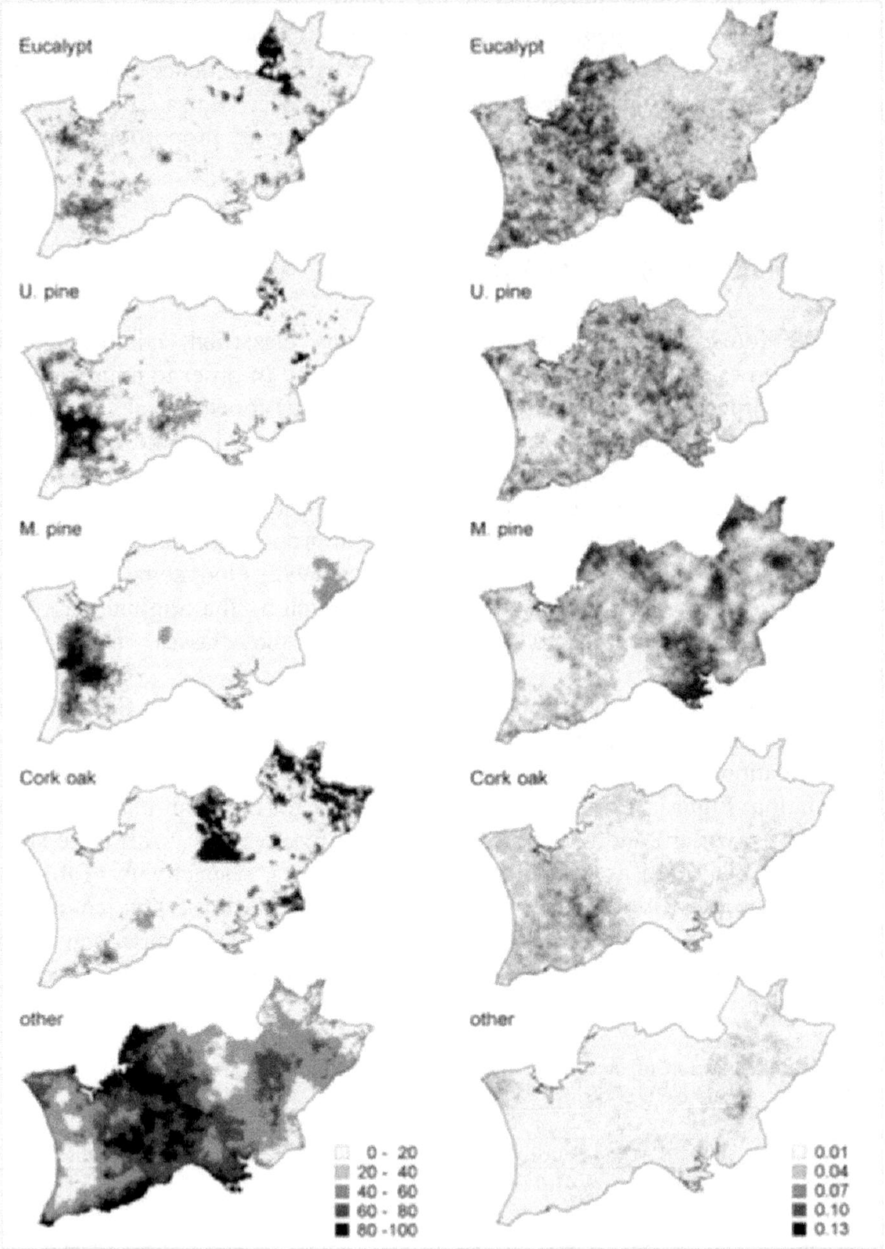

Figure 4. DSCoS coverage proportion estimates (left) and estimation variance (right) for the four forest species and other cover classes.

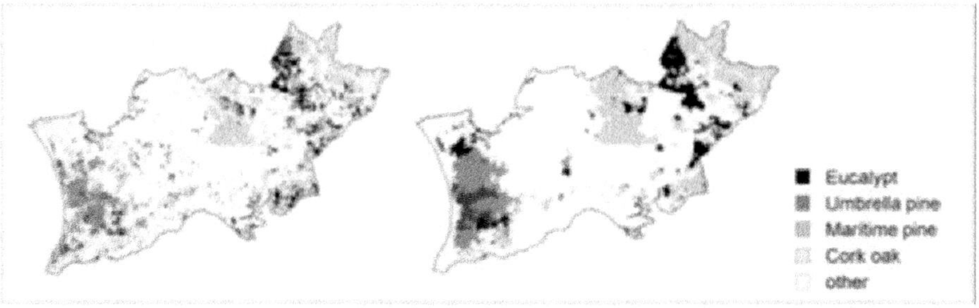

Figure 5. Forest cover classification obtained from the upscaled classified satellite image (left) and from the DSCoS (right).

To facilitate visual comparison of these two classification results, a separate map of classification differences is presented for each of the coverage classes (Figure 6). Differences comprise patches of apparent cover underestimation of the satellite image classification for both pine species. Differences in Cork oak classification mainly occur in the vicinity of areas dominated by this species' cover; they suggest the smoothing character of simulation in comparison to standard classification. The image of differences related to the sum of other classes appears to mirror the assemblage of the forest-species' images. Classification differences are correlated to species cover proportions, as differences occur mainly in regions where the cover has a high proportion, but show no positive correlation with local variances.

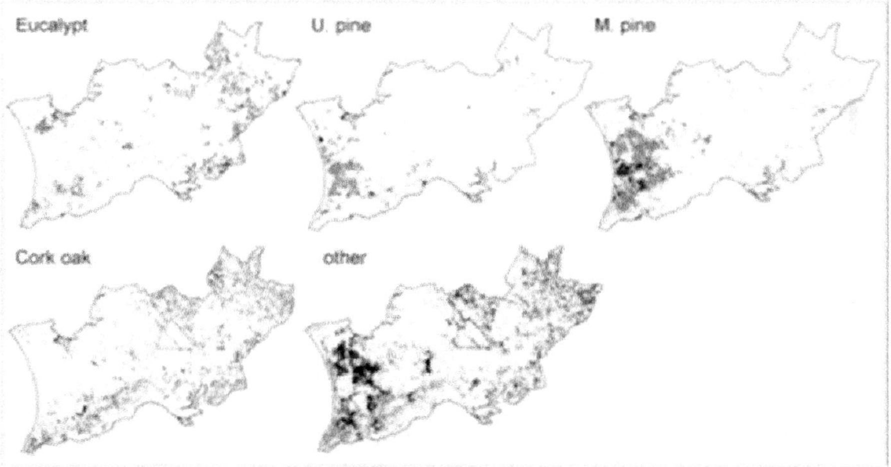

Figure 6. Difference between the DSCoS (DScl) and satellite image classification (SIcl); white: DScl = SIcl,: DScl yields pictured class and SIcl does not, black: SIcl yields pictured class and DScl does not.

Cross-validation was performed on the 156 training areas (used for the satellite image classification) within the study area that were not used as *hard data* for the geostatistical calibration procedure. Both satellite image classification and the applied co-simulation predict Eucalypt, Cork oak and other coverages well (>75% of correctly predicted grid cells). Overall SI classification classifies 80% of the training data area correctly, and DSCoS 75%. This is not surprising as SI classification was calibrated on these training data. Furthermore, SI classification appears to perform better than DSCoS classification for Eucalypt, Maritime pine and the other cover classes, whereas DSCoS performs better on Umbrella pine and Cork oak (Table 3). There are differences in classes attributed to the miss-classified training data; DSCoS attributes more data unduly to the class of other covers (*e.g.* for the pine species) than satellite image classification. Overall, SI classification correctly predicts more.

Table 3. Percentage of correctly (in bold) and miss-classified training data areas, for the satellite image (SI) and DSCoS classifications.

		Training data cover class				
		Eucalypt	U. pine	M. pine	Cork oak	Other
SI	Eucalypt	**87.0**	3.7	20.6	0.6	0.4
Class	U. pine	2.6	**52.3**	9.9	0.3	2.6
	M. pine	7.2	30.3	**55.9**	1.1	1.7
	Cork oak	1.8	0.8	8.8	**92.2**	6.0
	other	1.5	13.0	4.9	5.9	**89.4**
DSCoS	Eucalypt	**82.5**	2.1	21.7	0.0	3.5
class	U. pine	8.7	**59.5**	5.8	0.0	0.6
	M. pine	2.5	1.9	**39.8**	0.0	15.4
	Cork oak	0.0	0.0	8.4	**99.0**	3.2
	other	6.2	36.5	24.3	1.0	**77.2**

7. DISCUSSION AND CONCLUSIONS

This study proposes geostatistical satellite image calibration based on the stochastic Direct Sequential Co-Simulation (DSCoS) procedure and on local co-regionalisation models. This is an alternative approach to spatial forest species characterisation from satellite imagery and field data using co-located co-kriging (Pereira *et al* 2000). Given our data, consisting of frequently fragmented forested areas, co-located co-kriging would produce excessively continuous surfaces. The applied simulation technique provides a more close-to-reality image of forest cover. DSCoS enabled estimations of forest cover probabilities for the Setúbal peninsula as well as local uncertainty assessment, based on a satellite image classification covering the whole area and on forest cover observations collected on the ground.

In comparison to the satellite image classification, DSCoS yielded smoother distributions of forest species cover probabilities. Satellite image classification failed to reproduce the coverage observed at many of the training data used for its calibration and on many of field observations we used as *hard data*. The latter were inevitably perfectly reproduced by the simulations, which was conditioned to them.

DSCoS and satellite image classification performed differently for the different cover classes in the cross-validation performed on part of training data, previously used for the satellite image classification. In comparison, DSCoS performed better than SI classification for Eucalypt and Cork oak and worse for Umbrella and Maritime Pine. Overall, both methods failed to classify numerous training data of both pine species correctly. On the other hand, most Eucalypt and Cork oak areas were correctly classified, probably because these two forest resources have more continuous distributions on the Setúbal peninsula, unlike Umbrella and Maritime Pine, which occur mainly scattered and intermingled.

Limitations to this study were posed by the quality of the collected hard data, as sampling was not stratified over all classes and therefore not representative of the studied area. Furthermore, the aggregation of the original satellite image classification map to the lower study spatial resolution implies a loss of information, with penalization of less frequent and more scattered species. During upscaling each new grid cell was assigned to the most frequent class of the nine underlying 30×30 m grid cells, penalizing less frequent cover classes. This may partly explain the simulation methods failure to improve on Eucalypt and Maritime pine classification – species that frequently cover small areas. Upscaling of the posterior probabilities for these grid cells might have been a more correct alternative.

Another constrained is that simulations are based on the satellite image classification as *soft data*, which is not always optimal. During the five years between the collection of the training data used for satellite image calibration and for the validation of SI and DSCoS classification will some land cover may have changed, accounting for part of the misclassified training areas.

Future research should focus on improvements in the *soft* and *hard data*. In this context, the local uncertainty assessment provided by the simulation approach will allow the identification and classification of areas that need to be re-sampled and monitored for a better planning of those forest resources. Validation should be performed on an independent data set.

REFERENCES

1. Caers, J., 1999, Adding local accuracy to direct sequential simulation. Stanford Center for Reservoir Forecasting, Annual Meeting 12, v. 2.
2. Caers, J., 2000, Direct sequential indicator simulation. In: Kleingeld, W.J. and Krige, D.G. (eds.), Geostats 2000 Cape Town, vol. 1, p. 39-48.
3. Fouquet, C. and Mandallaz, D., 1993, Using geostatistics for forest inventory with air cover: an example. In: A. Soares (ed.), Geostatistics Troia'92, vol. 2, p. 875-886.
4. Goovaerts, P., 1997, Geostatistics for natural resources characterization. Oxford University Press, New York, 483 p.
5. Goovaerts, P., 2000, Geostatistics approaches for incorporating elevation into the spatial interpolation of rainfall. Journal of Hydrology, in press.
6. Journel, A.G., 1994, Modelling uncertainty: some conceptual thoughts. In: Dimitrakopoulos, R. (ed.), Geostatistics for the Next Century: Kluwer Academic Pub., Dordrecht, The Netherdlands, p. 30-43.
7. Nunes, M.C., Sousa, A.J. and Muge, F.H., 2000, The use of remote sensing imagery to update forest cover. In: W.J. Kleingeld and Krige, D.G., (eds), Geostats 2000 Cape Town, vol. 2, p. 559-570.
8. Pereira, M.J., Soares, A. and Rosário, L., 2000, Characterization of forest resources with satellite SPOT images by using local models of co-regionalization. In: Kleingeld, W.J. and Krige, D.G. (eds.), Geostats 2000 Cape Town, vol. 2, p. 581-590.
9. Soares, A., Pereira, M.J., Branquinho, C. and Catarino, F., 1997, Stochastic simulation of lichen biodiversity using soft information from remote sensing data. In: Soares, A., Gómez-Hernández, J. and Froidevaux, R. (eds.), GeoENV I – Geostatistics for Environmental Applications, Kluwer Academic Pub., p. 375-387.
10. Soares, A., 1992, Geostatistical estimation of multi-phase structures. Mathematical Geology, Vol. 24 (2), p. 149-160.
11. Soares, A., 2001, Direct Sequential Simulation and Cosimulation. Mathematical Geology, Vol. 33 (8), p. 911-926.

USE OF FACTORIAL KRIGING TO INCORPORATE METEOROLOGICAL INFORMATION IN ESTIMATION OF AIR POLLUTANTS

H. Caetano[1], M. J. Pereira[1] and C. Guimarães[2]
[1]*Environmental Group of the Centre for Modelling Petroleum Reservoirs, CMRP/IST,Av. Rovisco Pais, 1049-001 Lisbon, Portugal.* [2] *CVRM- Geossystems Center. Av. Rovisco Pais, 1049-001 Lisbon, Portugal.*

Abstract: A monitoring campaign for several airborne pollutants was conducted in Portugal, covering the entire country. Static samples (diffusion tubes) of SO_2, NO_2 and O_3 were collected in a regular grid of 20□20 km. In this paper, we present a methodology, using factorial kriging and morphological kriging concepts, to incorporate local information about wind directions in the spatial estimation of air pollutant concentrations. Pollutant concentration $Z(x_0)$ at location x_0 can be interpreted as a linear combination of independent components $Z^i(x_0)$ driven by wind direction and velocity i. Local spatial components $Z^i(x_0)$ are estimated through factorial kriging, based on variograms $\gamma_i(h)$ with local anisotropies determined by wind directions. Wind direction and velocities measured during the period of the sampling campaign were inferred for the entire country. Local estimates of main wind histogram classes were used to weight different spatial components of pollutant concentration $Z^i(x_0)$. Maps of SO2, NO2 and O3, estimated under the influence of local wind characteristics, were obtained for entire country.

1. INTRODUCTION

Following the European directive of Air Quality (1999/30/CE) a monitoring campaign for several airborne pollutants – SO_2, NO_2 and O_3 – was conducted in Portugal, covering the entire country with a regular sampling grid of 20×20 km.

X. Sanchez-Vila et al. (eds.), geoENV IV – Geostatistics for Environmental Applications, 55-65.

In a first phase, maps of ordinary kriging estimates of those elements were calculated, giving first spatial dispersion images for these pollutants. There are, however, some external factors – like, for example, topography, wind direction and speed – that are known to affect the spatial dispersion of the pollutants. Some geostatistical approaches have been incorporating the effects of the wind in the estimation or simulation of pollutant concentrations: Soares et al., 1993, transform the anisotropy ellipse, according to the wind direction of a given day, to estimate pollutant concentrations; Pereira et al., 1997 use a deterministic dispersion model (Gaussian plume) which is fed with the wind regime, among other parameters, to preview, in certain spots, the pollutant concentrations prior to simulation to the entire spatial domain; Nunes et al., 1999, use the wind rose for a given period to deform the spatial reference before simulating SO_2 concentrations.

In this paper, we present a methodology, using factorial kriging, to incorporate regional information about wind directions and speed in the spatial estimation of air pollutant concentrations. The estimation is performed for the entire country of Portugal. Given the regional scale of the sampling and of the resultant estimated maps, the influence of wind directions and speed are merely considered as main regional trends. Also, spatial locations of possible pollutant sources were not taken into account, mainly because most of them are unknown or diffuse. The proposed method was applied to produce maps of the mentioned airborne pollutants.

2. METHODOLOGY

Let us assume that the pollutant concentration at a given location, accumulated during a given short period of time, say one week, is an average concentration resulting from the influence of different wind regimes, in particular from wind direction and speed.

Considering that pollutant concentration $Z(x_0)$ at location x_0 can be interpreted as a linear combination of independent components $Z^i(x_0)$ driven by *Nd* wind-direction classes i:

$$Z\left(x_0\right) = \sum_{i=1}^{Nd} \delta_i Z^i\left(x_0\right) \quad i=1, Nd \tag{1}$$

Z(x) has a stationary global variogram $\gamma(h)$ and covariance $C(h)$ and, assuming that the global covariance can be decomposed into anisotropic structures corresponding to each main wind directions:

$$C(h) = \sum_{i=1}^{Nd} \delta_i C_i(h) \quad \text{with} \quad \sum_{i=1}^{Nd} \delta_i = C(0) \tag{2}$$

the weights δ_i correspond to the sills of each spatial component of covariance $C_i(h)$ and reflect the effects of each class of wind directions.

2.1 Estimating spatial components

At each location x_u the spatial component $Z^i(x_0)^*$ – equivalent to the pollutant concentration driven by wind characteristics of direction class i – can be estimated by factorial kriging, using the covariance model $C_i(h)$ or variogram $\gamma_i(h)$ (Goovaerts, 1997):

$$Z^i(x_0)^* = \sum_{\alpha=1}^{N} \lambda_\alpha Z(x_\alpha) \tag{3}$$

Ci(h) is the global covariance model corrected for the wind-direction characteristics at location x0. The dual representation of [3] is:

$$Z^i(x_0)^* = \sum_{\alpha=1}^{N} \sum \varphi_\alpha C_i(x_\alpha, x_0) \tag{4}$$

where φ_α are the dual weights, calculated with the global covariance model, associated to the covariances $C_i(x_\alpha x_0)$.

Traditional factorial kriging identifies different spatial components of a physical phenomenon with the structures of its global variogram or covariance. In this case, once the different spatial components have the same covariance models, with equal ranges, but with different spatial anisotropy directions, $Z^i(x_0)^*$ can be viewed as the morphological kriging estimate of $Z(x)$ with local anisotropy models (Soares, 1992). The total concentration at x_0 is:

$$Z(x_0)^* = \sum_{\alpha=1}^{N} Z^i(x_0)^*$$

2.2 Estimation of each spatial component variogram $\gamma_i(h)$

$\gamma(h)$ is the global variogram model estimated with the entire set of samples. $\gamma_i(h)$ is assumed to be equivalent to $\gamma(h)$ but corrected for local anisotropies determined by local wind directions. This means that $\gamma_i(h)$ is obtained by rotating the anisotropy ellipse with the angle of the wind direction at a given location. Remains the calculation of δ_i correspondent to the sills of each spatial component of variogram $\gamma_i(h)$, which reflect the effects of each class of wind directions (Soares et al, 1993). For example, in one location where a given wind class j does not exist, $\delta_j=0$, or if the wind just happens to be exclusively of class j $\delta_j=1$.

To calculate the values $Z_i(x_0)$ at any location x_0, according to (3)or (4), one needs to know the variogram sills for all wind-direction classes i at the spatial locations of pollutant samples x_α.

First an average value of wind direction for the entire sampling period is calculated at the location of monitoring stations of wind direction and speed. The mean values of the different monitoring stations are coded in a binary vector, depending on whether each station belongs to each one of the Nd classes. At location x of each monitoring station the following binary vector is created:

$I_i(x) = 1$ if monitoring station x belongs to class i, otherwise $I_i(x)=0$ $i=1, Nd$

$$\text{Ii(x)=prob}\{x \in \text{wind class i}\} \quad \text{i=1,Nd} \tag{4}$$

The stations' probabilities to belong to those Nd classes are calculated (i.e. spatially inferred by ordinary kriging of the vector) for all nodes of a regular grid where pollutant concentrations are to be estimated:

$$I_i(x_0)^* = \sum_\alpha \lambda_\alpha(x_0) I_i(x_\alpha) \tag{5}$$

These estimated probability values $prob\{x_0 \in wind\ class\ i\}^*$ are scaled to $C(0)$ by the following product

$$\delta i\ (x0) = Ii(x0)^*.\ C(0) \tag{6}$$

that can be identified with $\delta_i(x_0)$.

Note, that in the traditional factorial kriging approach the components variograms $\gamma_i(h)$, corresponding to the different structures of global $\gamma(h)$, are estimated just once and considered representative for the entire area. In the proposed approach $\delta_i(x_0)$, the sills of $\gamma_i(h)$ at x_0, are estimated at every location x_0, corresponding to the *Nd* wind direction classes.

3. CASE STUDY: SPATIAL CHARACTERIZATION OF CARBON DIOXIDE, NITROUS DIOXIDE AND OZONE

3.1 Experimental data-Measurements of airborne pollutants and climatologic data

A monitoring campaign for several airborne pollutants was conducted in Portugal, covering the entire country. Static samples (diffusion tubes) of SO_2, NO_2 and O_3 were collected in a regular grid of 20×20 km . Figure 1, a) shows the locations of the regular pollutant-sampling grid. During the campaign, which took about 15 days, wind direction and speed were also recorded in a set of meteorological monitoring stations. In Figure 1, b) the wind-directions frequency histograms for the entire set of monitoring stations are presented.

3.2 Global variograms

Global variograms were calculated for each pollutant. Figure 2 shows the anisotropic variograms of NO_2. A map of NO_2 values was estimated, through ordinary kriging using the global variogram model (Figure 3), which can be compared with the equivalent map estimated using the proposed methodology.

3.3 Calculation of component variograms $\gamma_i(h)$

Once the global variogram $\gamma(h)$ is known, the idea is to adapt the direction of the anisotropy ellipse, according to the proximity of the wind-direction measurements. According to the exposed methodology, the following sequence of steps was taken to calculate the different component variograms $\gamma_i(h)$.

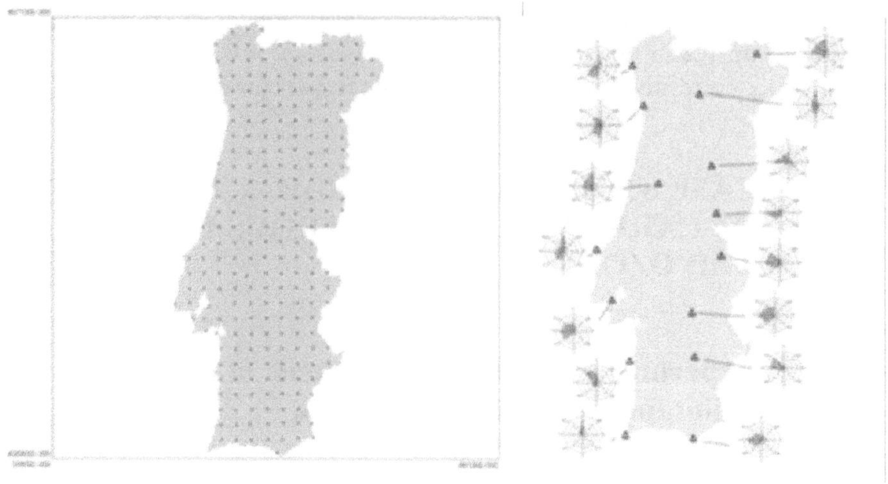

Figure 1. a) Spatial location of the regular sampling grid; b) wind-directions frequency histograms for the entire set of monitoring stations.

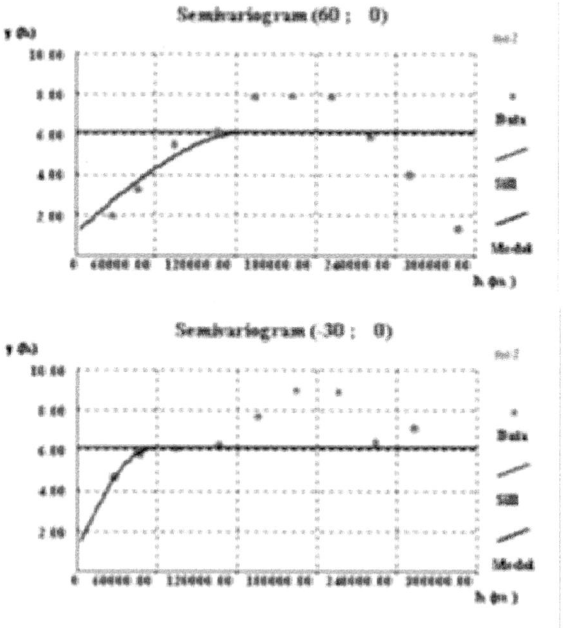

Figure 2. Variograms of NO_2 for the two main directions.

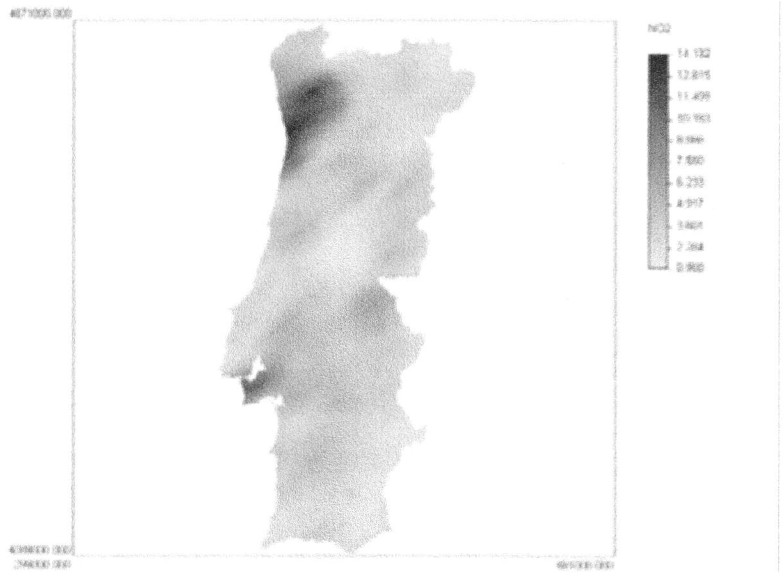

Figure 3. Estimated map of NO_2 (ordinary kriging) with one global model of variograms.

i) Calculation of the mean value of wind directions for each monitoring station presented in Figure 1, b).

A weighted average of wind direction vectors was computed with the wind speeds as weights, to account for the transport effect. The histogram of mean values of wind directions is shown in Figure 4. The following main wind direction classes were adopted:

	Min	Max	Mean	
C1	200.2423	222.3506	212.1230	S 32^0 7′ W
C2	252.0354	259.7979	255.9167	
C3	272.0298	288.3953	279.1886	
C4	315.5580	326.9635	322.5489	

(0° corresponds to the N/S direction)

Wind-direction mean values for each of the monitoring stations were coded in a binary vector:

$I_i(x) = 1$ if monitoring station x belongs to class i, otherwise $I_i(x)=0$ $i=1, 4$

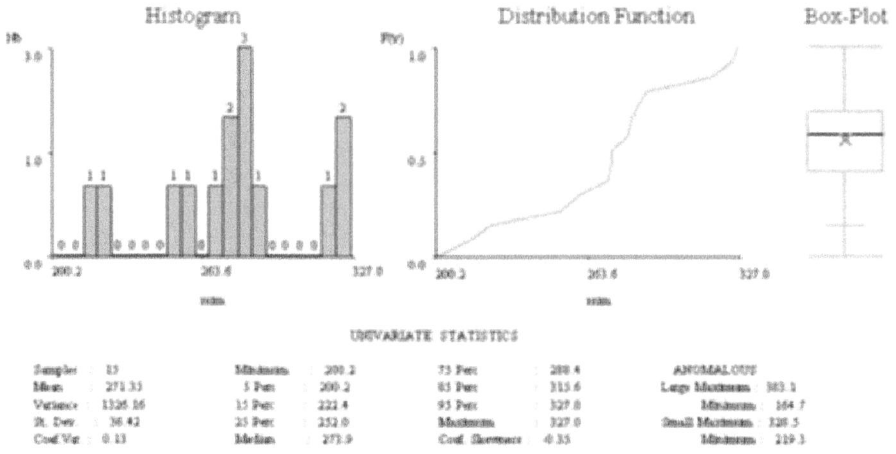

Figure 4. Histogram of mean values of wind directions.

ii) Spatial inference of $I_i(x)$ at a regular grid of points where pollutant concentrations are to be estimated.

We used ordinary kriging of the indicator vector (5)

$$I_i(x_0)^* = \sum_{\alpha} \lambda_\alpha(x_0) I_i(x_\alpha) \qquad i=1,4$$

After re-scaling the estimated probabilities $I_i(x)^*$ to the global variance $C(0)$, sills for the different variograms $\gamma_i(h)$ at location x_0 are obtained: $\delta_i\ (x_0) = I_i(x)^*.\ C(0)$

Figure 5 shows the inferred $I_i\ (x_0)$ for the 4 wind-direction classes, for the entire country, exhibiting the regional main trends of those classes of wind directions.

3.4 Estimation of pollutant concentrations: SO_2, NO_2 and O_3

Maps of the three pollutant elements were obtained for a regular grid of 1x1 km, applying the factorial kriging estimation procedure of equations (3) and (4). Figure 6 shows the estimated concentrations for SO_2, NO_2 and O_3.

4. FINAL REMARKS

The presented methodology aims at incorporating meteorological conditions – wind directions and speed – into the spatial estimation of airborne pollutants. A regional trend of the effects of these wind characteristics on the dispersion of tree airborne air pollutants is used in a factorial kriging approach.

The very promising results encouraged the promoters of this study to pursuit it for a second campaign focussed on the same pollutant elements.

We believe that the final model would be enriched with a narrower grid of meteorological measurements and with the incorporation of the main topographic patterns into the final model.

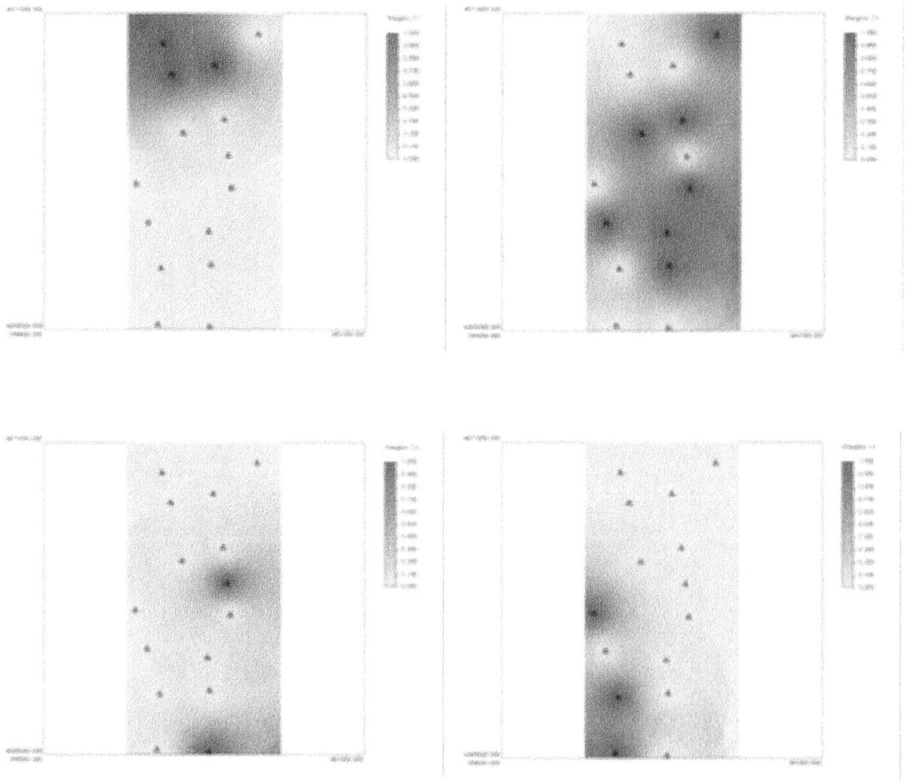

Figure 5. Estimated values of $I_i\ (x_0)^*$ for the 4 wind-direction classes.

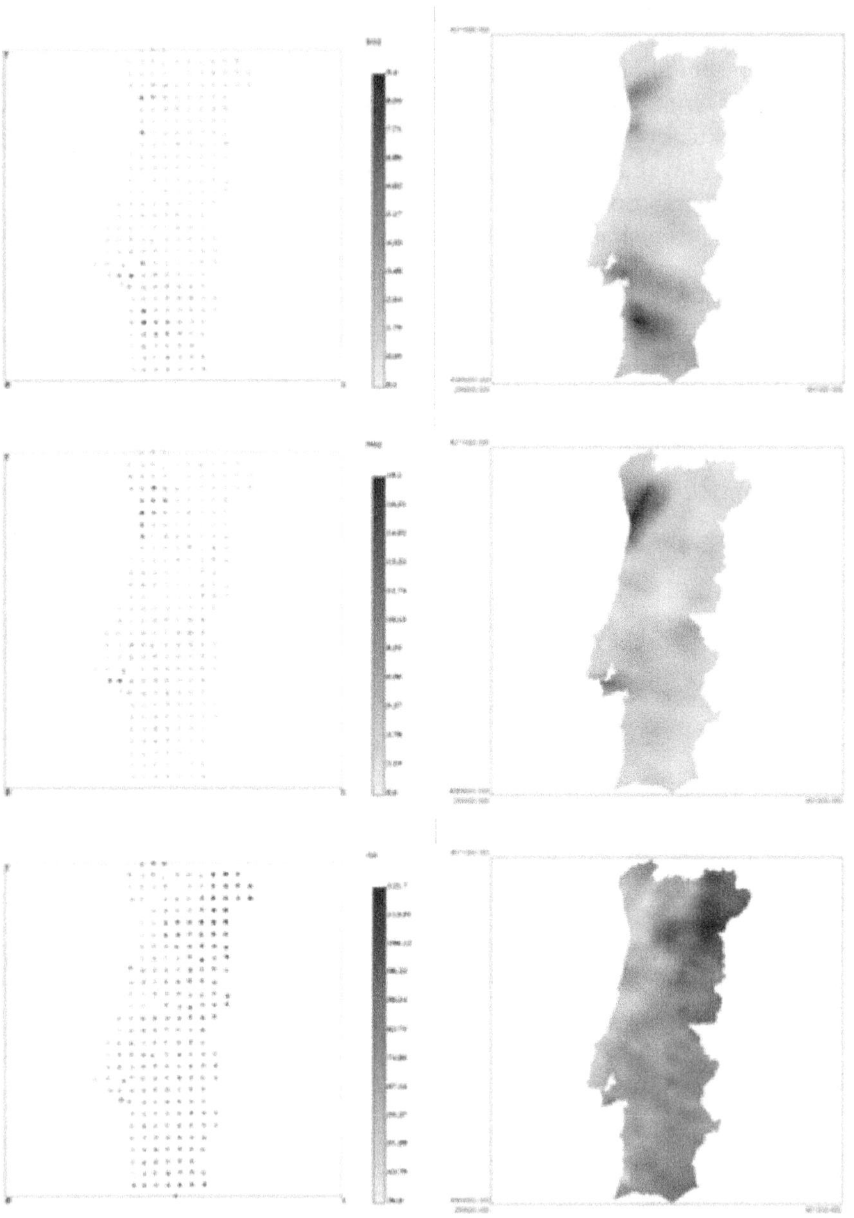

Figure 6. Estimated concentrations for SO_2, NO_2 and O_3.

REFERENCES

1. Goovaerts, P. (1997) Geostatistics for Natural Resources Evaluation. Oxford University Press, 483 p.
2. Nunes, C., Soares, A. and Ferreira, F. (1999) Evaluation of Environmental Costs of SO_2 Emissions using Stochastic Images. GeoENV II – Geostatistics for Environmental Applications. Gomes-Hernandez, J., Soares, A. and Froidevaux, R. (eds.), Kluwer Academic Publishers, pp. 113-125.
3. Pereira M.J., Soares, A. and Branquinho C. (1997) Stochastic Simulation of Fugitive Dust Emissions. Wollongong '96, 5th Geostatistics Congress, Baafi E. (ed.), Kluwer Academic Publishers, vol II, pp. 1055-1066.
4. Soares A. (1992) Geostatistical Estimation of Multi-Phase Structures", Mathematical Geology, 24 (2), pp. 153-164.
5. Soares A., Távora J., Pinheiro L., Freitas C. and Almeida J. (1993) Predicting Probability Maps of Air Pollution Concentration: A Case Study on Barreiro/Seixal Industrial Area". Geostatistics Troia'92, Soares, A. (ed.), Kluwer Academic Publishers, Holland, pp. 625-636.
6. Soares A. (2001) Sequential direct simulation and co-simulation. Mathematical Geology, 33(8), pp. 149-160.

HIGH RESOLUTION OZONE MAPPING USING INSTRUMENTS ON THE NIMBUS 7 SATELLITE AND SECONDARY INFORMATION

G. Christakos[1], A. Kolovos[1], M. L. Serre[1], C. Abhishek[1] and F. Vukovich[2]
[1]Center for the Advanced Study of the Environment (CASE), University of North Carolina-Chapel Hill, NC, USA; [2]Science Applications International Corp. (SAIC), Raleigh, NC, USA

Abstract: The high natural variability of ozone concentrations across space-time and the different levels of accuracy of the algorithms used to generate data from measuring instruments can not be confronted satisfactorily by conventional interpolation techniques. This work suggests that the Bayesian Maximum Entropy (*BME*) method can be used efficiently to assimilate salient (although of varying uncertainty) physical knowledge bases about atmospheric ozone in order to generate and update realistic pictures of ozone distribution. On theoretical grounds, *BME* relies on a powerful scientific methodology that does not make any of the restrictive modelling assumptions of previous techniques and integrates a wide range of knowledge bases. A study is discussed in which *BME* assimilates data sets generated by measuring instruments on board the Nimbus 7 satellite as well as uncertain measurements and secondary information in terms of total ozone-tropopause pressure empirical equations. The *BME* total ozone analysis eliminates major sources of error and produces high spatial resolution maps that are more accurate and informative than those obtained by conventional interpolation techniques.

Key words: Ozone, atmosphere, TOMS, SBUV, BME, spatiotemporal, geostatistics.

1. INTRODUCTION

Analyses of total ozone (TO_3) have been produced on a global basis using data from the Total Ozone Mapping Spectrometer (TOMS) since the late 1970s. The last decade, climatological analyses of Tropospheric Ozone

X. Sanchez-Vila et al. (eds.), geoENV IV – Geostatistics for Environmental Applications, 67-78.

Residual (TOR), which is an estimate of the tropospheric TO_3 and which was, in the initial work, the difference between TO_3 from *TOMS* and the stratospheric ozone determined from the Stratospheric Aerosol and Gas Experiment (SAGE) instrument, have been developed ([1]). *TOMS* data are collected globally on a daily basis, but the integration of years of *SAGE* data were required to provide a reliable analysis of stratospheric ozone on a global basis ([2]). Attempts have been made to develop daily maps of *TOR* using data from the Solar Backscatter Ultraviolet (SBUV) remote sensing system, which measures ozone in 12 Umkehr layers. However, one of the major problems in applying *SBUV* with *TOMS* data to develop *TOR* estimates is the differences in *spatial resolution*. For illustration, the locations of *TOMS* and *SBUV* measurements obtained on July 6, 1988 are shown in Fig. 1. The *SBUV* data gaps have been traditionally filled using

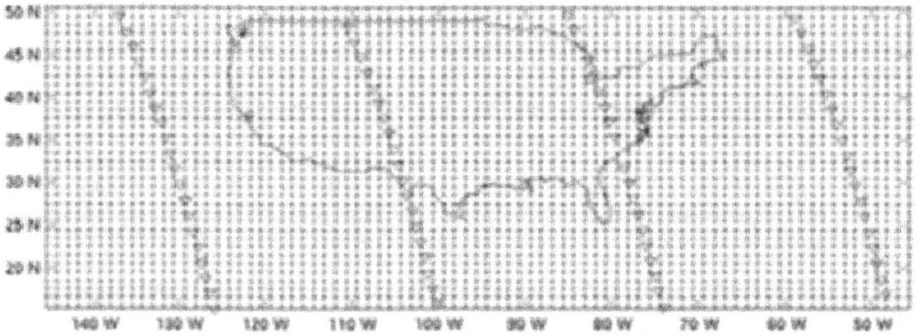

Figure 1. Grid coverage of satellite ozone measurements (July 6, 1988) for *TOMS* (plus markers) and *SBUV* (triangles) instruments.

conventional interpolation procedures so that stratospheric ozone from the *SBUV* instrument would be available at the data locations of the *TOMS* instrument. Poor correlation in Fig. 2 demonstrates the problem with using conventional interpolation procedures to fill the data gaps between orbital tracks for the *SBUV* data and points to a major source of error in *TOMS/SBUV TOR*. The high ozone variability across space-time together with the different levels of accuracy attributed to the instruments above, introduces considerable sources of uncertainty in the representation of ozone distribution when conventional interpolation procedures are used to fill *SBUV* data gaps ([3]). Many of the existing procedures –polynomial interpolation, basis functions, spatial regression, kriging, and neural networks [4], [5]– lack the scientific methodology to assimilate rigorously essential sources of physical knowledge and the conceptual organization to account for space-time variability effects. The underlying modelling assumptions are very restrictive (e.g., linearity, normality, overparameterization, and physical model-independence) and often lead to unrealistic representations of the actual characteristics of ozone variability.

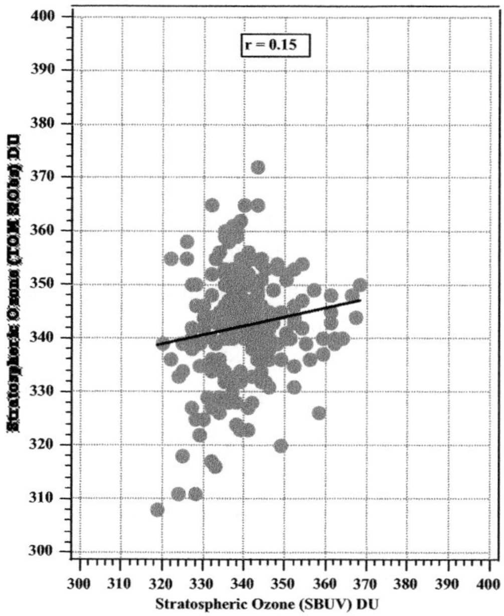

Figure 2. Plots of stratospheric ozone (=TOMS values minus tropospheric ozone from Wallops Island ozonesonde) vs. stratospheric ozone using interpolated SBUV values (1985-1989) .

Fig. 2, e.g., demonstrates the necessity for application of advanced mapping techniques that provide theoretical support and technical capabilities to adequately represent ozone variability and blend various knowledge bases (data collected at sparse *SBUV* measurement points, uncertain evidence, and secondary information). Such techniques must predict ozone concentrations at unsampled locations to fill data gaps with which analyses of stratospheric ozone can be generated that will have increased accuracy. A group of advanced techniques possessing these desirable features are provided by *Modern Spatiotemporal Geostatistics* (*MSG*; [6]). The paper includes a preliminary study of the application of the *Bayesian Maximum Entropy* (*BME*) method of *MSG* to predict ozone concentrations at unsampled locations across space to fill *SBUV* data gaps by means of high resolution maps. The *BME* theory does not make any of the restrictive assumptions of conventional interpolation techniques. TO_3 obtained at the *SBUV* measurement locations were analyzed over the continental U.S. Secondary information (empirical TO_3-tropopause pressure analysis) was processed and applied and the resulting improvement in TO_3 prediction was investigated. The results of the TO_3 analysis using *BME* were compared with the analysis of the complete set of *TOMS* data of the region.

2. MODEL DESCRIPTION

MSG ([6]) provides a powerful framework for generation of informative maps of atmospheric variables. The random field $X(\boldsymbol{p})$, $\boldsymbol{p}=(\boldsymbol{s},t)$, offers a mathematically rigorous and physically meaningful representation of TO_3-distributions across space-time. Atmospheric studies are generally concerned with the prediction of TO_3 at a network of points $\boldsymbol{p}_k$, given background knowledge and a set of site-specific data $\boldsymbol{\chi}_{\text{data}}$ at points $\boldsymbol{p}_{\text{data}}$. At points $\boldsymbol{p}_k$, either we have no observations at all or the available data are considerably uncertain and cannot be used as reliable predictions of the actual TO_3 values, $\boldsymbol{\chi}_k$. One then seeks to derive the probability density functions (pdf) that characterize $X(\boldsymbol{p})$ at every node of the mapping grid in light of the physical knowledge sources considered. TO_3 predictions $\hat{\boldsymbol{\chi}}_k$ at any set of grid nodes $\boldsymbol{p}_k$ are derived from the pdf at these nodes by means of a suitable criterion (the criterion choice is not unique, but it depends on the study goals). The unifying epistemic background of the *MSG* techniques includes two fundamental tenets:

(Ta) Consider general knowledge bases ($\mathcal{G}$-KB) such as physical laws, primitive equations, stochastic representations, and statistical moments (including multiple-point, non-linear and high-order statistics, if available) to define a space of plausible events and their respective pdf $f_{\mathcal{G}}$ (i.e., KB=$\mathcal{G}$ in this case) by means of a *teleologic* approach.

(Tb) Eliminate from consideration those otherwise plausible events that are physically or logically inconsistent with the available specificatory KB ($\mathcal{S}$; which may include hard data and uncertain observations). Then reassign probabilities to the remaining plausible events to be consistent with the $\mathcal{S}$-KB by means of a *logic* system leading to an updated pdf $f_{\mathcal{K}}$ ($\mathcal{K}=\mathcal{G}\cup\mathcal{S}$).

The $\mathcal{G}$-KB in Ta is transformed into a set of integral equations ($\mathcal{G}$–equations) of the corresponding pdf, which are solved teleologically in terms of a final cause expressed by the action principle. The solution form depends on the action principle one adopts regarding the events deemed plausible before the available data is considered. *MSG* often uses the action principle of *maximum expected information.* I.e., while in previous principles the action sought refers to concepts like energy or time, the *MSG* action refers to information which may involve, in particular, the Shannon measure properly extended in a space-time domain (another solution involves the Fisher information measure, etc.; [7]). In Tb, the $\mathcal{G}$-based solution, $f_{\mathcal{G}}$, is revised via application of a logic system to yield a revised pdf model, $f_{\mathcal{K}}$. The *MSG* theory is very general allowing the use of different assimilation frameworks, including statistical inductive inference (e.g., Bayesian conditionalization rules) and stochastic deductive inference (e.g., material biconditionalization principles). *MSG* offers a long list of spatiotemporal analysis and modeling techniques [7]. The present TO_3 study will focus on the *BME* technique which is, perhaps, the most widely used at present. A step-by-step description of *BME* can be found in the relevant *MSG* literature.

3. MODEL APPLICATION

3.1 Modeling Methodology

BME produces high resolution TO_3 analyses with better accuracy than conventional techniques, since it provides an improved representation of spatial variability, does not require the limited modelling assumptions of conventional techniques, and can incorporate various kinds of uncertain data not used in conventional techniques. *BME* analysis consists of 3 stages:

a) Section 3.2: The variation of the random TO_3 field across space (July 6, 1988) is represented by a random field. Instead of using *SBUV* measurements directly as hard data, *TOMS* data closest to the *SBUV* measurement locations were selected as hard data (thus, differences in the level of accuracy between the two instruments need not be accounted for).
b) Section 3.3: Soft information, which relates TO_3 to tropopause pressure, was generated via application of an empirical physical equation.
c) Section 3.4: Conventional interpolation provided TO_3 at the data gaps between sub-satellite points based on *TOMS* data selected at the *SBUV* data locations (Approach 1). Then, *BME* predicted the TO_3 values at all grid nodes by assimilating hard data together with soft information. The results from Approaches 1 and 2 were compared with the complete set of *TOMS* measurements in the area defined in Fig. 1.

3.2 Spatial Variations of Total Ozone

Fig. 3 shows the actual TO_3 map generated using the entire *TOMS* data set (Fig. 3 serves as the reference map for comparison purposes). In the context of $\mathcal{G}$-KB, the TO_3 distribution is represented by the spatial random field $TO_3(s) = \overline{TO_3}(s) + X(s)$, where $X(s)$ is a zero-mean homogeneous random fluctuation, and the spatial trend $\overline{TO_3}(s)$ is determined by a moving window averaging of TO_3 data ([7]). Given TO_3 and $\overline{TO_3}(s)$ at each data point

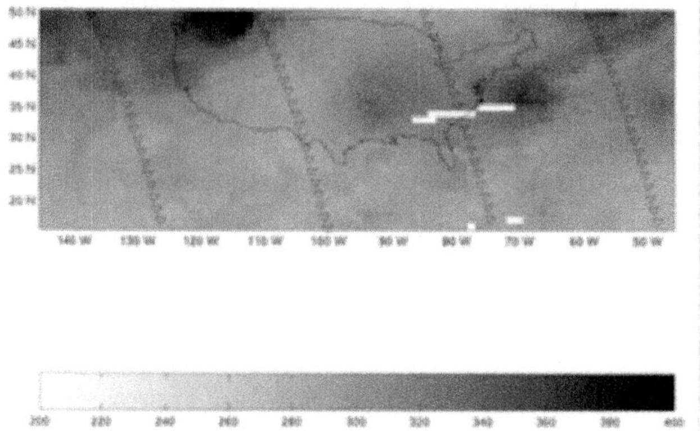

Figure 3. TO_3 map (in DU) obtained from *TOMS* instrument (July 6, 1988; blanc strips indicate areas where data was not available).

(triangles in Fig 1), the residual ozone $X(s)$ hard data are calculated from the equation above. The following covariance model (with nested exponential and gaussian components) is part of the $\mathcal{G}$ -KB about the residual ozone distribution, $c_x(r_{ij}) = c_1 e^{-3r_{ij}/a_1} + c_2 e^{-3r_{ij}^2/a_2^2}$, where $r_{ij} = |s_i - s_j|$. This theoretical model is fitted to the experimental covariance values (obtained from χ_{hard}) so that $c_1 = 75$ (DU^2), $a_1 = 15$ (degrees), $c_2 = 75$ (DU^2), $a_2 = 9$ (degrees). Each component of the covariance model accounts for half of the total variance (150 DU^2). The exponential component represents processes with somewhat high spatial variation over a long range of about 15 degrees (approx. 1667 *km* on the Earth's surface), while the Gaussian component represents smoother processes with a shorter range of about 9 degrees (approx. 1000 *km* on the Earth's surface).

3.3 Soft (Secondary) Information

Tropopause pressure data (P_t) are model-generated data (National Center for Atmospheric Prediction, NCAP). Starting from a phenomenological law, [8], the following formulation, which relates TO_3 with P_t, is obtained, $TO_3 = a_0 + a_1 \log P_t$, where $a_0 = TO_{3,0} + aH_{t,0} - aH_0 \log P_0$ and $a_1 = aH_0$

are estimated by experimental data fitting (a is linear rate of TO_3 decrease); P_0 is surface pressure, H is height, and H_0 is the scale height of the atmosphere (approx. 7 *km*). The a_0 and a_1 are viewed as random variables representing such factors as uncertainties due to ΔO_3-fluctuations and perturbations in the atmosphere. For each P_t-value, a soft pdf is derived that represents the distribution of TO_3 values and offers a physical basis for producing soft information. Fig. 4 is a typical scatterplot of TO_3 vs. P_t at

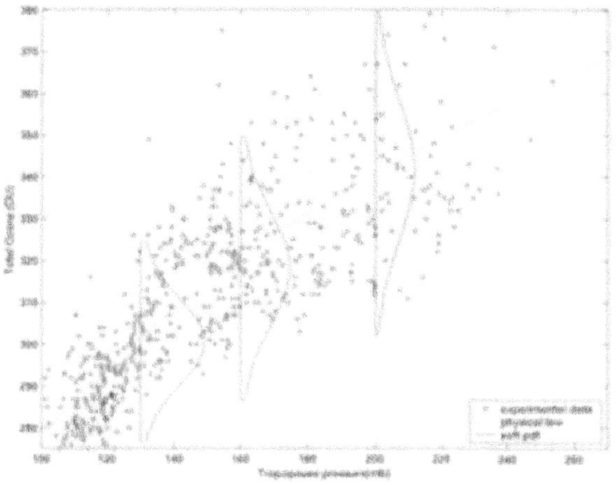

Figure 4. Scatter plot of TO_3 measurements vs. tropopause pressure.

concurrent points. Useful probabilistic representations of the uncertainty in the TO_3 values are generated: The data are divided into classes of contiguous, non-overlapping P_t-intervals. For each class the TO_3 mean and variance are derived as well as their pdf. Based on this procedure a probability datum for TO_3 is assigned to each P_t data point. Densities for 3 selected classes are plotted in Fig. 4, for illustration. *BME* can be applied in to its fullest ability by improving the physical relationship of pressure vs. height using additional information sources (e.g., potential vorticity data and temporal information in the data). Moreover, *BME* is an efficient tool for mapping various types of informative variables (categorical, etc.; [9], [10]).

3.4 Modeling Results

First, we assumed that S-KB consists solely of the hard TO_3 data set at the *SBUV* measurement points. A linear spatial regression technique is used to predict TO_3 in the remaining region. Note that this technique can be derived as a limiting case of the general *BME* theory under restrictive modelling conditions –a fact that demonstrates the generalization power of *BME*. The corresponding TO_3 map is shown in Fig. 5a. The prediction error std. deviation of the TO_3 prediction at location $\boldsymbol{s}_k$ is calculated as $\sigma_e(\boldsymbol{s}_k) = [c_x(0) - \sum_{i=1}^{M} \lambda_i c_x(r_{ik})]^{1/2}$, where M is the number of TO_3 data used, and λ_i are weights calculated from the regression system (the σ_e-map is plotted in Fig. 5b).

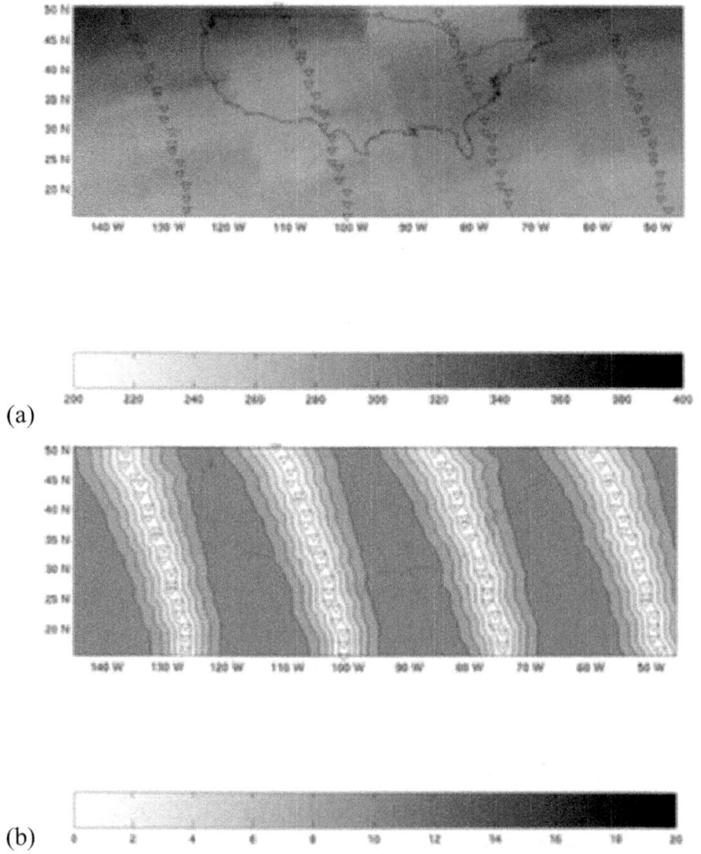

Figure 5. (a) TO_3 map (DU) using hard data at *SBUV* points; (b) map of prediction error std deviations. Approach 1.

Next, for Approach 2 the soft information (Section 3.3) was assimilated in addition to the hard data set to predict TO_3 in the *SBUV* data gaps by *BME*. Just as for Approach 1, *TOMS* data closest to the *SBUV* measurement locations were used as hard data. The resulting *BMEmean* map (i.e., the prediction at each point is the mean of $f_{\mathcal{K}}$) is plotted in Fig. 6a. The associated map of prediction error std. deviation at any point $\boldsymbol{s}_k$ (Fig. 6b) was obtained using the expression $\sigma_{\mathcal{K}}(\boldsymbol{s}_k) = [\int d\chi_k\,(\chi_k - \overline{x_k})^2\, f_{\mathcal{K}}(\chi_k)]^{1/2}$ (the mean value of the TO_3 fluctuation at prediction point $\boldsymbol{p_k}$ is $\overline{x_{\boldsymbol{k}}} = 0$). Comparisons of the maps in Figs 5a and 6a with the map in Fig. 3 show that the map of Fig. 5a clearly misrepresents the spatial TO_3 variation in certain areas and exhibits poor accuracy away from

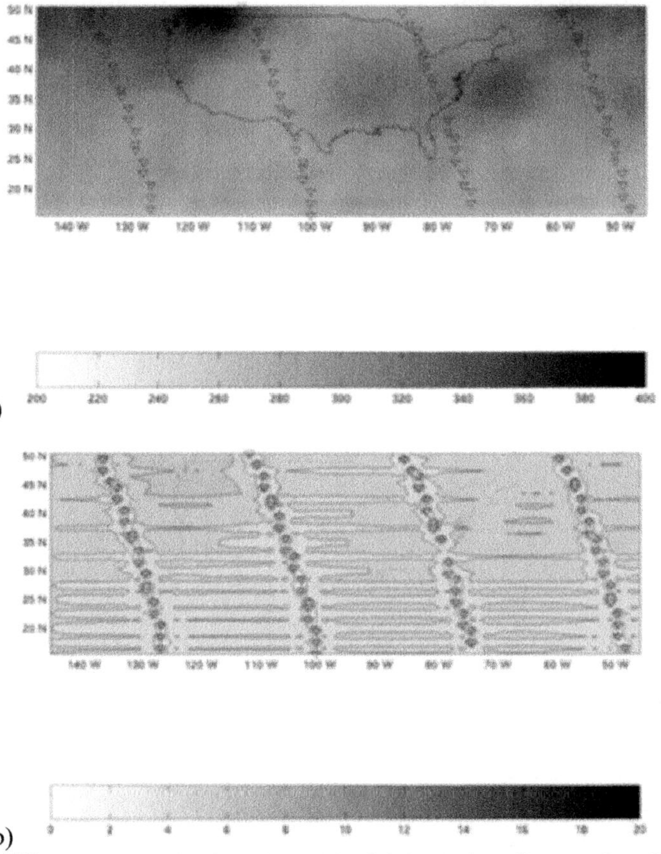

Figure 6. (a) TO_3 map (DU) using hard data and soft information; (b) map of prediction error std deviations. Approach 2.

hard data points, whereas Approach 2 offers a much more realistic representation of the spatial TO_3 variation, leading to noticeable improvements in prediction across space. The prediction error std deviation maps indicate that Approach 2 (Fig. 6b) offers a significant improvement over Approach 1 (Fig. 5b). In particular, Fig. 5b shows that the σ_e-errors are rather small along the satellite paths but increase considerably away from the paths, reaching their maximum values along the axis inbetween the paths.

The $\sigma_{\mathcal{K}}$-error map (Fig. 6b) depicts a more realistic distribution (error does not increase dramatically away of the satellite paths, etc.). On theoretical grounds, Approach 1 is a linear predictor; σ_e is independent of data values and, as a consequence, is the subject of criticism [11]. Approach 2 is a non-linear predictor, in general, and the $\sigma_{\mathcal{K}}$ depends on the specific data set offering an adequate prediction error assessment when the shape of $f_{\mathcal{K}}$ is not very complicated (e.g., if the underlying law is Gaussian, the probability that x_k lies in the interval $\hat{\chi}_{k,mean} \pm 1.96\sigma_K$ is 95%). In cases where $f_{\mathcal{K}}$ has a complicated shape, a realistic assessment of the analysis error is achieved using *BME* confidence sets.

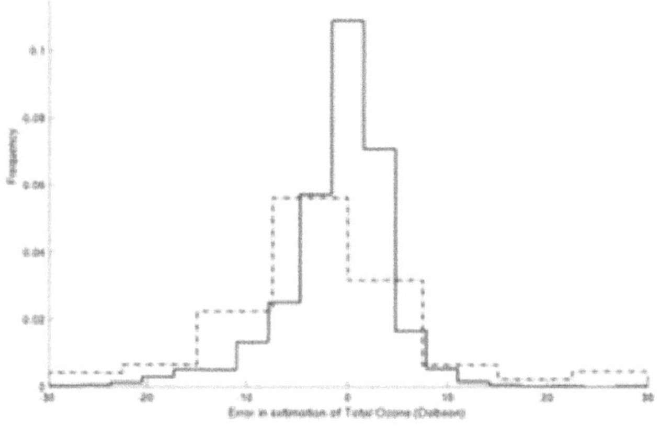

Figure 7. Histogram of spatial TO_3 prediction errors by Approach 2 (plain line) and by Approach 1 (dotted line).

To further compare the accuracy of Approach 1 vs. Approach 2, we calculated the differences between predicted TO_3 values (Figs. 5a and 6a) and actual values (Fig 3) at all points at which TO_3 values are available from *TOMS*. Histograms of prediction errors are shown in Fig. 7. Approach 2 has a sharper peak than Approach 1 around zero prediction error, which implies that *BME* produced more accurate TO_3 predictions at a much higher frequency than the conventional technique. The mean square error (*MSE*) drops from 106.50 DU^2 (Approach 1) down to 30 DU^2 (Approach 2), i.e., a substantial accuracy improvement of 72% in favor of Approach 2. Another measure of error indicating bias is the mean error (*ME*). In Fig. 7, $ME = -3.0$ *DU* for Approach 1 (indicating slight bias), whereas $ME = -0.9$ *DU* for Approach 2; i.e., a difference in accuracy of 69 % in favor of Approach 2.

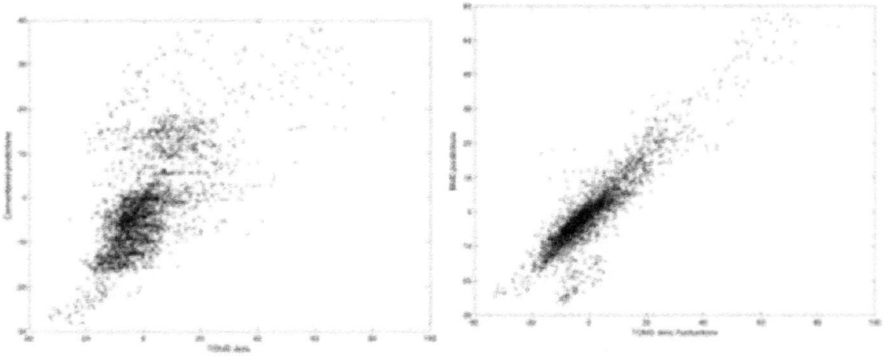

(a) (b)

Figure 8. Fluctuation scattergrams of *TOMS* data vs. predictions of (a) Approach 1and (b) Approach 2

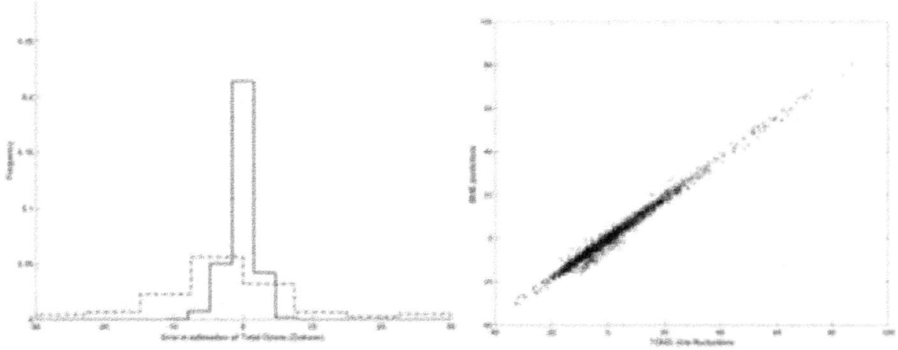

(a) (b)

Figure 9. (a) Histogram of TO_3 prediction errors by Approach 2 (plain line) and Approach 1 (dotted line). (b) Fluctuation scattergram of *TOMS* data vs. predictions by Approach 2. In addition to previous data arrangement, soft TO_3 data are available at map nodes.

Next, we assumed that soft data were also available at prediction nodes themselves. Soft means varied randomly within intervals including the *TOMS* data at the nodes. Although the soft data was intentionally contaminated by error, *BME* made optimal use of the situation and generated improved results. Fig. 9a compares the histogram of the prediction errors by Approaches 1 and 2. The improvements obtained by *BME* vs. conventional interpolation are significant: a sharper histogram around zero error; smaller

ME and *MSE* statistics [*ME*= -0.2 *DU* (Approach 2), = -3.0 *DU* (Approach 1); and *MSE*=3.40 DU^2 (Approach 2), =106.50 DU^2 (Approach 1)].

The corresponding scattergram of *TOMS* data fluctuations vs. *BME* predictions of the same fluctuations shows an almost perfect fit (Fig. 9b). As was mentioned, *BME* theory can be used to its fullest ability: *(i)* by performing a composite space-time ozone analysis and focusing on using uncertain *SBUV* data sets to construct maps of the ozone profile throughout the Earth; and *(ii)* by assimilating other soft knowledge sources, like potential vorticity data and temporal information in the data. The *(i)* and *(ii)* are important research and development issues, which will be the topics of future publications.

ACKNOWLEDGEMENTS

This work has been supported by a grant from the National Aeronautics and Space Administration (60-00RFQ041).

REFERENCES

1. Fishman, J., C. E. Watson, J. C. Larsen, and J. A. Logan, 1990. "Distribution of tropospheric ozone determined from satellite data." *J. Geophys. Res.*, 95, 3599-3617.
2. Vukovich, F. M., V. G. Brackett, J. Fishman, and J. E. Sickles II, 1996. "On the feasibility of using the tropospheric ozone residual for nonclimatological studies on a quasi-global scale." *J. Geophys. Res.*, 101, 9093-9105.
3. Fishman, J., and P. K. Bhartia, 2002. Personal communication, NASA Langley Research Center, Hampton, VA.
4. Cherkassky, V., and F. Muller, 1998. *Learning from Data*. J. Wiley & Sons, New York, N.Y.
5. Stein, M. L., 1999. *Interpolation of Spatial Data*. Springer, New York, N.Y.
6. Christakos, G., 2000. *Modern Spatiotemporal Geostatistics*, Oxford University Press, New York, N.Y.
7. Christakos, G., P. Bogaert, and M. L. Serre, 2002. *Temporal GIS: Advanced Functions for Field-Based Applications*, Springer-Verlag, New York, N.Y.
8. Wallace, J. M., and P. V. Hobbs, 1977. *Atmospheric Sciences – An Introductory Survey*. Acad. Press, San Diego, CA.
9. D'Or, D., and P. Bogaert, 2002. "Combining categorical information with the BME approach" *4th European Confer. on Geostatistics for Environm. Applic.*, Barcelona, Spain, 27-29 November, 2002.
10. Bogaert, P., 2002. "Spatial prediction of categorical variables: The BME approach" *4th European Confer. on Geostatistics for Environm. Applic.*, Barcelona, Spain, 27-29 November, 2002.
11. Goovaerts, P., 1997. *Geostatistics for Natural Resources Evaluation*. Oxford Univ. Press, New York, N.Y.

GEOSTATISTICAL DIGITAL IMAGE MERGING

J. Delgado-García[1], M. Chica-Olmo[2] and F. Abarca-Hernández[2]

[1]*Dpto. Ingeniería Cartográfica, Geodésica y Fotogrametría. Universidad de Jaén. c/ Virgen de la Cabeza, 2. E-23071 Jaén, Spain. e-mail: jdelgado@ujaen.es*
[2]*Dpto. Geodinámica. Universidad de Granada. Avda. Fuentenueva s/n. E-18071 Granada, Spain. e-mail: mchica@goliat.ugr.es, fabarca@goliat.ugr.es*

Abstract: The merging of multisensor image data is becoming a widely used procedure because of the complementary nature of various data sets. Ideally, the method used to merge data sets with high-spatial and high-spectral resolution that should not distort the spectral characteristics of the high-spectral resolution data. The paper presents a geostatistical image merging method. The approach takes into consideration important aspects like, for example, support of information and makes a real image integration using a sequential gaussian conditional cosimulation-based procedure. All the parameters of the integration (basically, the weights corresponding to the images to merge) are extracted from the images themselves, providing additional information of them like variability structures that can be used in other digital image processes like filtering, texture determination and classification.

Key words: Digital image merging, Geostatistics, Remote Sensing

1. INTRODUCTION

The digital image usage in environmental and cartographic applications is very frequent. Nowadays there is a wide range of systems that provide environmental and cartographic images in digital format. These images are classified in order to its spatial –ground sample distance, GSD– and spectral

X. Sanchez-Vila et al. (eds.), geoENV IV – Geostatistics for Environmental Applications, 79-90.

resolution. Unfortunately, in most case both resolutions are in opposition. The high-resolution sensors have a low resolution whereas the multispectral sensors have a good spectral resolution but a bad spatial resolution that limits their use in some environmental detailed applications.

The problem is solved using the digital image merging procedures. The main objective of these methods is obtaining synthetic images that combine the advantage of the high spatial resolution of one image with the high spectral resolution of another image.

These merged images have important environmental applications like land-use, vegetation, lithological photointerpretation and cartography and process monitorization (for example, pollution control). Applications that need to combine the multispectral information with a good spectral resolution that allows the cartographical product generation in adequate scales.

The geometric registration is a straightforward process. However, the mixing of information into a single data is not straightforward. Ideally, the method used to merge data sets with high-spatial resolution and high-spectral resolution should not distort the spectral characteristics of the high spectral resolution data. Not distorting the spectral characteristics is important for calibrating purposes and to ensure that targets that are spectrally separable in the original data are still separable in the merged data set (Chavez et al., 1991).

The objective of this paper is to present the preliminary results of a geostatistical merging image methodology. The method has been used to merge the information contents of Landsat-7 ETM+ (GSD=30m) and aerial orthoimage (GSD=3m). The method is compared to other well-known classical (non-geostatistical) merging procedures.

2. THE DATA SET

2.1 Study area

The study area used in this work covers a 60x60 km area localized in the Granada province (S of Spain), just at the northern border of Sierra Nevada mountains (figure 1), approximately 80km from Granada. In this zone, there was a very important iron open-cut mine (Alquife mine) that had a 3.3 Mt/yr. iron ore production in the mid-nineties. Actually the mine is closed and the old open-cut appears like a lake (figures 2 and 3).

2.2 Images

The data set used for this application are basically composed by the following images:

a) Aerial orthoimage: the higher spatial resolution image is obtained from the digital differential rectification of 1:40000 aerial photograms using a digital photogrammetric system. This 8-bits image has a 3m GSD (Figure 2).
b) Landsat-7 ETM+ images: this sensor provides 8 bands, 3 visible (ETM1: Blue, ETM2: Green, ETM3: Red), 1 NIR (ETM4), 2 MIR (ETM5 and ETM7), 1 TIR (ETM6) and 1 panchromatic (PAN). In this work, only ETM3, ETM4, ETM5 (bands that present maximum OIF, Chavez et al, 1982) are used (Figure 3).

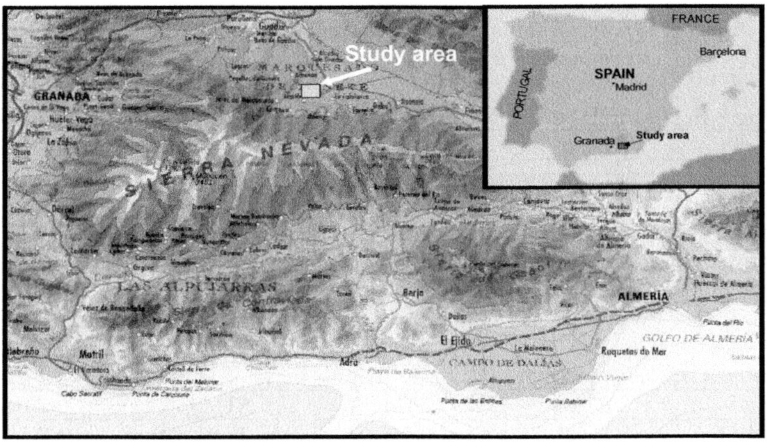

Figure 1. Localization map.

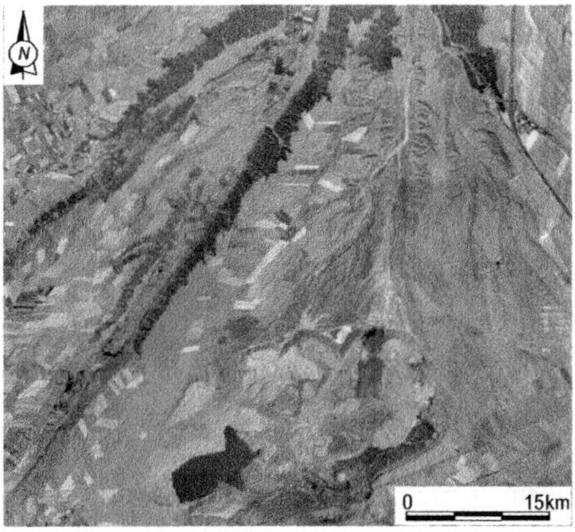

Figure 2. Digital aerial orthoimage. GSD=3m; Image Size=2000x2000 pixels (Grayscale representation from digital number 0 –black– to 255 –white–).

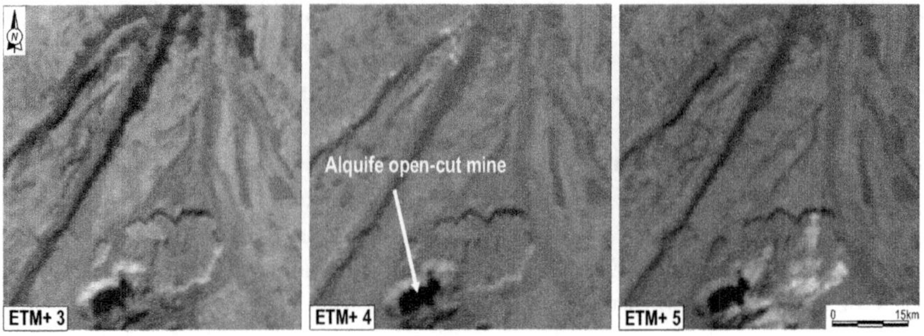

Figure 3. Landsat ETM+ images: GSD=30m; Image size: 200x200 pixels (Grayscale representation: from DN=0 –black– to white DN=255 –white–).

Table 1. Basic statistics of the images (data are integers ranging from 0 to 255).

Image	GSD(m)	Mean	Std.Dev.	Minimum	Maximum
ETM+3	30	73.55	15.14	25	139
ETM+4	30	63.00	10.09	18	136
ETM+5	30	66.88	15.38	14	185
Ortho	3	118.79	33.32	0	255
Ortho-30*	30	118.82	33.19	0	255

*Image obtained from the mean value of the corresponding 3m-orthoimage pixels

Table 2. Correlation Matrix.

	ETM3	ETM4	ETM5	Ortho-30*
ETM+3	1.000	0.557	0.450	0.375
ETM+4		1.000	0.635	0.282
ETM+5			1.000	0.185
Ortho-30*				1.000

*Image obtained from the mean value of the corresponding 3m-orthoimage pixels

3. METHODOLOGY AND APPLICATION

3.1 Geostatistical merging method

The geostatistical merging method presented in this paper consists basically in a series of geostatistical techniques. The general schema can be shown in the figure 4.

The process begins with a normal score transformation of the image data is applied in order to use a unique statistical distribution for all the images, previously a despiking process in the sense proposed by Verly (1986) was applied avoiding the clusters in the original data distribution that could generate problems in the transformation.

Once the data are transformed into a gaussian distribution, the structural analysis has been made. The gaussian image variograms are computed in column and in row directions. The obtained variograms present an isotropic behavior so the omnidirectional variograms were used for the adjustment to the models. The experimental variograms and theoretical model parameters are presented in the Figure 5 and Table 3.

All images present variograms models quite similar. The nested model is composed by 2 structures: a) Exponential, short range (around 60-150m) that represents 70-80% of total variance and b) Exponential, medium range (375-600m for visible bands and 2500-4500m for infrared bands) that represents 20-30% of total variance.

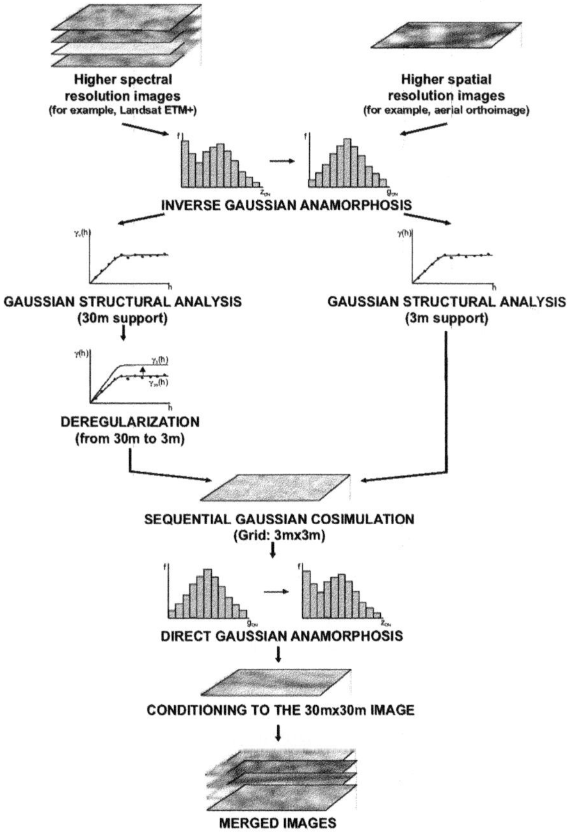

Figure 4. Flowchart of the geostatistical digital image merging procedure.

One of the most important advantages of the geostatistical approach to the image merging is its capability to consider correctly the information support problem in the down sampling process from the original 30m to the final 3m. Using the previous models it is possible to obtain the 3m variograms models parameters applying an iterative deregularization process based in the expressions presented in Clark (1977). The deregularization variograms parameters are given in the Table 4.

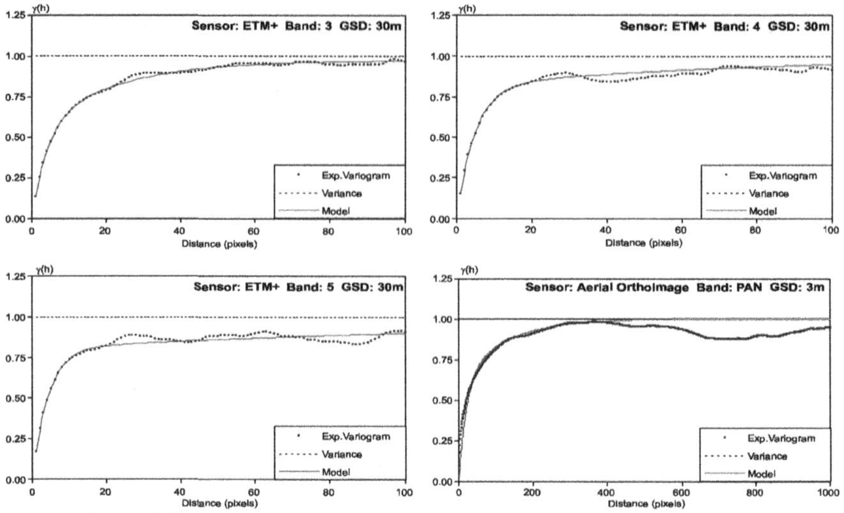

Figure 5. Experimental variograms and theoretical models for the image data.

Table 3. Variograms model parameters for ETM+ and Ortho-30 images.

Image	Struct.1	Sill1	Range1	Struct.2	Sill2	Range2
ETM3	Expon	0.540	120.00	Expon	0.430	660.00
ETM4	Expon	0.810	144.00	Expon	0.190	2400.00
ETM5	Expon	0.800	126.00	Expon	0.200	4500.00
Ortho-3	Expon	0.620	60.00	Expon	0.380	375.00

The distance parameters must be multiplied by 3 to obtain the approximate correlation ranges. Range units in meters.

Table 4. Deregularizated variograms model parameters for the ETM+ images.

Image	Struc.1	Sill1	Range1	Struc.2	Sill2	Range2	Sill1+Sill2	C(v,v)
ETM3	Expon	0.735	96.0	Expon	0.425	630.0	1.16	0.199
ETM4	Expon	1.100	102.0	Expon	0.290	3000.0	1.39	0.316
ETM5	Expon	1.082	99.0	Expon	0.300	5400.0	1.31	0.302

Range units in meters

These parameters are the basic input information in order to obtain the simulated data sets on 3m-pixel size. The simulated images preserve the variability of the deregularizated variables and can be obtained using the geostatistical simulation methods. The used method has been the sequential gaussian cosimulation due to its simplicity and its speed (due to the huge volume of data that it is necessary to simulate, that can be millions of data). The cosimulated images were obtained with the GSLIB SGSIM version 2.9.02 program (Deutsch and Journel, 1997) following the procedure described in Goovaerts (1997).

The sequential gaussian cosimulation method allows integrating the information provided by a secondary variable (3m ORTHO image data) through a collocated cokriging process. The non-conditional simulated images are shown in Figure 6 and their basic statistics in Table 5.

Figure 6. Geostatistical non-conditioned cosimulated images. Image size: 2000x2000 pixels. (Greylevel representation from digital number 0 –black- to 255 –white-).

Table 5. Descriptive statistics of non-conditional cosimulated images.

	ETM+3		ETM+4		ETM+5	
	Mean	Variance	Mean	Variance	Mean	Variance
Geostat	0.0210	1.0506	0.0224	1.3119	0.0251	1.2358

The last operation in the geostatistical image merging is the conditioning of the cosimulated images. For this operation the original 30x30m ETM+ images have been used. Using these images, it is possible to obtain the correction factors for each image and pixel in order to ensure the coincidence between the original digital numbers and the mean values of the 10x10 pixels group of 3m cosimulated images. The final merged images are presented in figure 7 and their corresponding basic statistics in Table 5.

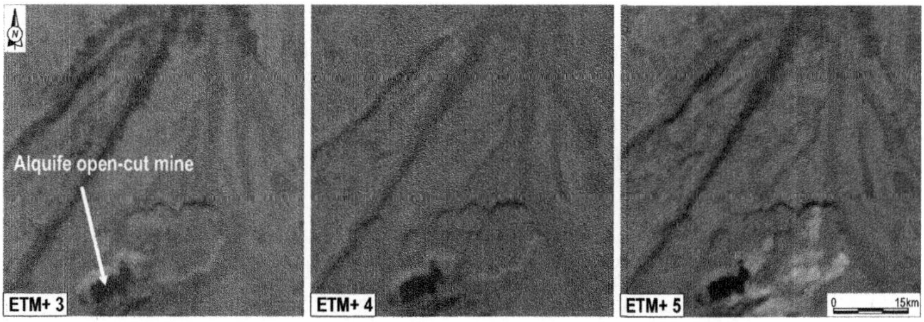

Figure 7. Geostatistical merged images. Image size: 2000x2000 pixels. (Greylevel representation from digital number 0 –black- to 255 –white).

Table 6. Descriptive statistics of merged images using the geostatistical approach.

	ETM+3/ORTHO		ETM+4/ORTHO		ETM+5/ORTHO		Correlation Coefficient		
	Mean	Var	Mean	Var	Mean	Var	¾	3/5	4/5
Geostat	73.05	18.12	62.50	13.98	66.38	20.34	0.67	0.60	0.78

3.2 Classical (non-geostatistical) merging methods

In order to compare the obtained results several non-geostatistical image merging methods have been applied. The methods used were: a) Hue-Intensity-Saturation; b) Principal Component Analysis; c) High-Pass Filter and d) Color Normalized.

3.2.1 Hue-Intensity-Saturation (HIS)

HIS is one of the most often used methods to merge multisensor image data. Haydin et al. (1982) merged using this procedure Landsat MSS with Return Beam Vidicon and Heat Capacity Mapping Mission data. HIS method is widely used to merge Landsat TM and SPOT-P data (Chavez et al., 1991). The method uses three bands of the lower spatial resolution image and transforms these data into HIS space. The higher spatial resolution image is constant stretched in order to adjust the mean and variance to the intensity one. The stretched image replaces the intensity component image before the images are retransformed back into the RGB space (Figure 8A, Table 7).

3.2.2 Principal Component Analysis (PCA)

The PCA method is similar to the HIS one. The higher spectral resolution images are used as input to a forward principal component procedure. As the HIS procedure, the high-resolution image must be stretched to have approximately the same mean and variance as the first principal component –PC1–. The results of the stretched image replace the PC1 band and the data are retransformed back into the original space (Figure 8B, Table 7).

3.2.3 High-pass filter (HPF)

The HPF procedure is based in a data compression and reconstruction technique described in Schowengert (1980). In the HPF method, the higher spatial resolution data have a small high pass filter applied. The results of the small high pass filter contain the high-frequency component/information that is related mostly to spatial information. The spatial filter removes most of the spectral information. The HPF results are added, pixel-by-pixel, to the lower spatial resolution data set (Figure 8C, Table 7).

3.2.4 Color-normalized (CN)

The color normalized method (Vrabel, 1996) uses a mathematical combination of the color image and high-resolution data to merge the higher spatial and higher spectral resolution images. Each band in the higher spectral image is multiplied by a ratio of the higher resolution data divided by the sum of the color bands. The function automatically resamples the three-color bands to the high-resolution pixel size using a nearest neighbor, bilinear, or cubic convolution technique. The output RGB images will have the pixel size of the input high-resolution data (Figure 8D, Table 7).

3.2.5 Main drawbacks and advantages of the classical methods

The non-geostatistical methods have from a geostatistical point of view several drawbacks:

a) Do not take into consideration the information support of the merging data. Usually, they need a resample process based in split the original pixel into a smaller size. The statistics of the obtained images preserve mean and variance of the original ETM+ images, which do not have sense from a geostatistical point of view (variance must increase with the pixel size reduction).
b) HIS and PCA are not really merging methods. They consist in the substitution of the high-spectral images with a high-spatial resolution image based on the correlation of the both data sets. The correlation level does not modify the process but it has influence in the final results. These methods only can be applied into triplets of bands.
c) Do not provide any additional information about the images (spatial variability, scale of variation,...) and the merging process is not controlled by the user (black-box automatic process).
d) The characteristics of the images obtained from the different methods have important differences in their appearance and basic statistics. These differences can have an important influence in the image interpretation and classification.

Nevertheless, all of the presented classical methods are very easy to apply and are implemented in the most popular remote sensing software, like for example, ERDAS-Imagine, ER-Mapper and ENVI.

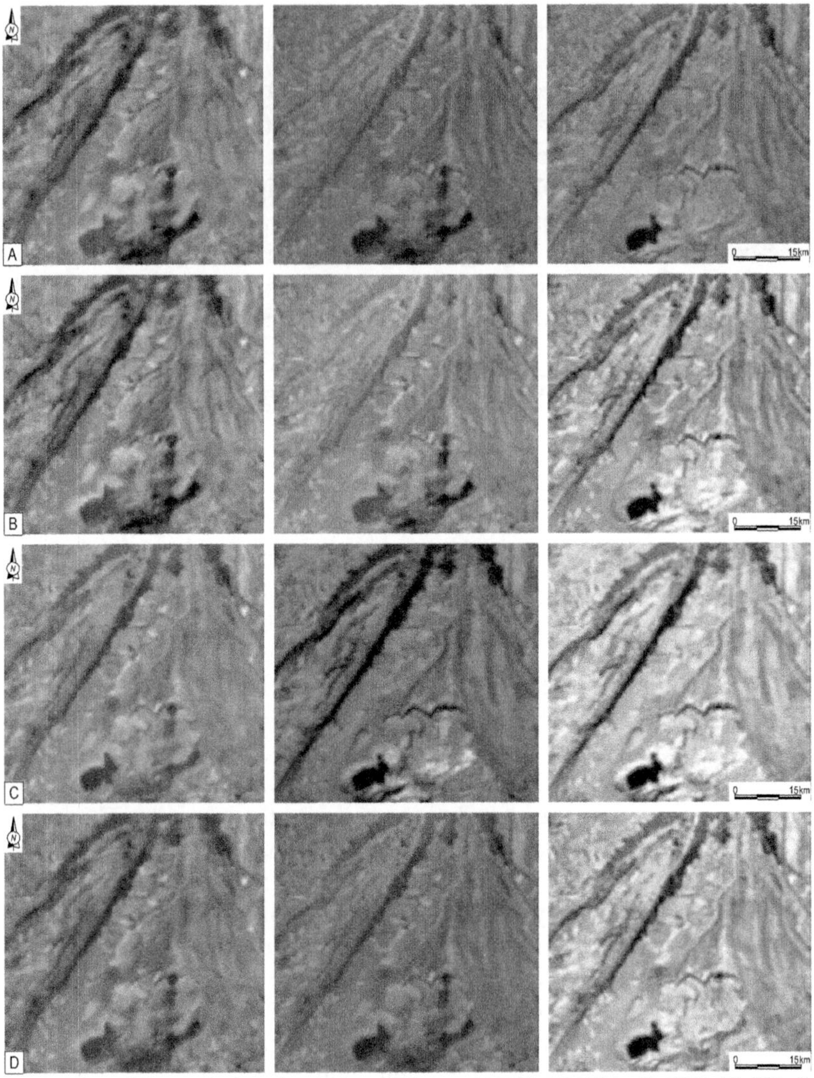

Figure 8. Merged images using classical procedures. A) HIS, B) PCA, C) HPF, D) CN. Left: ETM+3, Middle: ETM+4, Right: ETM+5.

Table 7. Descriptive statistics of merged images using classical methods

	ETM+3/ORTHO		ETM+4/ORTHO		ETM+5/ORTHO		Correl.Coeficient		
	Mean	Var.	Mean	Var.	Mean	Var.	3/4	3/5	4/5
HIS	73.38	15.94	62.90	11.84	66.10	13.37	0.59	0.48	0.70
PCA	73.55	15.62	63.00	12.28	66.87	13.15	0.40	0.55	0.60
HPF	172.99	48.52	92.32	18.98	121.92	23.62	0.69	0.77	0.98
CN	42.73	13.84	36.33	10.43	38.26	11.29	0.83	0.76	0.87

4. CONCLUSIONS

This paper has demonstrated that it is possible the digital image merging is possible through a geostatistical approach considering fundamental aspects like support effect and spatial variability of the images. The merged images using this procedure preserve the spectral characteristics of the higher-spectral resolution images.

The visual aspect of the geostatistical-merged images is quite different from the images obtained with classical methods. These differences are produced by the lower weight that the geostatistical method applies to the higher spatial resolution image. It is very important take into consideration that the geostatistical procedure makes a real integration of the images instead of the substitution made by the classical approaches. This is an important factor where it is necessary to work with non-visible spectral bands, which are low correlated with higher spatial resolution images that usually are panchromatic.

The main drawback of the geostatistical approach is its complexity. The method needs an important geostatistical background and suitable software. This software must be designed and optimized for large volume of data treatment.

Future works will focus on developing further approach in the deregularization and change support models and conditioning procedures. Additionally, other data sets will be examined using this procedure.

ACKNOWLEDGEMENTS

This work are part of the activities of the "Sistemas Fotogramétricos y Topométricos" and "Geoestadística, Teledetección y Sistemas de Información Geográfica" research groups of the Universities of Jaén and Granada (Andalusian Research Program). It was partially support by the R+D+I National Program of the Spanish Science and Technology Ministry under grant REN2001-3378/RIES. We also would like to thank both of the anonymous referees in providing positive comments.

REFERENCES

1. Chavez, P.R.; Sides, S.C. and Anderson, A. (1991). Comparison of three different methods to merge multiresolution and multispectral data: Landsat and SPOT Panchromatic. Photogrammetric Engineering & Remote Sensing, 57(3), 295-303.
2. Chavez,P.S.; Berlin, G.L. and Sowers, L.B. (1982). Statistical method for selecting Landsat MSS ratios. Journal of Applied Photographic Engineering, 8(1), 23-30.
3. Clark, I. (1977). Regularization of semivariogram. Computer & Geosciences, 3, 341-346.
4. Deutsch, C.V. and Journel, A.G. (1997). GSLIB: Geostatistical Software Library and User's Guide. Oxford University Press, 2nd edition, New York, 369 p.

5. Goovaerts, P. (1997). Geostatistics for Natural Resources Evaluation. Oxford University Press, New York, 483 p.
6. Haydn, R.; Dalke, G.W. and Henkel, J. (1982). Application of the IHS color transform to the processing of multisensor data and image enhancement. Proceedings, International Symposium on Remote Sensing of Arid and Semi-Arid Lands, Cairo, Egypt, 599-616.
7. Schowengerdt, R.A. (1980). Reconstruction of multispatial, multispectral image data using spatial frequency contents. Photogrammetric Engineering & Remote Sensing, 46(10), 1325-1334.
8. Verly, G. (1986). Multigaussian kriging – A complete case study. In: R.V. Ramani, editor, Proceedings of the 19th International APCOM Symposium, 283-298, Littleton, CO. Society of Mining Engineers.
9. Vrabel, J. (1996). Multispectral Imagery Band Sharpening Study. Photogrammetric Engineering & Remote Sensing, 62(9), 1075-1083.

ON THE EFFECT OF POSITIONAL UNCERTAINTY IN FIELD MEASUREMENTS ON THE ATMOSPHERIC CORRECTION OF REMOTELY SENSED IMAGERY

N. Hamm, P.M. Atkinson and E.J. Milton

Department of Geography, Southampton University, Southampton SO17 1BJ, United Kingdom. n.hamm@soton.ac.uk

Abstract: The empirical line method (ELM) is widely used to atmospherically correct airborne remotely sensed imagery. It is based on a simple linear regression between remotely sensed measurements of radiance and field based measurements of reflectance. To construct the regression, spatially coincident field and airborne measurements are paired. This research investigates the impact of uncertainty in the location of the field measurements on the outcome of the regression. First, block kriging was used to aggregate the field measurements to the same support as pixels. It was shown that large positional uncertainty gives a small effect on estimation of parameters of the ELM. Second the co-located field and pixel values were combined. It was shown that low positional uncertainty introduces variability in to parameter estimation for the ELM and that this is likely to concern the practitioner.

Key words: positional uncertainty, support, Empirical Line Method

1. INTRODUCTION

For optical remotely sensed data to be of lasting quantitative value it is important that the pixel values be defined in units of reflectance. However, remotely sensed data are typically provided in units of at--sensor radiance,

X. Sanchez-Vila et al. (eds.), geoENV IV – Geostatistics for Environmental Applications, 91-102.

which are affected by atmospheric conditions. Some form of atmospheric correction is required to transform the at-sensor radiance values to at-surface reflectance. There are three broad methods for atmospheric correction (Schott, 1997): (i) methods that use information available in the image itself; (ii) physically-based approaches where radiative transfer schemes are used to model the interaction of radiation with the atmosphere and; (iii) empirical relationships between radiance and reflectance. These techniques are reviewed and discussed elsewhere (Schott, 1997; Smith and Milton, 1999). The research presented in this paper seeks to characterise and model the uncertainty involved in implementing the empirical line method. Specifically, the objective of this research was to quantify the uncertainty involved when the location of field measurements is not known perfectly.

2. THE EMPIRICAL LINE METHOD

The empirical line method (ELM) is based on a simple, first-order linear regression model, where field-based measurements of reflectance, measured on a pseudo-point support, are the dependent variable and remotely sensed measurements of radiance, defined on a pixel-sized support, are the predictor variable. The field-based measurements are made using a radiometer or spectrometer to characterise the reflectance of a predefined selection of ground targets. These are combined with spatially coincident airborne measurements of radiance and the data set is used to estimate the parameters of the regression model. The estimated parameters are then used to predict at-surface reflectance over the remainder of the image.

Ground targets for use in the ELM, should conform to several criteria (Smith and Milton, 1999), as follows:

1. The targets are identifiable in the image and on the ground;
2. The targets cover a range of reflectance values from bright to dark within each wave band;
3. Targets should be near Lambertian;
4. Targets should be spatially homogeneous in the spectral domain;
5. The histogram of the reflectance measurements should be normally distributed in each band for each target;
6. All targets should be at the same altitude;
7. Targets should be larger than the effective resolution element of the sensor.

However, in any given exercise, it may not be possible to find targets that conform to all of these requirements. In particular, it may not be possible to find targets that are spatially homogeneous or normally distributed.

Typically, field-based measurements are taken at a number of random and *unrecorded* locations within each ground target and the sample mean and variance are estimated. The remotely sensed radiance value of each target is characterised by the mean and variance of several pixels within each target. The mean values of at-sensor radiance and at-surface reflectance then form a data pair, which are used to parameterise the regression model. Hence, the number of data pairs used in the regression model is the same as the number of ground targets. However, there are several problems with this approach:

1. There is no way of precisely linking specific field measurements with spatially coincident remotely sensed measurements.
2. The objective is to predict reflectance over pixel-sized supports, but the model is parameterised over a support of several pixels. Theoretically, however, the form of the model and the value of the model parameters may be different for different supports (Heuvelink and Pebesma, 1999).
3. No explicit effort is made to ensure that the at-sensor radiance and at-surface reflectance values used in the parameter estimation phase are defined on the same support. Hence the input data are inconsistent.
4. Mean values are used to construct a data pair for each target, thus artificially reducing the variance of the estimated parameters and the estimation of the variance parameter in the ELM.

These problems lead to practical and conceptual difficulties with implementing and applying the ELM. From the practical perspective, it means that targets that have spatial structure or that are not normally distributed (criteria 4 and 5) should not be used. Conceptually, the approach is flawed, particularly where uncertainty in the atmospheric correction needs to be quantified. The latter three problems mean that the variance in prediction of reflectance will not be evaluated properly.

3. METHODOLOGY

3.1 Framework

The discussion in section 2 highlights the need to adopt more suitable procedures if the practical and conceptual difficulties described are to be overcome.

Previous research (Hamm et al., 2002) has addressed the above issues by adopting a spatial sampling strategy (a nested grid) and recording the location of each field measurement (see Section 3.3). This allowed each field measurement to be linked to its spatially coincident pixel value. Under this scheme, the number of data pairs yielded is the same as the number of field samples used. Furthermore, this allowed relaxation of the criteria on spatial structure and allowed non-stationarity to be dealt with. This approach was shown to be essential to allow accurate parameter estimation, especially when there is spatial structure in the reflectance of the ground targets.

Block kriging and block conditional simulation were used to aggregate the field measurements to the same support as the image pixels, for input into the regression model. This was required to ensure that both variables that were input into the regression were defined on the same support. This was not essential for estimating the expectation of the slope and intercept parameters. However, it was required to gain a realistic estimation of the variance. Furthermore, it was shown that the variance in the regression model gained by using a conditionally simulated surface was higher than that obtained using the kriged surface. This was expected from theory.

The previous work, outlined above, advocated that the locations of field measurements need to be recorded, even if no geostatistical analysis is to be performed. Provision of surveying equipment and Global Positioning Systems (GPS) receivers mean that location can often be recorded in an efficient and cost effective manner. Hence it is useful to understand the implication if location is recorded less rigorously than it was in this study. The impact of positional uncertainty on geostatistical analysis and prediction is also of broader interest (Atkinson, 1996; Gabrosek and Cressie, 2002).

3.2 Field Site

Thorney Island in West Sussex, south east England contains a disused airfield with a range of surface cover types. These include asphalt, concrete and cropped grass, which are considered to be "typical" ground targets for use with the ELM (Smith and Milton, 1999). On 24th July 2001 the site was overflown by the Natural Environmental Research Council (NERC) aircraft which carried the Itres Instruments compact airborne imaging spectrometer (*casi*). Data were aquired at an altitude of approximately 1000 m on a north-to-south flight line oriented along the centre of the main runway. The field measurements were taken close to the centre of the image swath.

3.3 Method

Field measurements were taken on a nested square grid (Figure 1) using a Milton Multiband Radiometer (MMR) operating in dual beam mode (Milton, 1987). The instruments were inter--calibrated prior to and after use, allowing straightforward processing from radiance to reflectance. The MMR samples the electromagnetic spectrum in four broad wavebands that are designed to correspond to the first four bands of the Landsat Thematic Mapper (TM) sensor. The broad wavebands of the MMR mean that it is not ideal for operational atmospheric correction of *casi* data. However, the rapid measurement time of the MMR made it ideal for acquisition of a spatially distributed large sample (approximately 250 samples per target) and the data gained were suitable for tackling the key research questions outlined in this paper. For this paper, analysis and discussion focuses on deriving broad-band reflectance for the first waveband of the MMR (approximately 420-530 nm, blue). The corresponding band used from the image data is the first waveband of the *casi* (approximately 440-560 nm).

The location of each measurement was surveyed and recorded relative to UK Ordnance Survey (OS) trigonometric points. The *casi* data were also geometrically corrected to the OS National Grid. This allowed each field measurement to be located within the image. Careful attention was given to ensuring that the location of the field measurements were recorded rigorously and precisely, both relative to each other and to the OS National Grid. In an operational situation, it might not be possible to record location with such rigour and precision and this research seeks to examine the implication of that. However, it is assumed that the geometric correction of the imagery is perfect.

4. RESULTS AND ANALYSIS

4.1 Data exploration

Data summaries are presented in Figure 1 (assuming no positional uncertainty). Figure 1 suggests that the distributions of the data for asphalt and grass approach normality. However, the histogram for concrete suggests that the data are not drawn from a normal distribution. This is problematic for the "typical" implementation of the ELM, since this violates the criteria of normality stated in Section 2. However, Hamm et al. (2002) showed that criteria can be relaxed if co-located pixels and reflectance data are used.

Omnidirectional sample variograms (Matheron, 1963), assuming no positional uncertainty, are shown in Figure2. These give further evidence of the spatial structure in the reflectance of the surface that is implied by Figure 1. To address the issue of non-normality in the concrete data set, the concrete ground target was segmented to remove the area north and east of the (467116, 101765) OS National Grid co-ordinate.

4.2 Scaling up

Some formal procedure is required to predict values of at-surface reflectance on the same support as the field measurements. It is possible to use block kriging (Bierkens et al., 2000), using a grid defined by the pixel locations. The blocks can then be paired with co-located pixels for use in the regression model.

This paper forms part of a larger project, which is adopting a Bayesian framework, hence the model-based approach to geostatistics (Diggle, et al., 1998) is adopted. Furthermore, the model-based approach addresses a fundamental criticism of the classical approach, which is that it does not take account of the uncertainty involved in estimating the parameters of the assumed covariance function. Under the model-based framework, kriged predictions and conditionally simulated surfaces were obtained at regularly spaced points on a square grid. The grid was set up such that the points were also regularly spaced within each co-located pixel. The value of the block was then defined as the mean of all the predicted data points within each pixel. This follows the approach of Bierkens et al. (2000), who show that a variety of upscaling rules may be applied, depending on the circumstances.

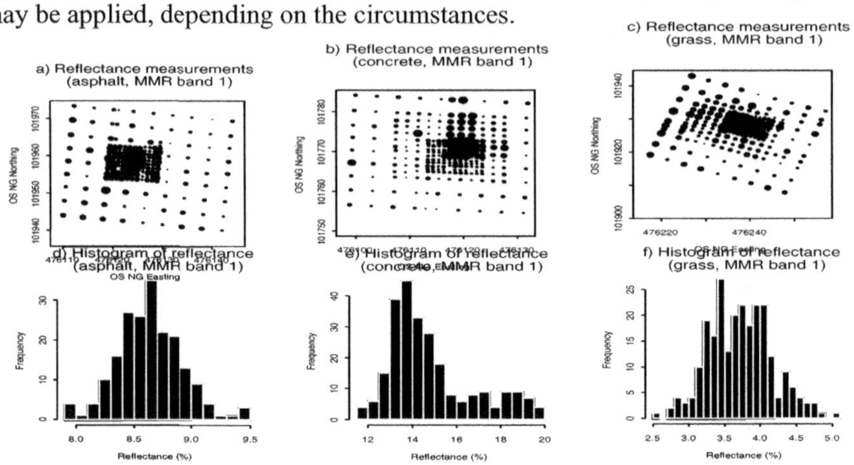

Figure 1. Data summaries for (a, d) asphalt; (b, e) concrete and (c, f) grass. The top row (a, b, c) shows the location of the field measurements (UK Ordnance Survey co-ordinates). The size of the points is proportional to the magnitude of the the reflectance value. The bottom row (d, e, f) shows a histogram for each target.

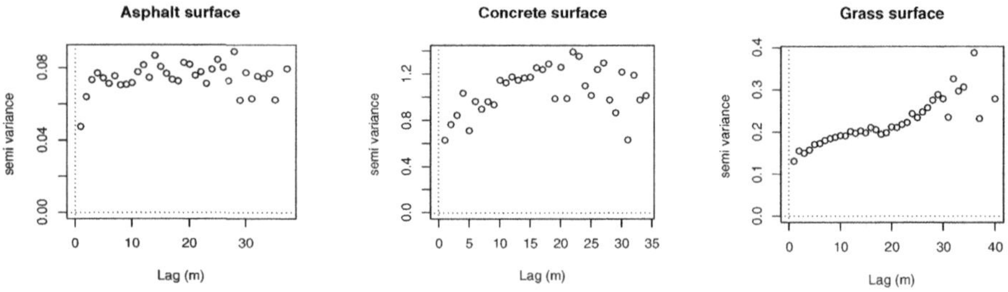

Figure 2. Sample variograms for the three surfaces.

4.3 Model-Based Geostatistics

Under the basic form of the model-based framework a Gaussian process is assumed (Diggle et al., 1998). Data are given in the form $(Z,\mathbf{x})_i$ where $\mathbf{x}_i$ is a location within the study region, and Z_i is the measurement. The existence of an unobserved stochastic stationary Gaussian process, $S(\mathbf{x}_i)$, $E[S(\mathbf{x}_i)] = \mu$ and $\rho(\mathbf{h}) = Corr[S(\mathbf{x}),S(\mathbf{x}-\mathbf{h})]$, is assumed. The exponential model for $\rho(\cdot)$ was used, since it gave a good fit to the data and is mathematically and computationally convenient. The joint distribution of $\boldsymbol{Z}$ is multivariate Gaussian:

$$\mathbf{Z} \sim MVN(\mu\mathbf{1}, \tau^2\mathbf{I} + \sigma^2\mathbf{R}) = MVN(\mu\mathbf{1}, \mathbf{V}) \quad (1)$$

where $\mathbf{1}$ denotes a vector of ones, $\mathbf{I}$ is the identity matrix, $\mathbf{R} = R(\phi)$ and $\mathbf{V} = \tau^2\mathbf{I} + \sigma^2\mathbf{R}$. The data are conditionally independent, given $S(\cdot)$:

$$Z_i | S \sim N(\mu(x_i) + S(x_i), \tau^2) \quad (2)$$

Bayes' formula is:

$$\pi(\mu, \sigma^2, \phi, \tau^2 | z) \; \alpha \; L(z; \mu, \sigma^2, \phi, \tau^2)\pi(\mu, \sigma^2, \phi, \tau^2) \quad (3)$$

Posterior α Likelihood × Prior

where $L(\mathbf{z}; \mu, \sigma^2, \phi, \tau^2)$ is multivariate Gaussian. The posterior predictive distribution for $\mathbf{z}_{new}$ is obtained:

$$f(z_{new} | z) \;=\; \int f(z_{new} | \theta) L(z; \theta) d\theta \quad (4)$$

where $\theta = (\mu, \sigma^2, \phi, \tau^2)^T$. In terms of the variogram μ is the mean, ϕ is the range, τ^2 the nugget variance and $\tau^2 + \sigma^2$ is the sill.

The "geoR" package for R was used for parameter estimation and for prediction (Ribeiro and Diggle, 1999). The Bayesian inference scheme implemented in geoR allows for simultaneous parameter estimation and prediction. The choice of priors is recognised as a delicate issue in Bayesian inference and non-informative priors were adopted for μ and σ^2 and discrete priors for ϕ and τ^2. The posterior distributions are then obtained using a Monte Carlo inferential strategy. The reader is referred to Diggle and Ribeiro (2002) for further information.

Posterior distributions for the parameters are obtained (Equation 3), so the modelled variograms may be summarized using the mode, median and the mean. Under the model-based approach, the multivariate distribution for all data points is used. Hence the model is *not* fitted to the experimental variogram and is not directly comparable to it. A maximum lag, to which the model is to be fitted, is not imposed. Similarly, a global neighbourhood approach is used for prediction. This is different to approaches commonly taken in classical gcostatistics.

4.4 Positional uncertainty

In order to simulate positional uncertainty a random error term, ε, was added to the Easting and Northing of each field location, as follows:

$$x_i = Easting + \varepsilon_{x_i}$$

$$y_i = Northing + \varepsilon_{y_i} \quad (5)$$

where i refers to each individual field location. It was judged reasonable to model ε as being drawn from a Normal distribution, $\varepsilon \sim N(0,\cdot)$. Taking $\varepsilon \sim N(0,0.25)$ reflects the case where the operator has a high degree of confidence in their positional accuracy. However, if the user had not been able to give such attention to recording location, $\varepsilon \sim N(0,4)$ is likely to be more sensible. By adopting this procedure, different realisations of the sampling scheme for each surface can be simulated and used to explore the effect of positional uncertainty. These are termed the "perturbed" data sets. The effect of positional uncertainty is then analysed by reference to the resulting modelled variograms and the effect on the implementation of the ELM.

Adopting this approach tackles the situation where a measurement is performed at the intended location but attributed to an incorrect location. This is the *resource model* described by Gabrosek and Cressie (2002). This is a realistic scenario, given the practical implementation of the sampling strategy. An alternative approach would be to adopt the *design model* where the measurements are taken at an incorrect location but attributed to the intended location. The design model is not considered in this paper.

4.4.1 The effect of positional uncertainty on the variogram

The modelled variograms for the original (i.e. unperturbed) data set and for the perturbed data sets are shown in Figure 3. From theory, it is expected that positional uncertainty will lead to an increase in the variogram only at short lags (Atkinson, 1996; Gabrosek and Cressie, 2002). This is borne out for the asphalt and concrete surfaces, but not for the grass surface, where there is a decrease in the sill. Although unexpected, latter result is not inconsistent with other *experimental* results (Atkinson1996).

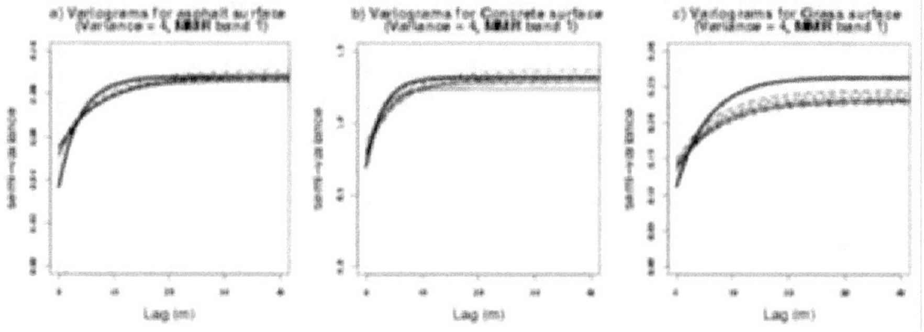

Figure 3. Modelled mean variograms for the three surfaces (a) asphalt, (b) concrete and (c) grass. The thick line is for the "original" (i.e. unperturbed) data set and the remaining lines are each for different perturbed data sets (in this case $\varepsilon \sim N(0,4)$).

4.4.2 The effect of positional uncertainty on the ELM

From the perspective of implementing the ELM, two questions need to be considered:

1. Does positional uncertainty lead to a change in the estimate of the slope, β, and intercept, α of the regression model?

2. Does positional uncertainty lead to a change in the estimate of the variance, λ^2, in the regression model.

The regression was performed using the co-located kriged blocks and pixels (see Section 4.2). Classical and Bayesian diagnostics were used for analysis of the regression. Non-informative priors were selected for the regression parameters, hence the point estimates were the same under both frameworks. The estimated parameters for the original (unperturbed data-set) are given in Table 1, together with examples of estimated parameters for the perturbed data sets (for both $\varepsilon \sim N(0,0.25)$ and $\varepsilon \sim N(0,4)$). The examples given are for illustration and are consistent with other results.

The results presented in Table 1 suggest that the introduction of positional uncertainty does not affect the estimate of β and λ. However, where ε is large, this can affect the estimate of α. For many remote sensing applications this change in α may be unimportant, since it introduces a small bias of less than 1% (Smith and Milton, 1999). The lack of sensitivity of β is encouraging. It should be realised that the blocks are derived from a kriged surface. Using blocks that are derived from conditionally simulated surfaces are likely to lead to a larger and more realistic estimate of the variance in the regression model (see Section 3.1). Current research is directed at incorporating the information contained in the conditionally simulated surfaces.

The above analysis was performed with a large data set that was time consuming and laborious to collect. Furthermore, the geostatistical analysis requires specialist expertise and may be time consuming. Hence the practitioner is likely to want to collect a smaller number of samples and adopt a more simple method for pairing the data. Previous research (Hamm et al., 2002) demonstrated that geostatistical analysis is not required to accurately estimate α and β, although it is necessary to pair the co-located point measurements of

Table 1. Estimated parameters for the ELM, implemented by pairing blocks and pixels. Orig. indicates the results for the unperturbed data sets. P.1, P.2 and P.3 indicate different realisations of the perturbed surface for $\varepsilon \sim N(0,0.25)$ and $\varepsilon \sim N(0,4)$.

	$\varepsilon = 0$	$\varepsilon \sim N(0,0.25)$			$\varepsilon \sim N(0,4)$		
	Orig	P.1	P.2	P.3	P.1	P.2	P.3
α	-0.76	-0.76	-0.75	-0.74	-0.75	-0.67	-0.71
β	0.0029	0.0029	0.0029	0.0029	0.0029	0.0029	0.0029
λ	0.53	0.53	0.53	0.53	0.53	0.53	0.52

reflectance and pixels. If the user does not require an accurate estimate of λ for their application, they might, feasibly, make a limited number of field measurements (perhaps less than 10 per target) and record the locations. This scenario was recreated by randomly selecting 1, 3 and 10 measurements for each target, pairing them with their co-located pixel value, and inputting the data pairs into the regression model. A large perturbation of the location may, therefore, lead to a field measurement being paired with an adjacent pixel and the objective was to investigate the implications of this.

Results of this analysis are shown in Table 2, for the cases of 3 and 10 measurements per target. The results for the case of 1 measurement per target are not shown, since classical and Bayesian diagnostics indicate that we would have low confidence that $\alpha \neq 0$ and $\beta \neq 0$ (95% confidence interval). Where 3 and 10 measurements are used it is sometimes found that $\alpha \neq 0$ (95% confidence interval). As before, this issue with α may not concern the practitioner (who might set $\alpha = 0$). Of much greater concern is the variability that is introduced into the estimation of β, since small changes in this parameter can lead to large changes in predictions of reflectance, the magnitude of which will vary with the brightness of the target. This effect is illustrated in Figure 4 and is likely to seriously concern the practitioner, especially if they are interested in targets that are bright or dark relative to the mean reflectance value. In addition to this variability in the point estimates of the parameters (α, β and λ), the positional uncertainty also increases the standard deviation of the estimated parameter. This effect leads to a decrease in the precision of predictions based on the ELM, as illustrated, for an analogous case, by Equation 4.

Table 2. Estimated parameters for the ELM, implemented by pairing field measurements and pixels. The top (bottom) set is for the scenario where 10 (3) measurements are taken for each target. Orig. indicates the results for the unperturbed data sets. P.1, P.2 and P.3 indicate different realisations of the perturbed surface for $\varepsilon \sim N(0,0.25)$ and $\varepsilon \sim N(0,4)$. The * indicates that the estimated parameter is not significantly different to 0 (95% confidence interval).

	$\varepsilon = 0$	$\varepsilon \sim N(0,0.25)$			$\varepsilon \sim N(0,4)$		
	Orig	P.1	P.2	P.3	P.1	P.2	P.3
α	-0.64*	-1.06	-0.73	-0.67	-0.08*	-1.03	-0.86
β	0.0029	0.0030	0.0029	0.0029	0.0027	0.0031	0.0030
λ	0.82	0.83	0.94	0.78	0.92	1.05	0.76
α	-0.45*	-1.33	-0.53*	-0.64*	-0.48*	-1.75*	-0.76*
β	0.0029	0.0031	0.0029	0.0029	0.0028	0.0033	0.0029
λ	0.65	0.90	0.51	0.83	0.88	1.51	0.54

5. CONCLUSIONS

The research discussed in this paper highlights several conceptual and practical issues. First, the effect of positional uncertainty on the variogram is demonstrated and the results are broadly consistent with theory and with the results of Atkinson (1996). Second, it implies that, given a large sample and when the field data are aggregated to the same support as the remotely sensed data, the estimation of the slope and intercept of the ELM is not sensitive to realistic errors in the location of the field measurements. This is an encouraging result, although a fuller assessment is required, by using the conditionally simulated blocks to more accurately quantify the variance in the regression. Finally, the practical and realistic scenario, where a relatively small sample is used and co-located field measurements and pixel values are combined was considered. It was shown that small perturbations of the locations of the field measurements introduce variability into the estimation of the slope of the ELM. This is likely to be of concern to remote sensing practioners who need to implement the ELM.

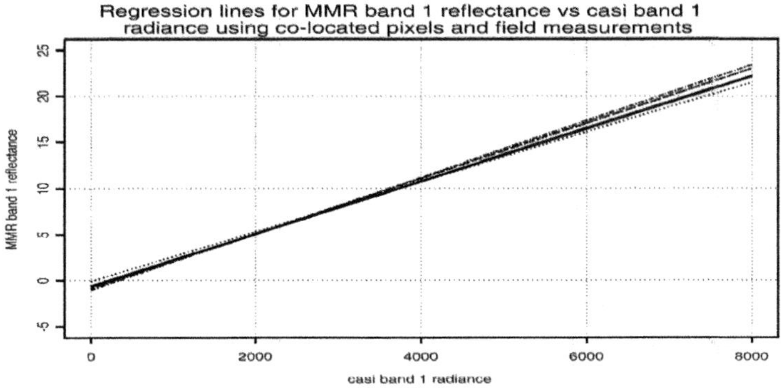

Figure 4. Implementation of the ELM (for co-located pixels and field measurements), illustrating the possible effects of positional uncertainty in a small sample (10 per target) on prediction.

ACKNOWLEDGEMENTS

This research is supported by a UK Natural Environmental Research Council (NERC) studentship (GT 04/99/FS/253) to N. Hamm. The NERC Airborne Remote Sensing Facility provided remotely sensed data. The NERC Equipment Pool for Field Spectroscopy provided field equipment and advice. The advice of P. Ribeiro and P. Diggle is gratefully acknowledged.

REFERENCES

1. Atkinson, P.M. 1996. Simulating locational error in field-based measurements of reflectance. In: A. Soares, J. Gomez-Hernandez and R. Froidevaux, eds, *geoENV I - Geostatistics for Environmental Applications*. London: Kluwer Academic Publishers, pp. 297-308.
2. Bierkens, M.F.P., P.A. Fink, P. de Willigen. 2000. *Upscaling and Downscaling Methods for Environmental Research*. London: Kluwer Academic Publishers.
3. Diggle, P.J., Jr.P. Ribeiro. 2002. Bayesian inference in model-based geostatistics. *Geographical and Environmental Modelling*. **6**(2):129-146.

4. Diggle, P.J., J.A. Tawn, R.A. Moyeed. 1998. Model-based geostatistics. *Applied Statistics*. **47**(3):299-350.
5. Gabrosek, J., N. Cressie, 2002. The Effect on Attribute Prediction of Location Uncertainty in Spatial Data. *Geographical Analysis*. **34**(3):262-284.
6. Hamm, N., P.M. Atkinson, E.J. Milton. 2002. Resolving the support when combining remotely sensed and field data: the case of the atmospheric correction of airborne remotely sensed imagery using the emiprical line method. In: G. Hunter and K. Lowell, eds, *Accuracy 2002*, Proceedings of the 5th International Symposium in Natural Resources and Environmental Sciences. Melbourne, pp. 339-347.
7. Heuvelink, G.B., E. Pebesma. 1999. Spatial aggregation and soil process modelling. *Geoderma*. **89**(1-2):47-65.
8. Matheron, G. 1963. Principles of geostatistics. *Economic Geology*. **58**:1246-1266.
9. Milton, E.J. 1987. Principles of field spectroscopy. *International Journal of Remote Sensing*. **8**(12):1807-1827.
10. Ribeiro, Jr.P., P. Diggle. 1999. geoR / geoS: A geostatistical software library for R/S-Plus. Technical report ST-99-09, Department of Mathematics and Statistics, Lancaster University, UK.
11. Schott, J.R. 1997. *Remote Sensing: The Image Chain Approach*. Oxford: Oxford University Press.
12. Smith, G.M., E.J Milton. 1999. The use of the empirical line method to calibrate remotely sensed data to reflectance. *International Journal of Remote Sensing*. **20**(13): 2653-2662.

GEOSTATISTICAL SPACE-TIME SIMULATION MODEL FOR CHARACTERIZATION OF AIR QUALITY

C. Nunes [1,2] and A. Soares[2]

[1] *Universidade de Évora, PortugaL. E-mail: carlanunes@alfa.ist.utl.pt*
[2]*Environmental Group of the Centre for Modelling Petroleum Reservoirs, CMRP/IST, Av. Rovisco Pais, 1049-001 Lisbon, Portugal.*

Abstract: The characterization of spatial uncertainty has been addressed in earth sciences using spatial models, based on stochastic simulation algorithms. Dynamic processes are characterized by two components – space and time. These usually have quite different levels of uncertainty: on the one hand, the heterogeneity of the static component – normally related to the space – can sometimes not be compared with the complexity of the dynamic part of the process; on the other hand, the available knowledge is usually quite different for these two components. This is possibly the main reason why the development of simulation algorithms for spatial processes with a time component is still at an early stage. The main goal of this study is to present a simulation model for the characterization of space-time dispersion of air pollutants. The objective of this model is to predict critical scenarios to support air quality control and management. This space-time simulation approach is applied to assess the particles contamination of Setúbal Peninsula (South of Lisbon – Portugal); a study, that is part of a project for the evaluation of regional air quality risk maps.

1. INTRODUCTION

This study presents a simulation process for the spatio-temporal characterization of air pollution dispersion, using simultaneous integration of spatial and temporal dispersion patterns. This process belongs to the family

X. Sanchez-Vila et al. (eds.), geoENV IV – Geostatistics for Environmental Applications, 103-114.

of geostatistical models for the characterization of spatio-temporal natural resources phenomena.

Geostatistical space-time models have been applied to several environmental areas, such as determination of space-time trends in the deposition of atmospheric pollutants (Eynon and Switzer, 1983; Bilonick, 1985, Kyriakidis and Journel, 1999[a)]), estimation of rain fall or piezometric head fields (Bras and Rodrigues-Iturbe, 1984; Rouhani and Wackernagel, 1990; Armstrong, Chetboun, and Hubert, 1993), spatial-temporal characterization of birds dispersion patterns (Santos et al, 2000), characterization of population dynamics in ecology (Hohn, Leibhold and Gribko, 1993) and design of sampling networks for monitoring spatiotemporal processes (Switzer, 1979).

In methodologic terms, the proposed method approaches the space-time referential as a finite collection of time series correlated in space (Solow and Gorelick, 1986; Kyriakidis and Journel, 1999 [b)]).

The goal of this methodology is not the inference of values in space and time, but the assessment of uncertainty using several critical scenarios to reproduce the spatial and temporal continuity and variability of the phenomena (Soares, Patinha and Pereira, 1996; Nunes, Soares and Ferreira 1999).

2. METHODOLOGY

To simulate several time series that have the same univariate, and bivariate statistics as the reality, a methodology composed by two main steps was developed (Nunes and Soares, 2002). In the first step a linear estimator of local cdf, for each monitoring station x_u is created, for the dispersion phenomena taking into account the spatial and the temporal previous occurrences. In a second step, a simulation process creates several time series based on the estimators defined in first step.

2.1 Estimation of bivariate distribution function $(Z(x_u,t),Z^*(x_u,t))$

Assuming a value $Z(x_u,t)$ of variable Z, measured in monitoring station x_u at time t – correlated with the concentrations measured in previous time periods at the same station and with concentrations measured at neighbouring monitoring stations at same time period, values of $Z(x_u,t)$ can be generate for all time periods and for all monitoring stations using the conditional distribution functions:

$$F\left(x_u,t;z\middle|z(x_u,t-1),z(x_u,t-2),\ldots,z(x_{u+1},t),z(x_{u+2},t),\ldots\right)=$$
$$=\mathrm{Prob}\{Z(x_u,t)\leq z|z(x_u,t-1),z(x_u,t-2),\ldots,z(x_{u+1},t),z(x_{u+2},t),\ldots\}$$

Note that in this case study the value $Z(x_u,t)$ is only correlated with the concentrations measured in previous time periods at the same station and with concentrations measured at neighbouring monitoring stations at same time period. So the space-time croos-covariances are neglected during estimation.

The main problem is how to estimate these *cdf* with a limited set of data – the time series of a few monitoring stations.

The idea of this paper is to calculate an approximation to these distribution functions using a linear combination of conditioning data:

$$F\left(x_u,t;z\middle|\sum_i \lambda_{ti} z(x_u,t-i)+\sum_j \lambda_{uj} z(x_{u+j},t)\right) \tag{1}$$

Based on historical data, bivariate histograms of $Z(x_u,t)$ and $Z^*(x_u,t)$, where $Z^*(x_u,t)$ is defined by:

$$Z(x_u,t)^*=\sum_i \lambda_{ti} Z(x_u,t-i)+\sum_j \lambda_{uj} Z(x_{u+j},t) \tag{2}$$

Can be estimated for each monitoring station and conditional distributions can be calculated from them. These weights can be computed by, for example, simple kriging.

With this approach, bivariate distribution functions $(Z(x_u,t), Z^*(x_u,t))$ can be estimated for each monitoring station from historical data, which allow the second step of this methodology, the simulation of $Z(x_u,t)$ values.

2.2 Simulation process

The values $Z(x_u,t)$ in each spatial location x_u at a time t are generated in a iterative simulation process and starts with the calculation of the conditioning data of [2], the estimated value $Z^*(x_u,t)$. Afterwards, the conditional distribution $F(Z(x_u,t)|\ Z^*(x_u,t)$ is retained from the estimated bidistribution function $(Z(x_u,t), Z^*(x_u,t))$. Finally, a valuc z is drawn from thc conditional distribution $F(Z(x_u,t)|\ Z^*(x_u,t)$. The process continues until all monitoring stations time series have been simulated.

The process can be summarized as follows. After defining the number of periods that are to be simulated in all monitoring stations, sequential simulation of the space-time process starts with a small set of seed values $Z(x_u,t-i)$. These are usually contiguous values taken from the historical data. The sequential procedure is illustrated in the following steps:

i- Define randomly the monitoring station location x_u to be simulated at time t.

ii- Estimate of $Z^*(x_u, t)$ according to [2] taking into account the seed values in the first steps and the previously simulated values afterwards.

iii- The conditional distribution $F(Z(x_u,t)| Z^*(x_u,t))$ is retained from the bi-distribution $(Z(x_u,t), Z^*(x_u,t))$ previously estimated for the monitoring station x_u.

iv- The simulated value $Z^s(x_u,t)$ of time t at x_u is drawn from $F(Z(x_u,t)| Z^*(x_u,t))$.

v- This value $Z^s(x_u,t)$ is added to data set. Return to step i) to simulated all monitoring stations for the same period t . When all monitoring stations, in time t, were simulated, return to i) with $t=t+1$, until all time series are simulated

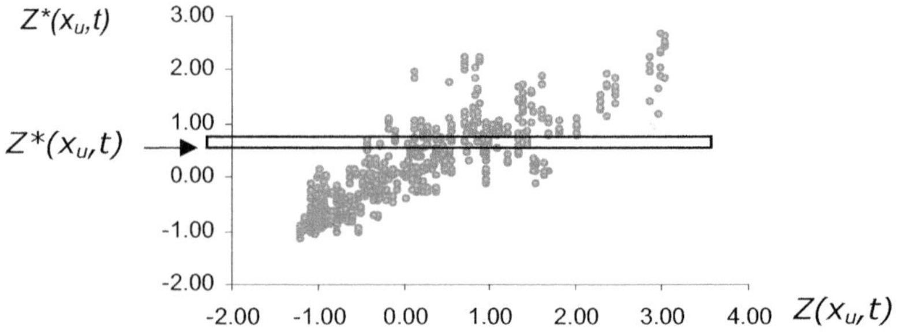

Figure 1. Sampling the the bidistribution $(Z(x_u,t), Z^*(x_u,t))$.

Generate $Z^s(x_u,t)$ values from $F(Z(x_u,t)| Z^*(x_u,t))$.

In step iii) a conditional distribution $F(Z(x_u,t)| Z^*(x_u,t))$ is defined or, i.e., the conditional histogram is retained from the global bi-histogram $(Z(x_u,t)| Z^*(x_u,t))$ (see fig.1).

To draw a simulated value z from the $F(Z(x_u,t)| Z^*(x_u,t))$, one follows the direct sequential simulation (dss) approach (Soares, 2000): the idea is to draw a value z from a portion of the conditional histogram $(Z(x_u,t)| Z^*(x_u,t))$, centered at the simple kriging estimator $Z^*(x_u,t)$ and with a local variance determined by the conditioning data of [2]. In dss algorithm local variance is given by simple kriging variance. In this case as the same conditioning pattern is used, the kriging variance values are the same. Hence, at each location x_u,t ,local variance $\sigma'^2(x_u,t)$ is calculated by the surrounding data - $Z^s(x_u,t-1)$, $Z^s(x_u,t-2)$,..., $Z^s(x_u,t-i)$, $Z^S(x_{u+1},t)$ $Z^S(x_{u+2},t)$, ..., $Z^S(x_{u+j},t)$ - and it is standardised by σ'^2 (x_u) which is the maximum σ'^2 (x_u,t) variance value observed for all times t at the same monitoring station x_u,.

$$\sigma^2(x_u, t) = \frac{\sigma'^2(x_u, t)}{\sigma'^2(x_u)} \quad (3)$$

Once it is defined the kriging estimate $Z^*(x_u,t)$ and local variance $\sigma'^2(x_u,t)$ a value z is drawn from the conditional distribution $F(Z(x_u,t)|\ Z^*(x_u,t))$ (annex 1).

The results of this simulation technique reproduce the statistics of observed data (real data): histogram and descriptive statistics. Theoretically, the spatial and temporal variograms are reproduced once the value z is drawn from the local distribution centered at the simple kriging estimator and with simple kriging variance (Journel, 1994). In this case [3] was used as a local variance. It is an approximation but temporal and spatial variograms succeed to be reproduced.

3. CASE STUDY

3.1 Air quality of Setúbal Peninsula

This case study aims at characterizing air quality in the Setúbal Peninsula.

Particulate emissions from three main non-diffuse sources – a cement plant, a power plant and a pulp mill – are periodically measured in a set of monitoring stations on daily average basis during 6 months (from 1/2/1997 to 31/7/1997) (Figure 2).

The global descriptive statistics are presented in Figure 3. Table 1 shows the descriptive statistics for each monitoring, during the referred period of time.

The most time consuming part of the data analysis and description is typically the description of spatial and/or temporal continuity. Though the variogram is the tool most commonly used by geostatiscians, it often suffers in practice from the combined effect of heterocedasticity and the preferential clustering of samples in areas with high values. In such cases, there are many alternatives that may produce clearer and more interpretable descriptions of spatial continuity. Of these alternatives, transformations of the original variables using local means and standard deviations are already quite common alternatives used by practioners.

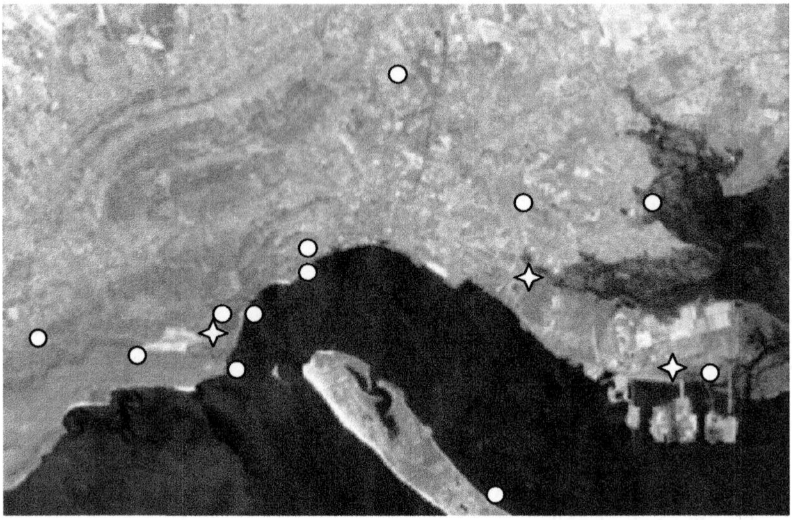

Figure 2. View of Setúbal peninsula with monitoring stations (0) and pollutant sources (◊).

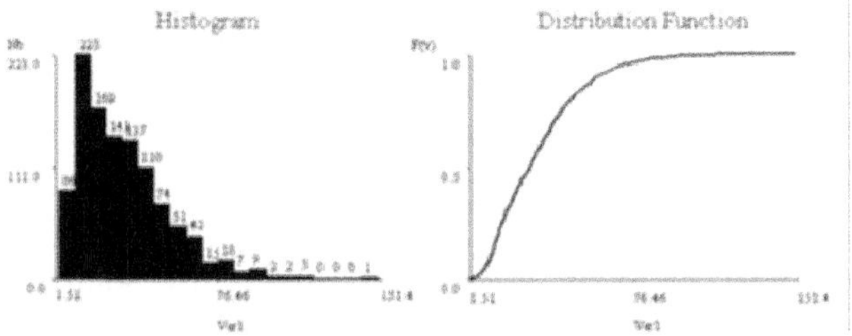

Mean	Variance	Min	Max	Skewness	P95
31.59	429	1.51	151.4	1.19	70.2

Figure 3. Histogram, distribution function and descriptive statistics of global data.

Table 1. Descriptive statistics of each monitoring station.

	PA	SF	SO	SE	SU	TR	P1	P2	P3	P4	P5	P6
Min.	13.28	8.01	8.67	5.15	1.51	2.31	11.00	5.40	10.80	11.00	18.10	3.10
Max.	83.22	79.58	71.43	57.44	62.92	38.02	97.50	151.40	120.80	114.30	107.50	86.50
Mean	36.41	32.21	26.38	25.57	21.85	12.53	48.69	37.94	55.41	58.75	52.64	35.28
SD	13.73	15.72	14.84	13.56	12.79	7.12	21.76	23.14	27.50	23.79	21.71	17.16

Global statistics and variograms of transformed experimental data are shown in Figures 4 and 5, respectively.

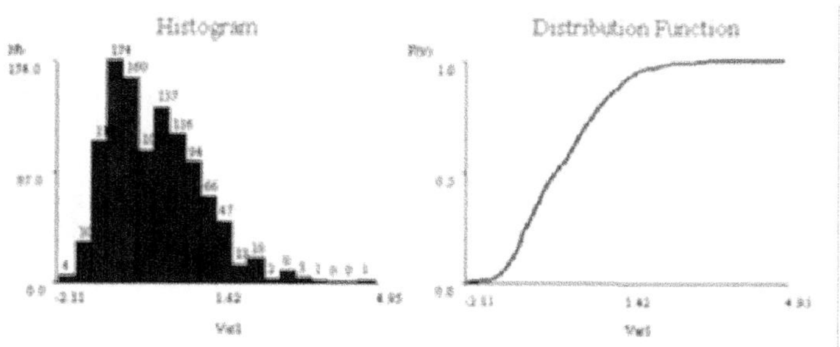

Mean	Variance	Min	Max	Skewness	P95
0	0.99	-2.11	4.95	0.71	1.65

Figure 4. Histogram, distribution function and descriptive statistics of transformed data.

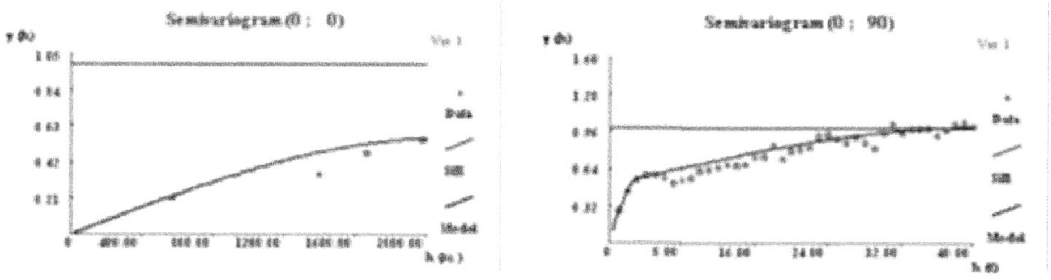

a) Spatial Variogram (isotropic) b) Temporal variogram

Figure 5. Spatial and temporal variograms of standardized data.

The methodology presented in section 2, was applied to these standardized data that were back transformed to the local original mean and standard deviation after the simulation process.

3.2 Space-time simulation of particulate concentration

The simulation process began with a small set of sequential values taken from the case study. Thirty different simulations were computed using the explained methodology. Global descriptive statistics of three detailed simulation examples, in standardized scale, are shown in Figure 6 and Table 2.

Comparing these results with the original parameters (Figures 4 and 5) one can see that this methodology reproduces the global statistics of the real data. Also the statistical parameters of these thirty simulations, for each monitoring station, have honored the statistics of the experimental data.

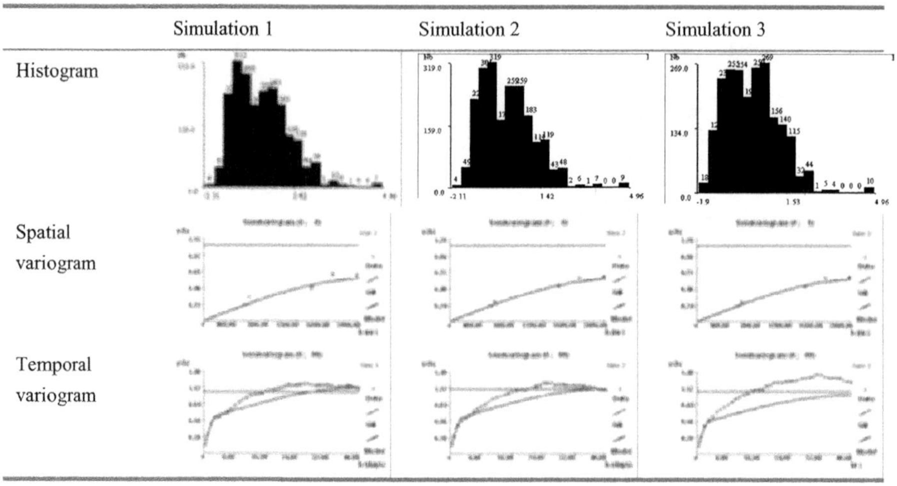

Figure 6. Histograms, spatial and temporal Variograms.

Table 2. Global descriptive statistics of the 3 detailed simulation examples.

	Mean	Variance	Min	Max	Skewness	P95
Ex. 1	0.1	1.14	-2.11	4.95	0.74	2.12
Ex. 2	0.07	1.1	-2.11	4.95	0.85	1.85
Ex. 3	0.16	1.05	-2.11	4.95	0.73	1.85

As mentioned, to reproduce the original phenomenon, it was necessary to apply a back-transformation for each monitoring station x_i:

$$Z''(x_i,t) = Z'(x_i,t) * \sigma_i + \mu_i$$

In figures 7, 8 and 9 the experimental data and two different simulations for three monitoring stations (SF, PA and P3, respectively) can be seen.

In the original scale, the histograms and the statistics descriptive observed in real data (figure 3) are honoured in all space-time simulations.

Note: Because of the heterocedascity observed in monitoring stations, spatial variograms where computed with standardized data by the local mean and variance of the monitoring stations. Simulations, computed with standardized data, reproduce the spatial patterns, as they are revealed with spatial variograms of fig 6. After the transformation to the original particulate concentration variable, local means and variances of the monitoring stations mask, as they do with the experimental data, the spatial pattern.

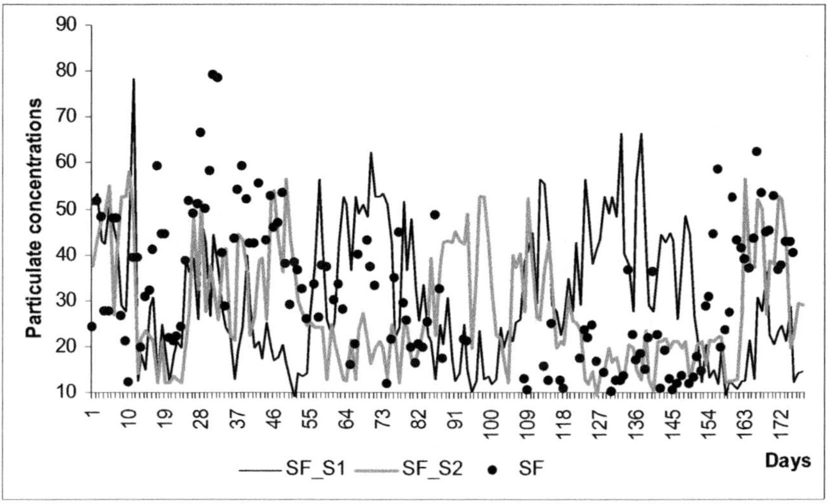

Figure 7. Particulates concentration measured at SF and two simulations (S1, S2).

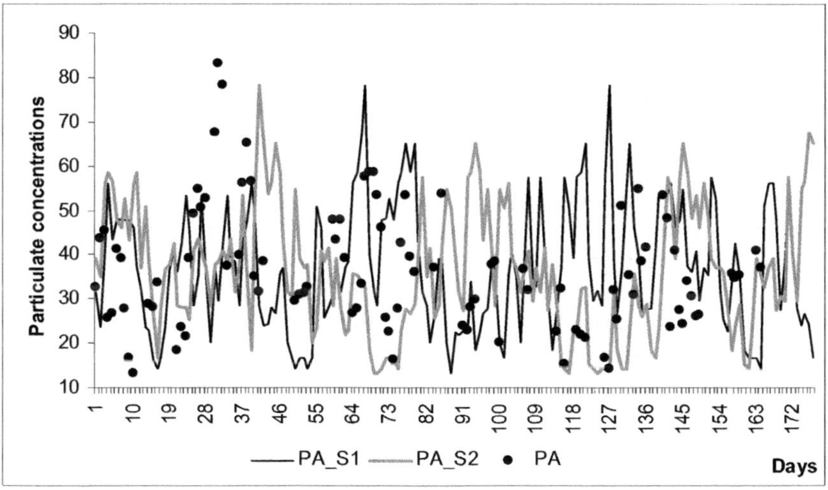

Figure 8. Particulates concentration measured at PA and two simulations (S1, S2).

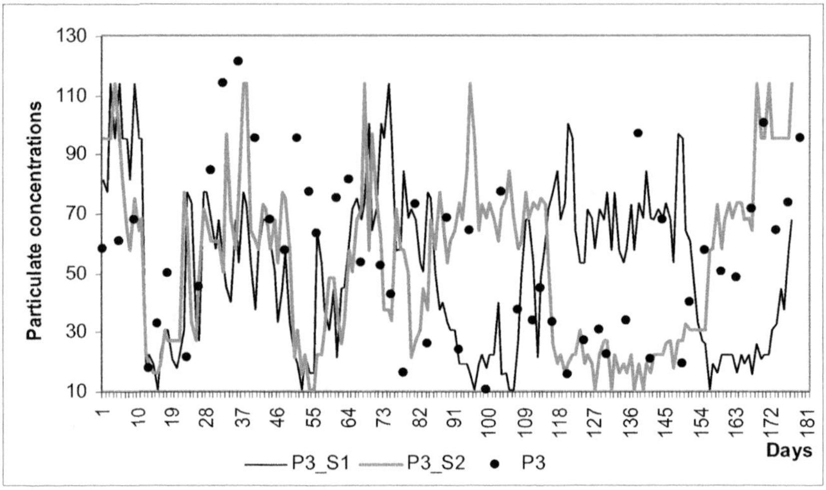

Figure 9. Particulates concentration measured at P3 and two simulations (S1, S2).

4. CONCLUSIONS

The proposed methodology of space-time simulation of natural phenomena showed very promising results for the present case study of particulate concentration characterization.

The spatial-temporal simulations, covering the monitoring stations localizations for all time t, can be used as conditioned data to a spatial simulation, for each time t, allowing the simulation of entire area, for any simulated time t.

The simulation methodology succeeded to reproduce the main space-time patterns as they are revealed by spatial variograms between monitoring stations and average time variograms of all monitoring stations.

The proposed space-time simulation model allows for the assessment of extreme and risk situations reproducing the impact of air quality on the neighbourhood eco-systems.

REFERENCES

1. Armstrong, M. Chetboun, G., Hubbert P., 1993, Kriging the rainfall in Lesotho, in A. Soares, ed., Geotatistics Tróia'92, Vol. 2: Kluwer Academic Publ., Dordrecht, p. 661-672.
2. Bilonick, R. A., 1985, The space-time distribution of sulfate deposition in the northeastern United States: Atmosph. Environment, v. 19, no. 11, p.1829-1845.
3. Bras, R.L., and Rodrigues-Iturbe, I, 1984, Randon Functions and hydrology: Addison – Wesley, Reading, MA, 559 p.
4. Eynon, B. P., and Switzer, P., 1983, The variability of rainfall acidity: Can. <Jour. Statistics, V. 11, no. 1, p. 11-24.

5. Hohn, M.E., Liebhold, A. M., and Gribko, L.S., 1993, Geostatistical model for forecasting spatial dynamics of defoliation caused by the gypsy moth (Lepidoptera: Lymantriidae): Environmental Entomology, v. 22, no. 5,p.1066-1075.
6. Kyriakidis, P. C. and Journel , A. J., 1999 [a)], Stochastic Modeling of spatiotemporal distributions: Application to sulphate deposition trends over Europe, in J. Gomez-Hernandez , A. Soares, and R. Froidevaux ed., geoENV II- geostatistics for environmental Applications,: Kluwer Academic Publ., Dordrecht, p. 89-100.
7. Kyriakidis, P. C. and Journel , A. J., 1999 [b)], Geostatistical Space- Time models: A review, Mathematical Geology, Vol.31, n0. 6.
8. Journel A., 1994, Modeling uncertainty: some conceptual thoughts, in Dimitrakopoulos, R., ed., Geostatistics for the next century: Kluwer Academic Pub., Dordrecht, The Netherlands, p.30-43.
9. Nunes C., Soares A. and Ferreira F., 1999, Evaluation of Environmental costs of SO_2 emissions using Stochastic Images, in J. Gomez-Hernandez, A. Soares, and R. Froidevaux ed., geoENV II- geostatistics for environmental Applications,: Kluwer Academic Publ., Dordrecht, p. 113-124.
10. Nunes C. And Soares A, Geostatistical Space-Time Simulation Model, paper submitted to Environmetrics, 2002.
11. Rouhani, S., and Wackernagel, H., 1990, Multivariate geostatistical approach to space-time data analysis: water resources Res., V. 26, no 4, p. 585-591.
12. Santos E., Almeida J., Soares A., 2000, Geostatistical Characterisation of the migration Patterns and Pathways of the Wood Pigeon in Portugal, Geostatistics 2000, W.J. Kleingeld, D.G. Krige Ed., vol.2, P. 615-622.
13. Soares A., 2001, "Direct Sequential Simulation and Co-Simulation", Mathematical Geology, Vol. 33, no.8, Pg. 911-926.
14. Soares, A., Patinha , P. J., and Pereira, M.J., 1996, Stochastic simulation on space-times series: Application to a river water quality modeling, in Srivastava, R.M., Rouhani, S., Cromer, M. V., Johnson, A.I., and Desbarats, a. J., eds., Geostatistics for Environmental and Geotechnical Applications: American society for Testing and Materials, west Conshohocken , Pa, p.146-161.
15. Solow, A. R., and Gorelick, S. M., 1986, Estimating monthly stream flow values by cokriging: Math. Geology, v. 18, no. 8, p.785-809.
16. Switzer, P., 1979, Statistical considerations in network design, Water Resources Res.,v. 15, no. 6, p.109-123.

ANNEX 1

A Gaussian distribution function can be used as a tool to sample the local bivariate distribution depending on the local variability information:

Suppose φ is the normal score transform of original $z(x,t)$ values:

$$y(x)= \varphi(z(x,t)) \quad \text{with} \quad G(y(x))=F_Z(z(x,t)) \tag{4}$$

The local estimate $Z^*(x_u,t)$ has the equivalent Gaussian value $y^*(x_u) = \varphi(z^*(x_u)$ which, together with the $\sigma^2(x_u,t)$ estimation variance, can define a Gaussian cdf $-G(y^*(x_u), \sigma^2\,(x_u,t))$.

To simulate a new value $z^s(x_u,t)$ the following sequence of steps is used:

- Generate a value p from a uniform distribution U(0,1)
- Generate a value y^s from $G\,(y^*(x_u), \sigma^2(x_u,t))$

$$y^s= G^{-1}(p) \tag{5}$$

- Finally, a simulated value $z^s(x_u,t)$ is obtained by the inverse transform φ^{-1}:

$$z^s(x_u,t)= \varphi^{-1}(y^s) \tag{6}$$

This means that $z^s(x_u)$ is sampled from intervals of $F_Z(z)$ defined by the local estimates $Z^*(x_u,t)$ and local variance $\sigma^2(x_u,t)$.

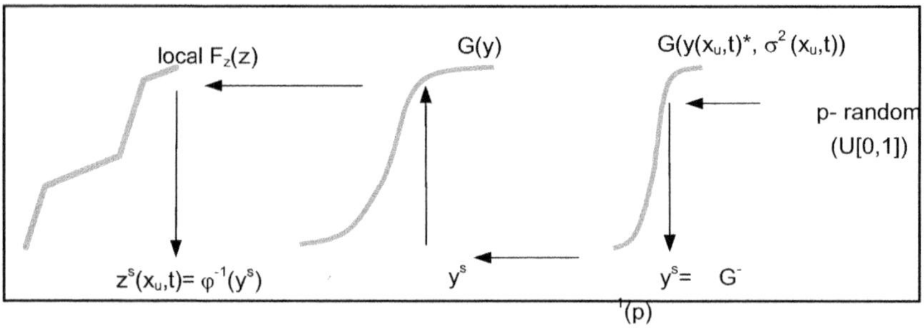

Figure 1. Illustrates the explained methodology applied to the $F(Z(x_u,t)|Z^*(x_u,t))$.

It is important to note that the Gaussian transformation is used solely for sampling intervals of the bidistribution $F_Z(z)$. It does not have any role in the estimation of local cdf, hence no Gaussian hypothesis of the original values is assumed. The entire sequential procedure is performed with the original variable $Z(x)$.

SOFT DATA SPACE/TIME MAPPING OF COARSE PARTICULATE MATTER ANNUAL ARITHMETIC AVERAGE OVER THE U.S.

M.L. Serre, G. Christakos and S.J. Lee
Center for the Advanced Study of the Environment (CASE), University of North Carolina-Chapel Hill, NC

Abstract: In the U.S., particulate matter (PM10) is considered an important criteria air pollutant and it is monitored throughout the country by means of a considerably dense network of stations. Because of the health risks associated with PM10, it is important to study carefully the spatiotemporal distribution of the air pollutant. In the last decade, the modern BME approach has emerged as an advanced function of temporal GIS (TGIS). The BME approach has certain powerful features and has been used for mapping PM10 and PM2.5 distributions in the U.S. and abroad. In this work we propose an approach to use available information to develop probabilistic soft data about the annual arithmetic average of PM10, and we use the BME framework to rigorously process that information and produce realistic spatiotemporal maps of PM10 distribution over the US. We apply the approach presented on a large PM10 dataset from the USEPA AIRS database covering the 1984 to 2000 period.

Key words: Particulate matter, space/time, mapping, Geostatistics, BME, soft data

1. INTRODUCTION

In the U.S., particulate matter of aerodynamic diameter less than 10 micrometer (PM_{10}) is considered an important criteria air pollutant and is monitored throughout the country by means of a considerably dense network of stations. PM_{10} is now referred to as the "coarse" particulate matter by comparison to the finer $PM_{2.5}$ criteria air pollutant. Because of the health risks associated with PM_{10} [1], it is important to study carefully its spatiotemporal distribution in the air. In the last decade, the modern

X. Sanchez-Vila et al. (eds.), geoENV IV – Geostatistics for Environmental Applications, 115-126.

Bayesian Maximum Entropy (BME) approach has emerged as an advanced function of temporal GIS (TGIS). The BME approach [2,3] has certain powerful features and has been used for mapping PM_{10} distributions in the U.S. and abroad [4,5,6]. In this work we are concerned with estimating the space/time distribution of the annual arithmetic average of PM_{10} over the United State. The annual arithmetic average provides a measure of the chronic exposure to PM_{10}, which is of concern for long-term human health effects [1]. A critical aspect of annual arithmetic average measures is that they have varying level of uncertainties depending on the number of observations available, and they display a high variability in both space and time. In this work we propose an approach to use available information to develop probabilistic soft data about the annual arithmetic average of PM_{10}, and we use the BME framework to rigorously process that information and produce realistic spatiotemporal maps of PM_{10} distribution over the US.

A spatiotemporal random field (S/TRF) $Z(p)$ is used to represent the randomness and correlation structure of the annual PM_{10} arithmetic average field across space and time. The vector $\boldsymbol{p} = (\boldsymbol{s}, t)$ defines a point in the space $\boldsymbol{s}$ and time t domain. Given certain general knowledge about the entire $Z(p)$ field (such as its mean trend and covariance structure), the BME method defines a space of plausible events, and then restricts this space to be consistent with available site-specific knowledge. The site specific knowledge includes hard data (accurate measures) and soft data (probabilistic description of the possible values for $Z(p)$ at some data points). After processing the general knowledge (mean and covariance) and employing a Bayesian conditionalization rule on the hard and/or soft data, BME yields a posterior probability density function (PDF) that characterizes $Z(p)$ at every point of a mapping grid, from which informative space/time maps of the annual PM_{10} arithmetic average are constructed. The *BMElib* package is used in this work, and readers are referred to the associated book [7] for more detailed information about the practical implementation of the BME method.

2. THE PM_{10} DATASET IN THE US

The PM_{10} data used in this analysis is based on PM_{10} measurements collected at 1168 monitoring stations distributed throughout the United States (Fig. 1). For each of the monitoring station the USEPA *AIRS* database [8] provides annual statistics over a 17-year period (1984-2000). Several of the 1168 monitoring stations were only in service for part of the 17 years so that annual statistics was available for only 7327 (or 37%) of the

Figure 1. Map of the location of the 1168 PM_{10} monitoring stations in the US.

1168*17= 19856 possible space/time data points. The annual statistics for a given data point available in this work consist only of the three following variables: (1) the number n_{obvs} of PM_{10} observations collected over the year (usually 24-hour average measurements), (2) the arithmetic average C_{ave} of these observations, and (3) their 95 percentile $C_{0.95}$ (i.e. the value that was only exceeded 5% of the time). Exploratory data analysis revealed that C_{ave} had a skewed distribution toward positive high values (top of Fig. 2), while log-transformed C_{ave} were approximately normally distributed (bottom of Fig. 2).

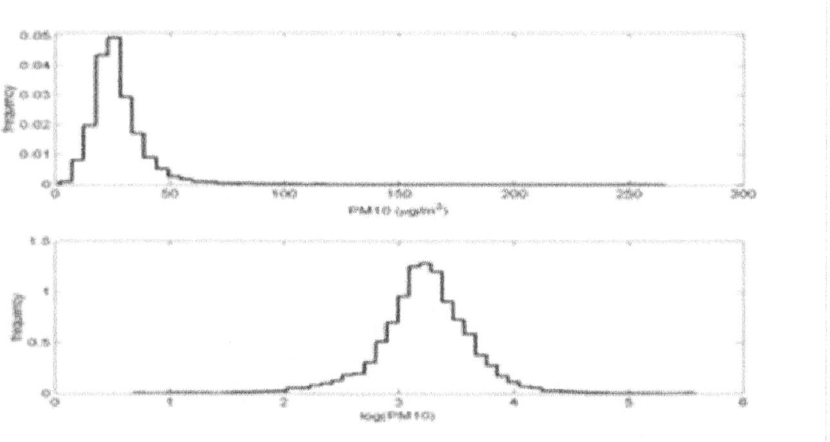

Figure 2. Frequency distribution (histogram) of C_{ave} (PM_{10} in μg/m^3) and log(C_{ave}).

Investigation of the *AIRS* PM_{10} dataset also revealed that n_{obvs} had a wide distribution of values (Fig. 3), with a mode of about 60 (i.e. the annual statistics for most data points was based on about 60 observations) but a minimum of 1 (i.e. several data points had only 1 observation) and a maximum in excess of 300. Furthermore the variance of the observation values, as indicated by $C_{0.95}$, varied significantly from data point to data point. The variation of the number of observations and variance leads to calculated arithmetic averages C_{ave} that have varying level of reliability, and it is critical that these different levels of reliability be incorporated in the analysis. For instance we found that $n_{\text{obvs}}=1$ for as many as 46 data points, leading to 46 calculated arithmetic average C_{ave} that are a lot less reliable than those obtained at data points with $n_{\text{obvs}}=60$. As a consequence we propose an approach that rigorously integrates the different levels of reliability in the calculated C_{ave} by treating them as soft data in the BME framework.

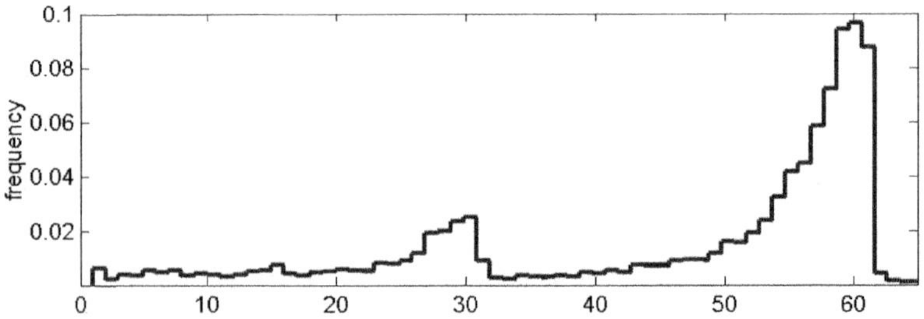

Figure 3. Frequency distribution of the number of observation n_{obvs}.for each data point.

3. GENERATING SOFT DATA

Let $Z(\boldsymbol{p})=Z(s,t)=T^{-1}\int_{u\in[t,t+T]} du\ C(s,u)$, where T=1 year, be the S/TRF representing the annual arithmetic average at spatial location $\boldsymbol{s}=(s_1,s_2)$ and time t of the instantaneous PM_{10} concentration $C(s,u);\ u\in[t,t+T]$. We assume that over the year $[t,t+T]$ for which the observations are collected at a data point $\boldsymbol{p}=(s,t)$ the expected value $\mu=\text{E}[C(s,u)];\ u\in[t,t+T]$, is constant, and that ergodicity applies so that μ is approximately equal to $Z(s,t)$. For that year the calculated arithmetic average $C_{\text{ave}}=1/\ n_{\text{obvs}}\ \sum_{i=1,n_{\text{obvs}}} C_i$ is an estimator of the expected value μ at $\boldsymbol{p}$, where C_i are the n_{obvs} observation values of PM_{10} concentrations over the year. Linear regression theory holds that $(\mu-C_{\text{ave}})/(s/\sqrt{n_{\text{obvs}}})$ is student-t distributed with n_{obvs}-1 degrees of freedom, where s^2 is an estimator of the variance of the n_{obvs} observation values assumed independent. In our work we had to use $C_{0.95}$-C_{ave} to obtain an estimate of s using $s \approx (C_{0.95}-C_{ave})/1.65$, but future work should use the classical $s^2=1/(n_{obvs}-1) \sum_{i=1,n_{obvs}} (C_i - Z_{ave})^2$ estimator when that information is available. Hence for each of the 7327 space/time data point $\boldsymbol{p}$ we use C_{ave}, n_{obvs} and $C_{0.95}$ to obtain a student-t probability density function (PDF) of $\mu \approx Z(\boldsymbol{p})$. Letting z be the random variable representing the S/TRF $Z(\boldsymbol{p})$ at a data point $\boldsymbol{p}$, and substituting μ with z under the ergodic assumption that they are equal over the averaging year corresponding to the data point $\boldsymbol{p}$, we get that $v=(z-C_{ave})/(s/\sqrt{n_{obvs}})$ is student-t distributed with n_{obvs}-1 degrees of freedom. The PDF of the variables z and v are related by the relationship $f_{S,z}(z)dz=f_v(v)dv$, from which we immediately obtain that the PDF for z is $f_{S,z}(z)=1/sn\ f_v(\ (z-C_{ave})/sn\ )$, where $sn=s/\sqrt{n_{obvs}}$, and f_v is the student-t PDF with $n_{obvs}-1$ degrees of freedom. This student-t PDF provides the soft probabilistic data that correctly prescribe different levels of reliability of our knowledge of the true (but unknown) value of $Z(\boldsymbol{p})$ at the data point $\boldsymbol{p}$ as a function of n_{obvs} and $C_{0.95}$-C_{ave} (i.e.. a small n_{obvs} or a large $C_{0.95}$-C_{ave} will yield a soft PDF with wide spread, i.e. high uncertainty of the actual arithmetic average $Z(\boldsymbol{p})$ at that data point).

4. SPACE/TIME VARIABILITY OF PM10

The S/TRF $Z(\boldsymbol{p})$ representing the annual PM_{10} arithmetic average was log-transformed to obtain the S/TRF $Y(\boldsymbol{p})=\log(Z(\boldsymbol{p}))$ with an approximately normal distribution (Fig. 2). The $Y(\boldsymbol{p})$ field was further decomposed into a mean trend function $m_y(\boldsymbol{p})$ and a residual S/TRF $X(\boldsymbol{p})$, such that $Y(\boldsymbol{p})= m_y(\boldsymbol{p})+X(\boldsymbol{p})$. The mean trend function was obtained with the *BMElib* package by using a moving window average of Y-data with an exponential space/time filter. This mean trend essentially "smoothes" the spatiotemporal fluctuations, and yields a residual field $X(\boldsymbol{p})=Y(\boldsymbol{p})-m_y(\boldsymbol{p})$ that is homogenous in space and stationary in time. The mean trend may be considered to be a deterministic function (i.e. a known function), while the residual field models all the uncertainties and variability associated with PM_{10} over the space/time domain of interest. The space time variability of $X(\boldsymbol{p})$ is described in terms of the space/time covariance function $c_x(r,\tau)=E[(X(s,t)-m_x(s,t))(X(s',t')-m_x(s',t'))]$, where $r=|s- \boldsymbol{s'}|$ is the spatial lag and and $\tau=|t-t'|$ is the temporal lag. Values of the covariance function $c_x(r,\tau)$ where estimated with the *BMElib* package for different classes of spatial and temporal lags. The experimental values estimated by the *BMElib* package using the log-transformed, mean trend removed X-data are shown in Fig. 4 as a function of spatial lag classes (top of Fig. 4) and temporal lag classes (bottom of Fig. 4). The theoretical covariance model selected to fit these experimental covariance values is a *non-separable* model that consists of the superposition of two exponential models with different spatial and temporal scales, as is shown in the following equation

$$c_x(r,\tau)=c_{01}\exp(-3\,r/a_{r1})\exp(-3\,\tau/a_{t1})+c_{02}\exp(-3\,r/a_{r2})\exp(-3\,\tau/a_{t2})$$

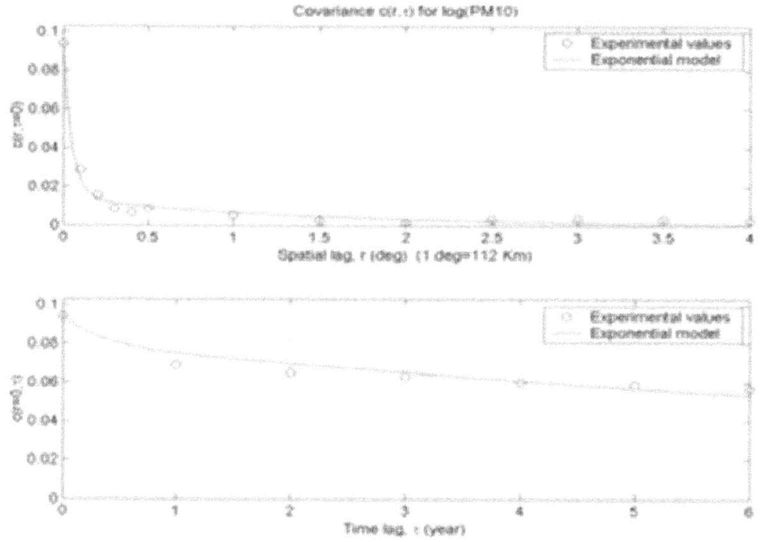

Figure 4. Covariance of *X(s,t)* as a function of spatial lag and temporal lag.

The first covariance component has a covariance sill of c_{01}=0.0141, a spatial range of a_{r1}=4 degrees (or approximately 448 *Km* on the Earth surface), and a temporal range of a_{t1}=1 year, while the second component has a sill of c_{02}=0. 0798, a spatial range of a_{r1}=0.15 degrees (or approximately 16.8 *Km*), and a temporal range of a_{t1}=45 year. We hypothesize that the first covariance component corresponds to PM_{10} fluctuations that are weather-related, with large spatial structures (e.g. large rainfall or wind events, cleaning airborne particulate over large spatial areas of a few hundred *Km* in size), but a short temporal scale of about 1 year (likely corresponding to seasonal fluctuations). Conversely we attribute the second covariance component to fluctuations that are caused by long-term human activities, with a spatial influence of just 10 to 20 Km (corresponding to the size of urban centers with high car traffic and the zone of influence of large point source of PM_{10}), and a temporal range of long duration of approximately 45 years. It is interesting to note the lasting effect of human activity on particulate matter pollution in the environment, which is an additional reason for the concerns associated with air pollution.

5. SPACE/TIME *BME* ESTIMATION RESULTS

In the space/time estimation of the annual PM_{10} arithmetic average we are concerned with obtaining a BME estimate of $Z(\mathbf{p})$ at any estimation point $\mathbf{p}_k$, and characterizing the associated estimation error by means of a confidence interval. The mean and covariance models and the set of probabilistic (soft) data provides the knowledge bases used in *BMElib* to calculate the BME posterior PDF of $Z(\mathbf{p}_k)$. The covariance model $c_x(r,\tau)$ provides the basis of the general knowledge of the log-transformed residual S/TRF $X(\mathbf{p})$. Hence we need to transform the student-t soft PDF for the Z-data to soft PDF's for log-transformed mean trend removed X-data. Once the soft data has been transformed and processed by *BMElib,* we obtain the BME estimate and confidence interval of $X(\mathbf{p})$ at the estimation point, which must be back-transformed to obtain the BME estimate and confidence interval of $Z(\mathbf{p})$. The steps to transform the Z-soft data and to back-transform the X-BME estimates are summarized as follow:

1. At each of the 7327 space/time data points we use Cave, nobvs and C0.95 to obtain the soft PDF for Z, fS,z(z)=1/sn fv((z-Cave)/sn), where fv is the student-t PDF with nobvs–1 degrees of freedom, sn=s/$\sqrt{n_{\text{obvs}}}$, and the standard deviation s is estimated from the 95 percentile C0.95 as s=(C0.95-Cave)/1.65 (as mentioned earlier one should use the classical variance estimator s2 if that information is available);
2. If at a monitoring station $C_{0.95}=C_{\text{ave}}$ for a given year (which happens when there are few observation values), then the standard deviation is conservatively taken as the largest standard deviation of all the recorded years at that monitoring station. Furthermore when $n_{\text{obvs}}=1$, the soft PDF is simply taken as the Gaussian PDF with mean C_{ave} and standard deviation equal to the largest recorded for that monitoring station.
3. The soft PDF for $Y=\log(Z)$ is given by $f_{S,y}(y) = z\, f_{S,z}(z)$, where $z=\exp(y)$;
4. The soft PDF for $X=Y-m_y$ is given by $f_{S,x}(x) = f_{S,y}(y)$, where $y=x+m_y$;
5. Using *BMElib* we obtain the BME posterior PDF for $X(\mathbf{p})$ at an estimation point $\mathbf{p}_k$, from which we get the median estimator $X_{k,\text{median}}$ (i.e. the value that has a probability of 0.5 to be exceeded), and the confidence interval $\text{CI}_x=[\, X_{k,l}\, ,\, X_{k,u}]$ for some predefined confidence level (e.g. the 95% confidence level).
6. The BME median estimate and confidence interval for Y are obtained by translation, i.e. $Y_{k,\text{median}}=X_{k,\text{median}}+m_y(\mathbf{p}_k)$ and $\text{CI}_y=\text{CI}_x+m_y(\mathbf{p}_k)$, where $m_y(\mathbf{p}_k)$ is the mean trend at the estimation point $\mathbf{p}_k$;

7. Finally the BME estimate and confidence interval for Z are obtained using the reciprocal of the log transform, i.e. $Z_{k,\mathrm{median}}=\exp(Y_{k,\mathrm{median}})$ and $CI_z=\exp(CI_y)=[\exp(Y_{k,l})\ ,\exp(Y_{k,u})]$.

(a)

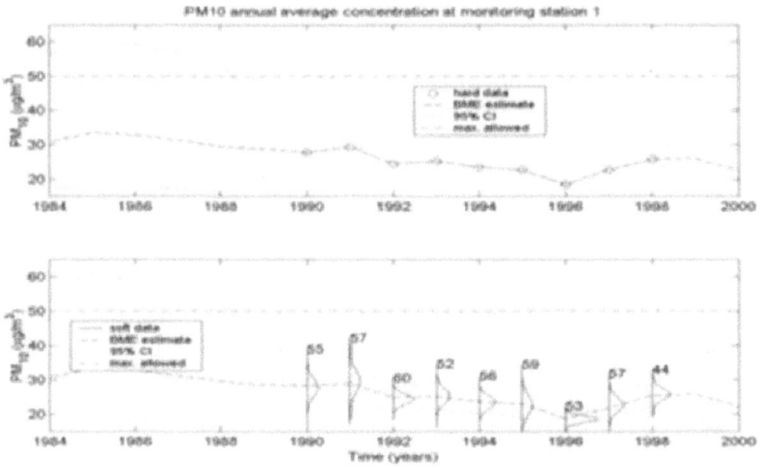

(b)

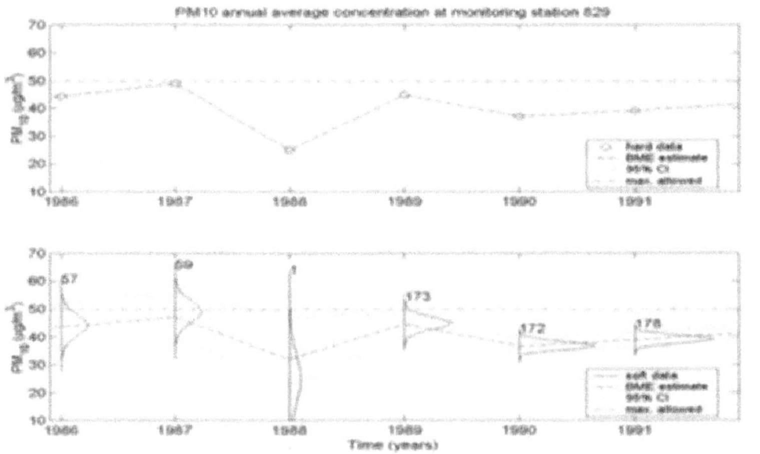

Figure 5. Annual PM_{10} arithmetic average BME estimates at (a) monitoring station 1, and (b) station 829. Results for hard data are on top plots and soft data on bottom plots.

In Figs. 5.a and 5.b we show the temporal plots of the annual PM_{10} arithmetic average at monitoring station 1 (Fig. 5.a) and monitoring station 829 (Fig 5.b). In each figure the top plot shows the estimation obtained if the calculated arithmetic average C_{ave} at each monitoring event was treated as hard data (i.e. deeming that Z= C_{ave} with a probability one), while the bottom plot shows the result obtained in this work where soft data is used to represent uncertainties associated with the calculated arithmetic average. The soft PDF $f_{S,z}(z)$ is shown vertically for each data point, and the number of observations n_{obvs} is written at the top of the soft PDF. As can be seen in the figures the spread (or uncertainty) of the soft PDF varies substantially from one data point to another, with wider spread (more uncertain data) associated with smaller number of observations n_{obvs} and with higher variance s^2 in the observation values. The BME median estimate of the annual PM_{10} arithmetic average is shown with a dashed line, while the confidence interval corresponding to the 95 % confidence level (i.e. the interval that contains the true $Z(\mathbf{p})$ with a probability of 0.95) are shown with dotted lines. As can be seen from these plots, the soft data approach presented in this work leads to results that are more realistic and physically meaningful than the classical approach of just taking C_{ave} as hard data. Consider for example the data point for year 1988 in Fig. 5 (b). The C_{ave} was obtained on the basis of only one observation, yielding a soft PDF with a wider spread than neighboring points, which is much more realistic than treating it as hard data. This results in an upper bound of the 95% confidence interval that is significantly higher than that obtained for hard data, which has a critical impact on any sort of risk assessment.

The BME median estimate for the soft data approach presented in this work can also be plotted as spatiotemporal maps. In Fig 6.a and 6.b we show the spatial map of the BME median estimate of the annual PM_{10} arithmetic average for years 1984 and 2000, respectively. These maps show the change over time in the spatial distribution of PM_{10} chronic levels in the US, with a clear general decline of the air pollutant. At each point of the map BME provides the full posterior PDF of the annual PM_{10} arithmetic average, from which can be extracted the estimation error variance normalized by the estimated value as shown in Fig 7 for year 2000. This map provides a measure of the estimation uncertainty of the map, which is lower at data points and increases as we go away from the data points. Using the BME posterior PDF we may also delineate areas that are such that the probability of Z being smaller than a critical values is at least equal to some acceptable confidence level (say. 0.95). For illustration purpose we show in Fig. 8.a and 8.b the areas not meeting this criteria, i.e. areas not attaining a probability of at least 0.95 that Z is less than the 50 μg/m^3 national standard (non attainment areas). These figures show that the non-attainment areas at the 95% confidence level have considerably diminished for PM_{10},, so that except for Southern California most of the US is attaining the national standard with a 95 % confidence level (however on-going work is showing that the situation is different when considering the fine particulate matter criteria air pollutant, $PM_{2.5}$).

(a)

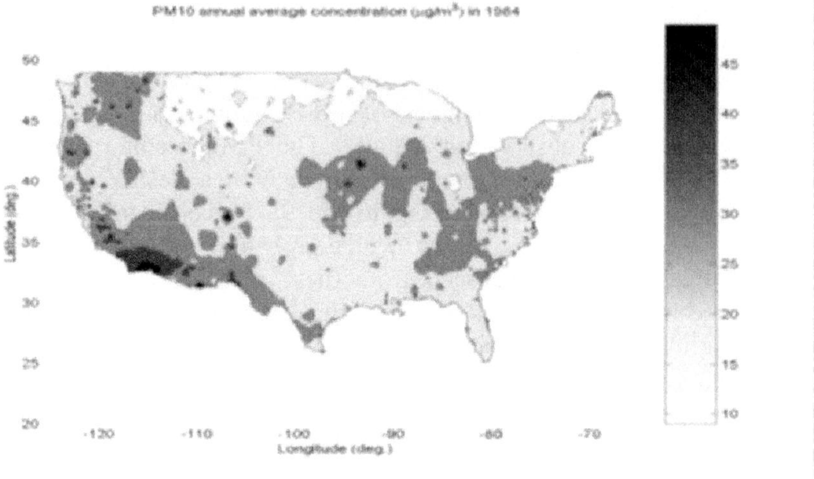

(b)

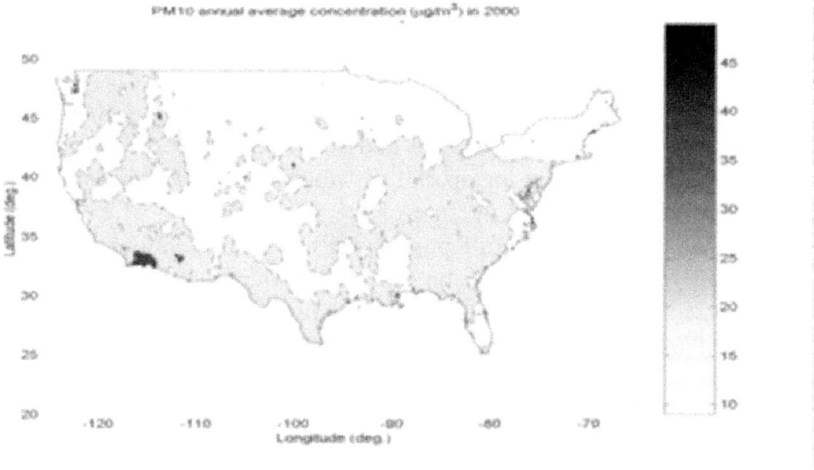

Figure 6. BME maps of annual PM_{10} arith. average (μg/m^3) in (a) 1984 and (b) 2000.

6. CONCLUSION

In this work we present an approach to model as soft information the annual statistics data available for PM_{10} over the US, and we use the BME framework to rigorously process that information and produce realistic spatiotemporal maps of the annual PM_{10} arithmetic average. The approach presented uses a student-t distribution that properly reflects varying levels of reliability of the soft data that depend on the annual number of observations and the variance of these observation values. Using the *BMElib* package we processed the large *AIRS* database of annual PM_{10} statistics for 1168 US monitoring stations over years 1984 to 2000, and we obtained spatiotemporal maps of annual PM_{10} arithmetic average distribution that are more realistic than those obtained with classical approaches not accounting for the composite space time effects and the uncertainties of the soft data.

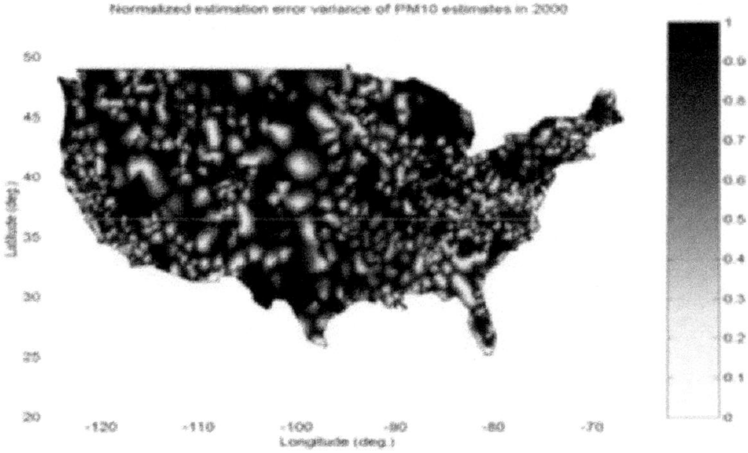

Figure 7. Normalized estimation error variance of the PM_{10} estimates in 2000.

ACKNOWLEDGEMENTS

This work has been supported by grants from the National Institute of Environmental Health Sciences (P42-ES05948 and P30-ES10126), and the Army Research Office (DAAG55-98-1-0289).

REFERENCES

1. USEPA , 1997. *National ambient air quality standards for particulate matter; Final Draft.* Federal Register 40 CFR Part 50, US Environmental Protection Agency, Washington D.C.
2. Christakos, G., 2000, *Modern Spatiotemporal Geostatistics*, Oxford University Press, New York, NY (3rd reprint, 2001).

3. Serre, M. L., and G. Christakos, 1999. Modern Geostatistics: Computational BME in the light of uncertain physical knowledge--The Equus Beds Study, *Stochastic Environmental Research and Risk Assessment*, 13(1), 1-26.
4. Serre, M. L., G. Christakos, and J. Howes, 2000. Powering an Egyptian air quality information system with the BME space/time analysis toolbox, In Proc. of *GeoEnv2000* (3rd Europ. Conf. on Geostatistics for Envir. Appl.), Avignon, France, Nov. 22-24.
5. Christakos, G. and M.L. Serre, 2000. BME analysis of spatiotemporal particulate matter distributions in North Carolina, *Atmospheric Environment*, 34, 3393-3406.
6. Christakos, G., M.L. Serre and J. Kovitz, 2001. Bayesian maximum entropy representation of particulate matter distribution in the state of California on the basis of uncertain measurements, *Journal of Geophysical Research-D*, 106 (D9), 9717-9731.
7. Christakos, G., P. Bogaert, and M.L. Serre, 2002. *Temporal GIS: Advanced Functions for Field-Based Applications*, Springer-Verlag, New York, N.Y. (CD Rom included).
8. USEPA *AIRS* database. http://www.epa.gov/air/data/index.html

(a)

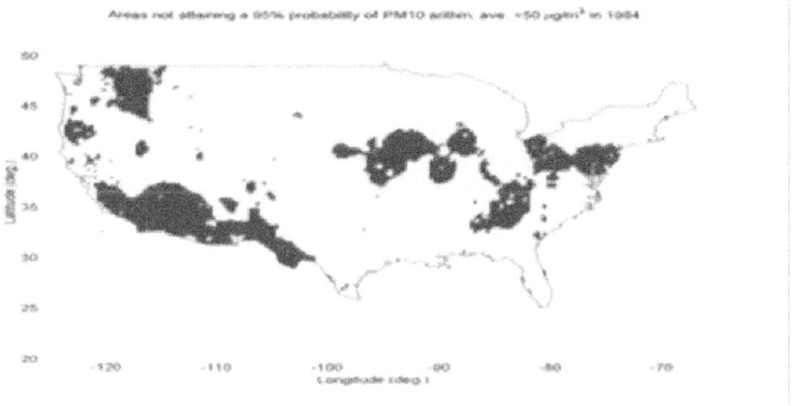

(b)

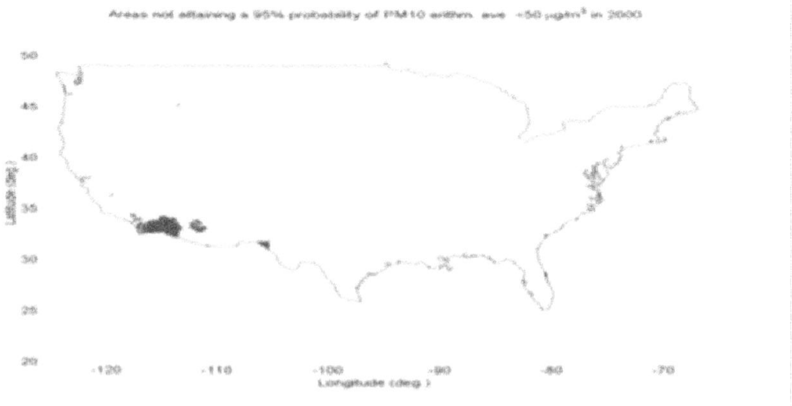

Figure 8. Areas not attaining a 95% probability of annual PM_{10} arithmetic average < 50 μg/m^3 for (a) year 1984 and (b) year 2000.

CHARACTERIZATION OF POPULATION AND RECOVERY OF IBERIAN HARE IN PORTUGAL THROUGH DIRECT SEQUENTIAL CO-SIMULATION

J. Almeida[1], E. Santos[2] and A. Bio[3]
[1]CIGA-FCT/UNL, 2829-516 Caparica, Portugal, ja@fct.unl.pt
[2]DGF, Av. João Crisóstomo, 28, 1069-040 Lisbon, emidio@dgf.min-agricultura.pt
[3]Environmental Group of CMPR/IST, Av. Rovisco Pais, 1049-001 Lisbon, Portugal, ana.bio@ist.utl.pt

Abstract: The objective of this work is to create a tool for demography characterization and management of the Iberian hare population in order to the yearly trend and evaluate recovery of the species in Portugal. The number of hunted animals caught in various hunting resorts is used as an indirect measure for the effective population size and distribution. The main output of the proposed tool consists of spatial maps illustrating the yearly abundance of the species in Portugal. Maps showing habitat carrying capacity, and the associated variance, which is largely due to local lack of information and to observation and sampling errors, are also presented.

Key words: Iberian hare, direct sequential co-simulation, carrying capacity, Gompertz curves

1. INTRODUCTION

Despite the controversy about which of three hare species actually occupy the Iberian Peninsula – *Lepus granatensis,* Rosenhauer, 1856 (Iberian hare), *Lepus europaeus,* Pallas,1778 (European hare) and *Lepus castroviejoi,* Palacios, 1976 (Broom hare) – *Lepus granatensis* has the largest distribution and is the only hare species present in continental Portugal. The Iberian hare prefers flat, sparsely-forested terrains and habitats

X. Sanchez-Vila et al. (eds.), geoENV IV – Geostatistics for Environmental Applications, 127-138.

related to extensive agriculture, and associated with vast agricultural landscapes, open forests and dispersed shrubs (Palacios and Meijide, 1979).

Until 1986, the Iberian hare populations were excessively hunted reaching the situation in which the population could not support the usual hunting pressure, so that it became necessary to restrict hunting permission to every second year. After 1986, recovery of the Iberian hare populations was actively promoted through two measures. Firstly, through the publication of the new hunting law, regulating the concession of demarcated hunting zones to hunting associations and hunting-related tourist industry. Secondly, with the elaboration and approval of management plans for sustainable hunting, essentially based on an average non-hunting period of three years followed by a careful exploitation that would not compromise the meanwhile recovered populations.

To validate the management of the parties involved, they are obliged to report hunting results on a yearly basis. These hunting results, geo-referenced and indexed to a surface, reflect the population's state at a given place and time and population change when a series of years' data are analysed. Considering that hunting legislation (and, therefore, pattern) for Iberian hare in Portugal did not change over the years under study, and assuming that hunting results are directly related to the species' population size in a given area, these hunting results will directly reflect species abundance. Consequently, the results presented here may be interpreted in terms of a familiar population dynamics models.

Population density can present various patterns of annual trends (Hedrick, 1984). When density is relatively low, as for the present species, it tends to evolve with positive growth. Population growth is dependent on factors intrinsic to the population, like birth and death rates, and extrinsic factors, like competition, predation and environmental limitations. Interaction of these factors will determine the population's future size and rate of change, until it reaches a state of equilibrium with the habitat's biophysical conditions, termed the habitat's carrying capacity *(k)*.

Considering intrinsic factors only, recruitment consists of births and immigration into the study area and losses consist of mortality and emigration. The Iberian hare's local mobility is believed to be small, yielding negligible gains or losses. Therefore population change is mainly determined by the balance between survival and mortality rates, with mortality caused either naturally or by hunting. As long as this balance is positive, growth continues to increase, following an S or sigmoid curve in time, until a threshold or limit determined by the given habitat resources is reached. The population's growth, initially exponential, reduces its speed as it approximates the state of generic balance with its environment evolving asymptotically towards the theoretical limit termed carrying capacity (*k*). This growth pattern can be modelled by several mathematical functions, for instance by Gompertz curves. These functions are characterized by: the lower asymptote being the population-growth starting level, the upper asymptote as the carrying capacity and the point of inflexion as the time of maximum growth. The Gompertz curve is given by the equation:

$$Y = A + Ce^{-e^{-B(x-M)}} \qquad (1)$$

where Y = dependent variable, x = time, A = the lower asymptote, C = the upper asymptote, M = the time of maximum growth and B = the growth rate.

Although the present study is based on evaluating the change of a factor, which causes mortality, and is measured using the number of animals hunted per 100 ha per year, the equation is assumed valid for population development as hunting pressure is directly related to the hare's population level. According to unpublished information and data collected in Portugal, the maximum carrying capacity is about 50 hares$\times$100 ha^{-1} for the best habitat, allowing for a maximum sustainable hunting pressure of about 25 hares$\times$100 ha^{-1}.

The main objective of the present study is the development of a carrying capacity map for the Iberian hare in Portugal, based on indirect population data (hunting resort reports), and the characterization of this species' population dynamics. A sequence of maps illustrating the pattern and yearly variations of the species' abundance is presented and the issue of one under-sampled year is addressed using available historical information and sequential co-simulation methods for its estimation. Uncertainties in the final carrying capacity map are evaluated in terms of estimation variance.

2. BACKGROUND OF DIRECT SEQUENTIAL CO-SIMULATION WITH A SET OF SECONDARY VARIABLES

To create a map of abundance for a given under-sampled year, both the data collected in that period as well as historical data, corresponding to previous time periods, will be considered. To accomplish that objective and map under-sampled years, a space-time geostatistical simulation model is proposed, which can be summarized in the following steps:

i) A set of spatial and time trend maps is built with historical data (Santos *et al.*, 2000). These trend maps are interpreted as spatial-temporal random fields and are inferred in space for fixed periods of time, namely years. Species abundances estimated (through ordinary kriging) for each year and the entire area, are considered as a significant spatial trend for that year i, i=1, Ny conditional to historical data.

ii) To map the abundance and respective uncertainty for a given under-sampled year, we propose simulation of the spatial dispersion of the hare population constrained to the historical trend maps. A direct sequential co-simulation algorithm is used, which calls for the local estimates of the abundance in the year t_j at location x_u, $z_1(t_j, x_u)$ – the primary variable – based on abundance of n neighbourhood values for the same year $z_1(t_j,x_\alpha)$ and on the secondary variables $z_2(t_i,x_u)$, i=1, Ny ($j \neq \{1...Ny\}$) obtained from the Ny yearly trend maps of abundance at location x_u. It is a co-located cokriging procedure with a multiple set of secondary variables.

This space-time model combines data from samples from one given year with multiple secondary data provided by several average maps from the recent past, conditioned to the local correlation between values of different time slices.

The proposed methodology is applied to obtain distinct local models of co-regionalisation between year t_j and each year of historical data t_i, i=1, Ny ($j \neq \{1...Ny\}$). This means, that the spatial pattern of year t_j can be correlated with different local areas at the same year and with the same local area at different years. The local models of co-regionalisation were computed on a moving window basis.

Direct Sequential Co-simulation with a set of secondary variables forms an extension of the algorithm proposed by Soares, 2001, and can be summarized as follows:

1. Define a random path visiting each node of a regular grid of nodes.

2. At each node x_u, simulate the value $z_1^s(t_j,x_u)$ using the Direct Sequential Simulation (DSS) algorithm:

   - Identify the local mean and variance of $z_1(x)$, $z_1(t_j,x_u)^*$ and $\sigma^2_{sk}(t_j,x_u)$, using the simple co-located kriging estimator with a multiple set of secondary variables:

$$z(t_j,x_u)=\sum_{\alpha=1}^{n}\lambda_\alpha z_1\left(t_j,x_\alpha\right)+\sum_{i=1}^{Ny}\lambda_i z_2(t_i,x_u) \qquad j\neq\{1...Ny\} \qquad (2)$$

Using the matrix formalism, the simple co-located kriging system with a

$$\left[\begin{array}{cccc|cccc}
1 & C^{t_j}_{12} & \cdots & C^{t_j}_{1n} & C^{t_jt_1}_{1u} & C^{t_jt_2}_{1u} & \cdots & C^{t_jt_{Ny}}_{1u}\\
C^{t_j}_{21} & 1 & \cdots & C^{t_j}_{2n} & C^{t_jt_2}_{2u} & C^{t_jt_2}_{2u} & \cdots & C^{t_jt_{Ny}}_{2u}\\
\vdots & \vdots & \ddots & \vdots & \vdots & \vdots & \ddots & \vdots\\
C^{t_j}_{n1} & C^{t_j}_{n2} & \cdots & 1 & C^{t_jt_{Ny}}_{nu} & C^{t_jt_{Ny}}_{nu} & \cdots & C^{t_jt_{Ny}}_{nu}\\
\hline
C^{t_jt_1}_{u1} & C^{t_jt_1}_{u2} & \cdots & C^{t_jt_1}_{un} & 1 & C^{t_1t_2}_{u} & \cdots & C^{t_1t_{Ny}}_{u}\\
C^{t_jt_2}_{u1} & C^{t_jt_2}_{u2} & \cdots & C^{t_jt_2}_{un} & C^{t_2t_1}_{u} & 1 & \cdots & C^{t_2t_{Ny}}_{u}\\
\vdots & \vdots & \ddots & \vdots & \vdots & \vdots & \ddots & \vdots\\
C^{t_jt_{Ny}}_{u1} & C^{t_jt_{Ny}}_{u2} & \cdots & C^{t_jt_{Ny}}_{un} & C^{t_{Ny}t_1}_{u} & C^{t_{Ny}t_2}_{u} & \cdots & 1
\end{array}\right]
\left[\begin{array}{c}
\lambda^{t_j}_1\\ \lambda^{t_j}_2\\ \vdots\\ \lambda^{t_j}_n\\ \hline \lambda^{t_1}_u\\ \lambda^{t_2}_u\\ \vdots\\ \lambda^{Ny}_u
\end{array}\right]
=
\left[\begin{array}{c}
C^{t_j}_{1u}\\ C^{t_j}_{2u}\\ \vdots\\ C^{t_j}_{nu}\\ \hline C^{t_jt_1}_{u}\\ C^{t_jt_2}_{u}\\ \vdots\\ C^{t_jt_{Ny}}_{u}
\end{array}\right]$$

multiple set of *Ny* secondary variables is defined as follows:

Where:

$C^{t_j}_{\alpha\beta}$ = Covariance between samples at locations x_α and x_β in year t_j

$C^{t_j t_a}_{u\alpha}$ = Cross-covariance between samples at location x_α in year t_a and location to estimate x_u in year t_j

$C^{t_a t_b}_{u}$ = Cross-covariance between years t_a and t_b at location to estimate x_u

$\lambda^{t_j}_{\alpha}$ = Weights of primary information

$\lambda^{t_a}_{u}$ = Weights of secondary information

$C^{t_j}_{\alpha u}$ = Covariance between samples x_α and location to estimate x_u in year t_j

$C^{t_j t_a}_{u}$ = Cross-covariance between years t_j and t_a at location to estimate x_u

with $\alpha = 1\ldots n$; $\beta = 1\ldots n$; $a = 1\ldots Ny$; $b = 1\ldots Ny$; $j \neq \{1\ldots Ny\}$

- Locally resample the histogram of $z_1(x_u)$, for instance using a normal score transform (φ_1) of the primary variable $z_1(x)$, and calculate $y(x_u)^* = \varphi_1(z_1(t_j,x_u)^*)$;
- Draw a value p from a uniform distribution $U(0,1)$;
- Generate a value y^s from $G(y(x_u)^*, \sigma^2_{sk}(x_u))$: $y^s = G^{-1}(y(x_u)^*, \sigma^2_{sk}(x_u), p)$;
- Return the simulated value $z_1^s(x_u) = \varphi_1^{-1}(y^s)$ of the primary variable.

3. Loop until all nodes are simulated.

Assuming Markov-type approximation, the cross-covariance function can be calculated using the following relation in terms of covariance or correlograms (Almeida and Journel, 1994), which calls only for the inference of the primary variable covariance function and the correlation index $\rho_{12}(0)$ between the primary and secondary variable.

The set of simulated images of hare abundance obtained for the entire area at time period t_j allows for calculation of the average species abundance and uncertainty assessment.

3. CASE STUDY

3.1 Estimation of yearly maps of hare population

The first objective of this study is to present a sequence of maps illustrating the pattern and yearly variations of the abundance of Iberian hare in Portugal. Comparing the number of hunting resorts that reported annual results, there has been a gradual increase in the amount of available data since 1989, with exception of the year 1998 (Table 1). In 1998, the number of reserves reporting the number of hunts, was exceptionally low, mainly in the resorts located in the south. To improve the estimation of the hare abundance for this under-sampled year, a spatial-temporal simulation algorithm was used, whose results are presented in section 3.2.

Table 1. Basic national statistics of collected data: number of hunting resorts (#); mean, variance (var), minimum (Min) and maximum (Max) of reported individuals per 100 ha.

Years	'89	'90	'91	'92	'93	'94	'95	'96	'97	'98	'99
#	118	327	322	570	1075	992	1293	1360	1557	904	1422
Mean	0.04	0.14	0.27	0.89	1.18	1.82	2.26	2.02	2.86	1.12	2.64
Var.	0.13	0.52	0.99	3.99	6.35	12.28	18.23	12.70	20.16	5.05	22.59
Min	0	0	0	0	0	0	0	0	0	0	0
Max	3.75	6.95	7.68	17.98	23.97	40.12	54.53	27.77	41.83	20.49	40.56

Experimental variograms were calculated for the available annual density data and were fitted with isotropic models with two spherical model structures. Figure 1 illustrates examples of experimental variograms and the theoretical models adjusted for the sequence of the two most recent years. When compared to the years of '99, the year of '98 shows a high continuity, due to the existence of large gaps of information in the southern zone and, simultaneously, because this zone is characterised by a high heterogeneity as can be observed looking at the remaining years.

Yearly estimated maps of hare abundance (Figure 2) were calculated by ordinary kriging. This sequence of maps clearly shows an increase of the abundance of the hare in Portugal, mainly in southern and eastern areas. When looking carefully at the image estimated for '98, smooth areas of high values are observable in the south, with an unrealistic propagation of the high values until the southern shoreline (Algarve).

Year '99: $\gamma(h)=0.6\ \mathrm{Sph}(a=10000)+0.4\ \mathrm{Sph}(a=90000)$

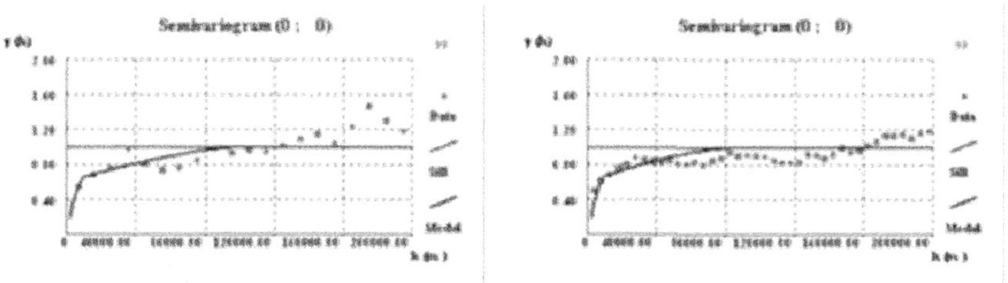

Year '98: $\gamma(h)=0.35\ \mathrm{Sph}(a=10000)+0.65\ \mathrm{Sph}(a=175000)$

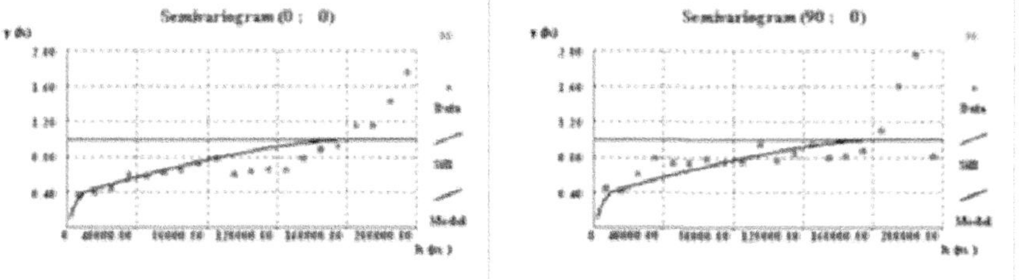

Figure 1. Experimental variograms for the number of individuals captured per area unit, for the sequence of years '99 and '98 and models fitted.

This is clearly a consequence of the reduced number of samples used in this particular area. Given that this area is the most significant in the evaluation of hare populations in Portugal, improvement of this year's estimation, taking also historical data into account, is more than justified. Thus, the following simulation methodology is proposed: direct sequential co-simulation with a multiple set of secondary information, as introduced above.

3.2 Improving inference for the '98 hare population map taking into account historical data

The objective of this part of the work is the inference of a map for '98 using the available historical information next to the data from year '98. Local correlation maps were computed between data of that year and of the remaining years as historical information and, also, between all the remaining years themselves. These maps of correlation were computed using moving windows of adaptable size, in order to include always a significant set of data (20 samples) that allows the calculation of a correlation coefficient. Given the 11 years of historical information available, 55 local correlation maps were constructed covering all possible combinations.

Detailed observation of the local correlation maps relative to the year '98 shows that the correlation is higher for the southern and western parts of the country, and diminishes through the years. For example, in the southern region the correlation remains higher than 0.6 for the most recent years (after

'96). In the western region, the correlation remains always high, given that they are consecutive years of almost zero recorded abundances.

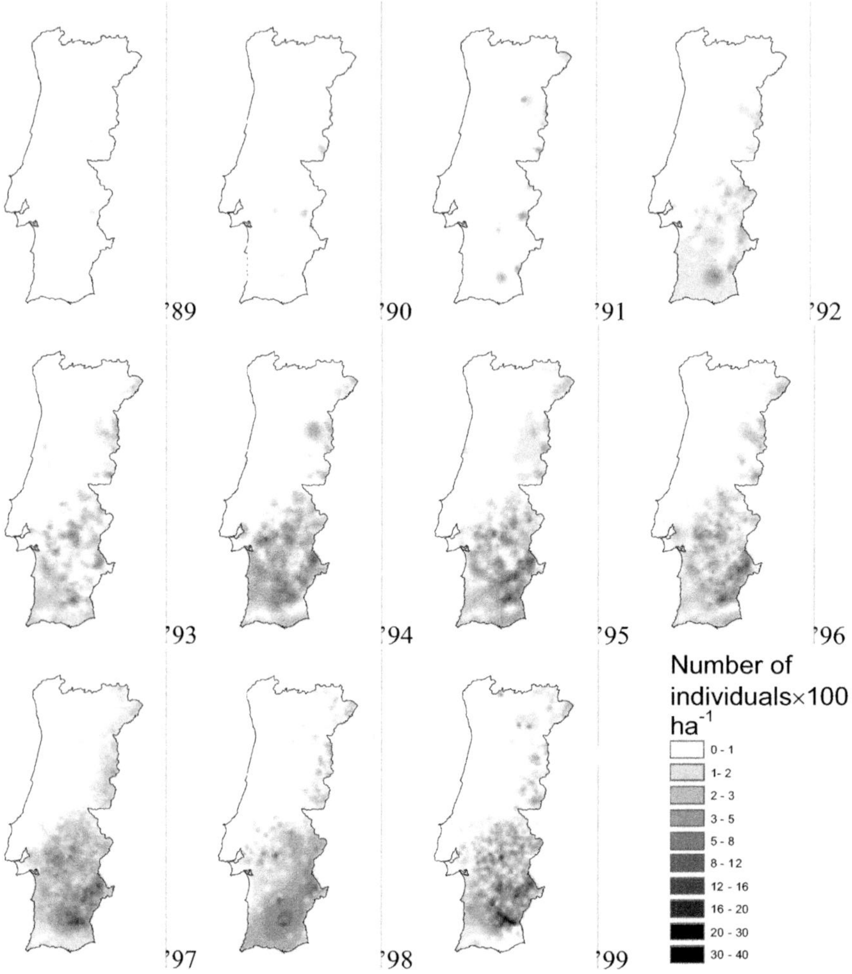

Figure 2. Estimated patterns of hare abundance for the sequence of years 1989–1999.

Using the yearly estimated maps as secondary information, the maps with the local correlation coefficients and data from the year '98, 10 simulated images of abundance have been generated for the year '98 (3 simulated scenarios are presented in Figure 3), using the methodology of direct sequential co-simulation, presented above. A simple average of the simulated images allowed for the construction of an average image for the year '98 (Figure 4). As can be observed, the resultant modelled distribution corresponds better with the data from previous and subsequent years, in comparison to the previous image of Figure 2.

Figure 4 shows that the highest values of abundance are observed in the southern zone and the residual values for shoreline of the Algarve are now in agreement with the historical data. This map takes into account the histogram of the experimental values for the year '98, whichas a maximum of 21 individuals per 100 ha. This is a relatively low limit in comparison to adjacent years, mostly caused by the gap in data reported from the south. This final image clearly shows the main patterns of high values found in the historical data.

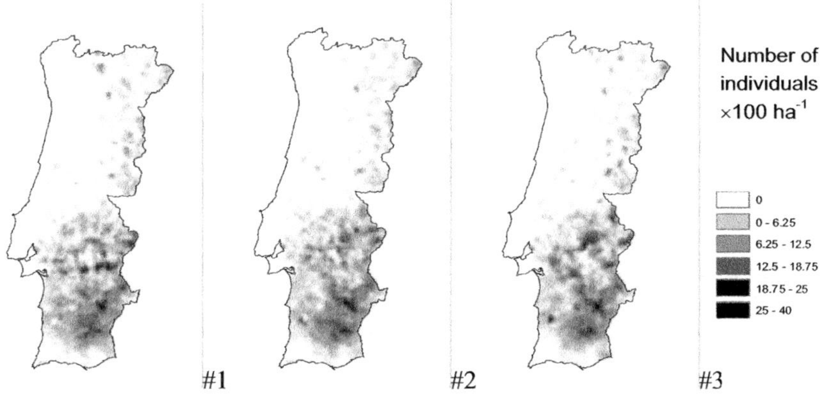

Figure 3. Three examples of simulated scenarios of hare abundance for the year '98.

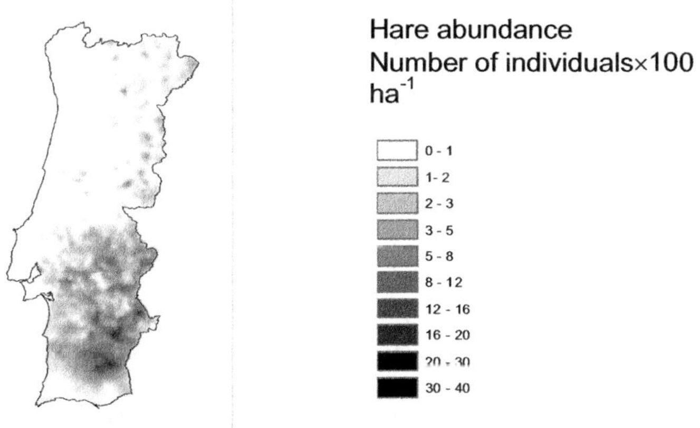

Figure 4. Proposed average map of hare abundance for the year '98.

3.3 Carrying capacity map

In this final part of the case study one additional map was constructed, illustrating the present-day trend in Iberian hare abundance. All hunting resorts were classified into six classes based on the total number of individuals captured during 1999. A categorical, nation-wide map (Figure 6, left) was constructed using the multiphase indicator kriging algorithm, after a transformation of probability values into categorical values based on the maximization criteria of local and global probabilities (Soares, 1992, Almeida *et al.*, 1993). All of the categories exhibit a characteristic spatial distribution related to habitat sustainability.

Sinusoidal trend curves of the Gompertz type were automatically adjusted to the data derived from each of the classes (excluding class 1 with zero values) (Figure 5). Converting the map of classes into carrying-capacity values, using the upper-level asymptote of each class, it is possible to visualize trend values on the national scale (Figure 6, right). This constitutes a forecast trend map based on experimental data reported in 1999 by the hunting resorts (categorical map) and regional fitted curves. One disadvantage of this map is the high influence of local values from '99, leading to high heterogeneity in several areas and, sometimes, hiding regional tendencies. Thus, a simulation procedure conditional to the soft data only is proposed to obtain a smooth map of the global tendency.

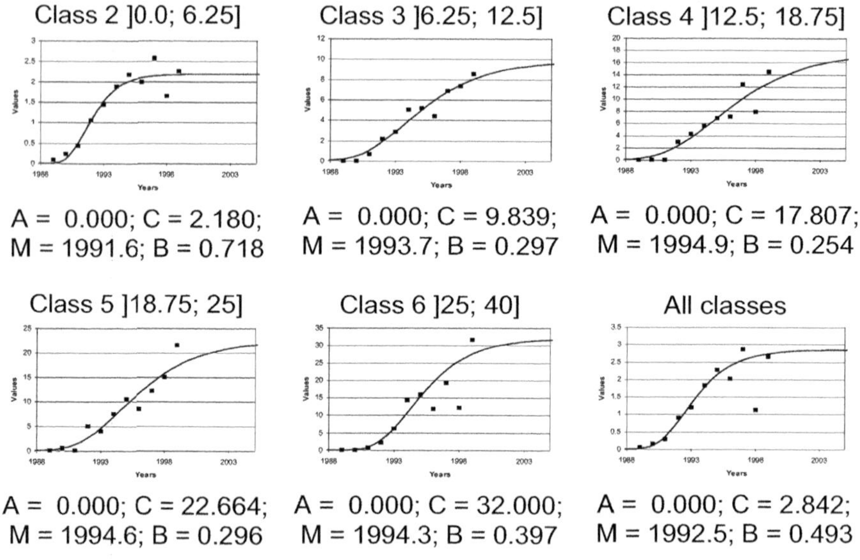

Figure 5. Gompertz curves fitted for each interval and parameters (See eq. 1).

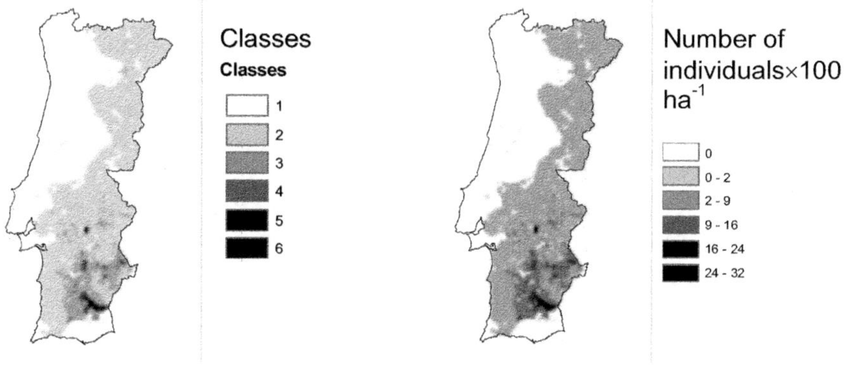

Figure 6. Left: map of interval classes. Right: class-driven trend values of carrying capacity.

If we admit that the carrying capacity map represents the limit of the abundance, the proposed method consists of a carrying capacity simulation, imposing the abundance histogram of 1999, the variogram of these same data and the map of correlations between the more recent, consecutive years (between '97 and '99, if we exclude '98). Therefore, the idea is to generate scenarios for the evolution of hare abundance, assuming that the local correlations will remain steady and that we reached a histogram representative of data in a limit situation. Thus, 10 simulated images of abundance were generated, resulting in a local average map of the trend (Figure 7, left) and a map of variance (uncertainty) (Figure 7, right).

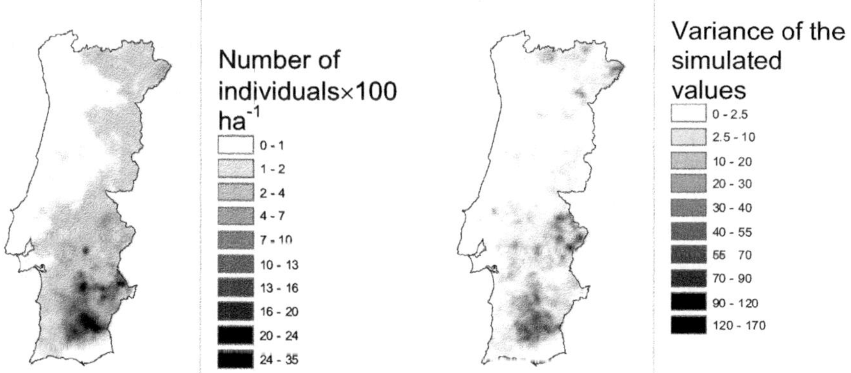

Figure 7. Left: local carrying capacity; right: local variance of simulated values.

4. FINAL REMARKS

There is evidence for a relationship between Iberian hare recovery and sustainable capacity of habitats in Portugal. Inference of missing information for the year '98, through integration of historical data series, was successful, allowing for integration of that under-sampled year in the annual trend used for the production of a carrying-capacity map. The application of non-conditional simulation to the carrying-capacity tendency map enabled the filtering of anomalous records, leading to a map that is more consistent with the observed regional variability.

Figure 7 left, identifies three major zones in Portugal: marginal areas (upper western half of the country and southern coast); areas of medium suitability (upper eastern half); and areas of high suitability (southern half with exception of the southern coast).

The respective variance map, however, points at the presence of instable areas that may not yet have reached their final carrying capacity and at anomalous values which may have generated entropies in the calculation process. Some potentially exceptional areas (e.g. in central south-east) are apparently still in an evolutionary phase of the process.

REFERENCES

1. Almeida, A. and A. Journel (1994). Joint simulation of multiple variables with a Markov-type corregionalization model. Mathematical Geology 26(5), p. 565-588.
2. Almeida, J., A. Soares and R. Reynaud (1993). Modelling the shape of several marble types in a quarry. Proceedings of the 24th International Symposium APCOM, Montreal, Vol. 3, p. 452-459.
3. Hedrick, P. W. (1984). Population biology – The evolution and ecology of populations. Jones and Bartlett publishers, p. 177-194. Boston.
4. Palacios, F. and M. Meijide (1979). Distribuicion geográfica y hábitat de las liebres en la Península Ibérica. Naturalia Hispanica N°19 p3-40. Instituto Nacional para la Conservacion de la Naturaleza, N° 19. Madrid.
5. Santos, E., J. Almeida and A. Soares (2000). Geostatistical characterization of the migration patterns and pathways of the Wood Pigeon in Portugal. Proceedings of the 6th International Geostatistics Congress, Cape Town, South Africa, p. 615-622.
6. Soares, A. (1992). Geostatistical estimation of multi-phase structures. Mathematical Geology, Vol. 24(2), p. 149-160.
7. Soares, A. (2001). Direct Sequential Simulation and Cosimulation. Mathematical Geology, Vol. 33(8), p. 911-926.

UNCERTAINTY MANAGEMENT FOR ENVIRONMENTAL RISK ASSESSMENT USING GEOSTATISTICAL SIMULATIONS

J. Deraisme[1], O. Jaquet[2] and N. Jeannée[3]
[1]Geovariances, 49bis av Franklin Roosevelt, 77212 AVON Cedex, France; [2]Colenco Power Engineering Ltd, Mellingstr. 207,5405 Baden, Switzerland; [3]Geovariances, 49bis av Franklin Roosevelt, 77212 AVON Cedex, France.

Abstract: Geostatistical simulations are very popular in the petroleum and mining industries as they address some issues where "kriging-like" techniques fail. The multiple capabilities of geostatistical simulations have also proven to be of major interest to the environmental sciences. Non-linear estimation techniques such as disjunctive kriging, uniform conditioning or conditional expectation may be convenient for solving problems like the estimation of the probability of exceeding thresholds and contaminated volumes. But if these problems involve multiple point statistics or non-stationary cases, these techniques are not sufficient. Besides, in many situations, the multivariate aspect of the problem cannot be ignored and co simulation methods turn out to be the most efficient solution. The powerful contribution of geostatistical simulation methods to environmental issues is illustrated with applications in the domains of air pollution, soil contamination and hydrogeological modelling. The first example shows how simulations can quantify the risk of exposure of a city's population to air polluted with NO_2. The second example deals with soil contamination with poly-cyclic aromatic hydrocarbons at former industrial sites. The last example is taken from a national program for the storage of nuclear waste. When faced with the complexity of today's environmental risk assessment issues, optimal decision making requires knowledge of the prevailing uncertainties. Geostatistical simulations provide an assessment framework as well as solutions to achieve this goal.

1. INTRODUCTION

Geostatistical simulations have been received with success because of their capabilities to answer complex questions related to environmental

X. Sanchez-Vila et al. (eds.), geoENV IV – Geostatistics for Environmental Applications, 139-150.

issues. The simulations belong to the group of non-linear methods because they aim at reproducing the actual spatial distribution. Also, they provide a powerful means to characterize uncertainty. In this paper we want to illustrate the first point and particularly the fact that simulations can do more than other non-linear techniques (conditional expectation, disjunctive kriging, uniform conditioning, etc.) by demonstrating three examples of case studies made using the Isatis software (Bleines and al., 2001). The advantage of simulations arises from different aspects: relationships of any kind between the variables and known factors, no prerequisite of stationarity, taking into account support effect imaging of the heterogeneities of the medium to provide an input model for complex simulations of transfers.

2. FIRST EXAMPLE: RISK OF POPULATION EXPOSURE TO AIR POLLUTION

2.1 Probability of exceeding pollutant threshold

Air quality is regulated by European directives that prescribe a set of limits to be respected. This example, taken from a geostatistical study made on behalf of Air Normand, concerns the evaluation of the risk of pollution by NO_2 in the agglomeration of Rouen. The risk considered here involves the probability of exceeding a threshold of 40 $\mu g/m^3$ on a yearly basis.

The annual average of NO_2 concentration can be estimated on the basis of measurements from 89 diffusive samplers. Sufficient spatial coverage over the agglomeration allows to map correctly the pollutant by kriging techniques. A correlation with a synthetic co-factor, combining pollutant emissions and population data has led to an improved estimation by means of collocated co-kriging techniques (see Figure 1).

A systematic bias arises from the smoothing effect of kriging when we use kriging as an input to non-linear calculations like, in the present case, the application of a threshold. Geostatistical conditional simulations also using the synthetic co-factor, provide a consistent solution essentially because they reproduce the actual variability. But the probability of exceeding the threshold (see Figure 2) can be more easily calculated by the conditional expectation based on an assumption of multigaussian distribution of the pollutant (after normal score transform). By counting the collocated co-kriging values above the threshold we find a "polluted" area of 39 300 hectares. If now we use the conditional expectation as well as the simulation to calculate the polluted area with a probability of 50 % (close to what kriging says), we find a surface 25 % larger. In fact, by doing so we have just verified the bias previously mentioned. However, simulations are not really necessary to calculate the probability of pollution or to determine the geographic boundaries of the polluted area. Simulations become more interesting when introducing a spatial relationship with another parameter like the population density in order to evaluate the proportion of the population that is exposed to a pollution in NO_2 above 40 $\mu g/m^3$.

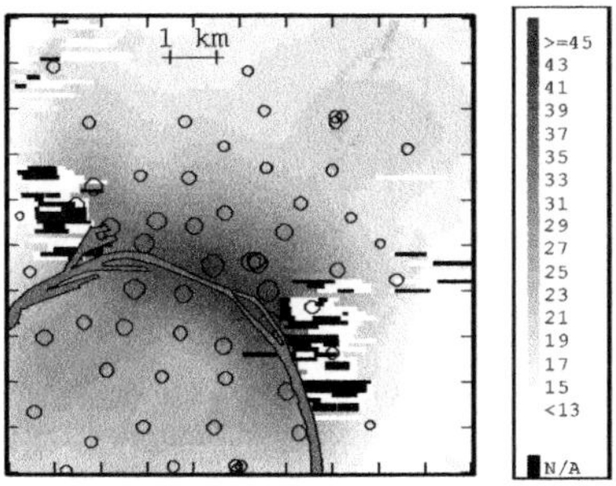

Figure 1. Collocated co-kriging of NO2 from diffusive samplers.

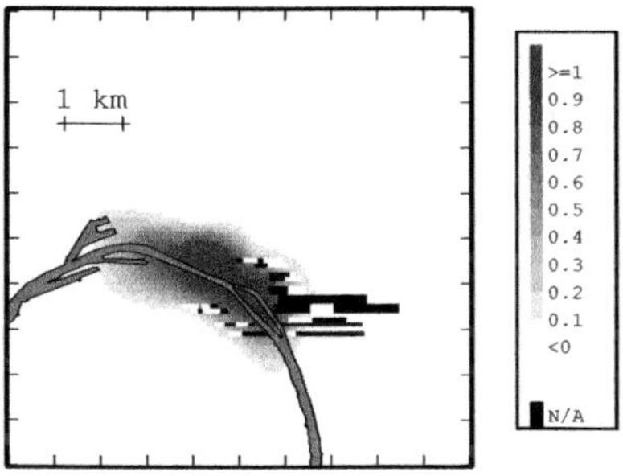

Figure 2. Probability map of exceeding a threshold in NO2 concentration.

2.2 Consequence of the pollution on the population

A first simple solution consists in multiplying the probability calculated above by the population density. We then find that there is a probability of 50 % for 8 % of the population to be exposed and a probability of 90 % for 3 % of the population to be exposed. This is not correct, because this solution ignores the variability in the position of the polluted area as illustrated for two simulations in Figure 3. It shows the overlapping area of the two zones which are interpreted as those of highest risk in two separate simulations. In fact, the overlapping area is centered on the Seine river with an important traffic axis but no population!

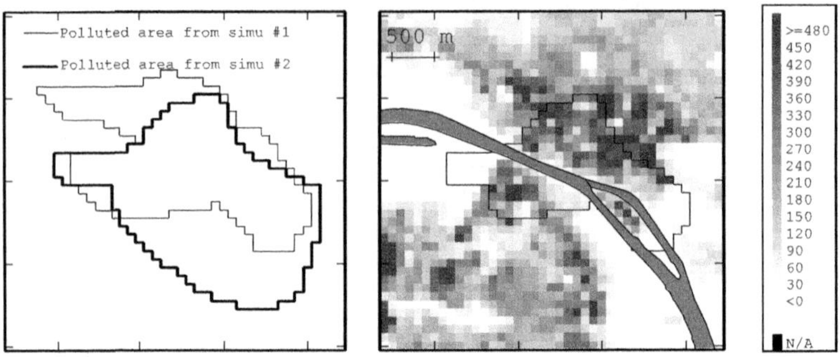

Figure 3. Superposition of the "polluted" areas from two simulations. On the right, the overlapping "polluted areas" are displayed as an overlay on the population density.

Let's suppose that the population density is very heterogeneous, meaning for instance that within the most probable polluted area the population is scarce because of the proximity to emissions from motorway traffic. In that case the "naïve" calculation will grossly underestimate the risk for the population, basically because the pollution and the population are not independent, their correlation being used when simulating with the co-factor. The right thing to do is, after calculating for each simulation the population concerned, to estimate the distribution and derive statistics and probabilities.

Applied to this case we find that instead of 3 % of the population (about 350 000 inhabitants) being exposed to the pollution with a probability of 90 %, we get 6.3 % by using 100 simulations (see Figure 4).

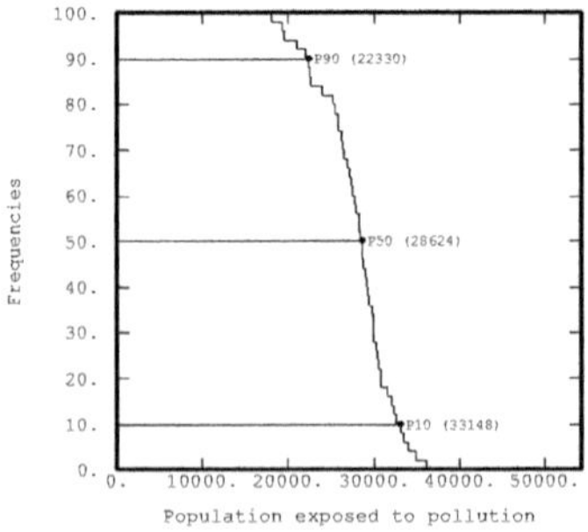

Figure 4. Distribution of the population exposed to pollution from 100 simulations.

In this example simulations appear to be a powerful method to estimate a quantity like a product of a variable (here the population density) and an indicator of a second factor (here NO_2>40 µg/m^3). A too simplistic response may cause significant errors.

3. SECOND EXAMPLE: HAZARD ANALYSIS AND DELINEATION OF POLLUTED AREAS

3.1 Context

Cleaning up a polluted site requires the delineation of areas where the concentration is above a critical level. True grades are always unknown, and kriging gives only estimated values of these grades. In order to take into account the (lack of) precision of the estimator, it is useful to add to the kriging estimate the probability with which the unknown variable exceeds a given level. Furthermore, the knowledge of mean grades above the threshold are of real interest, the cost of the remediation being directly influenced by the level of contamination of the areas that need to be treated.

Interest could be put on "punctual" grades or on their mean on blocks of a given size. For a given threshold, support effect implies that the proportion of blocks with a concentration above the threshold varies with the size of the block. Consequently, taking into account the size of the support used in the remediation step – which depends on the chosen remediation technique and the future use of the site - is necessary. Conditional expectation and disjunctive kriging are able to estimate the probability to exceed a given threshold and the mean grade over the threshold. As soon as a change of support model is required, these methods require stationarity, not only local stationarity, and their application in a multivariate framework is not easy.

With the availability of an auxiliary variable densely known over the field and correlated to the pollutant of interest, we show how hazard analysis might be obtained from conditional collocated co-simulations. By hazard analysis, given a threshold and the size of the blocks of interest, we mean the computation of probabilities that the threshold is exceeded, the corresponding volumes that have to be remediated, and the mean grades of pollutant over the threshold.

3.2 Data

We are interested in the pollution of a former coke plant in northern France with Polycyclic Aromatic Hydrocarbon (PAH) compounds - dataset provided by the Centre National de Recherche sur les Sites et Sols Pollués (Douai, France). We focus here on the benzo(a)pyren (BaP), a five cycles non-volatile, non-soluble and highly carcinogenic PAH. 52 points have been sampled on a regular square grid with an interval of 10 m and local narrowings at 5 m with a 1 m deep core drill. The mean concentration in BaP equals 37.8 mg kg^{-1} with a standard deviation of 96.2 mg kg^{-1} (Jeannée, 2001). The usual remediation value for this pollutant being 10 mg kg^{-1}, at least some areas will have to be cleaned up, which is confirmed by linear kriging techniques. Regarding the historical information, two pools of coal tar are located on the sampled area; they have been excavated and one of them, located in the south, has been filled in with non-polluted material; backfill coming from the excavation of the north coal tar has been dumped in the Northwest of the site (see Figure 5).

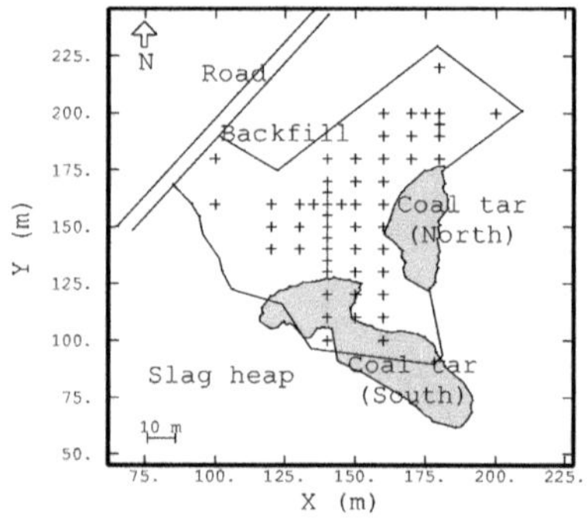

Figure 5. Configuration of the site, location of PAH sampling points.

Qualitative characteristics of samples have also been observed on a refined grid of 5 by 5 m: presence/absence of coal, coal tar, smell, limestone grains, stonework pieces, greenish color of the sample, dross, etc. A correspondence analysis has been performed in order to synthesize their information. This factorial analysis technique reduces the high number of variables to a mere few of non-correlated factors containing the information about the data. Here, the first factor (called "auxiliary factor" hereafter) distinguishes backfilled materials and soil in place and is correlated to the BaP grades. The use of the auxiliary factor, known on the 5 by 5 m grid, will therefore improve the knowledge of the grade pollutant in places where no PAH analysis has been performed.

3.3 Results

After a gaussian transform of BaP grades and auxiliary factor, a bivariate variogram model is fitted on the experimental variograms, and 200 collocated conditional block simulations are performed using the Turning Bands method (Chilès and Delfiner, 1999) on 5 by 5 m blocks.

We are interested in the remediation of areas where the BaP grade exceeds the usual intervention level of 10 mg kg^{-1}. Figure 6 shows the importance of the use of auxiliary information correlated to the variable of information if we want to assess the contaminated volumes. Indeed, taking into account this information leads to a ca. 15 % decrease of the polluted volumes. The probability that the BaP grade exceeds 10 mg kg^{-1} on the blocks is derived from the simulations.

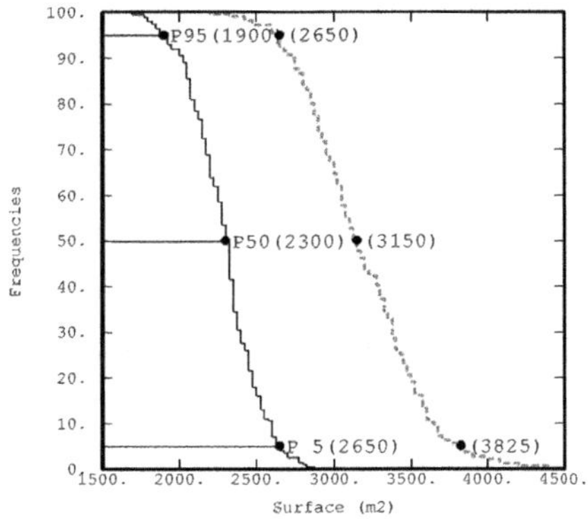

Figure 6. Cumulate histograms of the surfaces where the pollutant concentration exceeds 10 mg kg^{-1} for univariate (dotted line) and collocated (solid line) simulations.

These estimates are used to consider several remediation scenarii, corresponding to financial and sanitary choices. For instance, Figure 7 shows the mean BaP grades above 10 mg kg^{-1} on 5 by 5 m blocks, together with the delineation of the area where the probability to exceed 10 mg kg^{-1} in BaP is greater or equal to 0.2.

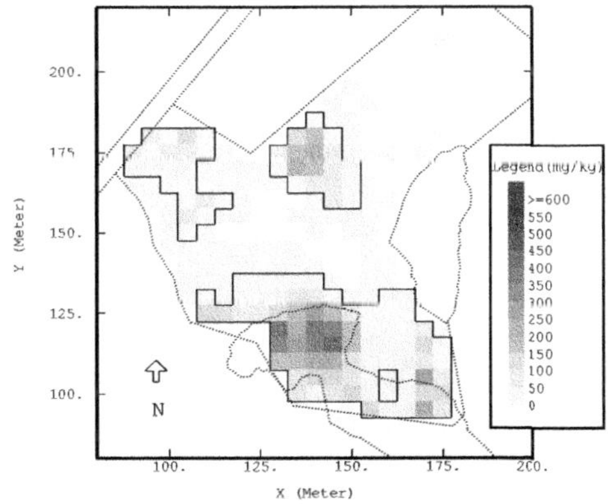

Figure 7. Mean grades above 10 mg.kg^{-1} on 5 by 5 m blocks. Outline of the area where P[BaP > 10 mg kg-1] ≥ 0.2 (solid lines). Contour of the site and location of coal tars (dotted lines).

4. THIRD EXAMPLE: CASE STUDY IN HYDROGEOLOGY

4.1 Objectives

This study (Jaquet and Siegel, 2000) has been performed for SKB (Swedish Nuclear Fuel and Waste Management CO) within the framework of the assessment of the long-term safety of a deep repository for spent nuclear fuel. The objectives of this modelling study were (a) the enhancement of the understanding of subglacial groundwater flow due to basal ice melting and (b) the evaluation of the impact of subglacial groundwater flow on a repository in terms of its relative position with respect to the ice margin of the glacier. The achievement of these goals has required a probabilistic description of the hydraulic conductivity using geostatistical simulations which are then used as input for the numerical modelling of glaciation effects.

4.2 Issue and approach

The modelled domain whose size is 250 * 10 * 4 km^3 comprises the Äspö region (see Figure 8). The host rock considered is of granitic type and contains major fracture zones. When assessing host rock capabilities, one key parameter is the hydraulic conductivity. Because of its spatial variability, the characterization of the hydraulic conductivity is a major issue when modelling hydrogeological processes using numerical (deterministic) methods. Due to the complexity of the spatial behavior of the hydraulic conductivity and the little amount of available data, a probabilistic approach is chosen for the spatial description of the hydraulic conductivity.

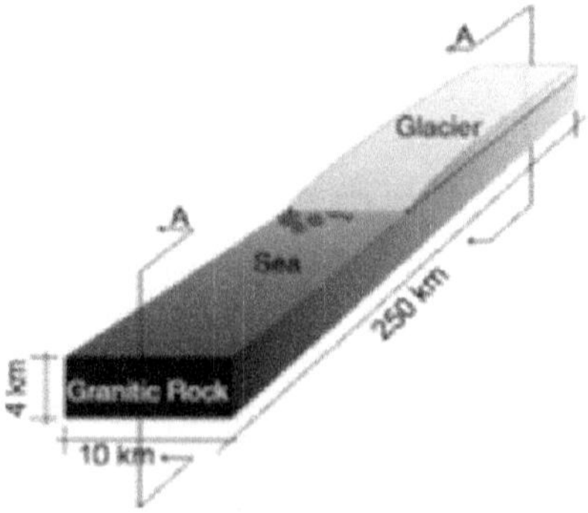

Figure 8. Model domain with glacier location; the ice margin is placed right above Aspö (A-A: location of cross-section).

4.3 Data

The spatial variability of the hydraulic properties of the rock mass and the major fracture zones at Äspö was characterized by Rhén et al. (1997). The rock mass was divided into four hydrogeological units for which statistical parameters are available (cf. Table 1). The hydraulic conductivity is assumed to follow a log-normal distribution (i.e., the log-conductivity distribution is Gaussian). The isotropic range was estimated using experimental variogram calculations performed for the regional scale. For this modelling study, the range is assumed to remain constant for the four hydrogeological units considered. A Gaussian random function with an exponential variogram was then selected for the geostatistical simulation of the spatial variability of the rock-mass log-conductivity.

Table 1. Rock mass hydraulic parameters (after Rhén et al., 1997).

Depth [m]	Geometric mean of hydraulic conductivity[1)] [m/s]	Standard deviation [log 10]	Range [m]
0 – 200	$1.3 \cdot 10^{-7}$	0.96	825
200 – 400	$2.0 \cdot 10^{-7}$	0.65	825
400 – 600	$2.6 \cdot 10^{-7}$	0.79	825
600 – 2000[2)]	$4.7 \cdot 10^{-8}$	0.72	825

1) Equal to the mean of the log-conductivity values.
2) In the model these statistical parameters are assumed valid from 600 to 4000 m depth.

4.4 Geostatistical simulation

The characterisation of the log-conductivity of the rock mass requires the generation of a realisation of the Gaussian random function with an exponential variogram (cf. section 4.3). The log-conductivity is simulated in 3 dimensions using the Turning Bands method (Chilès and Delfiner, 1999; Lantuéjoul, 2002) implemented in the NAMMU package (Marsic et al., 2001). The result is a Gaussian normalised simulation of the log-permeability (i.e., with zero mean and unit variance). This geostatistical simulation is then scaled according to the parameters related to the hydrogeological units defined in Table 1 (cf. section 4.3).

The porosity is calculated from the simulated conductivity using a deterministic correlation (i.e., a power function) which was fitted between porosity and conductivity data (Rhén et al. 1997). Finally, conductivity and porosity values are assigned to each finite element for the mesh of the numerical model.

Figure 9 illustrates the horizontal log-conductivity and porosity fields for a central segment of the model's bottom. The presence of the major fracture zones can be observed; their hydraulic influence is implicitly reproduced using the IFZ (Implicit Fracture Zone) method (Marsic et al., 2001). The corresponding statistics of the realisation are given in Table 2.

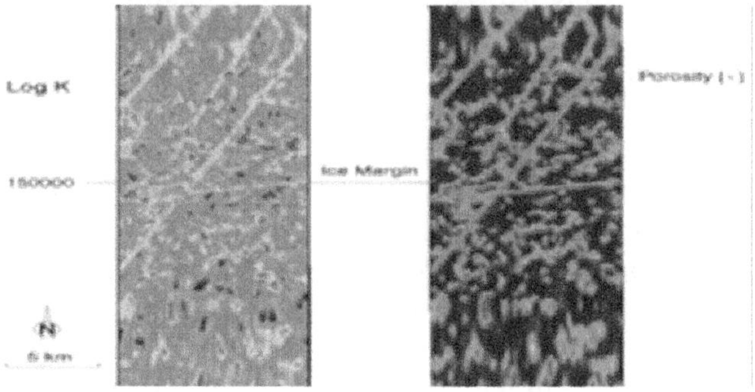

Figure 9. Geostatistical simulation of log-conductivity and porosity fields (horizontal cut at a depth of 4000 m for a portion of the model domain).

Table 2. Geostatistical realization: mean values of hydraulic properties.

Hydrogeological unit	Geometric mean of conductivity[1)] [m/s]	Arithmetic mean of porosity [-]
2: 0 – 200	$1.7 \cdot 10^{-7}$	$9.8 \cdot 10^{-4}$
3: 200 – 400	$2.3 \cdot 10^{-7}$	$6.3 \cdot 10^{-4}$
4: 400 – 600	$2.2 \cdot 10^{-7}$	$8.1 \cdot 10^{-4}$
5: 600 – 4000	$4.8 \cdot 10^{-8}$	$2.3 \cdot 10^{-4}$

1) Equal to the mean of the log-conductivity values.

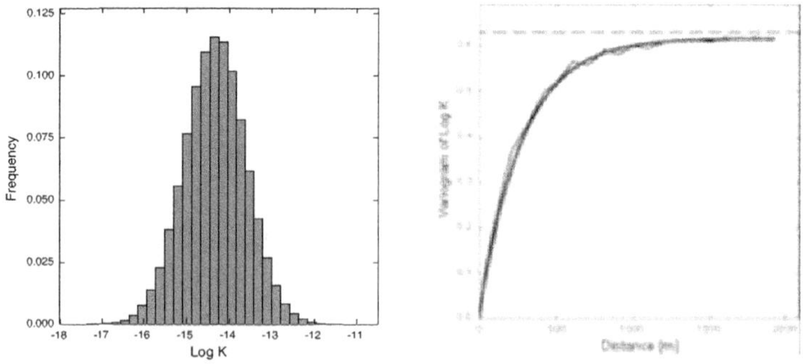

Figure 10. Log-conductivity (i.e., Log K): histogram of unit 5; experimental and exponential (bold line) variograms calculated for a portion of the model domain: 10 * 10 * 4 km^3 (horizontal dashed line at the level of the log-conductivity variance).

The discrepancies between the geometric means of the realisation and their input values (cf. Table 1) are related to the usual statistical fluctuations associated with a given realisation. A verification of the characteristics of the model (i.e., a Gaussian distribution with an exponential variogram) is performed. The results are shown in Figure 10. The symmetric shape of the log-conductivity histogram for the lower hydrogeological unit is of the Gaussian type. The experimental variogram calculated for a portion of the model domain can be fitted with an exponential model with a practical range of 825 m. The input statistical parameters of the geostatistical realisation (for the hydraulic conductivity) can be reproduced; ergodicity problems are thus avoided when simulating the hydraulic conductivity. Then, the effects of the spatial variability of the hydraulic properties on modelling results are assessed using a single realisation.

4.5 Input for hydrogeological modelling

This 3-dimensional simulation of the hydraulic conductivity and porosity then serves as input for the numerical modelling of density-driven flow induced by the variable salinity of the groundwater (see Figure 11); the freshwater input is provided through subglacial groundwater flow due to basal ice melting. The required flow and transport equations are solved using the package NAMMU (Marsic et al., 2001). Finally, the resulting site performance measures (e.g., travel time from potential repository location to the surface) obtained through numerical modelling integrate the characteristics of the host rock; i.e., the effects of spatially variable hydraulic parameters are propagated into numerical modelling results.

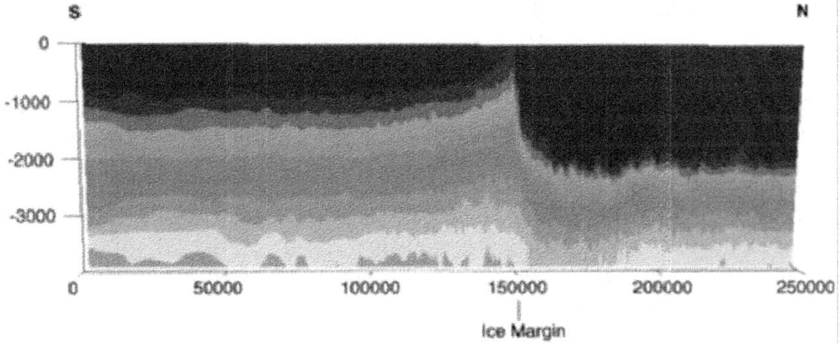

Figure 11. Numerical modelling result: salinity at time 122 years. The fingering effects are due to the spatially variable conductivity; mixing and salt transfer processes are enhanced (cross-section A-A : cf. Figure 9)

5. CONCLUSIONS

The three applications presented have shown how anthropogenic or natural phenomena with complex spatial variability can be characterised using geostatistical simulations. These methods provide a powerful contribution where the application of solely deterministic approaches could not deliver answers when dealing with complicated environmental issues. Emphasis has also been put on the advantage of simulation approaches compared to direct estimation methods, even non-linear geostatistical ones.

The simulations provide an indispensable tool for studying problems involving correlations between different factors as well as complex processes like transport and flow phenomena. Even a limited number of simulations can provide acceptable solutions avoiding large errors coming from "classical" methods.

These geostatistical solutions constitute the foundation for risk assessment studies aiming for the determination of consequences on humans. Furthermore, the inherent uncertainty due to the spatial variability can be estimated and propagated into numerical (deterministic) modelling results for predictive purposes. Thus, the management of this uncertainty will allow for optimal decision-making by authorities and stakeholders when faced with today's environmental concerns.

REFERENCES

1. Bleines C. et al, 2001. ISATIS Software Manual, 3rd Edition, Géovariances Fontainebleau, 585 pp
2. Jeannée N. 2001. Caractérisation géostatistique de pollutions industrielles de sols. Cas des HAP sur d'anciens sites de cokeries. Thèse de Doctorat en Géostatistique, Ecole des Mines de Paris.
3. Chilès J.P. and Delfiner P. 1999. Geostatistics: modelling spatial uncertainty, Wiley Series in Probability and Mathematical Statistics, 695p.
4. Jaquet O., Siegel P. And Klementz W. 2002. Groundwater flow and transport modelling during a glaciation period, SKB Report to be published.
5. Rhén I., Gustafson G., Stanfors R., and Wikberg P. 1997. Äspö HRL – Geoscientific evaluation 1997/5. Models based on site characterisation 1986-1995. SKB Technical Report TR-97-06.
6. Lantuéjoul C. 2002. Geostatistical simulation: models and algorithms, Springer, 256p.
7. Marsic N., Hartley L., Jackson P., Poole M. and Morvik A. 2001. Development of hydrogeological modelling tools based on NAMMU, SKB Report R-01-49.

A SPATIAL ASSESSMENT OF THE AVERAGE ANNUAL WATER BALANCE IN ANDALUCIA

K. Vanderlinden[1], J.V. Giráldez[1], and M. Van Meirvenne[2]

[1]*Dept.of Agronomy, Córdoba University, Avda. Menéndez Pidal s/n, P.O. 3048, 14080 Córdoba, Spain.*
[2]*Dept Soil Management & Soil Care, Ghent University, Coupure 653, B-9000 Ghent, Belgium.*

Abstract: A simple methodology to assess spatially the average annual water balance in the Region of Andalusia is presented, taking advantage of previously produced maps of average annual precipitation (*P*) and reference crop evapotranspiration (ET_o). Using a simple bucket model, daily series of actual evapotranspiration and total runoff were calculated, from which average annual actual evapotranspiration, *E*, and total runoff, *Q*, were obtained. Considering average annual values, the water balance problem of a homogeneous land area reduces to the question: How is *P* split up between *E* and *Q*? Budyko's empirical relationship offers an answer to this question, relating the index of dryness, $R=ET_o/P$, to the ratio between *E* and *P*. Similar relations are used to transform a map of *R* into first estimates of *E* and *Q*. These maps are consecutively used as local mean maps in simple kriging with varying local means (SKlm) or as an external drift variable in kriging with an external drift (KED). The cross-validation statistics show larger errors for the *Q* estimates, due to its skewed distribution, but only small differences are observed between SKlm and KED. Finally, block kriging is used to produce maps of *E* and *Q* with both methods.

Key words: average annual water balance, simple kriging with varying local means, kriging with an external drift, Budyko diagram, Andalusia

1. INTRODUCTION

Lack of available natural water resources is an important matter of concern among scientists and decision makers in regions where the occurrence of precipitation is irregular and seasonal, as is the case for the Region of Andalusia in

X. Sanchez-Vila et al. (eds.), geoENV IV – Geostatistics for Environmental Applications, 151-162.

southern Spain. This region is subject to a Mediterranean climate, in which rainy periods are succeeded by large, dry periods with high temperatures and consequently large evapotranspiration rates. A detailed study of the average annual water balance and its spatial distribution may shed a light on this problem.

During a shower, the rainfall water reaches the soil surface, infiltrates and moistens the underlying soil layers, according to their properties and their moisture content. This infiltrated water may be transpired by the plants, evaporate from bare soil or move downwards to the underlying aquifer. On the other hand, when the rainfall rate exceeds the infiltration capacity of the soil, runoff is produced and a first division occurs at the soil surface, depending principally on the soil moisture conditions: the division of rainfall in evapotranspiration and runoff. This division has been one of the main issues of Hydrology. Soil moisture is also a factor that affects a second division at the soil surface. The energy that the soil surface receives from the sun is split up into latent heat, used to evaporate the available water, and sensible heat, which is essential to numerous chemical reactions that take place in living organisms on the earth surface.

Traditionally, very simple bucket-type models have been used to calculate the components of the soil water balance. Rainfall is added and evapotranspiration is subtracted from a soil water store at monthly or daily intervals, and when the maximum storage capacity is exceeded, runoff is generated (Boughton, 1968; Alley, 1984). These models are usually applied to large areas or entire catchments, using monthly or daily data. The difference between these models consists of how the dependence of evaporation on soil moisture is described. Milly (1994b) developed a model that provides an approximate description of the water balance problem of large areas, starting from the hypothesis that the long-term water balance depends only on the local interaction, attenuated by soil moisture storage, of fluctuating water supply (rainfall) and demand (evapotranspiration). The author is able to explain 88% of the spatial variability of the observed average annual runoff in the U.S., east of the Rocky Mountains. Other models are based on resolving the Richards equation using numerical techniques (Kroes et al., 2000) or using approximate analytical solutions (Broadbridge and White, 1988), but are not suited for regional studies because they require too many observed input variables.

The aims of this study are: (1) the analysis of the average annual soil water balance in the Region of Andalusia, using a simple bucket model, and (2) the development of an adequate spatial interpolation methodology for actual evapotranspiration and total runoff.

2. MATERIALS AND METHODS

2.1 The Average annual water balance

Considering the law of mass conservation, the total water balance of an area or region can be expressed as (Brutsaert, 1982, § 1, 11):

$$\frac{dw}{dt} = (p - e)A + q_i - q_o \qquad (1)$$

with p [LT^{-1}] the rainfall rate and e [LT^{-1}] the actual evapotranspiration rate within the area A [L^2], q_i [L^3T^{-1}] the surface and ground water inflow rate, q_o [L^3T^{-1}] the surface and ground water outflow rate, w [L^3] the water volume stored in the area A, and t [T] the time. The average actual evapotranspiration rate over the area A can then be calculated as:

$$e = p + \left[q_i - q_o - dw/dt\right]/A \,. \tag{2}$$

Since it is very difficult to measure q_i, q_o and w, application of equation (2) is usually restricted to the case of annual values of e, E. If an annual time interval is considered, it can be assumed that the average dw/dt value is zero and ground water flow can be neglected if a very large area is considered. Moreover, if the considered area is a natural catchment, there is no surface water inflow and only the surface water outflow remains. Taking this into account, equation (2) can be written as:

$$E = P - Q \text{, or } P = E + Q \,, \tag{3}$$

with P [L], Q [L], and E [L] the annual totals of precipitation, total runoff, and actual evapotranspiration. Equation (3) quantifies the partitioning of the received rainfall into actual evapotranspiration and total runoff. This relationship has made the inference of empirical relationships between E and P, or between Q and P possible. Bailey (1979), Eagleson (1981) and Brutsaert (1982) give an overview of these relationships. Especially the equations proposed by Budyko (1974), Lettau (1969) and Lettau and Baradas (1973) are useful to understand how this partitioning occurs. Usually an index of dryness is used that is defined as:

$$R = \frac{R_n}{\lambda P} = \frac{ET_o}{P} \,, \tag{4}$$

where R_n [ML^2T^{-2}] is the total annual net radiation, λ [MLT^{-2}] is the latent heat of vaporisation of water, and ET_o [L] the total annual reference crop evapotranspiration.

Budyko (1974) tried to fit the following equations to data from a large number of watersheds around the world:

$$\frac{E}{P} = 1 - \exp(-R) \tag{5}$$

$$\frac{E}{P} = R \tanh\left(\frac{1}{R}\right), \tag{6}$$

and observed that the vast part of the data lay between these two curves, what led him to propose the geometric mean of the latter two curves:

$$\frac{E}{P} = \sqrt{R \tanh\left(\frac{1}{R}\right)\left[1 - \exp(-R)\right]} \tag{7}$$

Budyko used to write these equations in terms of the annual net radiation, expressed in equivalent evaporation height. In the work by Milly, this value is approximated by ET_o. These equations represent the Budyko diagram, which shows the relationship between E/P and R, and which characterises the average annual water balance. E/P is a measure for the average annual water balance, given that it quantifies the repartition of rainfall in evapotranspiration and runoff. On the other hand, R, is a climatic index (Bailey, 1979, Figure 3.1). Values of R exceeding 1 represent dry and arid climates, where the annual water balance is characterised by a limited water supply. Small values of R (<1) correspond with humid climates, where the annual water balance is characterised by a limited energy supply. This distinction corresponds to the fact that annual evapotranspiration approximates to annual rainfall in regions where the annual energy supply to the earth surface exceeds largely the required quantity for vaporising the annual precipitation.

2.2 The Milly model

Milly (1993, 1994a, 1994b) explored the possibility to explain the annual soil water balance using a simple model with a limited water storage and infinite infiltration capacity. The soil volume considered is bounded above by the soil surface and has a depth of 1m, which is an approximation of the average plant root depth. It is assumed that the vegetative cover is dense enough to neglect direct evaporation from the soil surface and that the horizontal dimensions of the control volume are large as compared to the horizontal water flow in the root zone, due to soil heterogeneity and local topography (approximately 100 m). The water balance of this control volume can be expressed as:

$$\frac{dw}{dt} = i - e - q\,, \tag{8}$$

with i [LT^{-1}] the infiltration rate. A complete description of the assumptions made can be found in Milly (1994b). Taking these assumptions into account equation (8) can be written as:

$$\frac{dw}{dt} = \begin{cases} 0 & p > et_o \text{ and } w = w_0 \\ 0 & p < et_o \text{ and } w = 0 \\ p\text{-}et_o & \text{otherwise} \end{cases} \tag{9}$$

Moreover, e and q are simply obtained from:

$$e = \begin{cases} et_o & \text{when } w > 0 \\ 0 & \text{when } w = 0 \end{cases} \tag{10}$$

$$q = \begin{cases} 0 & when\ w < w_o \\ p - et_o & when\ w = w_o\ and\ p > et_o \end{cases} \tag{11}$$

This model is applied to completed meteorological time series (Vanderlinden, 2002) with variable length, from 160 meteorological observatories within the Region of Andalusia, corresponding to time periods between 1920 and 1998.

2.3 Location of the study region and data description

In this study we focus on the Region of Andalusia, situated in the south of Spain. Figure 1 shows the geographical situation of this region and the locations of the 160 meteorological stations, where estimates of w and daily series of p and et_o were available (Vanderlinden, 2002) to calculate the daily water balance using Milly's model.

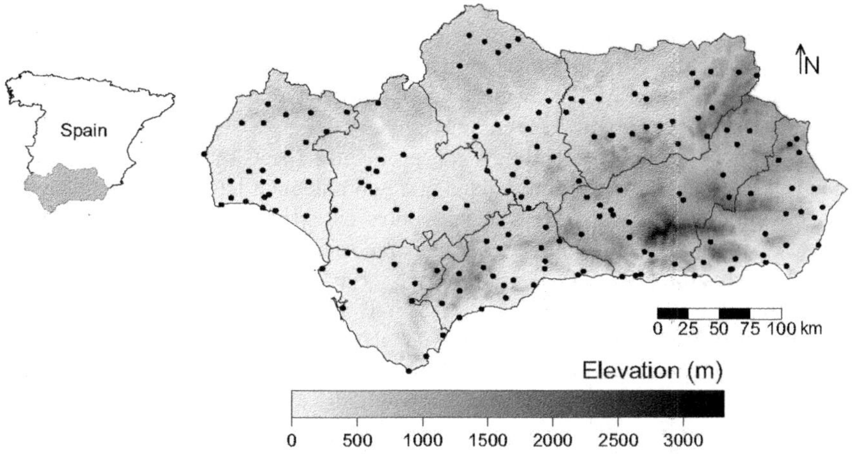

Figure 1. Geographical situation of the study region and location of the 160 meteorological stations projected on an elevation map of Andalusia.

Figure 2 shows a map of the index of dryness, R, generated from previously produced maps of P and ET_o. It can be observed that R is only smaller than 1 in the mountainous areas of Andalusia, which is an indication of the semi-arid or even arid character of a large part of this area. Anyhow, in order to give an exact description, the characteristics of the intra-annual climatology and its influence on the annual water balance should be taken into account as well, but these are not included in the index of dryness. The data on w_o were obtained from a previously produced map (Vanderlinden, 2001), where point data on w_o were calculated using pedotransfer functions (PTF's) of Schaap et al. (2001).

2.4 Geostatistical framework

Applying the previously mentioned concepts and model to the available daily data, an estimation of the daily water balance can be made at the available meteorological stations and daily values of actual evapotranspiration, e, and total runoff, q, can be obtained. From these series, average daily, monthly, seasonal or annual values can be calculated. However, we are not only interested in the temporal

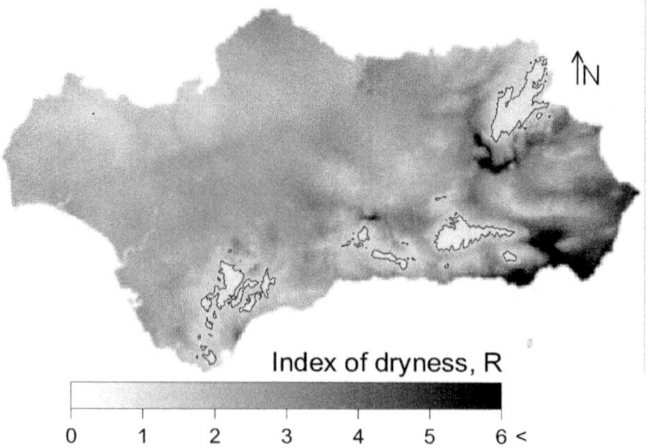

Figure 2. Map of the index of dryness, R=ET$_o$/P. The contour corresponds with a value of R=1.

evolution of these variables, but also in its spatial variability at each of these temporal resolutions. In order to simplify the analysis, here we will focus only on the spatial distribution of the average annual water balance within the Region of Andalusia, using geostatistical techniques.

The choice of the methodology for producing these maps should take into account issues related with "Interpolate first and calculate later" (IC) or "Calculate first and interpolate later" (CI) working routes (Stein et al., 1991; Heuvelink and Pebesma, 1999), and with the use of exhaustive secondary information provided by the Budyko diagram. Since the water balance model is highly non-linear, it could be argued that the IC approach is more suited (Heuvelink and Pebesma, 1999; Addiscott and Tuck, 2001). Since the model is run on a daily time scale, this methodological route would require daily maps of p and et_o, which would be very labour intensive, or require spatiotemporal methods, which are beyond the scope of this study. Moreover, the use of interpolated input data could lead to error accumulation in the output. So, for rather practical reasons we preferred the CI route.

Valuable exhaustive secondary information, in terms of R, can be obtained from previously produced maps of P and ET_o, for which the elevation (DEM) was used as a secondary variable (Vanderlinden, 2002). Finally, in a similar way as in the Budyko diagram, R can be related to E and Q.

A straightforward way to incorporate this secondary information into the spatial interpolation scheme is using simple kriging with varying local means (SKlm) (Goovaerts, 1997, §6.1.2):

$$z^*_{SKlm}(x_o) = m(x_o) + \sum_{i=1}^{n} \lambda_i^{SKlm} \left(z(x_i) - m(x_i) \right), \tag{12}$$

with $m(x_o)$ and $m(x_i)$ respectively the previously calculated local mean value at the estimation point, x_o, and at the n neighbouring points, x_i, with data values $z(x_i)$ and corresponding weights, λ_i^{SKlm}. This estimator requires the variogram of the residuals. An alternative easy way to incorporate secondary information is using kriging with an external drift (KED) (Goovaerts, 1997, §6.1.3), where the local mean is modelled locally as a linear function of the secondary or external variable:

$$m(x) = a_o(x) + a_1(x)\,y(x), \tag{13}$$

with $y(x)$ the secondary variable, and $a_o(x)$ and $a_1(x)$ coefficients that are supposed to be constant within the search neighbourhood. Both methods were implemented using GSLIB (Deutsch and Journel, 1998) and the required variograms were calculated and modelled using VARIOWIN (Panatier, 1996).

3. RESULTS AND DISCUSSION

3.1 Model performance and basic statistics

Average annual values of the components of the soil water balance in Andalusia, E and Q, were calculated from the daily output of the Milly model. The basic statistics of these data are shown in Table 1. It can be seen that the variance of E increases as w_o is duplicated and that the average of E increases with nearly 13 %.

Table 1. Basic statistics of the average annual components of the soil water balance at 160 meteorological stations within the Region of Andalusia, using the Milly model with the original w_o values or $w_o \times 2$.

n=160	M	med	Min	max	s	s^2	$skew$	$kurt$
				Milly, original w_o				
E (mm)	408.6	416.8	208.4	610.4	75.0	5627.6	-0.4	0.5
Q (mm)	164.1	120.4	0.0	1418.7	188.8	35665.7	3.4	15.6
				Milly, $w_o \times 2$				
E (mm)	463.9	469.4	208.7	753.8	100.7	10139.4	-0.2	0.4
Q (mm)	107.5	58.5	0.0	1302.8	167.9	28186.8	3.9	20.3

The Q data show highly positively skewed distributions, due to the large number of locations where annually only a small amount of runoff is generated, while there exist only a few locations where runoff is very high, due to a limited w_o or due to a high P. A duplication of w_o reduces the variance of Q and decreases the average of Q with nearly 35 %. The calculated values are represented in the Budyko diagram in Figure 3. It can be seen that they are situated below the empirical relationships of Budyko (equation (5)) or Lettau (1969) and an exponential relation, similar to equation (5) is fitted to the data.

The important differences between the calculated values and these curves are due to the important seasonality of the Andalusian climate, where the annual signals of rainfall and ET_o are completely out of phase (Milly, 1994b) and to the relatively low w_o values that were used in the calculations ($\overline{w}_o$=110 mm). Milly (1993, 1994a, 1994b) used values of approximately 150 mm. It can be shown that the differences with the Budyko diagram diminishes as w_o is increased. Comparison of the average values of E and Q with those obtained in a national study by the Ministerio de

Medio Ambiente (1998) on the Spanish water resources shows that values of w_o between 150 and 170 mm should be used in our approach in order to obtain comparable results.

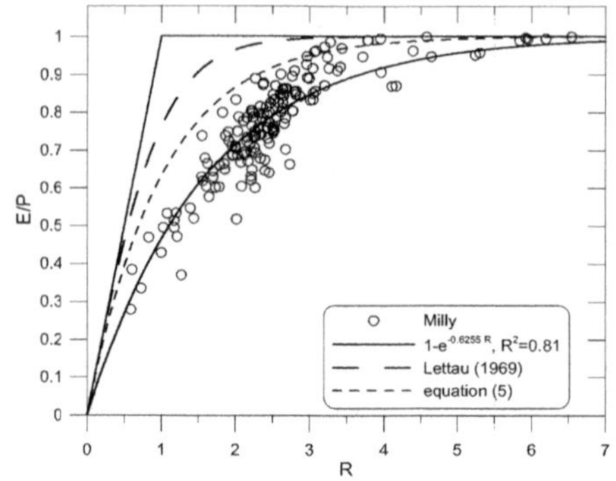

Figure 3. Representation on the Budyko diagram of the average annual water balance data, calculated with Milly's model.

3.2 Exhaustive secondary information and variography

Figure 3 shows that R can be used to predict E/P, or E and Q, which are more relevant in this context. In turn of using the fitted exponential relationship from this figure, we preferred to fit directly a curve to the R-E and R-Q data. In both cases an exponential model with two parameters is fitted, as shown in Figure 4. For $R>4$, Q is practically insensitive to R, because almost no runoff is generated, but becomes highly sensitive to R for $R<3$. In addition there are less data points for this interval, which makes it difficult to obtain a reliable fit. These relations are then used to produce maps of E and Q (Figure 5) from the map of R (Figure 2). These maps constitute a first estimation of the spatial distribution of these variables and can be considered as exhaustive secondary information that can be incorporated in the spatial interpolation procedure using SKlm or KED.

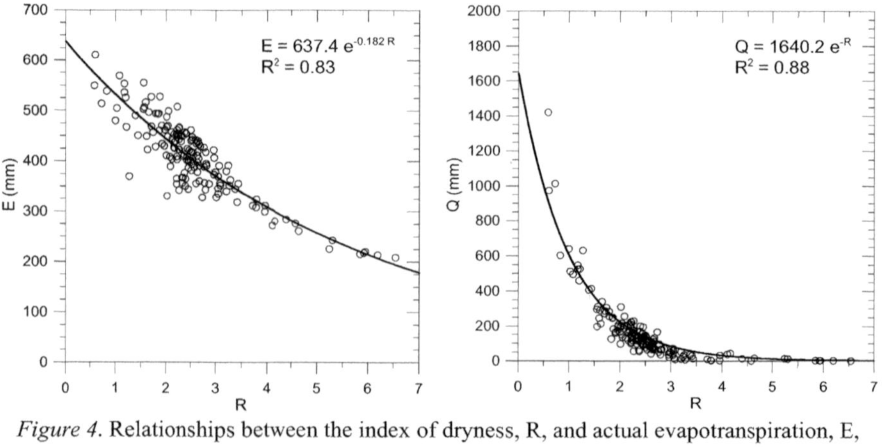

Figure 4. Relationships between the index of dryness, R, and actual evapotranspiration, E, and total runoff, Q.

In the first case these maps are considered as varying local means and interpolation is actually done with the residuals, requiring a residual variogram. In the second case the maps from Figure 5 are considered as an external variable that is supposed to be related linearly with the primary variables. In this case a directional variogram is used for the direction of mayor continuity or the direction in which the drift is less apparent. Also the omnidirectional variogram is often used, since it can be argued that the drift is usually not observed at the first lags of the variogram. Notice that both interpolation methods deal with the non-stationary conditions of the skewed *Q* data. Both maps of Figure 5 show a similar spatial pattern, alike to that of *P*, because it is this variable that conditions the water balance in the mayor part of Andalusia.

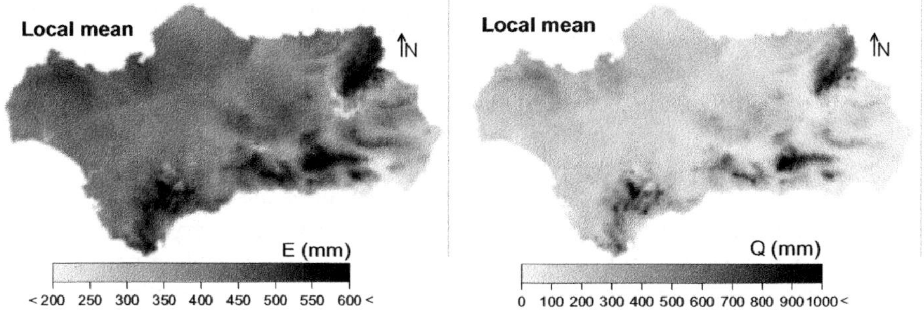

Figure 5. Local mean maps for E and Q, calculated from the map of R (Figure 2) and the relations from Figure 4.

From the maps in Figure 5 the residuals can be obtained and residual variograms can be calculated. These, together with the original variograms are represented in Figure 6. Exponential models were fitted with a zero nugget effect, except for the residual variogram of *E*, where a spherical model was chosen. The ranges of the original variograms for *E* and *Q* are 92 and 76 km, respectively, and 36 and 48 km for the residual variograms. The large difference between the sills of the original and the residual variograms indicates that the maps from Figure 5 explain a large part of the variability in the *E* and *Q* data, but do not capture the small scale variability, since the residual variograms still show a strong spatial correlation structure.

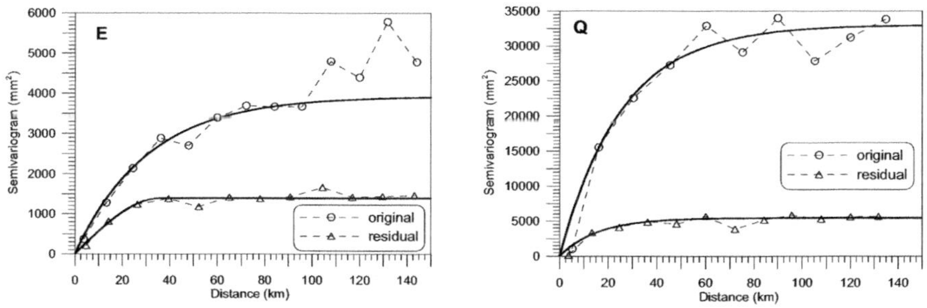

Figure 6. Variograms of E and Q for the original data and the residuals, with a fitted model.

3.3 Cross-validation and Spatial interpolation

The cross-validation statistics in Table 2 show that the difference between KED and SKlm is very small. In the case of *E*, KED performs slightly better and for *Q*,

SKlm is superior, especially in terms of MRE. The bias (ME) is in all cases negligible and the correlation coefficient is larger than 90 %.

Table 2. Statistical parameters of the cross-validation of E and Q, comparing SKlm and KED.

	*ME**	*MAE*	*RMSE*	*MRE*	*R*
			SKlm		
E (mm)	-2.8	23.9	32.6	0.06	90.3
Q (mm)	1.5	39.1	68.6	0.70	93.6
			KED		
E (mm)	-0.4	21.5	31.0	0.05	91.1
Q (mm)	-1.3	40.5	67.5	1.06	93.3

* *ME*: Mean Error, *MAE*: Mean Absolute Error, *RMSE*: Root Mean Square Error, *MRE*: Mean Relative Error, *R*: correlation coefficient

Although differences between both interpolation methods were small, according to the cross-validation results, *E* was interpolated using KED and SKlm was used for *Q*. Figure 7 shows the corresponding maps which were produced using block kriging with blocks of 1 km^2, a search neighbourhood radius of 80 km and a number of neighbouring points between 4 and 16. These maps are very similar to those presented in Figure 5, but show more local detail. Cross-validation only evaluates the goodness of the spatial estimation methods and not the entire methodology. This requires the comparison of total estimated runoff values with observations at gauging stations along the fluvial network of the different basins in the region. The general spatial pattern of *E* and *Q* corresponds well with those presented by the Ministerio de Medio Ambiente (1998, Figure 86 and 89). Since the interpolation errors for *Q* are larger, this map can also be produced from the maps of *E* and *P*, having in mind that *Q* = *P*-*E*.

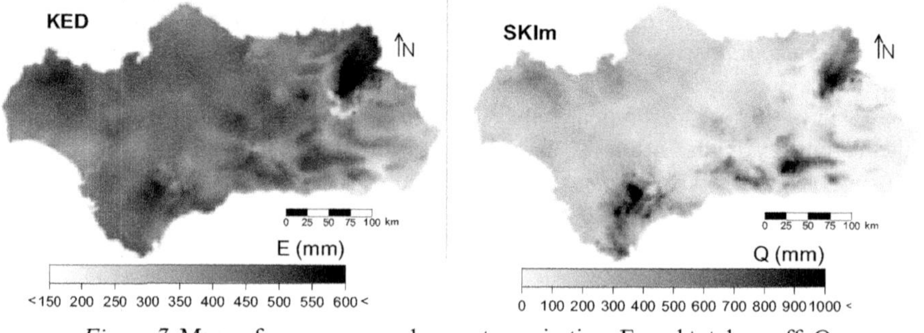

Figure 7. Maps of average annual evapotranspiration, E, and total runoff, Q.

4. CONCLUSIONS

We analysed the average annual soil water balance at 160 meteorological stations within the Region of Andalusia, using a simple bucket model. Modelled variables were actual evapotranspiration and total runoff. Basic statistics were calculated for these variables and an adequate interpolation methodology was established. The Budyko diagram offers the possibility to infer a relationship between the index of dryness and both variables. Since the annual signals of precipitation and reference evapotranspiration are out of phase, and since a relative small water storage capacity was used, the data did not fit well to the empirical

Budyko curve, for which an extra fitting parameter had to be used. Once these relationships were inferred, the index of dryness map was transformed in maps of both variables, which were used as exhaustive secondary information in Simple kriging with local varying means and kriging with an external drift. The cross validation shows that both methods give similar results. The proposed method constitutes a simple alternative for average annual large scale spatially distributed hydrological modelling as a used by the Ministerio de Medio Ambiente (1998). Further research will focus on the average seasonal behaviour of soil water balance and on the comparison of the total runoff of the Guadalquivir watershed, a mayor watershed in the Region, with stream flow data in order to obtain an overall evaluation of the method.

REFERENCES

1. Addiscott, T.M. and G. Tuck, 2001. Non-linearity and error modelling in soil processes. *Eur. J. Soil Sci.*, 52:129-138.
2. Alley, W.M., 1984. On the treatment of evapotranspiration, soil moisture accounting, and aquifer recharge in monthly water balance models. *Water Resour. Res.*, 20:1137-1149.
3. Bailey, H.P., 1979. Capítulo 3. Semi-Arid Climates: Their Definition and Distribution. In: A.E. Hall, G.H. Cannell and H.W. Lawton (Eds.), *Agriculture in Semi-Arid Environments*, Springer-Verlag, Berlin.
4. Boughton, W.C., 1968. A mathematical catchment model for estimating run-off. *J. Hydrol. N.Z.*, 7:75-100.
5. Broadbridge, P and I. White, 1988. Constant rate rainfall infiltration: A versatile nonlinear model. 1. Analytic Solution. *Water Resour. Res.*, 24:145-154
6. Brutsaert, W., 1982. *Evaporation into the Atmosphere. Theory, History, and Applications.* D. Reidel Publishing Company, Dordrecht.
7. Budyko, M.I., 1974. Climate and life, Academic Press, New York.
8. Deutsch, C.V. and A.G. Journel, 1998. *GSLIB. Geostatistical software library and user's guide*, 2nd edition. Oxford University Press, New York.
9. Eagleson, P.S., 1981. Dynamic hydro-thermal balances at macroscale. In: P.S. Eagleson (Ed.), *Land surface processes in atmospheric general circulation models.* Papers presented at the World Climate Research Programme Study Conference, Greenbelt, Maryland, Cambridge University Press, pp 289-357.
10. Goovaerts, P., 1997. *Geostatistics for natural resources evaluation.* Oxford University Press, New York.
11. Heuvelink, G.B.M. and E.J. Pebesma, 1999. Spatial aggregation and soil process modeling. Geoderma, 89:47-65.
12. Kroes, J.G., J.G. Wesseling and J.C. van Dam, 2000. Integrated modelling of the soil-water-atmosphere-plant system using the model SWAP 2.0. An overview of theory and application. *Hydrol. Proc.*, 14:1993-2002.
13. Lettau, H., 1969. Evapotranspiration climatonomy. 1. A new approach to numerical prediction of monthly evapotranspiration, runoff and soil moisture storage. *Mon. Wea. Rev.*, 97:691-699.
14. Lettau, H.H. and M.W. Baradas, 1973. Evaporation Climatonomy II: Refinement of parameterization, exemplified by application to the Mabacan river watershed. *Mon. Wea. Rev.*, 101:636-649.
15. Milly P.C.D, 1993. An analytic solution of the stochastic storage problem applicable to soil water. *Water Resour. Res.*, 29:3755-3758.
16. Milly P.C.D.,1994a. Climate, interseasonal storage of soil water, and the annual water balance. *Adv. Water Resour. Res.,* 17:19-24.
17. Milly P.C.D, 1994b. Climate, soil water storage, and the average annual water balance. *Water Resour. Res.,* 30:2143-2156.

18. Ministerio de Medio Ambiente, 1998. El libro blanco del agua en España, M.M.A., Madrid.
19. Panatier, Y., 1996. *VARIOWIN: Software for spatial data analysis in 2D*. Springer Verlag, New York.
20. Schaap, M.G., F.J. Leij and M. Th van Genuchten, 2001. Rosetta: a computer program for estimating soil hydraulic parameters with hierarchical pedotransfer functions. *J. Hydrol.*, 251:163-176.
21. Stein, A., I.G. Staritsky, J. Bouma, A.C. van Eijnsbergen and A.K. Bregt, 1991. Simulation of moisture deficits and areal interpolation by universal cokriging. *Water Resour. Res.*, 27:1963-1973.
22. Vanderlinden, K., 2001. Spatial estimation of the soil water storage capacity for water balance modelling in southern Spain. Communication presented at Pedometrics 2001, 4th conference of the Working Group on Pedometrics of the International Union of Soil Science, September 19-21, 2001, Gent, Belgium.
23. Vanderlinden, K, 2002. Análisis de procesos hidrológicos a diferentes escalas espacio-temporales. PhD thesis, Departamento de Agronomía, Universidad de Córdoba, Córdoba, Spain (in Spanish).

MODELING PHYTOPLANKTON: COVARIANCE AND VARIOGRAM MODEL SPECIFICATION FOR PHYTOPLANKTON LEVELS IN LAKE MICHIGAN

L.J. Welty and M.L. Stein

Department of Statistics, University of Chicago. 5734 South University Avenue, Chicago, IL. USA

Abstract: Algae and phytoplankton are crucial elements of marine ecosystems and of the global carbon cycle, which engenders widespread interest in better understanding their spatial and temporal variability. In situ fluorometry provides detailed measurements of phytoplankton levels; appropriate statistical models are necessary in order to elicit information about the distribution of phytoplankton biomass from this data. Challenges associated with such a data analysis include covariance model specification for processes in which variation in the vertical and horizontal directions differ greatly. Though the ideas presented here were developed with an eye to understanding phytoplankton dynamics, they may be helpful in developing models for other geophysical and environmental processes measured along vertical and horizontal dimensions.

1. INTRODUCTION

Phytoplankton, unicellular algae found primarily in oceans and lakes, are important components of marine ecosystems. They are at the base of the food chain, so insufficient numbers mean that few other species can survive. Excessive algal growth, common in nutrient rich polluted marine waters,

X. Sanchez-Vila et al. (eds.), geoENV IV – Geostatistics for Environmental Applications, 163-173.

may squeeze these same species out. Furthermore, most of the carbon fixation that occurs in the oceans is due to phytoplankton respiration and accounts for a sizable portion of global carbon fixation Falkowski (1994).

Chlorophyll is found in algae and phytoplankton, and because it may be measured using a variety of methods, it is often tracked as an indicator of algal biomass. Determining chlorophyll content by discrete water samples alone is an accurate but inefficient method for obtaining detailed descriptions of phytoplankton dynamics. A more efficient method for measuring chlorophyll is by chlorophyll fluorescence.

In situ chlorophyll fluorescence measurements capture chlorophyll levels over large spatial scales in real-time. When exposed to blue light near 430 nm, chlorophyll emits red light near 680 nm (Falkowski et al., 1985). The strength of the emission is roughly linearly related to chlorophyll. In situ fluorometers provide a real-time voltage output (approximately one observation per second, for instance), and when towed through lake or ocean waters provide a more complete picture of chlorophyll dynamics than water samples alone. The voltage output is calibrated to chlorophyll level by water samples taken along with fluorescence measurements.

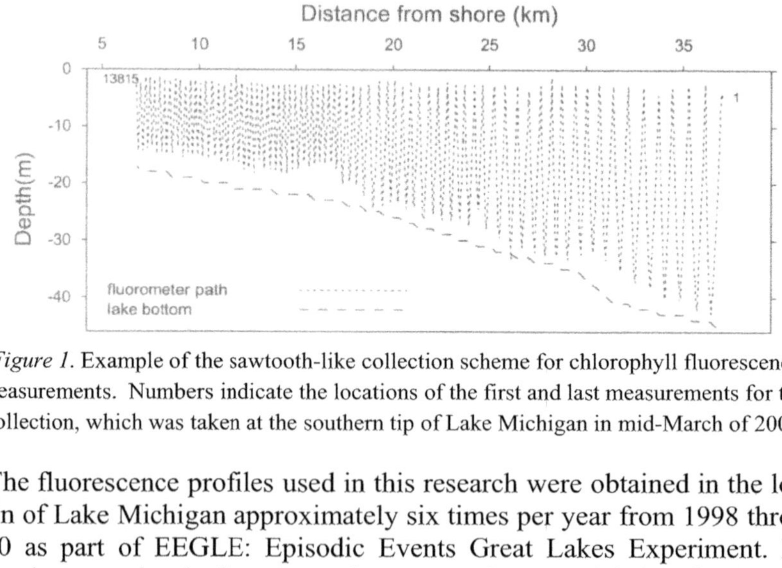

Figure 1. Example of the sawtooth-like collection scheme for chlorophyll fluorescence measurements. Numbers indicate the locations of the first and last measurements for this collection, which was taken at the southern tip of Lake Michigan in mid-March of 2000.

The fluorescence profiles used in this research were obtained in the lower basin of Lake Michigan approximately six times per year from 1998 through 2000 as part of EEGLE: Episodic Events Great Lakes Experiment. Data collections consisted of towing a fluorometer in an undulating fashion from surface to bottom and bottom to surface repeatedly for approximately 25 kilometers along transects extending from shore toward the lake's center (Figure 1).

2. AXIAL DEPENDENCE OF PHYTOPLANKTON

Unlike the most commonly used spatial models that assume isotropy, any tenable model for phytoplankton or similar marine measurements must account for the distinctly different processes along the horizontal and vertical axes. Variables affecting phytoplankton (temperature, water mixing, nutrient levels, to name only a few) change differently depending on if one moves roughly parallel to the surface/bottom or toward the surface/bottom. For example, during the spring and summer months, the surface layer in Lake Michigan is warmed by solar radiation and becomes thermally isolated from the deeper and cooler waters. The temperature differential inhibits water mixing between the two layers, so that the chlorophyll levels may vary significantly through the water column. During this time, a five meter change in vertical position may result in a much greater change in chlorophyll level than a kilometer change in the horizontal.

3. VARIABILITY IN THE HORIZONTAL AND VERTICAL DIMENSIONS

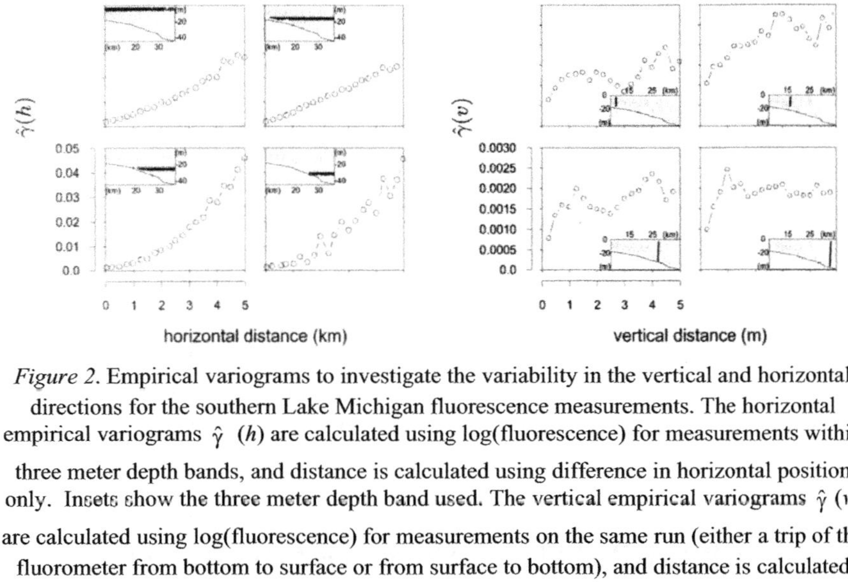

Figure 2. Empirical variograms to investigate the variability in the vertical and horizontal directions for the southern Lake Michigan fluorescence measurements. The horizontal empirical variograms $\hat{\gamma}$ (h) are calculated using log(fluorescence) for measurements within three meter depth bands, and distance is calculated using difference in horizontal position only. Insets show the three meter depth band used. The vertical empirical variograms $\hat{\gamma}$ (v) are calculated using log(fluorescence) for measurements on the same run (either a trip of the fluorometer from bottom to surface or from surface to bottom), and distance is calculated using difference in vertical position only. Insets show the run of measurements used for the corresponding empirical variogram.

We consider 13,815 fluorescence measurements collected in mid-March of 2000 at the southern tip of Lake Michigan. Temperature measurements taken simultaneously are nearly constant in the vertical, and decrease from

4.5°C to 2.5°C as we move away from shore. The sawtooth-like collection scheme (Figure 1) prevents us from completely separating the horizontal variability from the vertical variability in our measurements, so as an approximate method for investigating the variability in each direction we subdivide the data into regions with small range in depth and regions with small range in distance from shore. We calculate empirical variograms based on distance from shore for observations that are no more than three meters apart vertically, as well as empirical variograms based on depth for observations that are in the same run – either a pass up from bottom to surface or a pass down from surface to bottom (Figure 2). In addition, we calculate the average across all runs of the empirical variograms based on depth (Figure 3).

The empirical variograms show that there is considerably greater variability along the horizontal dimensions than the vertical. The process along the horizontal appears nonstationary with no discernible sill, while the process along the vertical appears stationary with a sill near 0.002 (see Figure 3). The horizontal process appears to have range greater than 5 km while the vertical process appears to have range near 1 m. There is no obvious evidence that the horizontal variogram depends on depth or the vertical variogram on distance from shore, so an intrinsic model of order zero may be appropriate.

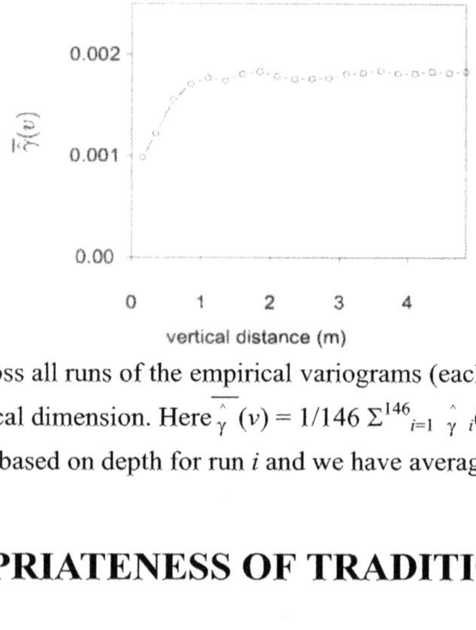

Figure 3. Average across all runs of the empirical variograms (each calculated as $\hat{\gamma}(v)$ in Figure 2) for the vertical dimension. Here $\bar{\hat{\gamma}}(v) = 1/146\, \Sigma^{146}_{i=1}\, \hat{\gamma}_i(v)$, where $\hat{\gamma}_i(v)$ is the empirical variogram based on depth for run i and we have averaged over all 146 runs.

4. INAPPROPRIATENESS OF TRADITIONAL MODELS

We consider our log(fluorescence) measurements as observations from a random field $Z(h,v)$ on $\mathbf{R}^2$, where h is horizontal distance from shore (measured in kilometers), v is depth (measured in meters), and our specific region of $\mathbf{R}^2$ is the approximately triangular area bounded by the water surface and lake bottom. For notational simplicity, we do not consider any nugget effects in what follows. (The empirical variograms suggest that our model should include a nugget of approximately 0.0009).

We have observed that $Z(h,v)$ is a badly anisotropic process. Linear transformation of one or more of the dimensions also fails to produce tenable variogram or covariance models for $Z(h,v)$. If the variogram for $Z(h,v)$ had the form

$$\gamma((h_1,v_1),(h_2,v_2)) = f\left(\sqrt{h^2+\alpha v^2}\right)$$

where $|h_1-h_2| = h$, $|v_1-v_2| = v$, and where f is conditionally negative definite, then $\gamma((h1,v_1),(h_2,v_1)) = f(h)$ and $\gamma((h_1,v_1),(h_1,v_2)) = f(\alpha v)$. Since f must have the same sill in both directions, this model cannot possibly describe the variograms shown in Figures 2 and 3.

5. SPECIFYING VALID MODELS

Before formulating more complex models, we consider what functions make valid covariance and variogram models. Having $K(\mathbf{x})$, $\mathbf{x} \in \mathbf{R}^d$ positive definite is a necessary and sufficient condition for $K(\mathbf{x})$ to be the covariance function of a weakly stationary random field on $\mathbf{R}^d$. If $K_1(x)$ and $K_2(y)$ are valid covariance functions for $x,y \in \mathbf{R}$, then for $a,b > 0$, $aK_1(x) + bK_2(y)$ and $K_1(x)K_2(y)$ are valid covariance functions on $\mathbf{R}^2$.

Analogously, $\gamma(\mathbf{x})$ is a valid variogram model on $\mathbf{R}^d$ if it is conditionally negative definite on $\mathbf{R}^d$. If $\gamma(\mathbf{x})$ is a valid variogram model, then so is $b\gamma(\mathbf{x})$ for $b>0$ Cressie (1993). If $\gamma_1(\mathbf{x})$ and $\gamma_2(y)$ are valid variogram models on $\mathbf{R}$, then $\gamma(x,y) = \gamma_1(x) + \gamma_2(y)$ is valid as well (this follows directly from the conditional negative definiteness). It is generally *not* the case that $\gamma_1(x)\gamma_2(y)$ will be a valid variogram. As an illustration, suppose W is a stationary Gaussian random field on $\mathbf{R}^2$ with variogram model $\gamma(x,y) = b\gamma_1(x)\gamma_2(y)$, where b is nonnegative and $\gamma_1(0) = \gamma_2(0) = 0$. Then

$$\begin{aligned} Var\{W(x,y)-W(x,0)-W(0,y)+W(0,0)\} \\ &= 4b\gamma(x,0)+4b\gamma(0,y)-4b\gamma(x,y) \\ &= -4b\gamma_1(x)\gamma_2(y), \end{aligned}$$

which is nonnegative if and only if $b = 0$.

6. POTENTIAL MODELS

Given the axial dependence of the fluorescence measurements, it seems natural to consider a tensor product approach with covariance function for $Z(h,v)$ of the form

$$\text{Cov}\{Z(h_1,v_1),Z(h_2,v_2)\} = K_H(h)K_V(v) + K_H(h)$$

where K_H and K_V are the covariance models for the horizontal and vertical directions respectively. Then Cov$\{Z(h_1,v_1),Z(h_2,v_2)\}$ is positive definite as long as K_V and K_H are. This form allows for most of the variation to occur in the horizontal dimension as we require, but cannot be well specified when the variation along one dimension (in our case, the horizontal) is nonstationary.

One might also consider the model

$$\gamma\left((h_1,v_1),(h_2,v_2)\right) = \gamma_H(h) + \gamma_V(v)$$

which allows for different (possibly nonstationary) variogram models in the horizontal and vertical dimensions. As Chilès and Delfiner discuss, this model treats the process as exactly additive, *i.e.*

$$\begin{aligned} &\text{Var}\{Z(h,v) - Z(h,0) - Z(0,v) + Z(0,0)\} \\ &\quad = 4\gamma_H(h) + 4\gamma_V(v) + 4\left[\gamma_H(h) + \gamma_V(v)\right] \\ &\quad = 0, \end{aligned}$$

taking $\gamma_H(0) = \gamma_V(0) = 0$ Chilès et al. (1999). It seems unwise to use the above model unless one is quite sure that $Z(h,v)$ is exactly additive.

7. REVISED POTENTIAL MODELS

Modifications to the above approaches do suggest tenable variogram models. First, consider

$$\gamma\left((h_1,v_1),(h_2,v_2)\right) = \gamma_H(h) - K_H(h)K_V(v) + K_H(0)K_V(0) \tag{1}$$

where γ_H is a variogram model for the horizontal direction, and K_H and K_V are covariance models for the horizontal and vertical directions respectively (note that the two terms are not equivalent to $\gamma_H(h)\gamma_V(v)$). Taking both K_H and K_V positive definite results in $\gamma(h_1,v_1),(h_2,v_2))$ conditionally negative definite as required. The first term γ_H accounts for the nonstationarity in the horizontal direction, and the remaining terms for the interaction between horizontal and vertical variability.

Another reasonable model would be

$$\gamma\left((h_1,v_1),(h_2,v_2)\right) = \gamma_H(h) + \gamma_R\left(\sqrt{h^2 + \alpha v^2}\right). \tag{2}$$

Again the first term accounts for the nonstationarity in the horizontal direction, but here the second term requires that the remaining variability be attributed to geometric anisotropy. This model offers some flexibility and simplicity over (1) in that it does not require specification or existence of covariance functions in the vertical and horizontal directions. Model (1) is however quite different from model (2) in that it treats the process as locally nearly additive [Stein (1999),2.11].

8. MODEL COMPARISON FOR FLUORESCENCE MEASUREMENTS

We take (1) to have the form

$$\gamma(h,v) = \theta_0 1_{\{h+v>0\}} + \theta_1 h^{\theta_2} - \theta_3[1 - M_v(\theta_4 h)M_v(\theta_5 v)]$$

where θ_0 is the nugget effect, $M_v(z) = 2^{1-v} z^v K_v(z)/\Gamma(v)$, and K_v is a modified Bessel function Abramowitz (1965). The indicator function $1_{\{h+v>0\}}$ takes the value one if $(h+v)>0$ and zero otherwise. This model, which we will denote $M_T(h,v;\boldsymbol{\theta})$ for its tensor product like last term, treats $\gamma_H(h)$ in (1) as a power law variogram and K_H and K_V in (1) as covariance functions from the Matérn class Stein (1999). We take (2) to have the form

$$\gamma(h,v) = \phi_{0^1_{\{(h+v)>0\}}} + \phi_1 h^{\phi_2} - \phi_3\left[1 - M_v\left(\sqrt{\phi_4^2 h_2 + \phi_5^2 v^2}\right)\right]$$

where again we have added a nugget effect ϕ_0, $\gamma_H(h)$ in (2) is a power law variogram and γ_R in (2) is a Matérn class variogram. We denote this model by $M_G(h,v;\boldsymbol{\phi})$ for its treatment of the local behavior in the horizontal and vertical as geometrically anisotropic. We note that M_T and M_G are equivalent when h=0 or v=0.

Obtaining parameter estimates for M_T and M_G using exact likelihood methods is intractable given that each evaluation of the likelihood function requires $O(n^3)$ operations, and here n = 13,815. Our solution is to employ an approximate likelihood method similar to that proposed by A. Vecchia (Vecchia, 1988). For an observation vector $\mathbf{z} = (z_1,z_2,\ldots,z_n)$, Vecchia noted that it may be possible to approximate the likelihood

$$L(\theta|z) = p(z_1|\theta)\prod_{i=2}^{n} p(z_i|z_{i-1},\ldots,z_1,\theta),$$

where p denotes probability density, by considering the conditional density of z_i on some subset of $z_{i-1},\ldots z_1$ and hence reducing computation. Vecchia proposed conditioning on the m points in $z_{i-1},\ldots z_1$ nearest to z_i, with m generally much less than i–1 (*e.g.* m = 5). In order to account for long range spatial dependence, we alter Vecchia's scheme to condition on some points in $z_{i-1},\ldots z_1$ near z_i as well as some points in $z_{i-1},\ldots z_1$ that are much farther away from z_i. Estimating variogram parameters is our primary interest, so we extend Vecchia's idea to approximate the restricted log likelihood Kitanidis (1983) rather than the log likelihood. Forthcoming work by M. Stein, Z. Chi, and L. Welty details methodology and results for these extensions of Vecchia's work.

Table 1. Variogram parameter estimates for M_T and M_G obtained by maximizing the approximate restricted log likelihood $\tilde{R}$ for the log(fluorescence) observations $\mathbf{z}$. Estimates for parameters describing the long range horizontal dependence are quite similar, while estimates for parameters describing the local behavior of the process differ markedly. We consider M_G preferable to M_T as it has the larger approximate log likelihood.

	Parameter Estimates						$\tilde{R}$
M_T	θ_0	θ_1	θ_2	θ_3	θ_4	θ_5	25527.26
	3.93×10^{-4}	3.39×10^{-3}	1.22	1.43×10^{-3}	2.58×10^{2}	2.75	
M_G	ϕ_0	ϕ_1	ϕ_2	ϕ_3	ϕ_4	ϕ_5	25531.66
	3.49×10^{-4}	3.35×10^{-3}	1.25	1.47×10^{-3}	3.04×10^{2}	3.29	

We therefore take our log(fluorescence) observations $\mathbf{z}$ to be from a Gaussian random field with covariance structure given by $M_T(h,v;\theta)$ or $M_G(h,v;\phi)$, and let $\tilde{R}_T(\theta|\mathbf{z})$ and $\tilde{R}_G(\phi|\mathbf{z})$ represent the approximate restricted log likelihoods under the respective models. We order our observations $z_1,z_2,\ldots,z_{13,815}$ by the order in which they were collected (1). With appropriate adjustments for small values of i, we select our conditioning subset for z_i to consist of ten nearby previous points as well as ten roughly evenly spaced observations from more distant observations. We maximize $\tilde{R}$ using a conjugate gradient algorithm [Press et al., 1992, 10.6]. For initial computational simplicity, we do not maximize over ν, the smoothness parameter for the Matérn covariance function. Based on comparisons of the likelihood for $\nu = 0.5$, 1.0, and 1.5, as well as the shape of the empirical vertical variogram, we set $\nu = 1.0$. Results are shown in Table 1.

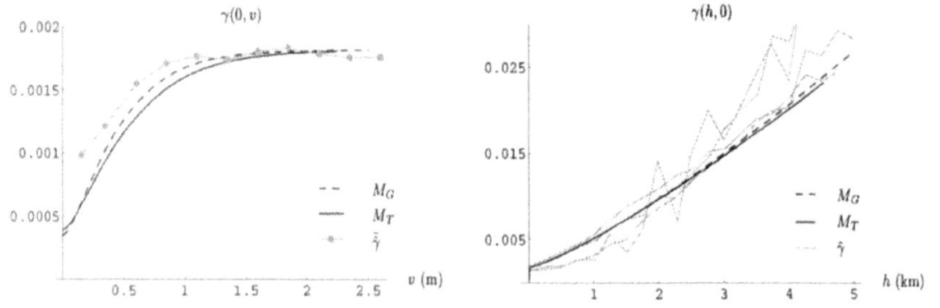

Figure 4. Implied horizontal and vertical variograms for $M_G(h,v;\tilde{\phi})$ and $M_T(h,v;\tilde{\theta})$ as well as empirical variograms $\hat{\gamma}(v)$ and $\hat{\gamma}(h)$, calculated as in Figures 2 and 3.

That $\tilde{R}_G(\phi|\mathbf{z}) > \tilde{R}_T(\tilde{\theta}|\mathbf{z})$ sugggets that $M_G(h,v;\phi)$ more reasonably describes the dependence structure of the log(fluorescence) measurements. The parameters describing the long range horizontal dependence are nearly the same for $M_G(h,v;\tilde{\theta})$ and $M_G(h,v;\phi)$; the largest discrepancies in estimates are for the nugget effect and fifth and sixth parameters, which describe the local variation of the process as well as the interaction between the horizontal and vertical dependence. One should note however that $1/\tilde{\theta}_4 \approx 1/\tilde{\phi}_4$ and $1/\tilde{\theta}_5 \approx 1/\tilde{\phi}_5$, so that the models do give similar estimates for the range of the vertical process and for the range of the local horizontal processes. The models predict nearly indistinguishable variograms in the horizontal direction for long ranges, and similar variograms along the vertical direction (Figure 4).

We note that the average empirical vertical variogram is slightly above either parameteric estimate (Figure 4), but that the difference is not necessarily an indication of model misfit. One possible reason for the discrepancy may be that $\hat{\gamma}(v)$ contains some horizontal variation (recall we calculated each $\hat{\gamma}(v)$ using points in the same run, which will vary slightly in horizontal coordinate). The difference may also be due to sampling variability in the empirical variogram. Points in empirical variograms are highly correlated, so empirical variograms contain significantly less information about processes than appearances suggest. For a more detailed discussion of this problem with empirical variograms and the advantages of using maximum likelihood, Stein (1999).

Figure 5 shows the models' distinct treatments of the interaction between the local horizontal and vertical variation. Significant differences near the origin illustrate the distinction in modeling the local process as geometrically anisotropic versus as a product form. As one moves away from the origin, M_G-M_T is nearly zero, suggesting little qualitative difference in the models for larger values of h and v.

9. CONCLUSIONS

We conclude that models of the form in (2) may more reasonably account for the axial dependence of our log(fluorescence) measurements, the distinctly different variability they exhibit along the axes, and the interactions of the variability along the horizontal and the vertical. The accuracy of interpolated values for cholorphyll fluorescence, which are crucial to calibrating the fluorescence values with water samples and hence to producing fields of predicted chlorophyll levels, may depend significantly on the relevance of our covariance model. In any case, it is important to consider alternative models to the usual isotropic or geometrically anisotropic suspects when data exhibit very different horizontal and vertical variability. One would expect such situations to arise not only for marine data, but also for geological data (where moving parallel to the the earth's surface is much different from tunneling toward its core), or for meteorological data (where variability through a layer of atmosphere may be much different than variability as one moves up through the atmosphere).

Comparing models like those we have suggested here may also provide additional understanding of such processes.

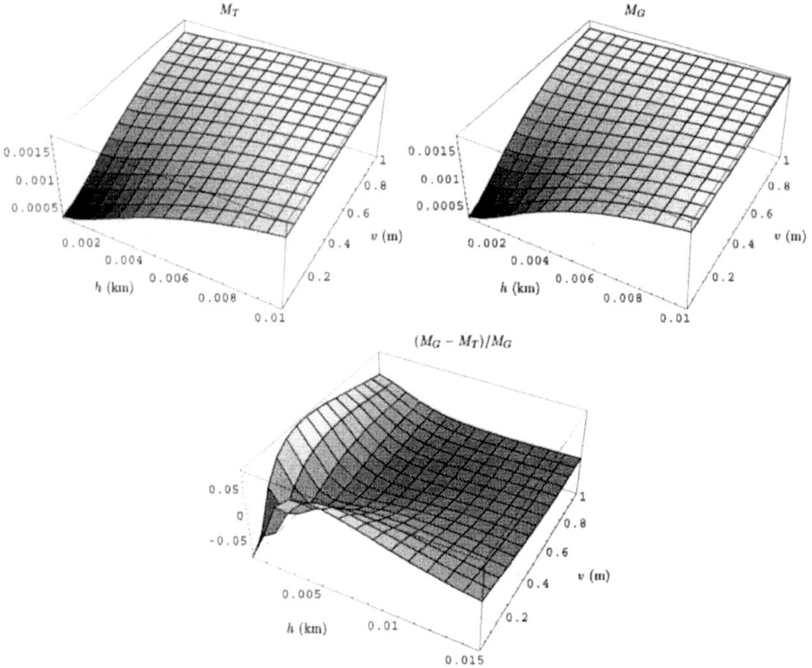

Figure 5. Estimated variograms $M_G(h,v;\tilde{\phi})$ and $M_T(h,v;\tilde{\theta})$, as well as their fractional difference, illustrating the distinct treatments of the short scale horizontal and vertical variation.

ACKNOWLEDGEMENTS

The chlorophyll fluorescence data were produced as part of the Episodic Events Great Lakes Experiment (EEGLE) Program, supported by the National Oceanic and Atmospheric Administration's Coastal Ocean Program and the National Science Foundation. The first author was supported by a National Science Foundation Graduate Research Participant Fellowship. The second author was supported by the United States Environmental Protection Agency through STAR Cooperative Agreement #R-82940201-0 to the University of Chicago. This research has not been subjected to the Agency's required peer and policy review and therefore does not necessarily reflect the views of the Agency, and no official endorsement should be inferred. The authors are greatly indebted to and wish to thank Barry Lesht of Argonne National Laboratory, Environmental Research Division, and Tom Johengen and Henry Vanderploeg of Great Lakes Environmental Research Laboratory, NOAA for providing access to and detailed information about the chlorophyll fluorescence data.

REFERENCES

1. Abramowitz, M. and Stegun, I. (1965). *Handbook of Mathematical Functions*, ninth ed. Dover, New York.
2. Cressie, N. (1993). *Statistics for Spatial Data*. Wiley, New York.
3. Falkowski, P.G., Kiefer, D.A. (1985). Chlorophyll *a* fluorescence in phytoplankton: relationship to photosynthesis and biomass, *Journal of Plankton Research*, **7**:715-731.
4. Falkowski, P.G. (1994). The role of phytoplankton photosynthesis in global biogeochemical cycles. *Phytosynthesis Research*, **39**:235-258.
5. Kitanidis, P.K. (1983). Statistical estimation of polynomial generalized covariance functions and hydrologic applications. *Water Resources Research,* **19**:909-921.
6. Press, W.H., Teukolsky, S.A., Vetterling, W.T., Flannery, B.P. (1992). *Numerical Recipes in C.* Cambridge University Press, Cambridge.
7. Stein, M.L. (1999). *Interpolation of Spatial Data: Some Theory for Kriging*. Springer, New York.
8. Vecchia, A. V. (1988) Estimation and identification for continuous spatial processes, *Journal of the Royal Statistical Society B*, **50**:297-312.

GEOSTATISTICAL INVERSE PROBLEM: A MODIFIED TECHNIQUE FOR CHARACTERIZING HETEROGENEOUS FIELDS

A. Alcolea[1], A. Medina[2], J. Carrera[1] and J. Jódar[1]
[1]*School of Civil Engineering. UPC, Dpt. of soil engineering and geosciences. Campus Nord. Bld. D2. c/ Jordi Girona, 1-3. 08034 Barcelona, Spain;* [2]*School of Civil Engineering. UPC, Dpt. of Applied Mathematics III. Campus Nord. Bld. C2. c./ Jordi Girona,1-3. 08034 Barcelona, Spain*

Abstract: This paper describes a modification of the self-calibrating method for generating equally likely realizations (conditional simulations) of the transmissivity field, that honour measurements of transmissivity and dependent variables (heads, concentrations, etc.). Soft data (e. g. geophysics) can also be included in the conditioning procedure as a external drift. Moreover, spatial variability patterns of the "real" field (as observed through field or lab experiments) are respected. The results of the algorithm are compared with those obtained by the most commonly used methods in groundwater, such as zonation and pilot points (conditional estimation methods). The performance of these geostatistical inverse approaches was compared on a synthetic data set, where the outcome is based on qualitative (resemblance between the obtained transmissivity fields and the 'real' one) and quantitative criteria (goodness of fit between computed and measured heads). Results show that the inclusion of head data in the conditioning procedure provides a better solution than the one obtained including only transmissivity data. Final comparison (simulations/estimations conditioned to both type of data) shows similar results. The choice of the best method depends on whether the modeller seeks small-scale variability (conditional simulation methods) or large-scale trends (conditional estimation methods).

1. INTRODUCTION

For many environmental applications, such as the selection of a waste disposal site, aquifer management or aquifer remediation, a good

X. Sanchez-Vila et al. (eds.), geoENV IV – Geostatistics for Environmental Applications, 175-186.

characterization of the aquifer properties is absolutely necessary. The heterogeneity of some of these properties is known to control the aquifer response. For instance, it is well known that the heterogeneity of the transmissivity field has a large impact on solute or gas transport through the geosphere. The representation of aquifer behaviour is, in a wide sense, referred to as numerical modelling.

The main objective of numerical modelling is to obtain a representation of the aquifer that 1) honour all available data, such as point transmissivity, heads and concentration measurements, geological/geophysical information, etc. and 2) respect spatial variability patterns as observed through field or lab experiments. For this purpose, geostatistical inversion approaches are ideally suited, and they can be classified in two groups: conditional estimation and conditional simulation methods. While the latter provides the 'best' estimate of the unknown field, the outcome of the former is a set of equally likely realizations that honour all available data.

Several approaches can be found in each one of the groups. Zonation (*Carrera and Neuman*, 1986), kriging and pilot points method (*Certes and de Marsily,* 1991) are the most frequent among those of conditional estimation. Among others, self-calibrating method (*Gómez-Hernández et al.,* 1997), Linearized Cokriging (*Kitanidis and Vomvoris,* 1983), Linearized Semianalytical (*Rubin and Dagan,* 1987), are included in the group of conditional simulation methods.

A good review of geostatistical inverse approaches is *McLaughlin and Townley* (1996). In that paper, a common theoretical framework and a theoretical comparison are presented. However, the major attempt to compare them numerically was given by *Zimmerman et al.* (1998).

In this work we present a modification of the self-calibrating method, with especial emphasis in the algorithm, as well as a numerical comparison on a synthetic example with methods of zonation, pilot points and kriging.

2. PARAMETERIZATION METHODS

Inverse procedures need to describe the spatial and temporal variability of unknown parameters, which is referred to as parameterization. We present here a brief description of the parameterization methods used in this paper. For further information, we address the reader to reviews such as *Carrera* (1987), *Yeh* (1986). Linear parameterizations can be expressed as:

$$p(\mathbf{x},t) = \sum_{j=1}^{n} p_j f_j(\mathbf{x},t) \qquad (1)$$

where p_j are scalars called models parameters (unknowns) and $f_j(\mathbf{x},t)$ are interpolation functions. Parameterization procedures differ according to these functions. The most commonly used in groundwater have been zonation (discretization) and pilot points, defined below.

- **Zonation**: A partition is made on the system. In every partition's zone, the function $f_j(\mathbf{x},t)$ has a predefined variation or a constant value (*Carrera and Neuman*, 1986).
- **Pilot points method**: The interpolation functions $f_j(\mathbf{x},t)$ are defined as kriging coefficients and p_j are the hypothetical parameters on a finite number of points, which are referred to as pilot points (*de Marsily*, 1978).

3. SUGGESTED APPROACH

The approach proposed here is a modification of the one by *Gómez-Hernández et al.* (1997). Unknown parameter (log-transmissivity in this case) is defined as the superposition of two fields: a deterministic drift and an uncertain component. The deterministic part (Y_{drift}) can be obtained through conditional simulation or kriging, depending on whether one seeks small-scale variability or large-scale trends, and therefore reproduces hard data (i.e. transmissivity measurements) and soft data (i.e. geophysical data can be included as a external drift). The uncertain part can be seen as a perturbation, such that the final field also reproduces data related to dependent variables (heads, concentrations, etc.). To overcome stability problems, this perturbation field is expressed in terms of a finite number of unknown perturbations (ΔY) at n points (similar to master locations at *Gómez-Hernández* work, but pilot points at *de Marsily*'s). Final expression of the parameterization for $\log_{10}$T field can be expressed as:

$$\mathrm{Y}(\mathbf{x}) = \mathrm{Y}_{\mathrm{drift}}(\mathbf{x}) + \sum_{i-1}^{n} \lambda_i(\mathbf{x})\Delta \mathrm{Y}_i \tag{2}$$

where λ_i are interpolation weights, which, in this case, are obtained through kriging (seven variants were implemented: simple kriging, ordinary kriging, kriging with locally varying mean, kriging with external drift, simple cokriging, ordinary cokriging and standardized ordinary cokriging). Given that the deterministic drift honors parameter data, we seek to determine a perturbation field such that the final field also honors data related to dependent variables (heads, concentrations, etc.). Next section describes the methodology to obtain the optimal values of the unknown perturbations at the master locations.

4. INVERSION PROCEDURE

The goal is to obtain optimal values of the perturbations such that the final field also honors dependent variable measurements. A common way to achieve it is to formulate the problem in terms of a 'performance criterion', expressing the difference between actual solution and what we know about the real system (measurements). This criterion is referred to as objective function and can be expressed as (only using head measurements):

$$J = \lambda_h \left(\mathbf{h} - \mathbf{h}^*\right)^t \underline{\mathbf{C}}_h^{-1} \left(\mathbf{h} - \mathbf{h}^*\right) + \lambda_{\Delta Y} \left(\Delta\mathbf{Y} - \Delta\mathbf{Y}^*\right)^t \underline{\mathbf{C}}_{\Delta Y}^{-1} \left(\Delta\mathbf{Y} - \Delta\mathbf{Y}^*\right) \quad (3)$$

where $\mathbf{h}^*$ is the vector of all head measurements, $\mathbf{h}$ are the corresponding computed heads, $\Delta\mathbf{Y}$ is the vector of $\log_{10}T$ perturbations at master points, $\Delta\mathbf{Y}^*$ their prior estimates. $\underline{\mathbf{C}}_h$, $\underline{\mathbf{C}}_{\Delta Y}$ are the corresponding covariance matrices and λ_h, $\lambda_{\Delta Y}$ are weighting coefficients.

The set of unknown perturbations that minimizes (3) makes the final field to honor all available data. It should be noticed that conditioning is enforced strictly, given that transmissivity measurements are honored by the deterministic part Y_{drift} and perturbation is zero at those points. Posed in this way, inversion becomes an optimization problem, performed by Levenberg-Marquardt's method.

One of the novelties is the inclusion of the plausibility term, accounting for the difference between prior and posterior estimations of transmissivity at the master points. Other works (e.g. *Capilla et al.*, 1997) calibrate the model only bearing in mind head or concentration measurements, obtaining solutions providing a good fit between calculated and measured values, but do not assure plausibility of estimates. This is a very important issue. As demonstrated by *Carrera and Neuman* (1986b) the inclusion of this term (regularization term in that paper) improves the conditioning of the inverse problem.

In our work, prior estimation of the perturbations at the master points are obtained through kriging, on the basis of transmissivity measurements. This formulation also improves the statistical consistency of the method. This issue will be discussed elsewhere.

5. SYNTHETIC EXAMPLE

In this section we present the comparison between 6 geostatistical inverse approaches, including the one proposed here. All of them were applied to a set of synthetic data, where the 'real' system was perfectly known a priori.

Flow domain is a square of 4000x4000 m^2 area, where inflows are prescribed to be 0.1 m^3/d at the left boundary and the head level was set to 0 m at the right boundary. Upper and lower boundaries are supposed to be impervious. There are also two internal sinks of 3 m^3/d in the middle part of the flow domain. (Figure 1)

Log-transmissivity is considered as a random field with a zero mean and a spherical isotropic covariance function, with a variance of 4.0 and a range of 1000 m (1/4 of the domain length). For the purpose of the transmissivity estimation/simulation, the domain is divided into 1600 squared blocks of 100x100 m^2 area.

Flow regime is transient with steady-state initial conditions; under steady conditions, no pump is assumed in the middle of the domain. Wells pump only during half part of the test. The storage coefficient was taken as constant and perfectly known, with a value of 10^{-5}.

This problem setup (see Figure 1) was considered as the model for the 'real' system and was used to derive the conditioning measurements (head and $\log_{10}T$ data) at 25 observation wells.

For the application of the pilot points method and the proposed approach, a uniform grid was generated, using three master points per correlation range, as suggested by *Gómez-Hernández et al.* (1997). This leads to a total number of 144 master points, a number large enough to reproduce spatial variability patterns, but small enough considering computational effort.

Six methods (summarized at Table 1) were applied to the set of synthetic data and evaluated both qualitatively (resemblance between the obtained transmissivity and head fields and the 'real' ones) and quantitatively, in terms of the errors in computed log-transmissivities.

Table 1. Summary of methods applied to the set of synthetic data

Group	Conditioning data	Method	Acronym
Conditional simulation	$\log_{10}T$	Suggested approach	CS-T
	$\log_{10}T$, h	Suggested approach	CS-Th
Conditional estimation	$\log_{10}T$, h	Pilot points	CETh-PP
	$\log_{10}T$	Ordinary Kriging	CET-K
	$\log_{10}T$, h	Kriging as drift + perturbation	CETh-MP
	$\log_{10}T$, h	Zonation	CETh-Z

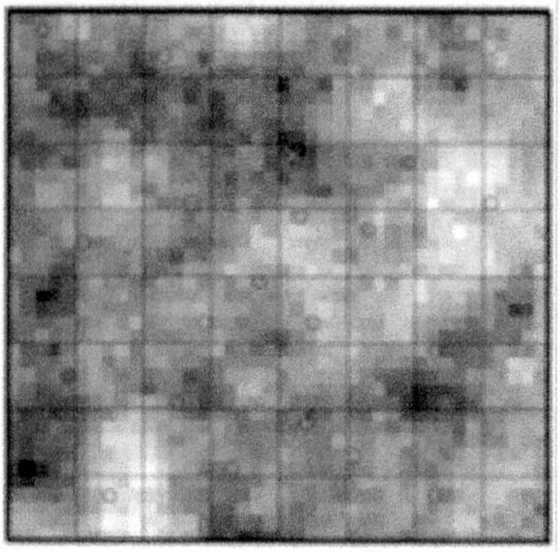

Figure 1. Synthetic example setup. "Real" transmissivity field, boundary conditions and position of the measurement points (circles).

An error vector $\mathbf{e}_j$ was defined for each simulation 'j', (unique in the case of conditional estimation):

$$\mathbf{e}_j^{\,i} = \mathbf{Y}_{calc,j}^{\,i} - \mathbf{Y}_{true,j}^{\,i} \quad i=1,N_b \quad j=1,N_S \tag{4}$$

where $\mathbf{Y}_{calc}$ and $\mathbf{Y}_{true}$ are the vector of calibrated and 'real' transmissivities of all blocks at simulation j.

Comparison is evaluated in terms of:

1. Mean error (ME): $ME = \frac{1}{N_s}\sum_{j=1}^{N_s}\frac{1}{N_b}\sum_{i=1}^{N_b}\mathbf{e}_{ji}$

2. Mean deviation error (MDE): $MDE = \frac{1}{N_s}\sum_{j=1}^{Ns}\left(\frac{1}{N_b}\mathbf{e}_j^t\mathbf{e}_j\right)^{1/2}$

3. Heads objective function: $\mathbf{J_h}^{\mathbf{j}} = \left(\mathbf{h}^{\mathbf{j}} - \mathbf{h}^*\right)^t\left(\mathbf{h}^{\mathbf{j}} - \mathbf{h}^*\right)$

where N_s is the number of conditional simulations and N_b is the number of transmissivity blocks (50 and 1600, respectively).

The first criterion measures the estimation biases and should be close to zero. The second one measures the difference between the true field and the obtained one, and (*Carrera and Glorioso*, 1991) should be smaller than the field variance (4 in this case). The third one measures the quality of the fit between calculated and measured heads at the observation points.

5.1 Visual comparison

Consider Figure 2, displaying the results of one of the realizations obtained by the proposed method. Comparing maps at column 1, one can observe that the simulation conditioned to $\log_{10}T$ data (b1) reproduces the large-scale patterns of the real field. However, there is still a large difference between the real field and the proposed one. This uncertainty can also be observed comparing maps (a2, 'real' heads) and (b2, predicted heads). Because head measurements were not included as conditioning data, measured heads do not have necessarily to be reproduced by the model, as shown at picture (b3).

This difference is reduced by adding the perturbation field (the one being calibrated on the basis of head measurements). Final solution is presented on row (c). The reduction of the uncertainty of the initial drift (conditioned only to $\log_{10}T$ data) can be observed in maps and pictures at row 'c'. Final field also reproduces large-scale patterns and is more alike than the initial drift. Also, head measurements are reproduced.

Figure 3 displays the results obtained by conditional estimation methods. The most important remark is that $\log_{10}T$ fields are inherently smooth. However, large-scale spatial patterns of the 'real' field are also honored, even in the cases where only $\log_{10}$ T data were used.

Considering rows (d) and (e) the similarity between true and calibrated fields is striking, if one seeks large-scale trends. Consider now map (c1), using pilot points method. One can see some singularities in the calibrated field, as measurements are fully respected. Row (e) displays the results obtained by the proposed method, using kriging as initial drift, jointly with the calibration of the perturbation field using the master points. This one does not present singularities on the final transmissivity map (e1), even though $\log_{10}T$ measurements are also respected.

Figure 4 displays a comparison between the average field of the 50 conditional simulations and the one obtained through zonation. As one can see, they are very similar. However, the average field is still sharp, probably because only 50 realizations were considered.

5.2 Numerical comparison

Table 2 displays the numerical aspects of the comparison. Considering mean error, all methods yielded similar results. Mean error was, in all cases, too high, but close to zero. So that, final solutions have a little bias.

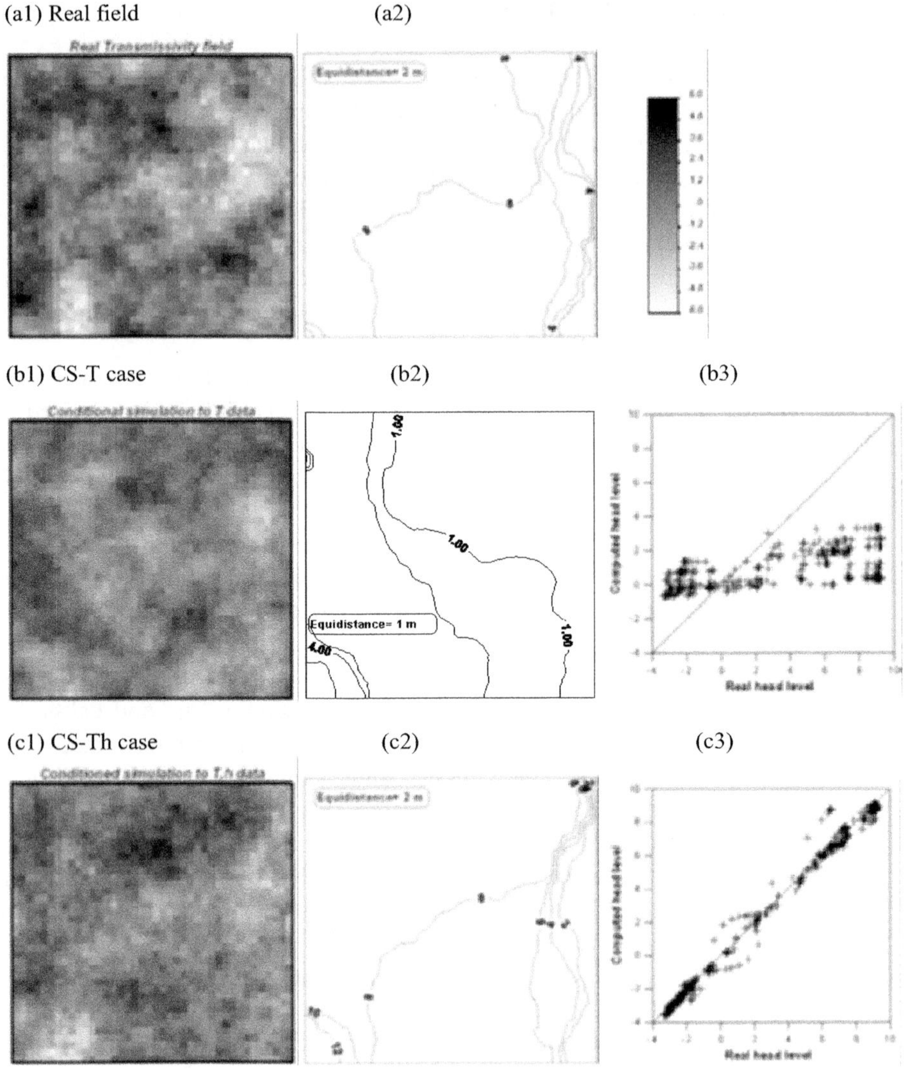

Figure 2. Results concerning conditional simulations. Row (a): 'real' $\log_{10}T$ field and 'real' head level field (steady-state). Row (b): conditional simulation to $\log_{10}T$ data. Row (c): Final field obtained with the proposed method, with field (b1) as initial drift. Column 1: $\log_{10}T$ maps. Column 2: head level map (steady state). Column 3: Plot of computed vs. measured head level.

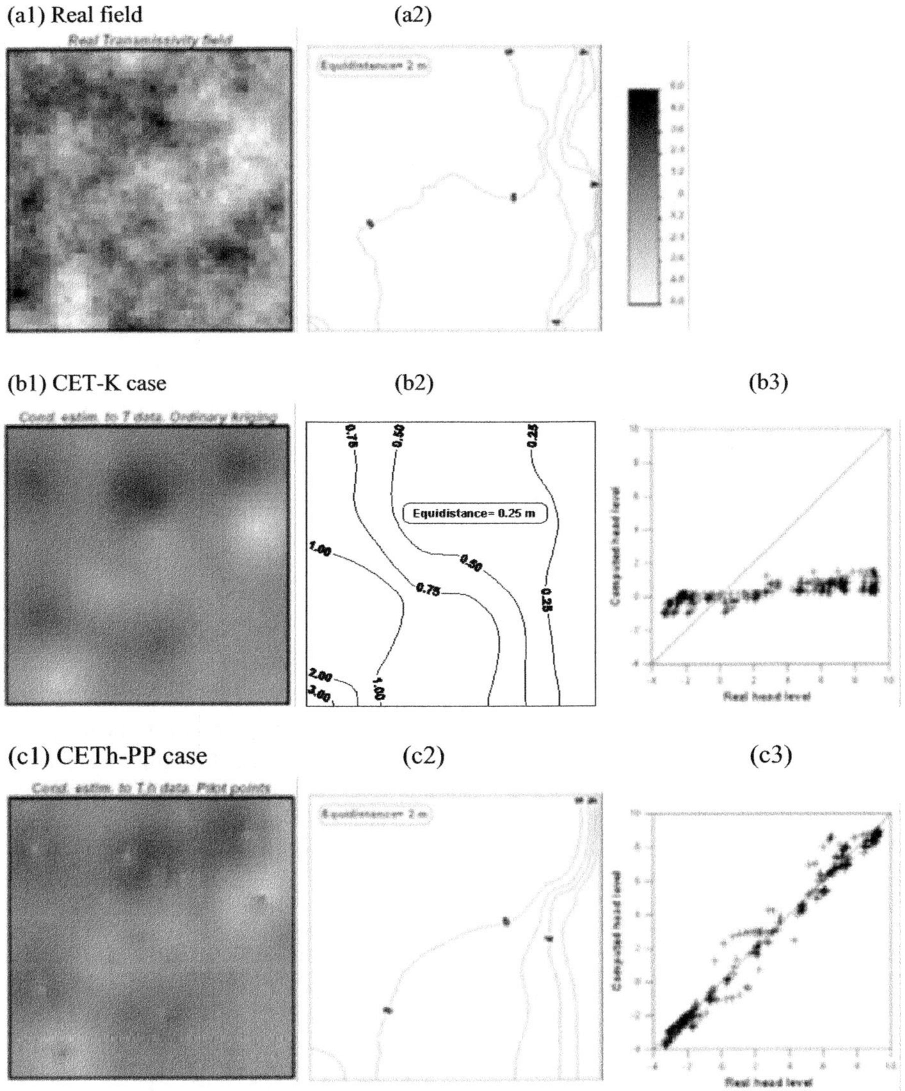

Figure 3a. Results concerning conditional estimation methods. Row (a): 'real' $\log_{10}T$ field and 'real' head level field (steady state). Row (b): conditional estimation to $\log_{10}T$ data using ordinary kriging. Row (c): conditional estimation to $\log_{10}T$ and head data using pilot points method. Column 1: $\log_{10}T$ maps. Column 2: head level map (steady state). Column 3: Plot of computed vs. measured head level.

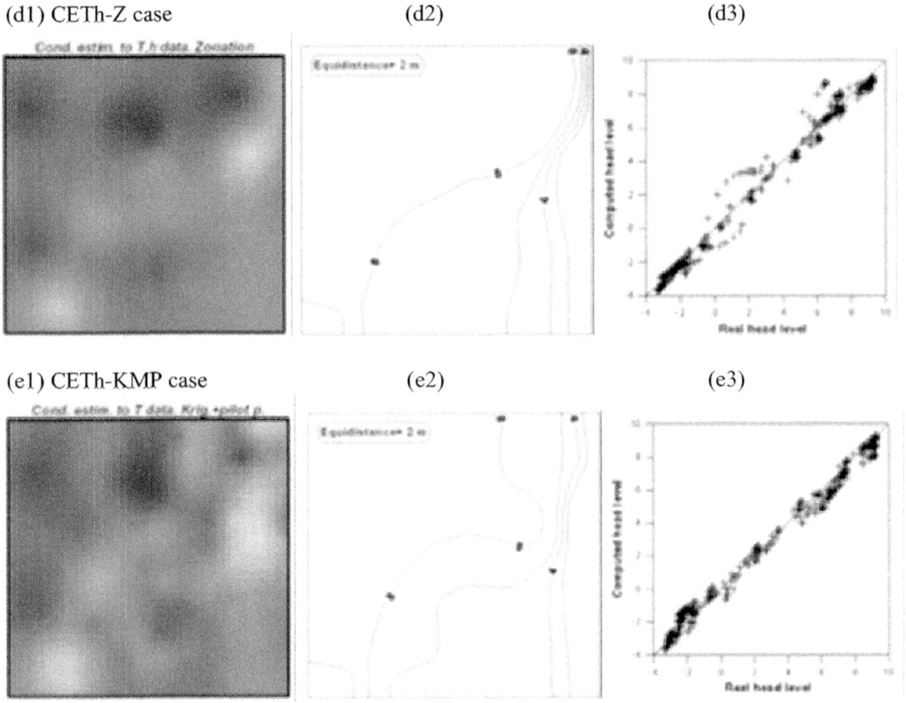

*Figure 3*b. Results concerning conditional estimation methods. Row (d): conditional estimation to $\log_{10}T$ and head data using the zonation approach. Row (e): conditional estimation to $\log_{10}T$ and head data using the proposed method, using a kriged field as initial drift. Column 1: $\log_{10}T$ maps. Column 2: head level map (steady state). Column 3: Plot of computed vs. measured head level.

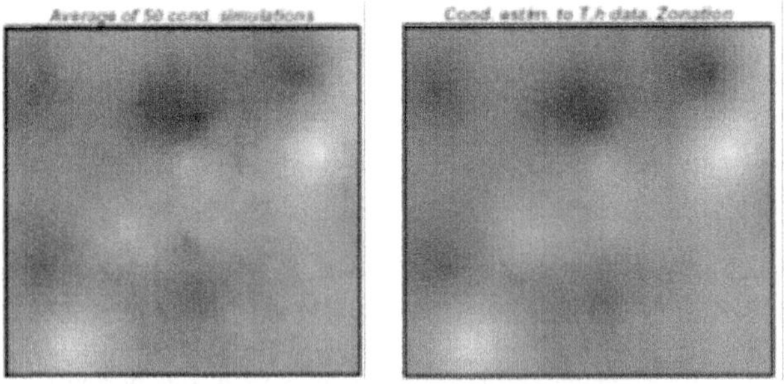

Figure 4. Comparison between the average field obtained with 50 conditional simulations to transmissivity and head level data and the $\log_{10}T$ map obtained through zonation (conditional estimation to $\log_{10}T$ and head data).

Considering mean deviation error, the suggested approach using kriging as initial drift displays a better behavior than the rest. All of the approaches yielded a mean deviation error under the standard deviation of the real field, showing, in general, a good performance.

The power of the suggested approach is shown considering head level fits (as calculated by J_h). In thirteen of the fifty conditional simulations, the suggested approach performed better than the zonation method, subjectively considered as the second best. Poor results of ordinary kriging are due to the fact of considering only $\log_{10}T$ measurements as conditioning data. In general, to consider both types of measurements as conditioning data improves the quality of the final estimation.

Table 2. Numerical comparison among the methods listed at Table 1.

Method	ME	MDE	J_h
CSTh (average 50 simul.)	0.33	1.87	282
Minimum CSTh	0.23	1.72	142
Maximum CSTh	0.44	1.97	791
CETh-KPP	0.10	1.53	206
CETh-PP	0.30	1.65	193
CET-OK	0.52	1.72	18200
CETh-Z	0.34	1.61	188

6. CONCLUSIONS

A modification of the self-calibrating method for generating equally likely realizations (conditional simulations) of the transmissivity field is presented. Final solutions honor measurements of transmissivity and dependent variables (heads, concentrations, etc.). Soft data (e.g. geophysics) can also de included in the conditioning procedure as an external drift.

Transmissivity field is defined as the superposition of a deterministic drift (obtained through kriging or conditional simulation), that honours log10T measurements

and reproduces spatial variability of the field being simulated and an uncertain perturbation field. The latter is optimized such that the final field also honours dependent variables measurements (heads in this work, although other type of measurements can be included easily).

Actual modifications consists of the addition of a penalty/regularization term in the objective function, considering plausibility of the model parameters, as well as the chance of using a kriged field as initial drift.

The algorithm is compared with the most frequently used conditional estimation methods (ordinary kriging, zonation and pilot points) on a set of synthetic data. The comparison is evaluated qualitatively and numerically. Both conditional estimation and conditional simulation approaches yielded good reproductions of the real system. The choice of the most appropriate method is somewhat subjective. It depends on whether the modeler seeks small-scale variability (conditional simulation) or large-scale trends (conditional estimation). However, single optimal estimate provided by conditional estimation should be used with caution for non-linear predictions. It is also (once more) corroborated that the inclusion of head measurements as conditioning data improves the quality (reduces the uncertainty) of the final estimation.

ACKNOWLEDGEMENTS

This work was funded by Spanish Nuclear Waste Management Company (ENRESA).

REFERENCES

1. Capilla, J. E., J. Gómez-Hernández and A. Sahuquillo, "Stochastic simulation of transmissivity fields conditional to both transmissivity and piezometric data- I. Demonstration on a synthetic aquifer", *Journal of Hydrology, 203, pp 175-188*, 1997.
2. Carrera, J. and S.P. Neuman, "Estimation of aquifer parameters under transient and steady-state conditions, 1, Maximum Likelihood Method incorporating prior information", *Water Resources Research, 22 (2), pp. 199-210,* 1986a.
3. Carrera, J. and S.P. Neuman, "Estimation of aquifer parameters under transient and steady-state conditions, 2, Uniqueness, Stability and Solution Algorithms", *Water Resources Research, 22 (2), pp. 211-227,* 1986b.
4. Carrera, J., " State of the Art of the Inverse Problem Applied to the Flow and Solute Transport Equations". Groundwater Flow and Quality Modeling (Custodio et al. Eds.), Riedel, Dordrecht, pp 549-583, 1987.
5. Carrera, J. and L. Glorioso, "On geostatistical formulations of the groundwater flow inverse problem", *Advances in Water Resources, 14, pp 273, 282 (1991)*
6. Certes, C. and de G. Marsily, "Application of the Pilot Point Method to the Identification of Aquifer Transmissivities", *Advances in Water Resources, 14, pp. 284-300 (1991)*
7. de Marsily, G., "De l'identification des systèmes en hydrogeologuiques (tome 1)", *Ph.D. Thesis, L'Univ. Pierre et Marie Curie-Paris VI, Paris, pp. 58-130 (1991)*
8. Gómez-Hernández, J., A. Sahuquillo and J.E. Capilla, "Stochastic simulation of transmissivity fields conditional to both transmissivity and piezometric data- I. Theory", *Journal of Hydrology, 203, pp 162-174*, 1997.
9. Kitanidis, P.K. and E.G. Vomvoris, "A geostatistical approach to the inverse problem in groundwater modeling (steady state) and one-dimensional simulations", *Water Resources Research 19(3), pp 677-690 (1983).*
10. McLaughlin, D. and LL. R. Townley, "A reassessment of the groundwater inverse problem", *Water Resources Research 32(5), pp 1131-1161, (1996)*
11. Yeh, W.W.G., "Review of parameter identification procedures in groundwater hydrology: The inverse problem", *Water Resources Research, 22 (2), pp. 95-108,* 1986.

2D PARTICLE TRACKING UNDER RADIAL FLOW IN HETEROGENEOUS MEDIA CONDITIONED AT THE WELL

C.L.Axness[1], J.J. Gómez-Hernández[1] and J. Carrera[2]

[1]*Departamento de Ingeniería Hidráulica y Medio Ambiente. Technical University of Valencia, 46071 Valencia, Spain. axness@upv.es, jaime@dihma.upv.es*
[2]*Department of Civil Engineering. ETSE Camins, Technical University of Catalunya, 03004 Barcelona, Spain. jesus.carrera@upc.es*

Abstract: The tracer test is one of the few experimental tools capable of estimating transport parameters at the local scale. Models used to estimate transport parameters (such as dispersivity) generally assume a homogeneous conductivity field. However, the distribution of the solute plume in heterogeneous media is primarily determined by the statistical nature of the hydraulic conductivity at the scale of the plume. We numerically simulate 2D transport of particles introduced into a steady injection well with prescribed head boundary conditions at the wellbore and at an exterior circle. We compute the travel time distribution of particles introduced into the well to points along a control circle of a given radius within a single transmissivity realization. In particular, we look at the effect of high, low, or average local wellbore transmissivity (as compared to the mean transmissivity of the domain) on the travel time distribution of each realization. We conclude that the difference between the logtransmissivity at the wellbore and the domain average logtransmissivity is likely to play an important role in the interpretation of dispersivity from conventional tracer tests.

X. Sanchez-Vila et al. (eds.), geoENV IV – Geostatistics for Environmental Applications, 187-198.

1. INTRODUCTION

The tracer test is the only means of directly obtaining an approximate in-situ measurement of the mass moment of a plume and estimating the associated transport parameters such as the dispersivity, yet the development of a solute plume under radial flow conditions near a well in heterogeneous media remains a poorly studied subject. The modeling of the hydraulic conductivity as a stochastic process or random space function (RSF) proved that plume moments (higher than second-order) were strongly influenced by heterogeneity at the scale of the plume. Methods used to estimate plume parameters from tracer tests generally assume spatial homogeneity. Recent work in the modeling of flow in heterogeneous media has established that the hydraulic head near a steady pumping well is strongly influenced (to first order) by the difference between the log-conductivity at the wellbore and the domain mean logtransmissivity (Axness and Carrera, 1999). In this article, we investigate the impact of this difference on the travel time distribution of particles instantaneously introduced into the wellbore of a steady injection well in a heterogenous porous media. We show this difference to be of first order importance in the computation of the spatial statistical moments of single realizations. We believe that this difference must be considered in the proper interpretation of tracer tests.

2. STATEMENT OF THE PROBLEM

We investigate two-dimensional mass transport in a heterogeneous medium using the method of particle tracking. The domain is a 2D fully saturated confined aquifer in which constant porosity is assumed. A constant head boundary condition, h_w, is prescribed at the wellbore, of radius r_w, and a smaller one, h_e at an exterior boundary along a circle of radius r_e. We assume a conservative solute, non-reactive with the medium. Other considerations made in this study include,

1. Particles are introduced directly at the edges of the discretization elements corresponding to the wellbore wall. Particles are not allowed to disperse back into the well in the case when local dispersion is applied.
2. The logtransmissivity is assumed multiGaussian and characterized by an exponential autocovariance function. The block transmissivity values are assigned as the point value of the transmissivity at the element centroid. This is discussed further in section 3.
3. The logtransmissivity at the wellbore is known.

A constant head on a circular outer boundary is prescribed for mathematical convenience in the computation of the hydraulic head and pore velocities. It could be criticized that the only real application in which this condition applies is that of a well in the center of a circular island. However, if the outer radius is sufficiently far from the well, the analysis of transport is relatively unaffected by such a boundary condition. The influence of boundaries on flow patterns under uniform flow conditions was investigated by Rubin and Dagan (1988). They found that under uniform mean flow conditions the hydraulic heads located two logtransmissivity correlation lengths from an imposed constant head boundary were approximately the same as those in an unbounded domain. This problem has also been investigated in Riva et. al. (2001), with similar conclusions. Although we have not done an exhaustive analysis to determine the distance at which the effect of the boundaries is negligible, we have set the control circle for transport three correlation lengths from the outside boundary. At this distance, the head distribution, as displayed in Figure 1, departs enough from the head at the outer boundary to consider that the boundary had a negligible effect at the control circle.

Under these conditions the transport equation is (Bear, 1972),

$$\frac{\partial C}{\partial t}+\nabla\cdot(\mathbf{v}C)=\nabla\cdot(\mathbf{D}\nabla C) \tag{1}$$

where $C(\mathbf{x},t)$ is the solute concentration, $\mathbf{D}$ is the dispersion tensor, and $\mathbf{v}$ is the pore velocity vector, with

$$\mathbf{D}=\begin{vmatrix} D_{xx} & D_{xy} \\ D_{yx} & D_{yy} \end{vmatrix} \tag{2}$$

where

$$D_{xx}=\alpha_L\frac{v_x^2}{v}+\alpha_T\frac{v_y^2}{v}+\mathbf{D}_m \tag{3}$$

$$D_{yy}=\alpha_T\frac{v_x^2}{v}+\alpha_L\frac{v_y^2}{v}+\mathbf{D}_m \tag{4}$$

and

$$D_{xy}=D_{yx}=(\alpha_L-\alpha_T)\frac{v_x v_y}{v}+\mathbf{D}_m \tag{5}$$

where $v=\sqrt{v_x^2+v_y^2}$ is the magnitude of the local pore velocity, v_x is the velocity component in the x-direction and v_y is the velocity component in the y-direction, α_L is the longitudinal dispersivity, α_T is the transverse dispersivity, and D_m is molecular diffusion.

As reported by Uffink (1985) and Kinzelbach (1990), in 2D, after the change of variables $v_x' = v_x + \partial D_{xx}/\partial x + \partial D_{xy}/\partial y$ and $v_y' = v_y + \partial D_{yx}/\partial x + \partial D_{yy}/\partial y$ the transport equation becomes the Ito-Fokker-Plank equation in two dimensions,

$$\frac{\partial c}{\partial t} + \nabla(\mathbf{v}'c) = \nabla^2(\mathbf{D}c) \tag{6}$$

which may be solved by particle tracking methods (Kinzelbach, 1990; Dagan, 1989; Wen and Kung, 1996). The particle position is given by

$$X(t+\Delta t) = X(t) + v'_x\,\Delta t + \frac{Z_1 v_x}{v}\sqrt{2\alpha_L v \Delta t} + \frac{Z_2 v_y}{v}\sqrt{2\alpha_T v \Delta t} \tag{7}$$

$$Y(t+\Delta t) = Y(t) + v'_y\,\Delta t + \frac{Z_1 v_y}{v}\sqrt{2\alpha_L v \Delta t} + \frac{Z_2 v_x}{v}\sqrt{2\alpha_T v \Delta t} \tag{8}$$

where $(X(t)\ ,Y(t))$ is the particle position at time t, Z_1 and Z_2 are two independent random deviates drawn from a Gaussian distribution with zero mean and unit variance.

The above development of the transport equation is general and applies to radial flow as well as Cartesian flow. In the case of radial flow, given that the average movement of the plume is either toward or away from the well, the longitudinal dispersivity is oriented along a radius extending from the center of the well while the transverse dispersivity is normal to this radius. In numerical simulations it is convenient to scale the elements so that elements far from the well are much larger than those close to it. For simulation of radial flow, the range of the local dispersivity near the well is greatly limited by the size of the elements and the fluid velocity in this region. Typically the local dispersivity is limited to be about 1/10 the element size in order to avoid non-physical behavior (particles backtracking) or numerical problems (particle jumping over elements). At distances far from the well, elements are large and the plume may be spread over a large area when the solute plume arrives at these elements. These are precisely the conditions under which stochastic theory and field experiments indicate an increasing dispersivity. In this case, the local dispersivity may compensate for a loss in spatial variability due to averaging logtransmissivity over larger element sizes. These conditions suggest the use of a local dispersivity that increases with element size (i.e., with radius from the well center). In this study we examine the influence of a local dispersivity that increases as a function of distance from the well. Specifically we use the model,

$$\alpha_i(r) = \alpha_{i1} + (\alpha_{i0} - \alpha_{i1})\exp[-(r - r_w)/l_i] \tag{9}$$

where i refers to the longitudinal (radial) or transverse (angular) local dispersivity type (L or T, respectively), α_{i0} is the i type local dispersivity at the wellbore, α_{i1} is the i type asymptotic local dispersivity ($r \rightarrow \infty$), l_i is a length scale in the i type direction r is the radius from the well center, and r_w is the well radius.

3. DESCRIPTION OF THE SIMULATIONS

Conditional realizations of logtransmissivity are drawn from a multiGaussian stationary random space function (RSF) of mean m_Y=-0.19, variance σ_Y^2=4.3, and isotropic stationary exponential covariance $C(\mathbf{s})=\sigma_Y^2 e^{-|\mathbf{s}|/lY}$, with a correlation length l_Y=15. The only conditioning datum is the logtransmissivity at the wellbore elements, the value of which is chosen to vary over several orders of magnitude coherently with the univariate distribution of the RSF. Each realization is characterized by the scalar value y_s which is defined as the difference between the logtransmissivity at the well and the spatial average of the transmissivity over the domain. Positive values of y_s indicate that the well is located in a zone of high transmissivity with respect to the rest of the domain, and conversely, negative values of y_s indicate that the well is located in a zone of low transmissivity.

Each realization is generated over a 2D annular domain of inner radius r_w=1 and outer radius r_e=100 (see, for instance, Figure 1) that has been discretized into 200 by 200 truncated sector elements in which the angle increment $\Delta\theta$ and the ratio of the outer to inner radius r_i+1/r_i are held constant. This discretization results in elements that are on the scale of r_ω/100 at the well to elements at the 10r_ω scale at the outer boundary. Given the non-uniformity of the grid, the generation of the field is carried out at the element centroids and the value generated at the centroid is assigned to the entire element regardless of its size. This approach maintains the statistical mean but introduces additional variability at the large outer elements when compared with the value that would have been obtained if the flow upscaled value had been computed. We have not investigated the impact of this increased variability for large elements, yet. Additionally, some artificial increase in the correlation scale at outer elements may be observed at the farthest distances from the well. Transport is analyzed only from the well up to elements of scale Δr=1.2. There is still a discrepancy in element sizes, which vary from Δr=0.01 to Δr=1.2, i.e., over two orders of magnitude. A more appropriate treatment of the generation of heterogeneous realizations over elements of varying support will be considered in our future work.

The conditioning value at the well has been chosen so that parameter y_s varies between -3 and 3 (in log units).

Flow is solved on each realization using the finite element code CFLOW (Axness, Carrera, and Bayer, 1998). Boundary conditions are prescribed heads h_e=0 at the outer radius, and h_w=10 at the well radius. Then, a modified version of the method by Cordes and Kinzelbach (1992) is used to obtain a mass conservative, continuous, velocity field in a sub-grid built after the division of each element in the original grid into four elements by the element diagonals. These velocities are bilinearly interpolated onto the simulation domain and used to track particles, with and without local dispersion using an adapted version of the constant displacement random walk code TRANSP by Wen and Kung (1996).

Particles are instantaneously introduced along the wall of the wellbore and tracked until they reach the control circle of radius r_c=55. The particle travel times to the control circle are recorded for all particles, and their mean, variance and distribution computed.

It should be noted that the travel time distribution is not the same as the concentration breakthrough curve at a single observation well. It is the distribution of particles arriving at a control circle rather than a single point.

4. DISCUSSION AND RESULTS

Figure 1 shows the entire annular problem domain including the transmissivity field, hydraulic head contour lines, and the particle tracks out to the control circle. The small white spot in the center of the annulus corresponds to the well. The set of black contour lines are the hydraulic head contour lines while the fine white lines are the particle paths for 200 particles introduced at the well. The transmissivity field statistics are discussed in the previous section. The transmissivity was conditioned at the well so that y_s=3, and no local dispersion was used in simulation (α_L=α_T=0). The effect of the heterogeneity is apparent in both the distortion of the head contour lines and the flow paths. Most notable is the fact that only a few particles enter low transmissivity zones and these particles themselves tend toward highly conductive paths. In this case, in which the well is conditioned to be in a highly transmissive zone, the hydraulic head contour lines are highly distorted and most of the drop in head occurs away from the well. Most of the particle transit time is in the area away from the well, which is of much lower transmissivity.

Figure 2 shows a plot of the transmissivity field, hydraulic head, and particle tracks in the area of the well for the same problem described above, again for a release of 200 particles. Note that in the lower left hand corner the particles avoid flowing through a low transmissivity region. Due to the correlation structure of the transmissivity field and the large gradient in the neighborhood of the injection well, the spatial distribution of particles in the well area is fairly uniform.

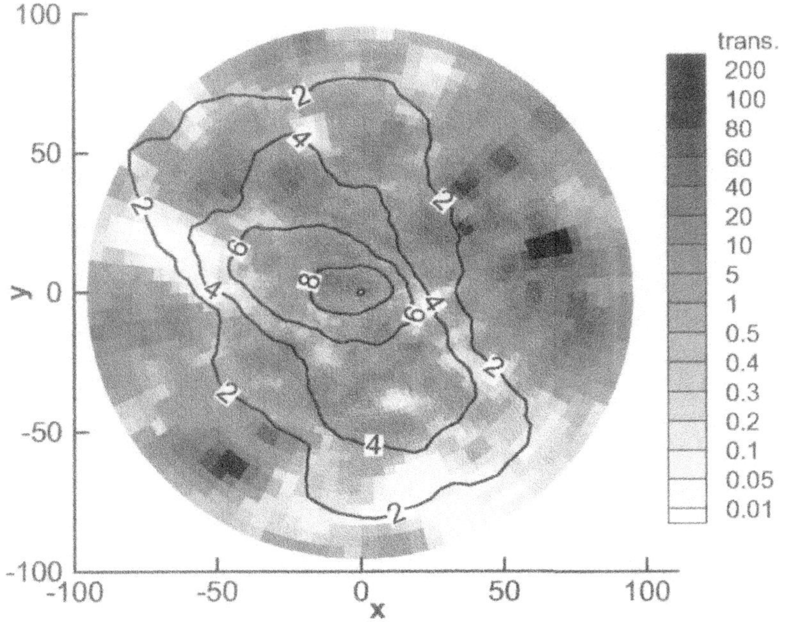

Figure 1. The entire domain including the transmissivity field, head contours and particle tracks for 200 particles. The domain is discretized into 40,000 sector elements, characterized by equal $\Delta\theta$, and increasing radial increments as the distance of the element from the well increases. Parameter values are y_s=3, r_w=1, r_e=100, h_w=10, and h_e=0 with the control circle at r_c=55. The realization was drawn from a RSF with statistics $\langle Y \rangle$=-0.19, σ^2_Y=4.3 and correlation length l_Y=15, conditioned to a wellbore value of 2.82. No local dispersion was considered in this simulation.

Figure 3 shows a plot of the particle tracks, transmissivity field and head contours in the area of the well for a realization with y_s= -3, that is, it has been conditioned to be much less transmissive at the well than the surrounding area. Additionally, radially-dependent local dispersion was added to the simulation. The parameters assumed for the radially-dependent dispersion were α_{L0}=0.01, α_{L1}=1, α_{T0}=0.01, α_{T1}=0.1, with l_L=l_T=10. Note that although the transmissivity field has the same spatial pattern for highs and lows as in the previous figure, the scale has changed about two orders of magnitude. Most of the head drop is now very close to the well, reflecting the lower (relative to the global mean) transmissivity in the area of the well. In this case most of the particle travel time is spent in the area of the well. The local dispersion tends to move the particles more erratically, with some particles now going through the low transmissivity zone in the lower left hand corner.

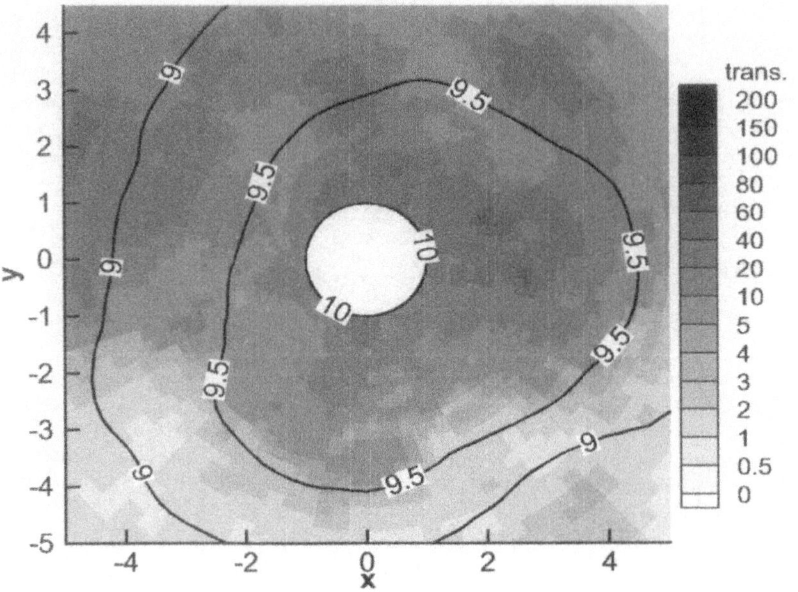

Figure 2. Close-up of the transmissivity field, head contours and particle tracks for 200 particles in the well area. The parameters for this figure are those given in Figure 1.

Figure 4 gives the travel time cumulative probability distribution function (cdf) for various values of the parameter y_s. The cdfs were computed from the travel times of 10,000 particles. We previously performed simulations (Axness et. al., 2002) that show that a few thousand particles are sufficient to converge to a stable cdf. The solid lines give the distribution in the case in which no local dispersion is considered while the dashed lines show results in which the previously described radially-dependent local dispersion is employed-although only for y_s=3 and y_s=-3. From the cdfs we observe that there is some tailing at the upper end of the probability distribution, similar to the tailing that is observed in tracer tests. It is of interest to note that this tailing is a product of only the heterogeneous behavior of the medium without the inclusion of matrix diffusion, which is often employed to explain tailing behavior. The behavior is more pronounced when the well transmissivity is lower than the domain mean (notice the logarithmic scale on the time axis), indicating that it is likely due to the injection of particles into zones of lower than average transmissivity in the area of the well. This leads one to speculate that tailing may be more pronounced in tracer tests conducted at wells with low local transmissivity as compared to wells with high local transmissivity as compared to the average field transmissivity.

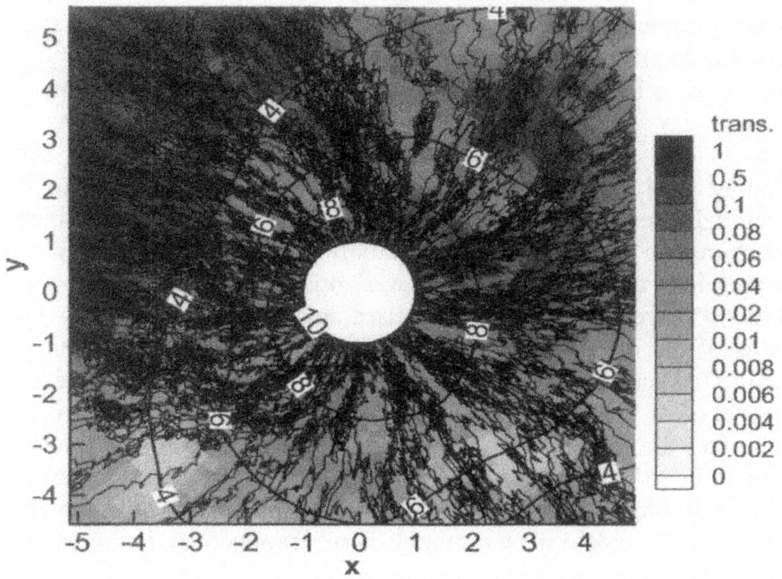

Figure 3. Close-up of the transmissivity field, head contours and particle tracks for 200 particles. The parameters of this figure are those of Figure 1 except that a radially-dependent local dispersion was assumed and the logtransmissivity value at the well was -3.19, so that y_s=-3. The radially dependent local dispersion parameters are given in the text.

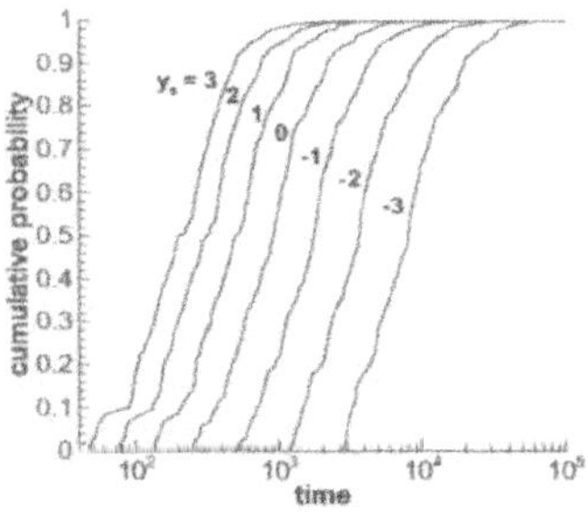

Figure 4. Comparison of the cumulative distribution function of travel times for simulations with different values of the parameter y_s. The solid lines correspond to simulations without local dispersion while the dashed lines represent the local radially dependent dispersive case with the parameter set of Figure 3.

The possibility of preferential channeling of the particles following paths of distinct transmissivities is apparent from Figure 4 for the case in which y_s>0 and no dispersion is introduced. Specially for y_s=2 and y_s=3, it is clear that the particles arrive from the well to the control circle in two distinct pulses. This multimodal characteristic of the cdfs is smeared out in the case dispersion is included and is less noticeable for the case in which y_s<0, which does not mean that particles cannot arrive to the control circle through distinct preferential areas in the domain, but with similar travel times through each preferential area.

Introducing local dispersion serves to smear out the curves but does not introduce any apparent bias. Within a regulatory context, the early mass arrival is one of the most critical parameters, our simulations show that including local dispersion does not modify the earliest particle arrivals, but substantially reduces the amount of mass arrived in the early times. Local dispersion also tends to increase the tailing of the curves, which is due to some particles dispersing into the low T zones,

Figure 5 gives the mean and standard deviation of particle travel times as well as of the inverse of the well discharge as a function of parameter y_s. Note the log scale used for the vertical-axes. The approximately exponential dependence of these three parameters on y_s over the range of y_s=-3 to y_s=1 emphasizes the importance of considering the relationship of the local well conductivity to that of the rest of the domain when modeling radial transport under non-uniform flow conditions. For larger values of y_s the dependence of the mean and standard deviation on y_s is sub-exponential, but remains strong.

5. CONCLUSIONS

We have developed a set of computer codes and modeling capabilities to explore the effect of heterogeneity on 2D radial transport from a steady-head injection well. We use these codes to explore the impact of y_s, i.e., the difference between the logtransmissivity at the wellbore, and the domain mean logtransmissivity, on particle travel time statistics to a control circle. We find an approximately exponential dependence of the mean travel time and standard deviation for wells that are less conductive than the effective mean logtransmissivity of the domain on y_s. This dependence is sub-exponential but remains strong for wells that are more conductive that the domain mean. We can conclude that the difference between the local conditions of conductivity at the well and the effective mean of the medium are important in radial flow transport problems, and we argue that it is likely that this parameter plays a key role in the interpretation of dispersivity from tracer tests.

We note that two dimensional transport realizations restrict the number of potential paths that the particles may take when compared to three dimensional transport. We also note, that for a better representation of a tracer test, the particles should be introduced into the well and the well modeled as a "mixing cell", instead of instantaneously distributing the particles at the outer wall of the wellbore. (The mixing cell approach will probably yield a smoother particle travel time distribution.) However, we believe that the main conclusions of this study regarding the influence of the parameter y_s on particle travel time statistics will hold in the case of a more realistic modeling of a tracer test.

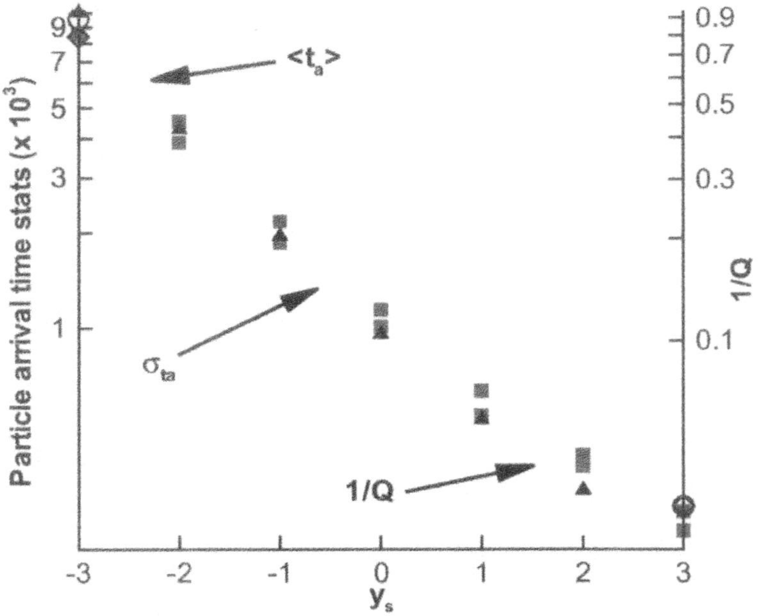

Figure 5. Mean, and standard deviation of travel times, and of the inverse of the well discharge, 1/Q, as a function of the parameter y_s. The solid line gives the mean travel time, the dashed line gives the standard deviation and the dash-dot line the inverse of the well discharge; all for the case of no local dispersion. The case of radially-dependent dispersion was solved only for the cases y_s=3 and y_s=-3, their means and standard deviations of travel time are represented by the open circles and diamonds, respectively.

Our future study will concentrate on the inclusion of a mixing cell model for the well, extension to pseudo-3D (stratified aquifer) geometries, the inclusion of spatially-variable retardation and porosity and change of scale. Our intention is to simulate as close as possible the impact that heterogeneity has on tracer tests both under reactive and non-reactive conditions.

REFERENCES

1. Axness, C.L., Gómez-Hernández J.J., Aliaga, R., Cassiraga, E.F., Guardiola-Albert C., Llopis C, Sahuquillo A., Capilla J.E., Rodrigo, J. (2002). Particle tracking in heterogeneous media under radial flow conditions: Presented at the *Gordon Conference on Flow and Transport in Permeable Media*, Boston, Mass., Aug. 4-9.

2. Axness C.L., Carrera, J. (1999). The 2D steady hydraulic head field surrounding a pumping well in a finite heterogeneous confined aquifer, *Math. Geology*, **31(7)**:873-906.
3. Axness, C.L., Carrera, J., Bayer, M. (1998). A new finite element formulation for solution of the hydrodynamic flow equation for convergent flow problems, *Computer Methods for Engineering in Porous Media Flow and Transport*, Sept. 28 - Oct. 2, Giens, France.
4. Bear, J. (1972), *Dynamics of fluid in porous media*, American Elsevier Publishing Company, Inc., NY.
5. Cordes, C., Kinzelbach, W. (1992). Continuous groundwater velocity fields and path lines in linear, bilinear, and trilinear finite elements: *Water Resour. Res.*, **28(11)**:2903-2911.
6. Dagan G., (1986), Statistical theory of groundwater flow and transport: pore to laboratory, laboratory to formation, and formation to regional scale. *Water Resour. Res.*, **22(9)**:120S-135S.
7. Dagan, G. (1989). *Flow and Transport in Porous Formations*, Springer-Verlag, NY.
8. Gelhar, L.W. (1993). *Stochastic Subsurface Hydrology*, Prentice-Hall, Englewood Cliffs, N.J.
9. Kinzelbach, W. (1990). The random walk method and extensions in groundwater modeling, in *Proc. Nordic seminar on groundwater modeling*, Randsvangen, Jevnaker, Norway, p. 1-25.
10. Riva, M.A., Guadagnini, A., Neuman, S.P., Franzetti, S. (2001). Radial flow in a bounded, randomly heterogeneous aquifer, *Transport in Porous Media*, **45**:139-193.
11. Rubin Y., Dagan, G. (1988). Stochastic analysis of boundaries effects on head spatial variability in heterogeneous aquifers, 1: constant head boundary, *Water Resour. Res.*, **24(10)**:1689-1697.
12. Sánchez-Vila, X., Axness, C.L., Carrera, J. (1999). Upscaling transmissivity under radially convergent flow in heterogeneous media, *Water Resour. Res.*, **35(3)**:613-622.
13. Uffink, G.J.M. (1985). A random walk method for simulation of macrodispersion in a stratified aquifer, *I.U.G.G. 18th Gen. Assem. Proc. Hamburg Symp*. I. A. H. S. Publ. **146**:103-114.
14. Wen X., Kung, C. (1996). A Q-basic program for modeling advective mass transport with retardation and radioactive decay by particle tracking, *Computers and Geosciences*, **21(4)**: 463-480.

COMPARISON OF GEOSTATISTICAL ALGORITHMS FOR COMPLETING GROUNDWATER MONITORING WELL TIMESERIES USING DATA OF A NEARBY RIVER

N. Barabás and P. Goovaerts

Department of Civil & Environmental Engineering, The University of Michigan, Ann Arbor, MI 48109-2125, U.S.A

Abstract: The benefit of river stage as secondary information in the kriging of groundwater level time-series is investigated for an unconfined, alluvial sand and gravel aquifer on Csepel Island in the Danube River. Factorial kriging analysis allows the filtering of the secondary information, which is then used in 3 forms: raw river data, the trend of the river data by itself or shifted by a well-specific lag time derived from river-well cross-correlograms. Cross-validation indicates that incorporation of river data using either kriging with an external drift or simple kriging with varying local means reduces the mean absolute error of prediction for 92% of the wells by an average of 18\% relative to ordinary kriging. The Danube's influence diminishes rapidly within the island, and two groups of wells are distinguished: one under the influence of the river and another, interior group. The kriging of the latter derives spurious benefit from the secondary information, possibly due to other seasonally varying influences.

X. Sanchez-Vila et al. (eds.), geoENV IV – Geostatistics for Environmental Applications, 199-210.

1. INTRODUCTION

Various physical models of stream-aquifer interactions have been developed and tested for alluvial aquifers (*Govindaraju* and *Koelliker*, 1994). However, detailed information on the hydraulic parameters of the aquifer are required for many of these models. When such information is inadequate, statistical models are an alternative that can enhance physical modeling. Physical interactions in the environment lead to statistical correlations in the data offering an opportunity to use better sampled variables in the estimation of missing values of the variable of interest even when no physical model is available. Csepel Island is a large 257 km^2 island in the Danube River, Hungary. In the north of the island, the Danube splits into two unequal branches, and groundwater flow direction changes according to river stage in the main branch. Well head at 38 monitoring wells (Figure 1) has been measured quarterly and river stage daily over 13 years. As well measurements do not coincide and are not sufficiently dense in time, data interpolation is necessary to assess groundwater flow frequency from various directions at each well (which will facilitate future identification of pollution sources).

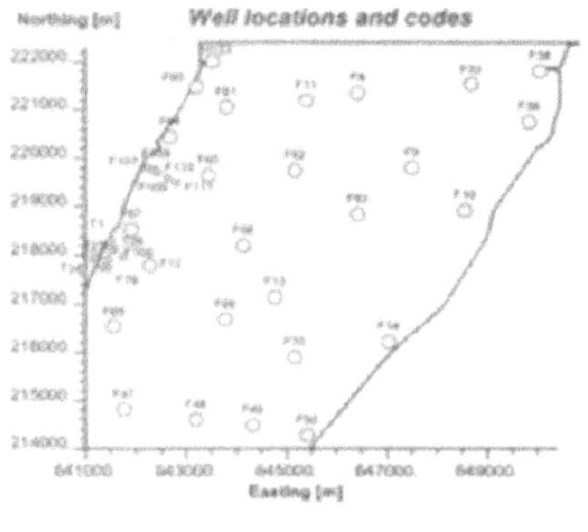

Figure 1. Csepel Island is located in the Danube River, just south of Budapest (Hungary). The site is in the upper half of the island and contains 38 monitoring wells.

Fluctuations in river stage induce corresponding responses in the aquifer that depend on diffusivity, porosity, and change in river stage and associated time frame (*Serrano* and *Workman*, 1998). While the oscillating signal may arrive almost unaltered at wells that are immediately adjacent, it will likely be dampened and delayed by the intervening aquifer matrix for more remote wells. In the extreme situation, some wells may fall completely outside the range of river influence. Therefore, different forms of river data may be best correlated with different wells: raw river data for wells next to the river, or, for remote wells, the trend of the river possibly shifted by a well-specific lag time.

The objective of this paper is to investigate how the correlation between groundwater levels and river stage can be used to interpolate groundwater level time-series at the wells. The first step is to retrieve from river data the information that is best correlated with well data. Factorial kriging analysis is a filtering technique used to estimate the trend of river data (*Wackernagel*, 1995), while river-to-well cross-correlograms serve to derive well-specific response lag times. *Rouhani* and *Wackernagel* (1990) adopted a similar approach and used cross-semivariograms and factorial kriging analysis to model the correlation between well time-series. Here the correlation of wells is assessed relative to a river instead. Simple kriging and kriging with an external drift then allow incorporation of the secondary information into the temporal interpolation of groundwater levels at the wells (*Goovaerts*, 1997). The most appropriate form of the secondary information likely varies with distance to the river. It is also possible that the river provide no benefit at all, in which case ordinary kriging of the well data alone may yield better predictions. This spatial variation of the results can help formulate criteria that define the influence of the river. With the ability to *a priori* choose the best approach for individual wells, the modeling of time-series over the rest of the island is facilitated.

2. SITE HYDROLOGY

The island consists of alluvial deposits that form a sand and gravel unconfined aquifer 3-5 m below finer sand and silt deposits. The aquifer itself ranges between 3 and 10 m in thickness under the study area. The main river branch (henceforth the Danube or the river) flows towards the south along the western edge of the island and has a mean annual volumetric flow rate of 2,380 m^3/s. The minor branch in the east has a substantially smaller flow rate at 3.5 m^3/s. This branch is controlled to maintain a stable stage at two dams at both ends of the island. While the stage in the Danube can fluctuate over several meters, the fluctuations in the minor branch do not exceed a few centimeters, and acts as a constant head boundary. Earlier studies, commissioned by the city waterworks have shown that the oscillations of the Danube have a strong influence on groundwater flow and cause its direction to change over a wide range.

River and groundwater data were made available by the waterworks of the city of Budapest and by the Hungarian Water Authority. River stage in the Danube was measured daily from 1984 to 1997. The hydrograph (Figure 2) shows that river stage fluctuates over a range of 8 m, and the maximum daily change was 1.9 m over the 14-year sampling period.

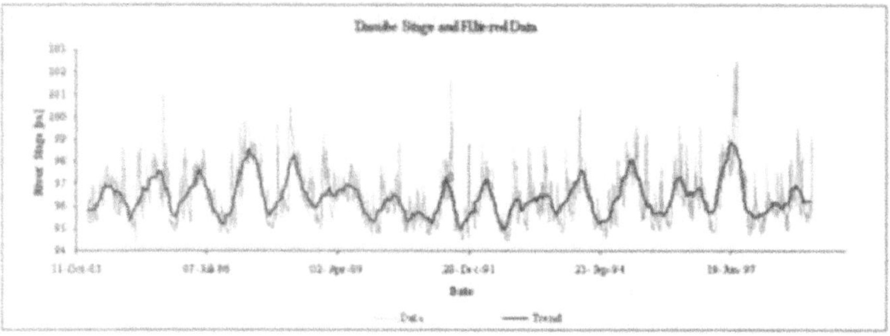

Figure 2. Hydrograph of river stage from 1984 to 1996. The black line is the trend of the hydrograph as estimated by kriging of the local mean.

In the same time period, 38 wells were sampled quarterly. Overall, average groundwater levels tend to increase towards the minor branch (stage constant at 96.5 m) and the north, while the well clusters near the main branch (west) display heterogeneity due to their proximity to extraction wells. The correlation between well and river data, $\bar{\rho}_D(0)$, ranges from 0.15 to 0.94 and, as expected, it decreases with distance to the river.

3. GEOSTATISTICAL ANALYSIS

The following sections describe the successive steps of the analysis: calculating the trend of the river data by kriging of the local mean, assessment of the lag times Δt with which the river stage signal arrives at the wells, and calculation of the residuals used in kriging at each well. Kriging of temporal data is closely related to other autoregressive models, e.g. the kriging system introduced here is the same as the Yule-Walker equations used in time-series modeling (*Bras and Rodriguez-Iturbe*, 1985).

3.1 Kriging of the Local Mean

The first step is the description of the temporal pattern of river stage data $D(t)$ and their decomposition into temporal processes on the basis of the nested semivariogram model $\gamma_D(\tau)$ (*Goovaerts*, 1997). The experimental semivariogram of stage data in Figure~3 shows strong oscillations and is modeled as a combination of an exponential (Exp), and a dampened hole-effect (HE) model (*Wackernagel*, 1995) defined as:

$$\mathrm{HE}\left(\frac{|\tau|}{a}, d\right) = 1.0 - \exp\left(\frac{-3\tau}{d}\right)\cos\left(\frac{\tau}{a}2\pi\right) \tag{1}$$

where τ is the lag time, a is the period, and the dampening effect d is the time in which the oscillation diminishes by 95%, with $d \geq 0$.

The complete expression for the Danube's semivariogram model, fitted using weighted least squares regression, is:

$$\begin{aligned} \gamma_D(\tau) &= \gamma_{\text{Exp}}(\tau) + \gamma_{\text{HE}}(\tau) \\ &= 0.98 \text{ Exp}\left(\frac{|\tau|}{77 \text{ days}}\right) + 0.35 \text{ HE}\left(\frac{|\tau|}{367 \text{ days}}, d = 44 \text{ days}\right) \end{aligned} \quad (2)$$

The period of 367 days reflects an expected agreement with the one-year cycle of seasonal variations.

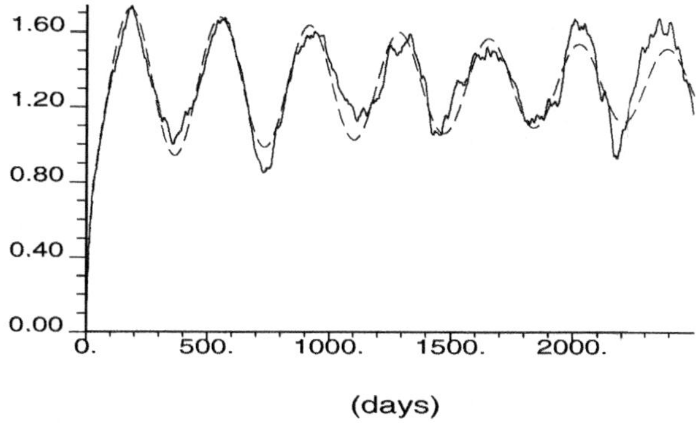

Figure 3. Experimental (solid line) semivariogram of the Danube data and the fitted model (dashed line).

Under model (2), river stage $D(t)$ can be decomposed as the sum of two independent, zero-mean, temporal processes D_{Exp} and D_{HE} corresponding to the basic semivariogram models, plus a trend component $m_D(t)$:

$$D(t) = D_{\text{Exp}}(t) + D_{\text{HE}}(t) + m_D(t) \quad (3)$$

In this paper, the well data are better correlated with the trend component m_D than with the two other processes D_{Exp} and D_{HE}. Hence, the following presentation is limited to the estimation of that component using kriging (*Matheron*, 1982, *Rouhani and Wackernagel*, 1990 and *Goovaerts et al.*, 1993). The mean at time t, $m^*_{\text{OK}}(t)$ is estimated as:

$$m^*_{\text{OK}}(t) = \sum_{\alpha=1}^{n(t)} \lambda_{\alpha m}^{OK}(t)\, D(t_\alpha) \quad (4)$$

where the kriging weights are the solution of the following system of ($n(t)$ + 1) linear equations:

$$\sum_{\beta=1}^{n(t)} \lambda_{\beta m}^{OK}(t)\,\gamma_{\mathrm{D}}(t_\alpha - t_\beta) + \mu_m^{OK}(t) \quad = \quad b_{\mathrm{Exp}} + b_{\mathrm{HE}} \qquad (5)$$

$$\alpha = 1,\ldots n(t)$$

$$\sum_{\beta=1}^{n(t)} \lambda_{\beta m}^{OK}(t) = 1$$

The right-hand side semivariance terms b_{Exp} and b_{HE}, in system (5) are the semivariogram sills, since the deterministic trend is uncorrelated with the data.

In practice, the kriging of aligned data may result in the so-called string effect (where "string" refers to one-dimensional data series such as time-series), whereby disproportionately high kriging weights are assigned to data at both ends of the search window (Figure 4). The reason is that these data don't have two contiguous neighbors, and thus they are considered as less redundant or more informative. Following *Deutsch* (1993), the correction involves wrapping the search window to a "circle" so that all data are equally redundant. So, after correction, the local mean is simply the arithmetical average of data within the search window.

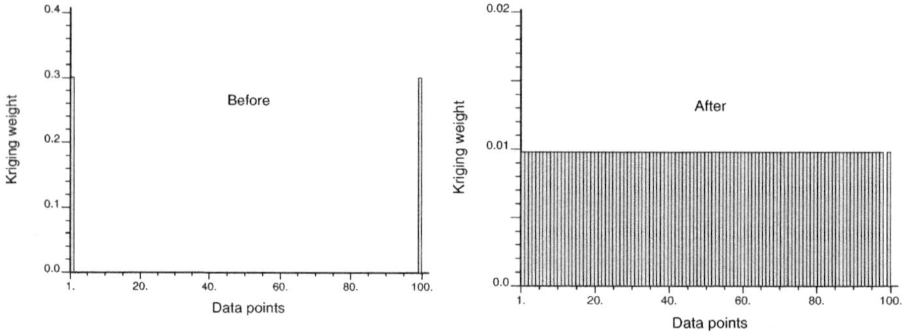

Figure 4. Kriging weights assigned to Danube data for the estimation of the trend before and after the string effect correction. Each bar represents the weight for one data point in a search window of 100 points, and the estimation is performed at the center of the window.

3.2 Calculation of well-lag times

When river stage changes, such as after precipitation events in the watershed or changes in dam operation, the river-aquifer head gradient changes and initiates a signal to which the aquifer responds. The signal then travels as a function of transmissivity through the aquifer. Within the aquifer, the wave is dampened and it dissipates or it is overcome by a possibly stronger signal. A statistical well-lag time can be calculated, which indicates how long such signals take to reach an individual well on average, given the history of river stage fluctuations over the observed period of time. A signal that is delayed will also be dampened, however, so the response is calculated relative to the trend of the river data rather than the daily data.

A delayed response to the influence of the Danube can be observed at several wells. The lag-time Δt for which the Danube trend $m_D(t - \Delta t)$ is the best correlated with the well data $z(t)$ is derived from the experimental cross-correlogram computed as:

$$\hat{\rho}_{m_D}(\tau) = \frac{\widehat{Cov}(\tau)}{\sqrt{\hat{\sigma}^2_{m_D} \cdot \hat{\sigma}^2_z}} \in [-1, +1] \tag{6}$$

with

$$\widehat{Cov}(\tau) = \frac{1}{N(\tau)} \sum_{\alpha=1}^{N(\tau)} m_D(t_\alpha - \tau) \cdot z(t_\alpha) - m_{m_D} \cdot m_z$$

$$\hat{\sigma}^2_{m_D} = \frac{1}{N(\tau)} \sum_{\alpha=1}^{N(\tau)} [m_D(t_\alpha - \tau) - m_{m_D}]^2 \qquad \hat{\sigma}^2_z = \frac{1}{N(\tau)} \sum_{\alpha=1}^{N(\tau)} [z(t_\alpha) - m_z]^2$$

$$m_{m_D} = \frac{1}{N(\tau)} \sum_{\alpha=1}^{N(\tau)} m_D(t_\alpha - \tau) \qquad m_z = \frac{1}{N(\tau)} \sum_{\alpha=1}^{N(\tau)} z(t_\alpha)$$

where $\hat{\sigma}^2_{\mathrm{mD}}$ and $\hat{\sigma}^2_{\mathrm{z}}$ are the variances of tail m_D-values and head z-values and m_{mD} and m_z are the respective means.

For example, at well F12 the correlation first increases and reaches a maximum of 0.55 at 49 days, which is a 1.34-fold improvement over the zero-lag correlation $\bar{\rho}_{\mathrm{mD}}(0)$, see Figure 5. Considering the well's distance to the river and the spatial trend of lag at other wells close to the river, the lag of 49 days is very long. Similarly high lags at two other wells near the river suggest that conductivity in the sand and gravel aquifer declines, slowing groundwater flow in a region approximately 1 km from the river.

Contrary to expectations, Δt does not exhibit a clear and consistent spatial pattern, except for wells within about 1-2 km of the river. Many wells in the middle of the island show no lag at all. This implies that the influence of the Danube is felt only a certain distance into the island beyond which lag-times may be spurious. Wells F13, F14, F69, F70, F62 and F63 show intermediate correlation (0.41-0.57) but no lag time. This may be due to other seasonally varying processes such as irrigation or evapotranspiration from an adjacent 8 km^2 forested area, which they enclose.

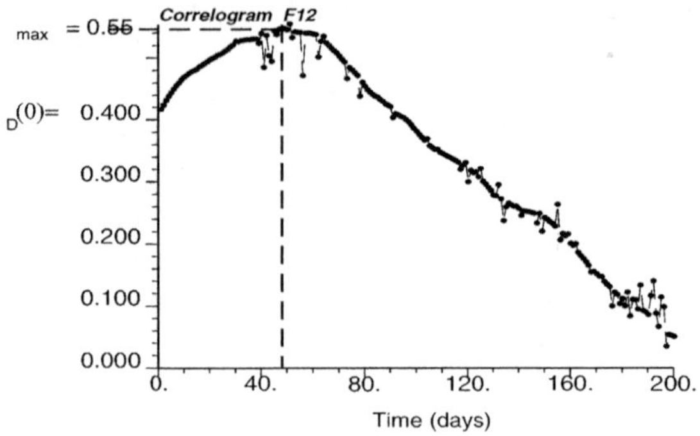

Figure 5. Cross-correlogram between the Danube trend and groundwater level at well F12, with a maximum correlation reached for a lag of 49 days.

3.3 Kriging of Time-Series

Various kriging algorithms are available to interpolate the well time-series. Three techniques are compared in this paper: ordinary kriging (OK) using only well data (reference technique), kriging with an external drift (KED) and simple kriging with varying local means (SKlm), both of which allow incorporation of Danube data. The theory of kriging is explained in detail in *Goovaerts* (1997), and only differences between the relevant kriging algorithms are stressed in this section.

All kriging estimates can be viewed as variants of the basic linear regression estimate:

$$z^*(t) - m(t) = \sum_{\alpha=1}^{n(t)} \lambda_\alpha(t)[z(t_\alpha) - m(t_\alpha)] \tag{7}$$

where $z^*(t)$ is the estimated variable at time t and λ_α is the weight assigned to the observation at time t_α. Kriging algorithms vary in their treatment of the trend $m(t)$ of the primary variable.

In OK, the mean is unknown and constant within the search window $W(t)$: $m(t) = m(t_\alpha)$: m unknown $\forall\ t_\alpha \in W(t)$. Since the kriging weights $\lambda_\alpha(t)$ sum to 1, the mean is filtered from the linear estimate (7), hence it is not directly involved in the estimation.

In both SKlm and KED the mean is modeled as a function of a secondary variable $y(t)$. In SKlm, this function $f(y(t))$ is determined prior to kriging, and its parameters are assumed globally constant. For example, a linear function leads to the following definition of the trend:

$$m_{SKlm}(t) = f(y(t)) = a + b\ y(t) \tag{8}$$

In the case of KED, the two trend coefficients a and b are estimated within each search window during kriging, allowing a local re-evaluation of the relationship between primary and secondary variables (*Wackernagel*, 1995):

$$m_{\text{KED}}(t) = f(y(t)) = a(t) + b(t)\,y(t) \tag{9}$$

with $a(t)$ and $b(t)$ constant within the search window. Unlike SKlm, the function $f(y(t))$ must be linear, which might require a prior transform of the data. For each well, three types of secondary information are considered here: the raw Danube data ($y(t) = D(t)$), and the trend of the Danube data ($y(t) = m_D(t)$), possibly shifted by a lag Δt, ($y(t) = m_D(t\text{-}\Delta t)$).

Both KED and SKlm, require a model for the semivariogram of residuals $r(t)$ between the primary variable $z(t)$ and its trend $m(t)$: $r(t) = z(t) - m(t)$. At each well, semivariograms were estimated and modeled for the groundwater levels and all possible residuals. For the groundwater levels, a hole-effect model with a yearly period was fitted at wells within 1-2 km of the Danube. Three wells in the interior of the island (F10, F69, F70) also showed a pronounced hole-effect, which, however, cannot be attributed to the Danube due to their distance (3-5 km). Other factors might be responsible, for example seasonal agricultural extraction. Figure 6 shows semivariograms for the raw data and two kinds of residuals at well F110. As expected, removing the Danube trend lowers the sill (smaller residual variance) and filters out oscillations. This pattern is observed at all other wells near the river.

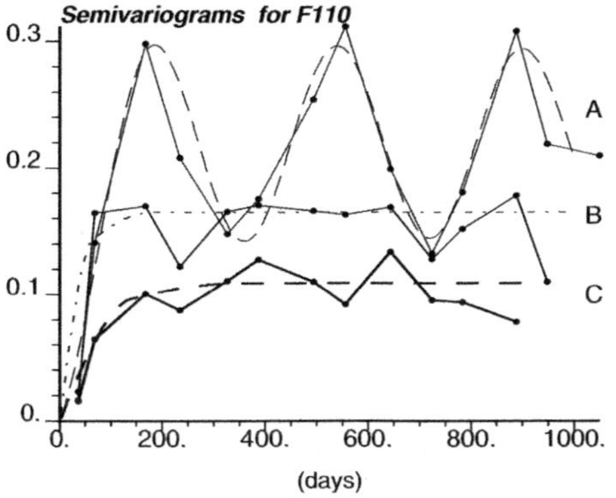

Figure 6. Experimental (solid lines) and model semivariogram (dashed lines) of the F110 data (A), and the residuals with the Danube (B) and the shifted Danube trend (*mD*(*t*-Δ*t*)) (C).

3.4 Performance comparison

The prediction performance of each combination of kriging technique and secondary information is assessed using the mean absolute error (MAE) of prediction obtained by cross-validation, whereby each observation is removed one at a time and is estimated using the remaining ones. The combination with the lowest MAE is then applied at each well to estimate daily groundwater levels. These estimates are in general quite close to the actual values, and the estimated time-series of all wells follow the sample time-series closely and reproduce the fluctuations. The magnitude of MAE in all cases, including OK, ranges from 0.06 m to 0.52 m, with a mean of

0.21 m. The histogram of MAE values over the site is bimodal, suggesting, again, the existence of two distinct groups. Not surprisingly, MAE is greater when the time-series variance increases (and the distance to the river decreases).

The benefit of the secondary information is measured by the following ratio: R = $(MAE_{OK} - MAE\text{min}) / MAE_{OK} \in [0,1]$, where MAE_{min} is the smallest MAE among OK and the 6 combinations of secondary information and kriging technique. Figure 7 (top graph) shows that all wells (except 3) benefit somewhat from the Danube data.

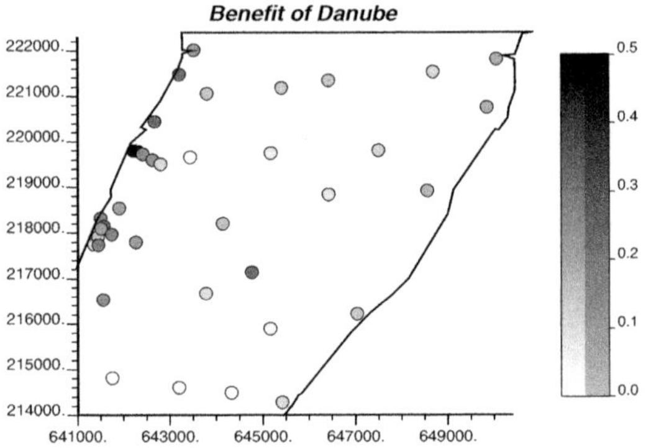

Figure 7. Kriging results: the benefit of incorporating secondary information as measured by *R*.

As expected, the gain is greatest near the Danube (e.g. best is well F107, with $R = 0.49$), and diminishes further inland showing a similar spatial trend as $\bar{\rho}_D(0)$, with an average value of 0.18 or 18% improvement over OK. Higher values are found along the minor river branch. The least improvement is seen in the southern wells. The three wells where OK performs best ($R = 0$) are scattered through the site. One of them (F65) is found quite close to the Danube and its semivariogram shows a pronounced hole-effect, yet its $\bar{\rho}_D(0)$ of 0.29 is very low. This could be due to low conductivity, geology, extraction activities nearby or well maintenance problems. There are two other wells in immediate proximity to the river (T24, T25) that do not show a great reduction in MAE with the inclusion of the Danube data. Their time-series have a high variance, yet the $\bar{\rho}_D(0)$ of 0.4 is weak as is the benefit the wells derive from the river data. The likely reason for this behavior is their proximity to extraction wells. Measurement error and well-maintenance problems (siltation) are also possible. F13 in the center of the island has a high benefit compared to surrounding wells. Good connectivity with the Danube is not a likely explanation since the lag-time is zero for this well. This result could be spurious, or it may indicate private extraction activities or other seasonal effects such as evapotranspiration. The best secondary information at each well is mapped in Figure 8, left graph. The Danube is best for adjacent wells, while the lagged trend is best for wells that are close (1-2 km) but not adjacent. This is expected: as the Danube signal travels through the aquifer, it is dampened and delayed. In the interior of the island, the Danube trend alone is best in most cases, but patterns are more random: there are wells that benefit most when the lag is included (F11, F12, F39) while others benefit most from the unfiltered river data. Thus, the benefit in the interior, though real, reflects geological heterogeneities or influences that also follow the seasonal cycle embodied by the Danube. From these results, we can conclude that the influence of the Danube diminishes rapidly indeed, giving rise to two groups as indicated by earlier observations as well: one near enough to the river to be influenced by it and another in the island's interior outside of the river's effect.

The best kriging technique tends to be SKlm in the north and KED in the south (Figure 8, right). The only known physical distinction between the northern and southern halves of the site is that the north is mostly urban in character, while the southern half is agricultural. This information is not enough to explain the north-south differentiation of best kriging technique. Nevertheless, even if there were a physical reason, when KED performs better, it does so by a smaller margin than SKlm: the average MAE improvement of KED over SKlm is 25%, while the gain of SKlm over KED is significantly greater at 40%.

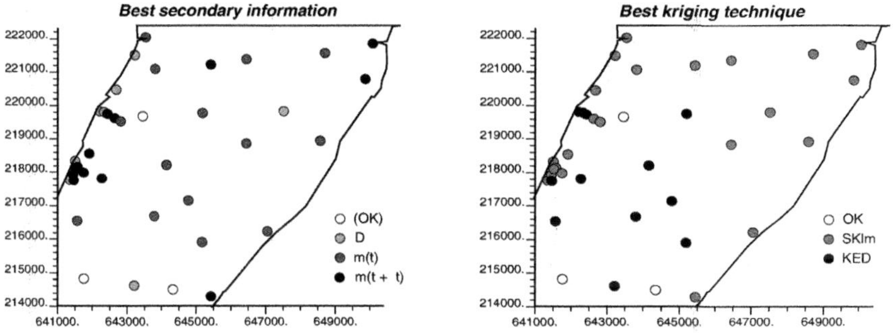

Figure 8. Kriging results: best secondary information (left) and best kriging technique (right) at each well as indicated by MAE_{min}.

4. CONCLUSIONS

Traditional models of stream-aquifer systems involve the solution of the Laplace or Boussinesq equations using the hydraulic parameters of the system (*Govindaraju* and *Koelliker*, 1994). In this paper, we show that a model of individual well time-series can be developed based solely on the statistical relationship between stream and aquifer data. Geostatistics provides tools to interpolate quarterly sampled time-series of groundwater levels using daily measurements of the stage of a nearby river as supplementary data. The first step is to process the river data (filtering, lagging) in order to extract the information best correlated with well data, which is then incorporated using kriging variants. Cross-validation shows that accounting for river data improves the prediction for 92% of the wells. Such "black box" modeling as presented here does not replace physical models, but offers an opportunity when hydrogeological data are insufficient or lacking. It can reveal hydrological connectivities that can also be applied as soft data or as elimination criteria in data fusion exercises as implemented by *Poeter* and *McKenna* (1995) and *McKenna* and *Poeter* (1995).

The results of statistical and geostatistical analyses as presented above could be integrated into a flow/transport model through refined hydrogeologic characterization (assignment of hydrologic conductivities and/or geologic categories in the model), by indicating which wells are in close hydrologic connection, and how far the river's influence extends.

REFERENCES

1. Bras, R.L. and Rodriguez-Iturbe, I., 1985. Random Functions and Hydrology. Addison-Wesley, Reading, Massachusetts.
2. Deutsch, C.V., 1993. Kriging in a finite domain. Math. Geol., 25: 41-52.
3. Goovaerts, P., 1997. Geostatistics for Natural Resources Evaluation. Oxford University Press, New York.
4. Goovaerts, P., P. Sonnet, and A. Navarre, 1993. Factorial kriging analysis of springwater contents in the Dyle river basin, Belgium. Water Resour. Res., 29 (7): 2115-2125.
5. Govindaraju, R.S., and J.K. Koelliker, 1994. Applicability of linearized Boussinesq equation for modeling bank storage under uncertain aquifer parameters. J. Hydrology, 157: 349-366.
6. Matheron, G, 1982. Pour une analyse krigeante de données régionalisées. Internal note N-732, Centre de Géostatistique, Fontainebleau.
7. McKenna, S.A., and E.P. Poeter, 1995. Field example of data fusion in site characterization. Water Resour. Res., 31 (12): 3229-3240.
8. Poeter, E.P., and S.A. McKenna, 1995. Reducing uncertainty associated with ground-water flow and transport predictions. Ground Water, 33 (6): 899-904.
9. Rouhani, S., and H. Wackernagel, 1990. Multivariate geostatistical approach to space-time data analysis. Water Resour. Res., 26 (4): 585-591.
10. Serrano, S.E., and S.R. Workman, 1998. Modeling transient stream/aquifer interaction with the non-linear Boussinesq equation and its analytical solution. J. Hydrol., 206: 245-255.
11. Wackernagel, H., 1995. Multivariate Geostatistics: An introduction with applications, Springer-Verlag, Berlin.

A GEOSTATISTICAL MODEL FOR DISTRIBUTION OF FACIES IN HIGHLY HETEROGENEOUS AQUIFERS

L. Guadagnini[1], A. Guadagnini[1] and D.M. Tartakovsky[2]

[1]*D.I.I.A.R., Politecnico di Milano. Piazza Leonardo da Vinci, 32, 20133 Milano (Italy)*

[2]*Theoretical Division, Los Alamos National Laboratory, Los Alamos, (USA)*

Abstract: We analyze flow in a heterogeneous aquifer composed of different geologic facies, whose hydraulic properties and internal geometries are uncertain. Our analysis employs random domain decomposition to derive robust moment equations for flow in composite porous media. The approach accounts explicitly for the multi-scale effects of material and geometric uncertainties on the ensemble moments of head and flux. We use an indicator-based geostatistical methodology to estimate the facies geometries and to quantify the corresponding uncertainty. We then apply our approach to a synthetic flow example, where stratigraphic and sedimentological data from a real aquifer are used to obtain the probabilistic facies distribution. We solve the equations for ensemble moments of hydraulic head and study the impact of unknown geometry of materials on the statistical moments of head.

1. INTRODUCTION

Uncertainty in hydraulic and transport parameters of natural geologic formations is conveniently accounted for by treating them as random fields. Consequently, flow and transport equations become stochastic. Much of the existing literature on stochastic hydrogeology deals with mildly heterogeneous formations, where variance of log-conductivity, σ_Y^2, is

X. Sanchez-Vila et al. (eds.), geoENV IV – Geostatistics for Environmental Applications, 211-222.

relatively small. While this assumption is crucial for closing the moment differential equations or for making Monte Carlo simulations manageable, it clearly limits the applicability of most such analyses. A recently proposed method of random domain decomposition [*Winter and Tartakovsky*, 2000, 2002] provides a general framework for modelling flow and transport in highly heterogeneous porous media consisting of multiple materials, by quantifying uncertainty in both spatial arrangement of geological facies and hydraulic properties within each facies. Since perturbation expansions are carried out within each facies separately, their accuracy and robustness remain high for most geological settings. The main unresolved challenges in applying the random domain decomposition (RDD) are the quantification of uncertainty (randomness) in a spatial arrangement of geologic facies from experimental data and the evaluation of random functional integrals. Specifically, so far there is no clear way to parametrize geometric uncertainty and to calculate integrals over random domains. Previous studies dealt with relatively simple material distributions [e.g., *Winter et al.*, 2002]. Here we use a typical data set from the alluvial aquifer system of the city of Bologna in Northern Italy and indicator geostatistical techniques to estimate the boundary between contrasting materials and to quantify the corresponding uncertainty. In particular, we utilize stratigraphic and sedimentological data to reconstruct the spatial extent of the aquitard that separates an upper contaminated aquifer from deeper aquifers. We then use this information as an input for a synthetic flow problem and solve the equations for ensemble moments of hydraulic head and study the impact of unknown geometry of materials on statistical moments of head.

2. METHODOLOGY

2.1. Composite medium model with RDD

Consider steady-state flow equation, $\nabla \cdot (K\,\nabla h) = f$, where K is (random) hydraulic conductivity, h is (random) hydraulic head and f is a (random) source function. It is common to use the Reynolds decomposition to represent random fields $\Re = \langle \Re \rangle + \Re'$ as the sum of their ensemble means $\langle \Re \rangle$ and zero-mean fluctuations $\Re'$. Then averaging the stochastic flow equations gives

$$\nabla \cdot [\langle K \rangle \nabla \langle h \rangle] - \nabla \cdot \mathbf{r} = \langle f \rangle \tag{1}$$

where $\mathbf{r} = -\langle K'\ \nabla h' \rangle$ is the residual flux. Deriving approximations for the residual flux is the crucial part of any stochastic analysis. One of the most popular approaches is to use perturbation expansions in σ_Y^2, variance of log-conductivity, $Y = \ln K$. Theoretically this limits the applicability of such solutions to mildly heterogeneous aquifers or to highly heterogeneous aquifers, in the presence of a large number of conductivity measurements. However, numerical simulations of *Guadagnini and Neuman* [1999b] demonstrated that the first-order perturbation approximations of hydraulic heads and fluxes remain robust for heterogeneous media with σ_Y^2 as large as 4. Random Domain Decomposition [*Winter and Tartakovsky*, 2000, 2002] extends the range of applicability of perturbation closures even further. RDD recognizes that high degree of spatial variability usually stems from the presence of different geological facies and explicitly accounts for it. RDD treats the porous medium (and its conductivity) as a doubly stochastic process, where both the facies geometries and their conductivities are random. This allows one to obtain the statistics of hydraulic head and Darcy flux in two steps. The first step consists of calculating the conditional statistics of the system states via perturbation approximations, e.g., $\langle h^{[1]}(\boldsymbol{x}\,|\,\Gamma)\rangle = \langle h^{(0)}(\boldsymbol{x}\,|\,\Gamma)\rangle + \langle h^{(1)}(\boldsymbol{x}\,|\,\Gamma)\rangle$, in powers of σ_{YMi}^2, variance of log-conductivity within the facies M_i. Here the superscript (i) denotes terms in the expansion proportional to the i-th power of σ_{YMi}^2 and the vertical bar denotes conditioning on the facies geometry, Γ. The zero- and first-order approximations of conditional mean hydraulic head are given by equations similar to Eqs. (6) – (9) of *Winter et al.* [2002]. The second step consists of calculating the corresponding statistics of the system states through the ensemble averaging over the geometry distribution $p(\Gamma)$, e.g.,

$$\langle h^{[1]}(\boldsymbol{x})\rangle = \int \langle h^{[1]}(\boldsymbol{x}|\Gamma)\rangle \; p(\Gamma)\; d\Gamma \; . \tag{2}$$

The conditional moment equations are solved numerically by the finite elements program of *Guadagnini and Neuman* [1999a], and (2) can conveniently be approximated by the law of large numbers. The first-order approximation of the hydraulic head variance, $[\sigma_h^2(\boldsymbol{x})]^{[1]}$, is calculated in a similar manner.

2.2. Identification of material distribution

This work is devoted to obtaining $p(\Gamma)$ from stratigraphic and sedimentological information that is used to characterize the alluvial aquifer system of the city of Bologna in Northern Italy. The 50 Km^2 area is part of the high–medium alluvial plain close to the city of Bologna - Regione Emilia Romagna, Northern Italy and is located within the Reno alluvial fan [Guadagnini *et al.*, 2002]. Three Plio-Pleistocenic age aquifers of fresh water have been identified, which are composed of (coarse and fine) alluvial and sea deposits. The coarse ones are essentially related to the fluvial activity of the Apenninic streams and of the Po River. Generally, these aquifers are separated by discontinuous horizons (aquitards) of variable thickness and lithology. The available hydro-geological data have been organized into an efficient database and used to reconstruct geological cross-sections, maps of basis and top surfaces of each recognized geological unit, their total thickness, and the volumetric fraction of permeable sediments. The latter is measured on the basis of the cumulative thickness of gravel (gravel and sand for the aquitards) divided by the total thickness. The Reno alluvial fan within the study area is wedge-shaped, becomes thicker in the North-direction, and tapers southward. It rests on sea clayey deposits with saline water. The three aquifer groups are separated by the two main aquitards, each about 20-30 m thick, as well as by other aquitards of lower standing. Here we concentrate on the reconstruction of materials' distribution within Aquitard *Alpha*. The latter is of particular concern for the local municipality, since it plays a major role as a separation element between the upper contaminated aquifers and the deeper aquifers that are currently heavily exploited for water supply [Guadagnini *et al.*, 2002]. Available 39 logs of geognostical boreholes and 183 well-logs reveal that the aquitard's thickness is highly variable, changing from 1 – 3 m in the vicinity of the peak of the alluvial fan to 8 – 12 m near the well fields, to even larger values in the northern part. The deposits are mainly silty-clayey, with local interbedding of coarser material. The quantity ✓*gr* + *sa*✓/*th*, representing the cumulative thickness of gravel (*gr*) and sand (*sa*) divided by the total thickness (*th*), is generally less than 0.2. However, it displays local peaks larger than 0.8, indicating possible discontinuities within the aquitard itself. In our analysis we categorize materials within Aquitard *Alpha* into two classes, i.e. low and high permeability facies, on the basis of available hydro-stratigraphic data. Presence of the high conductivity regions indicates possible connections between upper and lower aquifers. Following Ritzi *et al.* [1994], we use the indicator point kriging and probability cut-offs approach to reconstruct mean boundaries between the geologic facies. The procedure consists of the following steps: (*i*) Transform sedimentological data into an indicator function; (*ii*) Analyze the spatial correlation structure of the indicator function; (*iii*) Estimate the spatial distribution of the probability of occurrence of the low and high permeability facies. This corresponds to the spatial distribution of their local volumetric fractions; and (*iv*) Delineate the mean boundary between facies by introducing a probability (or local volumetric fraction) cutoff.

Step1. Let us introduce a (random) indicator function $I(\boldsymbol{x})$, such that $I(\boldsymbol{x}) = 1$ if the low conductivity facies is present at point $\boldsymbol{x}$, and $I(\boldsymbol{x}) = 0$ if the low conductivity facies is absent. This allows us to estimate the area over which the low permeability facies exists. Unlike Ritzi *et al.* [1994] who relied exclusively on conductivity data, we use both the sedimentological and stratigraphic data sets to assign values of the indicator $I(\boldsymbol{x})$. This is analogous to the approach used by Guadagnini *et al.* [2002]. Thus, a point in space is assigned to a low or high permeability material according to a suite of combinations of values of (a) local thickness of the aquitard (as estimated by stratigraphic analysis) and (b) percentage of coarse-grain materials integrated along the stratigraphic column within the identified thickness. The resulting indicator variable is of two-dimensional nature, identifying the planar distribution of materials within the investigated aquitard. Our analysis of raw data shows that the low-permeability facies is present at about 80% of the sampled locations and the spatial mean of the indicator is equal to 0.81.

Step 2. We use sample variograms to estimate the spatial correlation of $I(\boldsymbol{x})$. The directional variograms are computed using an angular tolerance of 30 degrees along the directions oriented at azimuths of 0, 45, 90 and 135 degrees from the North. Sample variograms exhibit no clear evidence of anisotropy. The isotropic exponential model with a nugget was fitted to the sample variograms, resulting in nugget = 0.08, sill = 0.11 and correlation scale = 350 m.

Step 3. We use the ordinary point kriging to compute the expectation of $I(\boldsymbol{x})$. The latter, of course, corresponds to the probability of encountering the low-permeability facies at a point $\boldsymbol{x}$ or, equivalently, to the volumetric fraction occupied by the low-permeability materials within a volume centered at $\boldsymbol{x}$ and corresponding to the vertical column over which sedimentological data have been integrated.

Step 4. The obtained two-dimensional kriging map of the indicator allows defining a probability level (i.e. a local volumetric fraction value) as a cutoff for delineating the mean boundary between the units [e.g., Johnson and Dreiss, 1989; Ritzi *et al.*, 1994]. To identify the proper volumetric fraction isoline, we followed the procedure of Ritzi *et al.* [1994] and compared the percentage of the total area covered by the low-permeability facies resulting from (a) the raw data and (b) contoured, in the kriged indicator map, by the volumetric fraction cut-off isoline equal to the global mean of the original indicator data. Demarcation of units resulting in the 81% coverage of low-permeability facies was insured upon using the cut-off $\langle I(\boldsymbol{x})\rangle = 0.81$. This results in a spatial distribution of the low- and high-permeability units that honors both the original data and the mean of the indicator data. We then obtain the gray-scale map of Figure 1, representing the spatial distribution of the local volumetric fraction of the low permeable unit. The solid line corresponds to $\langle I(\boldsymbol{x})\rangle = 0.81$ and represents our estimate (i.e. the mean) of the boundary between high- and low-permeability facies. Conductivity values are then assigned to grid cells depending on their location within the region.

The procedure outlined by Ritzi *et al.* [1994] provides a means to estimate mean boundaries between geological facies, without quantifying uncertainty. The approach we propose below fills this void:

- Assume that the selected value of the cut-off (in this case 0.81) defines a limiting value of the local volumetric fraction of low-conductivity materials; locations where this value is attained identify a contour line which constitutes the internal boundary separating regions occupied by the two materials.
- Use the mean and variance of I(x) computed by kriging at all points in space. Assuming that I(x) is a Gaussian field, this defines the full probability distribution and in particular the probability that the chosen cutoff value occurs at a given location in the aquifer.
- Draw probabilistic spatial distributions of the target cutoff isolines, thus identifying the probability levels associated with different spatial locations of the boundary between units.
- Assign weights to each realization of the spatial arrangement of units.

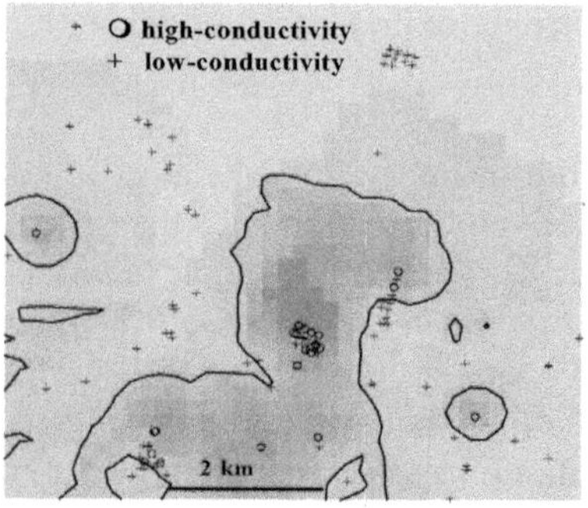

Figure 1. Spatial distribution of the local volumetric fraction of the low permeable unit (Grey scale: light – high volumetric fraction; dark – low volumetric fraction). The solid line corresponds to the mean boundary between high- and low-permeability materials. Data points are also shown.

Since I(x) is defined in the interval [0, 1], its distribution is normalized by the factor

$$A = \frac{1}{\sqrt{2\pi\sigma^2}} \int_0^1 \exp\left(-\frac{[\mathrm{I} - \langle \mathrm{I} \rangle]^2}{2\sigma^2} \right) d\mathrm{I} \qquad (3)$$

so that the weight of each realization of Γ in the probability space can be computed. These weights are then employed in a synthetic example below to derive the global (ensemble) moments of hydraulic head starting from the moments conditioned on various arrangements of material distribution.

3. SYNTHETIC FLOW PROBLEM

The probability map of facies' geometries constructed in the previous section can now be used to calculate the mean hydraulic head in Eq. (2). Since for this particular site the experimental data characterizing hydraulic properties within each geological facies are not available, we assume that within each facies the natural logarithm of hydraulic conductivity, $Y = \ln K$, is a statistically homogeneous Gaussian field. We set the mean log conductivities within the low and high permeability zones to $\langle Y_{\text{Low}} \rangle = 3.5$ and $\langle Y_{\text{High}} \rangle = 7.0$, respectively, when hydraulic conductivities are expressed in [cm/day]. We further assume that the conductivity of each facies has unit variance, the exponential correlation function with correlation scale, λ, and that conductivities of the two facies are uncorrelated. In contrast with deterministic trend models, the resulting conductivity field is essentially inhomogeneous, in that its (ensemble) mean, variance, and correlation function are all space dependent.

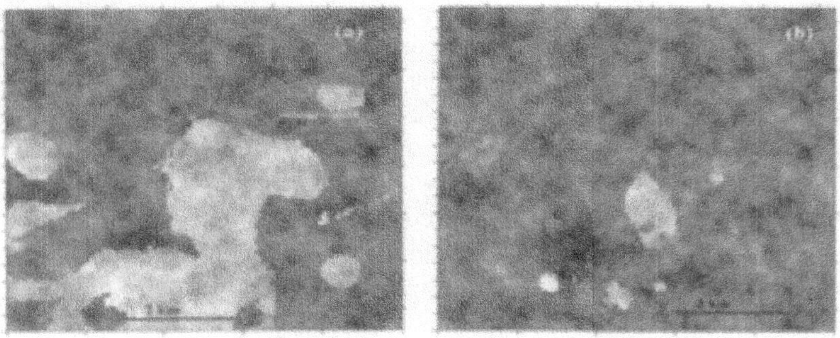

Figure 2. Conditional realizations of the composite domain corresponding to (a) the mean location of the boundary between the two materials and (b) a spatial distribution associated with a lower weight. Grey scale: light – high conductivity materials; dark – low conductivity materials.

Consider a rectangular flow domain of the size corresponding to the investigated area (*i.e.*, 7.2 × 6.8 km). For the sake of simplicity, we impose constant heads $H_A = 21.0$ m and $H_B = 1.0$ m on the left and right hand sides of the domain, while treating the remaining two boundaries as impermeable. This gives rise to the background hydraulic gradient of about 0.2%, the value representative of the field conditions. Pumping well is located in the middle of the field and operates at a steady-state flow rate of 100 m^3 / d. The domain is discretized by a grid of 19484 square elements (144 rows and 136 columns) of uniform size, $\Delta = 50$ m, with 5 points per correlation length of Y.

Figure 2 depicts two realizations of the composite flow domain, each one of them conditional on a particular location of the internal boundary between the materials. Figure 2a shows a logconductivity distribution corresponding to the mean boundaries between the two facies. Figure 2b shows a logconductivity distribution when the location of the internal boundary, defined by the 0.81 fraction of low-permeability material cut-off, has a different weight.

The correlation function for hydraulic conductivity of the composite medium is obtained by averaging the conditional correlation functions over all possible realizations of the materials distribution. Even though the two materials are assumed to be uncorrelated, there exists a transitional zone, where the points from the two materials are correlated. Within this zone the membership of a given point in a particular material is uncertain. Averaging over the boundary distribution smoothes the conditional correlation function of conductivity.

We obtain the conditional hydraulic head statistics (conditional mean and variance) by solving the RDD moment equations [Winter and Tartakovsky, 2000, 2002] with the stochastic finite element code of Guadagnini and Neuman [1999a]. The accuracy of the solutions of our moment equations is assessed through their comparison with Monte Carlo simulations. Guadagnini and Neuman [1999a, b] noted that a complete stabilization of the Monte Carlo statistics is not necessary for such a comparison to be meaningful. Therefore we limit the number of Monte Carlo simulations to 3000 for each of the log-conductivity fields. Since 21 realizations of the material distributions were considered in this study, we performed a total of $21 \times 3000 = 63{,}000$ Monte Carlo simulations. Overall agreement between the two solutions is excellent, except in the vicinity of the pumping well. This is in line with previous results of Guadagnini and Neuman [1999b]. Figure 3 shows the conditional mean and variance of hydraulic head for the material distributions of Figure 2a. The mean and variance of hydraulic head computed with RDD are shown in Figure 4. To ascertain the relative importance of the two sources of uncertainty (facies' geometry versus facies' conductivity), we show in Figure 5 the mean and variance of hydraulic head corresponding to the random facies geometry but deterministic (equal to their respective means) hydraulic conductivities. The comparison of Figures 4 and 5 reveals that this simplification leads to similar qualitative (but quantitatively different) spatial patterns of the mean drawdown and head variance.

4. COMPARISON WITH ALTERNATIVE MODELS

Next we compare the RDD approach with approaches that do not account explicitly for the presence of facies. Among these is a version of the dual permeability model, which expresses a local conductivity as a weighted sum of the conductivities of each facies,

$$K_{eq} = \text{Prob}[\text{I}(\boldsymbol{x}) = 1]\, K_{Low} + \text{Prob}[\text{I}(\boldsymbol{x}) = 0]\, K_{High} \tag{4}$$

The weights Prob[I($\boldsymbol{x}$) = 1] and Prob[I($\boldsymbol{x}$) = 0] = 1 – Prob[I($\boldsymbol{x}$) = 1] are determined by the kriging estimate of I($\boldsymbol{x}$). For each facies we generate 3000 log-conductivity fields with the same statistics used earlier and then use (4) to create realizations of the conductivity field. Contrary to the usual dual-continuum approach, which assumes that the volumetric fractions of the materials are constant over an entire flow domain, this approach results in a statistically inhomogeneous conductivity field.

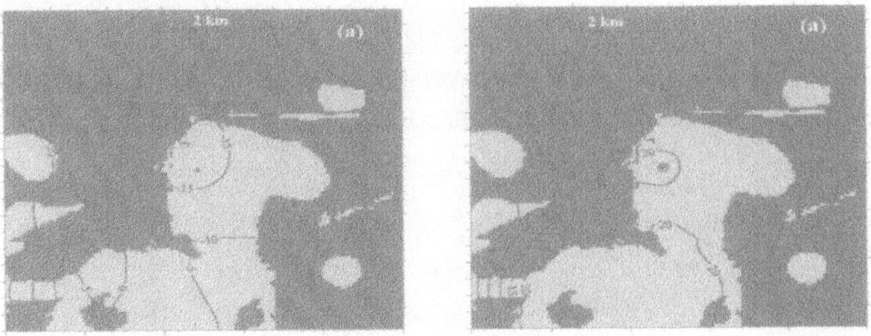

Figure 3. Conditional realizations of the mean and variance of hydraulic heads, superimposed on the corresponding mean (conditional) conductivity field.

Another approach is often referred to as a homogeneous approximation because it replaces a statistically inhomogeneous (stationary) conductivity field with the statistically homogeneous field whose statistics is determined as the mixture. For both approaches we use the Monte Carlo simulations to solve the flow equations. Figure 6 shows the hydraulic head statistics corresponding to the dual permeability distribution model. Figure 7 depicts the same quantities along the median transverse cross-section computed with all the models explored. The homogeneous approximation overestimates the drawdown and uncertainty (as quantified by head variance) at the well. The dual permeability distribution leads to the mean drawdown that is qualitatively and quantitatively similar to that obtained by considering only randomness in boundaries between materials and significantly underestimates head variance.

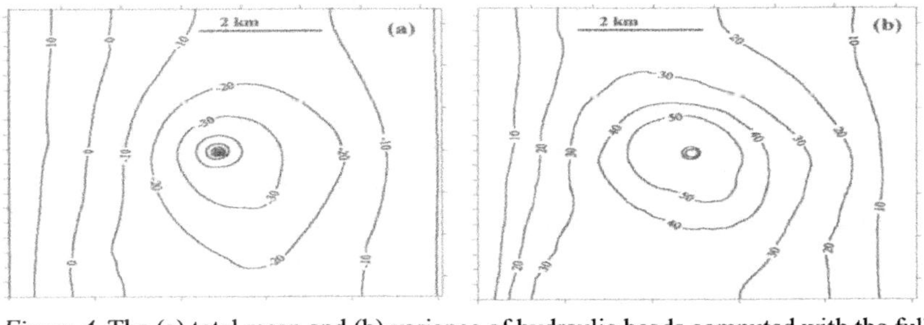

Figure 4. The (a) total mean and (b) variance of hydraulic heads computed with the full composite medium model with the full RDD.

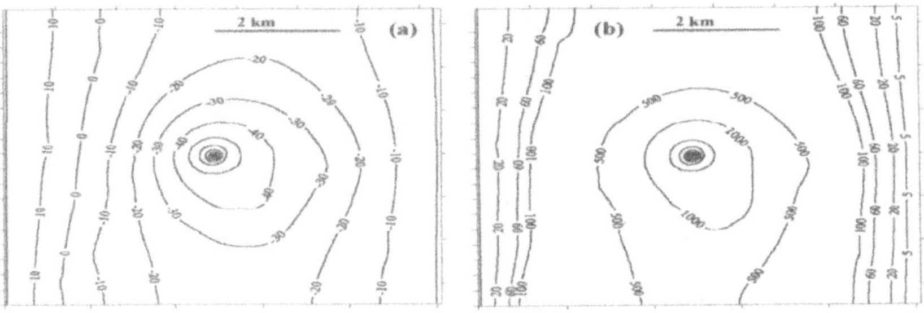

Figure 5. The (a) mean and (b) variance of hydraulic heads for random material distribution but deterministic (equal to their respective means) conductivities.

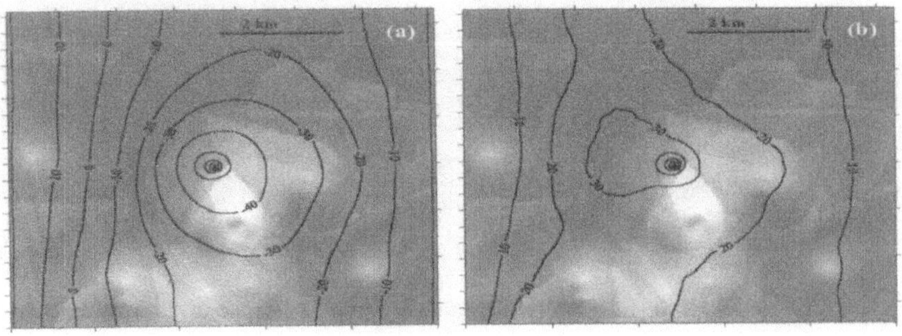

Figure 6. The (a) mean and (b) variance of hydraulic heads based on the dual permeability model. Superimposed is the corresponding mean conductivity field.

5. CONCLUSIONS

Our study leads to the following major conclusions:

1. One of the main difficulties in applying a random domain decomposition model is the identification of the spatial distribution of materials within a formation. We applied an indicator-based methodology to obtain an estimate of the spatial location of the boundary between contrasting materials as well as to quantify the associated uncertainty. The methodology is demonstrated on a synthetic flow example, where probabilistic material distribution is modelled using stratigraphic and sedimentological data from a real aquifer.
2. Our example emphasizes the qualitative and quantitative inadequacy of the homogeneous approximation for estimating the hydraulic head statistics. Similarly, replacing an essentially statistically inhomogeneous conductivity field with a model based on dual permeability concepts results in inaccurate solutions for the head statistics. The same holds for the models with deterministic trends in hydraulic conductivity, since they effectively disregard uncertainty in the facies geometry.
3. Our example emphasizes the qualitative and quantitative inadequacy of the homogeneous approximation for estimating the hydraulic head

statistics. Similarly, replacing an essentially statistically inhomogeneous conductivity field with a model based on dual permeability concepts results in inaccurate solutions for the head statistics. The same holds for the models with deterministic trends in hydraulic conductivity, since they effectively disregard uncertainty in the facies geometry.

4. The relative importance of uncertain geometry and uncertain conductivity was studied by comparing the case in which the material geometry is random, but the hydraulic properties of each material are fixed. Disregarding variability within materials leads to incorrect description of the statistical behavior of the system.

ACKNOWLEDGMENT

This work was supported by the European Commission under Contract No. EVK1-CT-1999-00041 W-SAHaRA. This work was supported in part by the U.S. Department of Energy under the DOE/BES Program in the Applied Mathematical Sciences, Contract KC-07-01-01. This work made use of shared facilities supported by SAHRA (Sustainability of semi-Arid Hydrology and Riparian Areas) under the STC Program of the National Science Foundation under agreement EAR-9876800.

REFERENCES

1. Desbarats, A. J., 1990: Macrodispersion in sand-shale sequence. *Water Resour. Res.*, 26(1), 153 – 163.
2. Johnson, N. M., and S. J. Dreiss, 1989: Hydrostratigraphic interpretation using indicator geostatistics. *Wat. Resour. Res.*, **25**(12), 2501-2510.

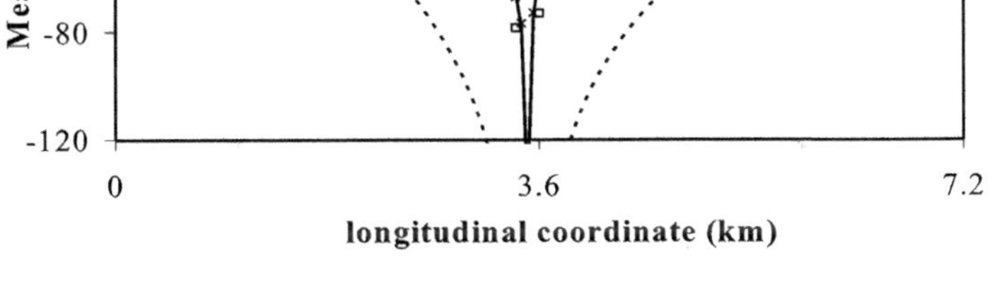

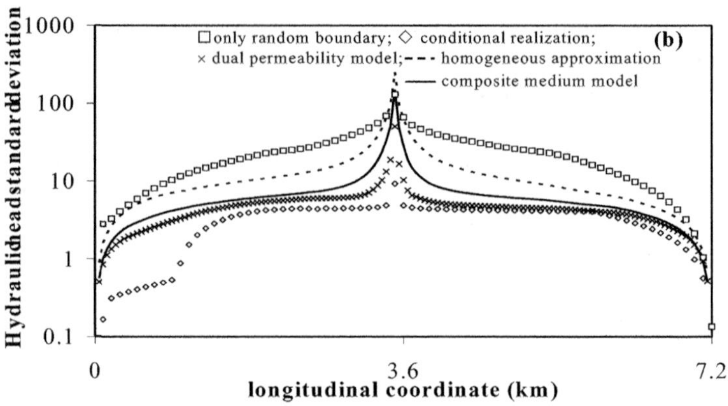

Figure 7. The (a) mean and (b) standard deviation of hydraulic head along the median transverse cross-section computed with all the models explored.

3. Guadagnini, A., and S. P. Neuman, 1999a: Nonlocal and localized analyses of conditional mean steady state flow in bounded, randomly nonuniform domains, 1. Theory and computational approach. *Wat. Resour. Res.*, *35*, 2999 - 3018.
4. Guadagnini, A. and S. P. Neuman, 1999b: Nonlocal and localized analyses of conditional mean steady state flow in bounded, randomly nonuniform domains, 2, computational examples, *Wat. Resour. Res.*, 35 (10), 3019-3039.
5. Guadagnini, L., M. Farina, S. Frontini, and M. Simoni, 2002: Geostatistical modelling of a heterogeneous alluvial aquifer, Proc. of the *International Conference on Calibration and Reliability in groundwater modelling*, ModelCARE2002, Prague, 167 – 171.
6. McLaughlin, D. and E. F. Wood, 1988: A distributed parameter approach for evaluating the accuracy of groundwater model predictions, 1. Theory. *Wat. Resour. Res.*, 24(7), 1037 – 1047.
7. Ritzi, R. W., D. F. Jayne, A. Z. Zahradnik, A. A. Field, and G. E. Fogg, 1994: Geostatistical modeling of heterogeneity in glaciofluvial, buried-valley aquifers. *Ground Water,* **32**(4), 666-674.
8. Winter, C. L., and D. M. Tartakovsky, 2000: Mean flow in composite porous media. *Geophys. Res. Lett.*, *27*, 1759 - 1762.
9. Winter, C. L., and D. M. Tartakovsky, 2002: Groundwater flow in heterogeneous composite aquifers. *Wat. Resour. Res.*, 38(8), 23.1 – 23.11.
10. Winter, C. L., D. M. Tartakovsky, and A. Guadagnini, 2002: Numerical solutions of moment equations for flow in heterogeneous composite aquifers. *Wat. Resour. Res.*, 38(5), 13.1 – 13.8.

INFLUENCE OF UNCERTAINTY OF MEAN TRANSMISSIVITY, TRANSMISSIVITY VARIOGRAM AND BOUNDARY CONDITIONS ON ESTIMATION OF WELL CAPTURE ZONES

H.J. Hendricks Franssen, F. Stauffer and W. Kinzelbach
Institut für Hydromechanik und Wasserwirtschaft, ETH Zürich, ETH Hönggerberg, 8093 Zürich, Switzerland

Abstract: The estimation of a drinking water well capture zone is uncertain due to spatial variability of transmissivity, among others. The spatial variable transmissivity is modeled by a Random Stochastic Function. It is common practice to fix the parameters that parameterize the adopted Random Stochastic Function Model. This paper presents a study that investigates to which extend the simulation results are influenced in case we do not fix these parameters and make them also random variables. The impact of this additional uncertainty is investigated both for forward models (only conditioning to transmissivity data) and inverse models (conditioning to transmissivity and head data). The results are compared with the impact of uncertainty in the boundary conditions. consequat

Key words: capture zones, stochastic, Bayesian.

1. INTRODUCTION

In order to maintain drinking water quality it is important to protect the zones around the drinking water wells from activities that may cause pollution. However, it is not desirable that large areas are excluded from activities that could yield important economical benefits. Therefore, it is crucial to characterize the groundwater flow around the drinking water well and more in particular, the zone from which contaminating particles could reach the drinking water well.

X. Sanchez-Vila et al. (eds.), geoENV IV – Geostatistics for Environmental Applications, 223-234.

The characterization of the well catchment is uncertain because of a (very) limited amount of measurement data, spatial variability of transmissivity and uncertainty on the spatio-temporal distribution of other parameters. The spatial variability of transmissivity is considered to be the most consequential one and quite some studies address its influence on the uncertainty of well capture zones (e.g. Franzetti and Guadagnini, 1996; Vassolo et al., 1998; van Leeuwen et al., 2000; Stauffer et al., 2002).

The common approach is to adopt a Random Stochastic Function (RSF) for the spatial variable transmissivity and to parameterize this RSF using the limited amount of measurement data. The estimated mean transmissivity and the transmissivity variogram are normally not subject to uncertainty and are fixed in the study. Full-Bayesian approaches take into account the uncertainty of these parameters (e.g. Woodbury and Rubin, 2000; Woodbury and Ulyrich, 2000). Feyen (2002) presents an application to the estimation of well capture zones. This paper illustrates how a more classical approach (the sequential Gaussian simulation to generate transmissivity fields and the sequential self-calibrated method for inverse conditioning) can also consider the mean transmissivity and the transmissivity variogram as random variables. In addition, a synthetic study investigates the impact of uncertainty of mean transmissivity and the transmissivity variogram and compares it with the influence of uncertainty of the boundary conditions. The influence of these sources of uncertainty is addressed in case only transmissivity data are used for conditioning and for the case that both transmissivity and hydraulic head data are used for conditioning.

2. METHODOLOGY

The spatial variable transmissivity field is modeled by a Random Stochastic Function. Sequential Gaussian simulation (Gómez-Hernández and Journel, 1993) is used to generate multiple equally likely transmissivity realizations, that are conditioned to transmissivity measurements. In the conventional approach, a mean transmissivity and a transmissivity variogram are supplied to the program. However, in order to consider uncertainty on the mean transmissivity and the transmissivity variogram, in this study the mean transmissivity and the transmissivity variogram may change from realization to realization. It means that we do not adopt one RSF, but a large series of RSF, the mean transmissivity and the transmissivity variogram being also random variables. However, in all cases a MultiGaussian distribution model is adopted.

In this synthetic study it was easy to build the probability density functions (pdf) of the mean transmissivity and the transmissivity variogram.

The reference transmissivity field was sampled 100 times (10 data) and from the 100 random data sets 100 different mean transmissivities and 100 different transmissivity variograms were estimated. In order to assure normality of the log transmissivity, the log transmissivity measurement data could be Normal transformed. However, in that case we would reproduce all the details of the experimental distribution (based on only 10 data) closely, neglecting the fact that the data set is only a random sample. The uncertainty in the mean transmissivity and the transmissivity variogram reflect the uncertainty for the case of 10 transmissivity data, and a moderately heterogeneous transmissivity field (variance lnT=1). Figure 1 gives the uncertainty of some of the input parameters for the generation of the log transmissivity fields.

In practice it is more difficult to build a model on the uncertainty of the mentioned hyper parameters (the parameters that parameterize the MultiGaussian model). This is also the weakest point in the Bayesian approaches, where subjective parameter values have to be introduced that characterize an unknown distribution. Here it is suggested that in case of a limited number of transmissivity data, formula from classical sampling theory can be used to estimate the variance of the mean transmissivity. Brus and De Gruyter (1994) show how also the uncertainty of the variogram can be estimated. However, in case of clustered measurement data or a very limited number of transmissivity data the uncertainty has to be addressed in a different way. An alternative is postulating a distribution for the mean transmissivity and the transmissivity variogram, and drawing values from these distributions. The distributions have to be broad enough to cover the believed uncertainty.

The impact of the uncertainty of the boundary conditions is studied by estimating the prescribed head values at the boundaries by 10 randomly sampled head data. A Monte Carlo approach has been used and for each realisation a different random data set is used. The 10 head data are used to estimate the head values at the boundaries by a second order polynomial. Because for each realization a different data set is used, also the estimated prescribed heads on the boundaries are different for each realisation. An alternative would have been to estimate the boundary heads by universal kriging, however, 10 head data did not allow to make a meaningful estimation of the variogram of the head residuals. In practice it is common to estimate the prescribed head values at the boundaries by interpolating or extrapolating head measurement data, because it is desirable to limit the domain that has to be studied in case no natural boundaries exist. In this study this practical behavior is imitated and it allows to quantify the impact of the uncertainty of the boundary conditions in a systematic way. However, one may argue that in practice the prescribed head values can be estimated in a more intelligent way, for example by using topographical information. One could argue that the applied approach here is a "worst case scenario" and that in practice the negative impact of not knowing the prescribed head values is less severe as in this study.

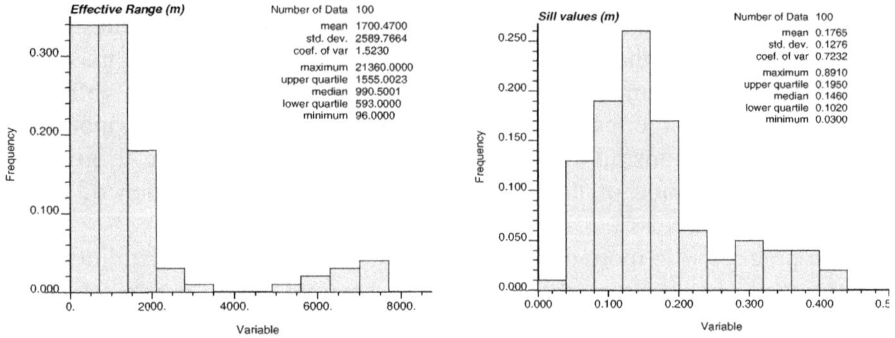

Figure 1. Histograms of (a) estimated range values and (b) estimated sill values of 100 variogram models that were estimated on the basis of 100 different data sets, randomly sampled from the reference log transmissivity field.

Section 3 gives details on the different scenarios that have been studied. For all of the scenarios, the following approach was followed. In case head data are available, realizations of log transmissivity are conditioned to both transmissivity and hydraulic head data, using the sequential self-calibrated method for the stochastic inverse modeling of groundwater flow (Gómez-Hernández et al., 1997) as implemented in the code INVERTO (Hendricks Franssen, 2001). Otherwise, only the forward flow and transport problem were solved. The approach consists of the following steps:

(1) 100 equally likely realisations of log decimal transmissivity are generated with GCOSIM3D (Gómez-Hernández and Journel, 1993). The realisations are conditioned to 10 transmissivity data. The log transmissivity mean and the log transmissivity variogram may be known, or may be estimated from the limited amount of transmissivity data.

(2) For each of the 100 realisations the groundwater flow equation is solved with the software INVERTO. In case only transmissivity data are available only the forward groundwater flow equation has to be solved, and the procedure continues with step 4. In case also hydraulic head data are available the measured heads are compared with the simulated heads and the following formula is evaluated:

$$J = \sum_{i=1}^{N_h} \xi_i (h_i^{SIM} - h_i^{MEAS})^2$$

where N_h is the number of head measurement locations, h_i the head at a measurement location, the weight ξ_i is chosen inverse proportional to the estimated measurement error (in our case all the measurement data have the same estimated measurement error (zero) which means that the weights are equal for all the data), *SIM* refers to simulated and *MEAS* to measured.

If J is smaller than a pre-defined tolerance value the measured heads are reproduced close enough. In case J is larger than the tolerance value the simulations continue with step 3.

(3) Because the head data were not matched close enough an iterative procedure starts that aims at matching the head data. Details on the optimisation procedure are given in Hendricks Franssen (2001). In this study, 400 master blocks parametrize the perturbation of the log transmissivity field and 50 master blocks the perturbation of the prescribed heads at the boundaries. In case of the boundary heads the maximum perturbation is arbitrarily set to 5.0 meters. After optimising the perturbations of the logtransmissivity field and (for some scenarios) the prescribed boundary heads, the procedure returns to (2) and the transmissivities and boundary heads are iteratively updated until the experimental heads are matched.

(4) The resulting solution of the groundwater flow equation is used to simulate the advective transport of particles. For each of the 100 realisations, one particle is released at the centre of each grid cell and tracked until it reaches a boundary of the system or the pumping well.

(5) Ensemble statistics are calculated over the 100 realisations. The following definitions hold:

$$AAE(X) = \frac{1}{N}\sum_{i=1}^{N}\left|\overline{X}_{SIM,i} - X_{REF,i}\right|$$

$$AESD(Z) = \frac{1}{N}\sum_{i=1}^{N}\sigma_{Z_i}$$

where *AAE* is the average absolute error, N the number of grid cells, X either log transmissivity, hydraulic head or particle capture probability (*CZ*), *SIM* the simulated value, *REF* the reference value and i a grid cell index. An overbar stands for ensemble average. With respect to the capture probability: $CZ(x,i)=0$ if a particle released from grid cell x for realisation i does not reach the pumping well and $CZ(x,i)=1$ if the particle reaches the pumping well. The average capture probability $CZ(x)$ for that grid cell is determined by averaging the 100 obtained $CZ(x,i)$. In the second equation Z stands for either log transmissivity or hydraulic head and σ is the ensemble standard deviation. The uncertainty with respect to the capture probability is given by:

$$AESD(CZ) = \frac{1}{N}\sum_{x=1}^{N}\min(CZ(x),\ \ 1 - CZ(x))$$

where *AESD*(*CZ*) is the domain averaged uncertainty with respect to the capture probability. For instance: if $CZ(x)=0$ or $CZ(x)=1$ the grid cell x does not contribute to *AESD*(*CZ*); if $CZ(x)=0.5$ the contribution is 0.5 (the largest contribution possible; the maximum uncertainty).

3. SYNTHETIC STUDY

The impact of the mentioned sources of uncertainty has been tested in a synthetic study. The studied 2-D domain has extensions of 5 x 5 km and is divided into 50 x 50 squared grid cells of size 100 m. The Northern and Southern boundaries are impervious and along the Western and Eastern boundaries fixed heads of respectively 0 m and 5 m are imposed. A pumping well is located 500 m West of the domain centre. The area receives a spatially uniform recharge of 363 mm/year. Steady-state groundwater flow in a semi-confined aquifer is simulated. A reference transmissivity field is generated with a mean transmissivity equal to –2.93 $\log_{10}(m^2/s)$ and an exponential variogram with a range of 500 m (1/10 of the domain) and a sill equal to 0.18861 $(\log_{10}(m^2/s))^2$ (ln$T$ variance=1). As a consequence, a water divide along the Eastern part of the area is present and the well pumps water from a considerable area located West of the water divide. Figure 2a gives the reference well catchment.

The impact of the uncertainty in the mean logtransmissivity and the logtransmissivity variogram are studied under two different conditions: (1) in case 10 transmissivity data are used to condition the transmissivity realisations, (2) in case both 10 transmissivity data and 10 head data are available and the transmissivity field is updated by inverse modelling. Also the impact of the uncertainty in the boundary conditions has been studied, with or without uncertainty in the mean transmissivity and the transmissivity variogram. For the case of uncertainty in the boundary conditions, the joint updating of the transmissivity field and the boundary conditions, by inverse modelling, is an additional simulation variant. See Table 1.

3.1 Influence of uncertainty in the mean transmissivity

Table 2 gives the scores on the evaluation criteria defined before. In this section we focus on the influence of the uncertainty in the mean transmissivity, without uncertainty in the transmissivity variogram and/or the boundary conditions. In case of forward modelling (scenario 1.1): the uncertainty of the mean transmissivity hardly affects the characterisation of the transmissivity field, the hydraulic head field and the well capture zone. However, the ensemble hydraulic head variance is much more affected by the uncertainty of the mean transmissivity. Compared with a similar case without uncertainty (scenario 0.1) an increase of the standardised ensemble variance (unconditional=100) from 88.4 (no errors) to 130.4 (error in mean logtransmissivity) occurred.

In case of inverse modelling (scenario 1.2): also in this case the uncertainty of the mean transmissivity yields slightly worse results. However, the impact of the uncertain mean transmissivity on the characterisation of the hydraulic head field is less than in scenario 1.1.

3.2 Influence of uncertainty in the transmissivity variogram

The effect of the transmissivity variogram uncertainty on the well capture zone characterisation is a bit larger than the effect of the uncertainty of the mean transmissivity, but also limited. The characterisation of the transmissivity and head fields is slightly worse than in the case of the correct

Table 1. Studied scenarios

Scenario	Uncertainty meanT?	Uncertainty variogram T?	Uncertainty Boundary conditions?	Conditioning data	Calibration BCS?
0.1	NO	NO	NO	10 T	n.a.
0.2	NO	NO	NO	10 T, 10 h	n.a.
1.1	YES	NO	NO	10 T	n.a.
1.2	YES	NO	NO	10 T, 10 h	n.a.
2.1	NO	YES	NO	10 T	n.a.
2.2	NO	YES	NO	10 T, 10 h	n.a.
3.1	YES	YES	NO	10 T	n.a.
3.2	YES	YES	NO	10 T, 10 h	n.a.
4.1	NO	NO	YES	10 T	NO
4.2	NO	NO	YES	10 T, 10 h	NO
4.3	NO	NO	YES	10 T, 10 h	YES
5.1	YES	NO	YES	10 T	NO
5.2	YES	NO	YES	10 T, 10 h	NO
5.3	YES	NO	YES	10 T, 10 h	YES
6.1	NO	YES	YES	10 T	NO
6.2	NO	YES	YES	10 T, 10 h	NO
6.3	NO	YES	YES	10 T, 10 h	YES
7.1	YES	YES	YES	10 T	NO
7.2	YES	YES	YES	10 T, 10 h	NO
7.3	YES	YES	YES	10 T, 10 h	YES

variogram. Surprisingly, the ensemble transmissivity variance and the ensemble head variance are lower in case the variogram is uncertain than in case the variogram is exactly known. It is found that the estimated average sill is below the true, unknown sill. This causes that the ensemble variances are underestimated. The lower variances give a false sense of security. Surprisingly, the well capture zone is better characterised in case the transmissivity variogram is uncertain. The explanation is thought to be the sampling error; if we would repeat the experiment for other reference fields, other results are expected.

In case hydraulic head data are available, the effect of the variogram uncertainty reduces and the results are closer to a similar case without variogram uncertainty. This also means, that the underestimation of the ensemble variances reduces or disappears.

3.3 Influence of uncertainty in both the mean transmissivity and the transmissivity variogram

In case both the mean transmissivity and the transmissivity variogram are uncertain, the results are still hardly affected by these sources of uncertainty. The characterisation of the hydraulic head field is the most affected. In case only logtransmissivity data are used for the conditioning, the AAE(h) decreases 39.8% as compared with the unconditional case. For the case that mean transmissivity and transmissivity variogram are known, the AAE(h) decrease is 50.3%. The differences are significant, but still not very large. However, the AESD(h) increases very significantly.

Again we observe that hydraulic head data are able to reduce the impact of the uncertainty in the transmissivity statistics.

3.4 Influence of uncertainty in the boundary conditions

In order to place the importance of the uncertainty of the mean transmissivity and the transmissivity variogram in a context, simulations were made in which the prescribed boundary heads were unknown.

Without uncertainty in the boundary conditions the 10 transmissivity data result in an AAE(h) reduction of 50.3%. With uncertainty in the boundary conditions this reduction is only 31.0%. The uncertainty of the hydraulic head field, as measured by AESD(h), is even more affected by the uncertainty of the boundary conditions. The characterisation of the capture zone, as measured by AAE(CZ) worsens only slightly; for the scenario without errors in the boundary conditions an AAE(CZ) reduction of 4.0% was achieved (as compared to an unconditional scenario), for the scenario with errors in the boundary conditions an AAE(CZ) increase of 4.5% occurs.

The smaller impact of the "wrong" boundary conditions on the well capture zone estimation can be explained by the fact that the estimated transmissivity field is not affected by the erroneous boundary conditions. The uncertain boundary conditions (scenario 4.1) affect more the capture zone estimation than the uncertain mean transmissivity (scenario 1.1) or the uncertain variogram (scenario 2.1).

For the case of inverse modelling with hydraulic head data (scenario 4.2), the negative impact of the uncertain boundary conditions on the hydraulic head field characterisation is less. While in scenario 0.2 (same amount of data, but no uncertain boundary conditions) the AAE(h) reduction is 72.0% (as compared with the unconditional case), it is 60.1% for scenario 4.2. This reduction of the impact of the uncertainty of the boundary conditions is also observed for AESD(h). Nevertheless, it is not the case for the characterisation of the well capture zone. Although in scenario 4.2 the head data yield an AAE(CZ) decrease of 9.1% as compared with an unconditional scenario, in case of error free boundary conditions (scenario 0.2) this decrease is 19.6%. The wrong boundary conditions limit in the inverse modelling the improvement of the characterisation of the transmissivity field. This is the reason why the impact of the erroneous boundary conditions on the well capture zone characterisation does not decrease in the inverse modelling.

Table 2. Scores on the evaluation criteria for the studied scenarios.

Scenario	*AAE*(*Y*)	*AESD*(*Y*)	*AAE*(*h*)	*AESD*(*h*)	*AAE*(*CZ*)
0.1	89.9	99.3	49.7	88.4	96.0
0.2	84.5	96.6	28.0	48.9	80.4
1.1	90.5	103.0	52.1	130.8	97.5
1.2	85.1	102.0	28.3	54.8	81.2
2.1	90.6	91.2	54.2	84.5	91.6
2.2	85.4	93.5	28.7	53.6	77.9
3.1	90.8	96.6	60.2	166.4	93.6
3.2	86.4	95.1	28.3	54.4	80.5
4.1	89.9	99.3	69.0	159.0	104.5
4.2	86.6	100.1	39.9	86.9	90.9
4.3	86.5	99.5	39.8	86.5	90.6
5.1	90.5	103.0	69.7	190.1	107.5
5.2	87.1	101.4	40.3	92.1	92.1
5.3	87.0	100.9	40.1	91.6	91.7
6.1	89.6	91.2	73.7	155.3	102.2
6.2	87.8	97.8	40.8	109.5	93.0
6.3	87.7	92.3	39.9	84.0	95.2
7.1	90.8	96.6	80.0	226.3	103.0
7.2	88.5	101.1	41.1	114.3	92.5
7.3	88.5	95.2	40.2	114.1	89.2

Also in case head data are available, and the transmissivities can be updated inversely, the uncertainty in the boundary conditions is much more consequential than the uncertainty of the mean logtransmissivity or the uncertainty of the logtransmissivity variogram.

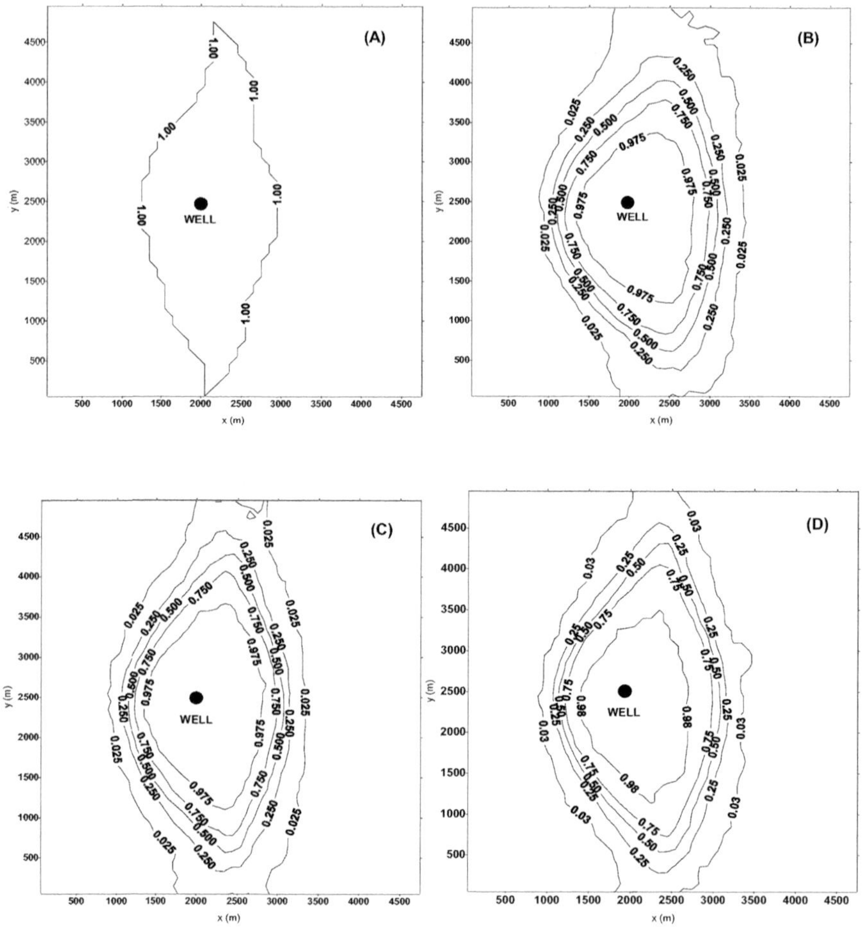

Figure 2. Ensemble averaged well capture zones with capture probabilities for (a) reference field, (b) unconditional simulations, (c) conditioning to 10 transmissivity and head data and no uncertainty in the mean transmissivity and the transmissivity variogram and (d) conditioning to 10 transmissivity and head data with uncertainty in the transmissivity variogram and the mean transmissivity.

The scenario with both head and logtransmissivity data is also repeated for the case that the modeller recognises the uncertainty/errors on the boundary head values and allows these values to be modified. The perturbation of the prescribed boundary heads (together with the logtransmissivities) results in only very slightly better simulation results.

3.5 Multiple sources of uncertainty

For the scenarios 5, 6 and 7 there are multiple sources of uncertainty. For the characterisation of the hydraulic head field, the logtransmissivity field and the well capture zone the negative impact of uncertain boundary conditions, uncertain mean log transmissivity and uncertain log transmissivity variogram, is approximately additive. As a result, the *AAE*(*h*), *AAE*(*Y*) and *AAE*(*CZ*) only increase slightly in case besides the uncertainty of the boundary conditions the other two sources of uncertainty are present.

The uncertainty of the estimation of the hydraulic head field, however, is much more affected in case of two or three sources of uncertainty. Because also in this case the impact of the multiple sources of uncertainty is approximately additive, in case of for example all the three sources of uncertainty the *AESD*(*h*) is more than double as big as in the unconditional case. Inverse modelling reduces largely the *AESD*(*h*), but in case of three sources of uncertainty the *AESD*(*h*) is always larger than in the unconditional case. On the contrary, the *AAE*(*h*) reduces until 40% of the *AAE*(*h*) in the unconditional case in case of three sources of uncertainty (and the use of hydraulic head data by inverse modelling).

4. CONCLUSIONS

This paper illustrates that it is possible to handle in a simple way uncertainty with respect to the mean transmissivity and the transmissivity variogram in the modelling of groundwater flow and mass transport.

At the same time it is found for this particular synthetic study that uncertainty on the mean transmissivity and the transmissivity variogram tend to have relatively limited consequences for the well capture zone estimation. Also in the Full Bayesian approaches in which these sources of uncertainty were addressed, the impact on the simulation outcomes was limited (Woodbury and Ulyrich, 2000; Feyen, 2002). The impact of the uncertainty of the mean transmissivity and the transmissivity variogram affects less the estimated ensemble averaged fields as the estimated ensemble variance of the fields. Inverse modelling helps to reduce the impact of the uncertainty of the mean transmissivity and the transmissivity variogram. Uncertainty in the transmissivity statistics has a much smaller impact on the well capture zone estimation than uncertainty in the boundary conditions. On the other side, even if boundary conditions, the transmissivity field, the transmissivity variogram and the mean transmissivity are uncertain, 10 hydraulic head data are able to yield a better characterisation of the well capture zone than in the unconditional case.

ACKNOWLEDGEMENTS

The study was performed within the European Research Project "Stochastic Analysis of Well Head Protection and Risk Assessment" W-SAHaRA. This project has been supported by the Swiss Federal Office for Education and Science (BBT).

REFERENCES

1. Brus D.J., de Gruijter J.J. Estimation of non-ergodic variograms and their sampling variance by design-based sampling strategies. Math. Geol. 1994; 26(4): 437-454.
2. Feyen L., *Stochastic delineation of well capture zones*. Ph.D dissertation. Department of hydrology and hydraulic engineering, Vrije Universiteit Brussel, 2002.
3. Franzetti S., Guadagnini A. Probabilistic estimation of well catchments in heterogeneous aquifers. Journal of Hydrology 1996; 174: 149-171.
4. Gómez-Hernández J.J., Journel A.G. Joint sequential simulation of multi-Gaussian fields. In *Geostatistics Troia'92 volume 1*, ed. Soares, A: 85-94, 1993.
5. Gómez-Hernández J.J., Sahuquillo A., Capilla J.E. Stochastic simulation of transmissivity fields conditional to both transmissivity and piezometric data. 1. Theory. Journal of Hydrology 1997; 1-4(203): 162-174.
6. Hendricks Franssen H.J. *Inverse stochastic modelling of groundwater flow and mass transport*. PhD dissertation. Technical University of Valencia, 2001.
7. Stauffer F., Attinger S., Zimmermann S., Kinzelbach W. Uncertainty estimation of well catchments in heterogeneous aquifers. Accepted for publication in Water Resources Research.
8. Vassolo S., Kinzelbach W., Schäfer W. Determination of a well head protection zone by stochastic inverse modelling. Journal of Hydrology 1998; 206: 268-280.
9. Van Leeuwen M., Butler A., te Stroet C.B.M., Tompkins J.A. Stochastic determination of well capture zones conditioned on regular grids of transmissivity measurements. Wat. Resour. Res. 2000; 36(4), 949-957.
10. Woodbury A.D., Rubin Y. A full-Bayesian approach to parameter interference from tracer travel time moments and investigation of scale effects at the Cape Cod experimental site. Wat. Resour. Res. 2000; 36(1), 159-171.
11. Woodbury A.D., Ulyrich T.J. A full-Bayesian approach to the groundwater inverse problem for steady state flow. Wat. Resour. Res. 2000; 36(8), 2081-2093.

EVALUATION OF DIFFERENT MEASURES OF FLOW AND TRANSPORT CONNECTIVITY OF GEOLOGIC MEDIA

C. Knudby and J. Carrera
Departament d'Enginyeria del Terreny i Cartogràfica, Universitat Politècnica de Catalunya, Barcelona, Spain

Abstract: In order to allow for reasonably exact modeling of fluid flow and contaminant transport in low permeable geologic media, it is of prime importance that the connectivity is represented with sufficient accuracy. Connectivity affects the effective large-scale value of hydraulic conductivity. It also affects the way in which porosity is accessed by solutes. Despite the apparent importance of connectivity, the use of parameters with a high level of information on connectivity is very limited in hydrogeology. We evaluate and compare several measures of flow and transport connectivity. Our results indicate that flow and transport connectivity are qualitatively different.

1. INTRODUCTION

Proper representation of connected features in geological media (high-conductivity flow paths and low-conductivity flow barriers) is of crucial importance when modeling flow and transport in geological media. Connectivity causes channeling (i.e. concentration of water flux along a small portion of the domain) which results in very significant reduction of solute travel time. The importance of a proper representation of connected features, which exist is many types of geological media frequently investigated in hydrogeology, has long been recognized (Fogg, 1986; Webb and Anderson, 1996). Nevertheless, very little attention has been paid to defining connectivity in a quantifiable manner in hydrogeological research.

This negligence can be partly attributed to the widespread use of the assumption that the distribution of hydraulic conductivities, K, in many geological media is approximately multilog-Gaussian. This assumption was

X. Sanchez-Vila et al. (eds.), geoENV IV – Geostatistics for Environmental Applications, 235-246.

originally based on the observations of Law (1944) and Davis (1969) who found that the point values of log-K in many natural media follow a Gaussian distribution. Considerable mathematical simplifications can be obtained by extending this assumption to one of multilog-Gaussianity. However, this involves an assumption on the spatial structure of the point values of K which is very rarely supported by data. In fact, multilog-Gaussianity implies minimal spatial correlation of extreme values (Journel and Deutsch, 1993; Gomez-Hernandez and Wen, 1998). Therefore, the widespread use of multilog-Gaussian K-distributions, which to a great extent is based on considerations of mathematical tractability rather than on available data, leads to a consistent underestimation of the connectivity. This consequence of the use of multilog-Gaussian K-distributions was analyzed by Sánchez-Vila et al. (1996). They found that for two-dimensional fields with higher correlation lengths for high-K zones than for low-K zones, the effective conductivity tends to be higher than the geometric average of the point values of K, K_G, which is the effective conductivity of an infinite multilog-Gaussian medium with isotropic correlation structure (Matheron, 1967).

One obvious step towards improved characterization and incorporation of connectivity would be the introduction of one or more measures of connectivity. The use of such parameters should make it easier to identify and quantify misrepresentation of connectivity. Also, when generating random fields using stochastic simulation, the connectivity measures could be used for conditioning. This would enable the generation of fields that are more realistic with respect to connectivity.

To the best of our knowledge, the only measure of connectivity used so far in hydrogeology stems from Percolation Theory (e.g. Stauffer and Aharony, 1991). The employed definition is purely topological in that a medium is connected only if a continuous path exists between two boundaries of the medium. As geological media never are completely impermeable the direct application of such a definition on a grain size scale would render all geological media connected. On the other hand, if the geological medium is treated as being composed of facies or indicators of K-intervals (e.g. Journel, 1983), then connectivity indicates whether or not certain facies or zones of K-values belonging to a certain interval percolate. In this case the measure plays a key role for the hydrological response of the system in question (Fogg, 1986; Fogg et al., 2001). Nevertheless, the use of such a simple definition of connectivity rules out the use of a significant part of the information potentially contained in a well-defined measure of connectivity. Two different media, one containing many connected features, the other one largely without connected features, might both be non-connected according to this definition despite the fact that their hydrological response is likely to be immensely different. In order to enable distinction between such media one needs a definition that allows a non-discrete quantification of the connectivity of a system.

Geostatistics is the natural choice for a framework into which measures of connectivity could be integrated. In geostatistics, one works with parameters such as the integral scale and the variance which both contain some non-discrete information on connectivity. However, standard geostatistical methods used in hydrogeology are based on the variogram that accounts for correlation as a function of distance between two points without consideration of what lies in between. Also, no consideration is taken to whether K is high or low except when indicator variograms are employed. Any reasonable measure of connectivity must consider "what lies in between" (e.g. by considering strings of high K-values, see fig. 1) and will therefore differ conceptually from variogram-based parameters. Western et al. (2001), who also call for quantification of connectivity, but in the field of surface hydrology, illustrate this nicely by use of the two fields presented in fig. 1.

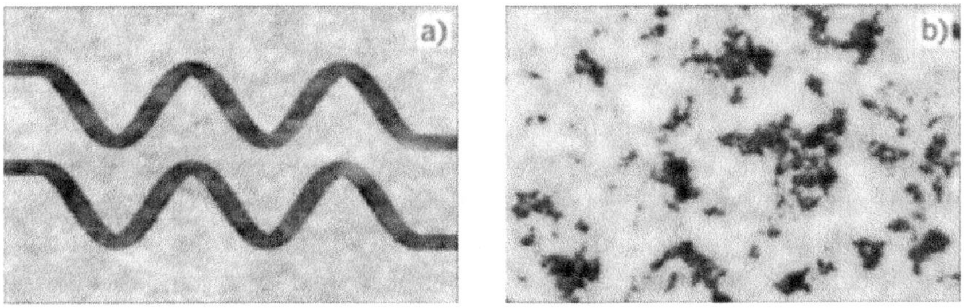

Figure 1. Conductivity fields with the same pdf's and omni-directional variograms, but with very different connectivity. From (Western et al., 2001)

The two fields have the same pdf and variogram, but are obviously very different with respect to connectivity. As connectivity conceptually is more closely related to channeling than any two-point correlation function used in geostatistics, one would have to go beyond standard geostatistical methods, or modify them, in order to address connectivity properly.

The objective of our work is to define easy-to-measure hydraulically based connectivity in order to make it easier to predict hard-to-measure transport based connectivity. In this paper we present and evaluate several measures of both flow and advective transport connectivity. It is not the intention of the paper to identify the best measure, but only to evaluate different measures, to test if they correlate, and to shed light on how the averaged response of flow and transport processes in geological media depend on connectivity.

This paper is organized as follows. First, we present a number of possible measures of "connectivity" which are based on different characteristics of flow and transport. Next, we describe the methodology that we have used to evaluate these measures. Subsequently we present the results of this evaluation. Finally, the outcome of the analysis is discussed and we conclude on which measures contain the most information on essential characteristics of the flow and transport in the media considered.

2. MEASURES AND CONNECTIVITY

Any valid measure of connectivity should meet the following requirements:

1. It should be quantifiable. This implies that it needs to be defined in terms of parameters which can be measured directly or estimated indirectly.
2. The value of the measure should contain information on the characteristics of flow and/or transport processes in the medium in question. This implies that it should be either related to or a function of parameters that exert control on flow and/or transport through the system.

Ideally, connectivity should be defined so that it can be derived directly from available field data such as, for example, pump tests. Nevertheless, in the present study, which deals with fully known synthetic fields only, it was considered useful to consider definitions of "connectivity" that allow for quantification of the parameter only when the exact distribution of hydraulic conductivities is known.

In the following, three flow-based measures of connectivity (CF_1, CF_2, CF_3), and two transport-based measures of connectivity (CT_1, CT_2) are presented.

2.1 Exponent for "Power Averaging" (CF_1)

The effective conductivity, K_{eff}, of a geological medium, can be estimated from power averaging of the point values of K (e.g. Renard and de Marsily, 1997). This involves estimating K_{eff} from

$$K_{eff} = \left(\frac{1}{V} \int_V K(\mathbf{x})^{CF_1} \, dV \right)^{\frac{1}{CF_1}} \tag{1}$$

where V is the volume in question, **x** is the location in space and CF_1 is an exponent. On the other hand, if K_{eff} is known (i.e. from a long pumping test (Meier et al., 1998)) then we can determine the value of the exponent CF_1 from (1). The value of CF_1 is an indicator of the connectivity of the field. For a layered medium in which the direction of flow is perpendicular to the layers (i.e. minimum connection), K_{eff} is equal to the harmonic mean of the point values of K. Thus, CF_1 assumes the value -1. Conversely, K_{eff} is equal to the arithmetic mean of the point values of K for a layered medium in which the direction of flow is parallel to the layers (i.e. maximum connection). In this case, CF_1 assumes the value 1. In between these extremes is the case of $CF_1 \rightarrow 0$. $\lim_{CF_1 \rightarrow 0}(K_{eff})$ corresponds to the geometric mean of the point values of K. Since the harmonic and arithmetic means are lower and upper bounds on the effective conductivity, respectively, CF_1 can be considered as an indicator of how the point values of K are organized within the medium in question - blocking the flow (CF_1 close to -1) or providing channels (CF_1 close to 1). In the two-dimensional case, CF_1 also indicates if the medium is more or less conductive than a multilog-Gaussian medium with the same pdf of K values (CF_1 smaller or greater than 0). Sanchez-Vila et al. (1996) discuss this issue further.

2.2 Ratio of K_{eff} to K_G (CF_2)

For the same reasons that CF_1 is a measure of connectivity, also the ratio

$$CF_2 = \frac{K_{eff}}{K_G} \tag{2}$$

is a measure of connectivity. In fact, the only difference between the two is the difference in the interval of values which the parameters can assume. Whereas CF_1 can assume values belonging to the interval [-1;1], CF_2 can assume any positive value. We test both in order to find out whether the scaling difference makes one measure easier to interpret than the other.

2.3 Ratio of the Critical Path Conductivity to K_G (CF_3)

The hydrogeological response of geological media which exhibit a high K-value variance will be closely related to the critical path conductivity, K_C, (Ambegaokar et al., 1971; Friedman and Seaton, 1998), also called the "extreme path conductivity" (Silliman and Wright, 1988). This is defined as the minimum conductivity along the path through the medium which has the highest minimum conductivity. A more intuitive measure of connectivity is therefore

$$CF_3 = \frac{K_C}{K_G} \tag{3}$$

2.4 Ratio between average arrival time and the arrival time of 5% of the solute (CT_1)

The breakthrough curve for solute being transported through a geological medium contains much information on the connectivity of the medium. If a very large proportion of the solute follows the same fast path, and as a consequence reaches the outlet shortly after injection, and over a small time interval, then the medium must be considered well connected. However, if the solute is spread out due to the lack of fast pathways, and therefore reaches the outlet late, and over a large time interval, then the medium must be considered badly connected. From these considerations it is clear that

$$CT_1 = \frac{T_{AVE}}{T_5}$$

where T_{AVE} is the average arrival time and T_5 is the time at which 5% of the solute has arrived at the outlet, can be considered a measure of transport connectivity.

2.5 Skewness of arrival times distribution (CT_2)

Based on the considerations outlined above, also the skewness of the breakthrough curve contains information on the connectivity. As a measure of connectivity, we have therefore also used the skewness of the breakthrough curve given by

$$CT_2 = \frac{1}{N_t}\sum_{j=1}^{N_t}\left(\frac{t_j - \bar{t}}{\sigma_t}\right)^3 \tag{4}$$

where N_t is the number of stream tubes used for the analysis, t_j are the average arrival times for the stream tubes, $\bar{t}$ is the overall average arrival time, and σ_t is the standard deviation of the arrival times. As CT_1 also CT_2 has the advantage of being based on data which are obtained from tracer tests.

3. METHODOLOGY

In order to evaluate the five measures of connectivity presented above, multilog-Gaussian fields were modified by rearranging K-values so that the connectivity was increased. The change in connectivity caused by the rearrangement was then analyzed and related to relevant hydrogeological parameters such as the effective conductivity and arrival times. Thus, the ability of the proposed measures of connectivity to describe the most important aspects of flow and advective transport could be evaluated.

The procedure was as follows:

1. The stochastic simulation program GCOSIM3D (Gomez-Hernandez and Journel, 1993) was used to generate several series of 50 two-dimensional Gaussian fields with the dimensions 64x64 cells. The original fields were generated used a spherical variogram with a variance of 4.0 and a range of 16 cell lengths.
2. All fields were modified by randomly choosing the location of a number of "fractures" - strings of cells with high K values - and subsequently interchanging the highest K values in the entire field with the values of the cells identified as fracture cells. This way, the histogram of K values remained unchanged. By ordering the two groups of values to be exchanged and placing the highest value from one group in the location of the highest value from the other group, and at the same time only interchanging a small fraction of the total number of K values, only very small changes in the variograms resulted from the modification of the fields. Fig. 2 shows two original and corresponding modified fields. The modification applied to the different series varied with respect to the number and length of connected features added to the fields.
3. Flow was solved for all fields using MODFLOW-2000 (Harbaugh et al., 2000). No-flow boundary conditions were imposed on two opposite sides. Two different constant heads were imposed on the remaining two boundaries.
4. Streamlines were determined by inverting both boundary conditions and conductivities and solving for flow. Fogg (1985) explains the procedure.
5. Values of all five connectivity measures listed in section 2 were calculated for all fields.

4. RESULTS

The six different series of 50 fields differed with respect to the number and length of the connected features generated by the modification of the K-distribution. In this paper, we only present the results from the first of the series. For this series, the original fields were modified by imposing the presence of four connected features of a maximum length of 32 cells. The connected features were allowed to fall partly outside the domain.

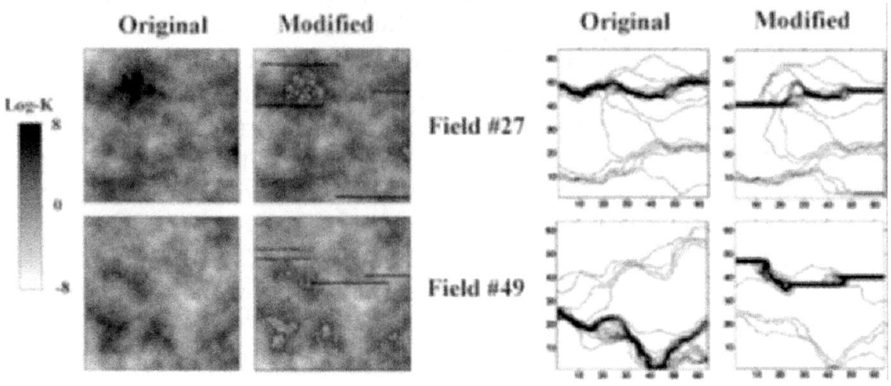

Figure 2. Two Gaussian fields before and after modification, and corresponding flow lines

It was found that the three measures of flow connectivity, CF_1, CF_2, and CF_3, contain the same information on flow connectivity. CF_1 and CF_2 are both functions of K_{eff} and are only different with respect to the scaling. In correspondence with the findings of Ambegaokar et al. (1971), we found that the effective conductivity of a system with high variance will be approximately given by K_C. Indeed, in one case (variance of log-K equal to 4), they correlate along a 1:1 line as shown in fig. 3. As a consequence, CF_3 differed only slightly from CF_2.

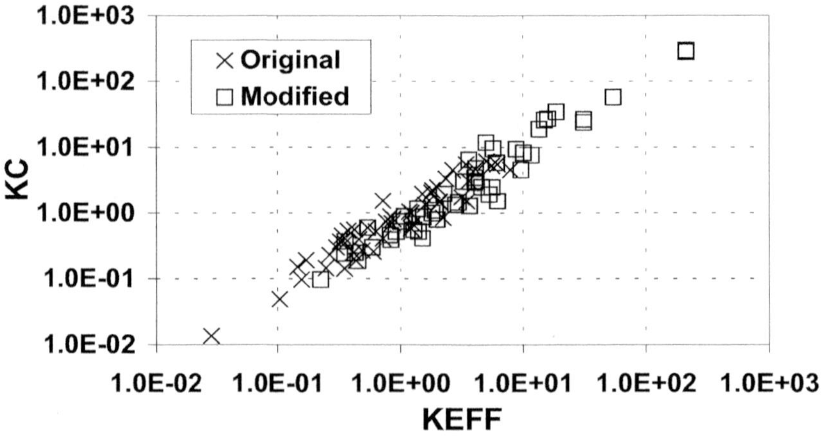

Figure 3. Relationship between the effective conductivity, K_{eff}, and the critical path conductivity, K_C.

The measure of transport connectivity based on the skewness of the breakthrough curve, CT_2, showed only a very minor increase as a consequence of the modifications of the fields. Thus, we chose to concentrate our analysis on CT_1 which changed more significantly as a consequence of the modifications.

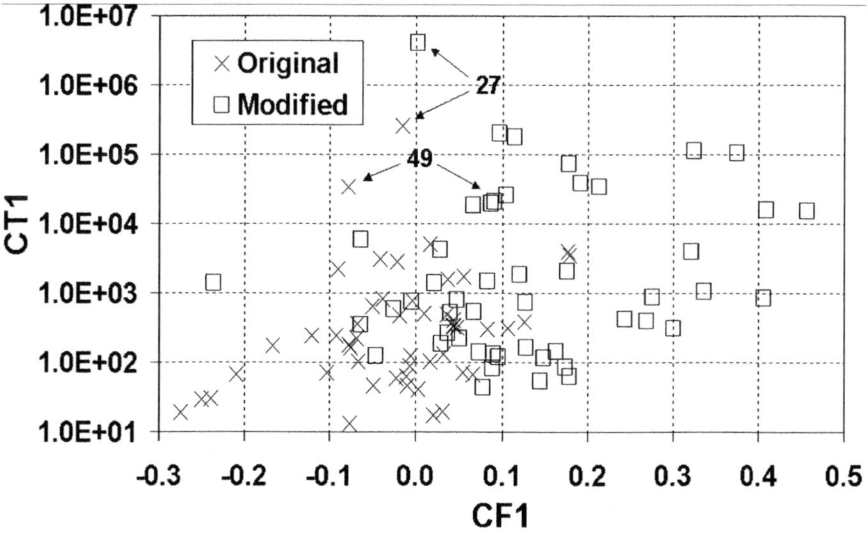

Figure 4. CF_1 vs. CT_1 for a series of 50 original and modified fields

Fig. 4 shows the relationship between CF_1 and CT_1 for a series of 50 fields. On average the rearrangement of the fields has caused a marked increase in both flow and transport connectivity, which suggests that they indeed measure what we perceive intuitively as connectivity. However, for some fields, the modifications caused an increase in one connectivity measure whereas the other measure remained constant or decreased. This indicates that there is a qualitative difference between connectivity of flow and transport as measured by CF_1 and CT_1. This corresponds with the conclusions of Scheibe and Yabusaki (1998) who showed that the power using for upscaling by power averaging depends on whether flow or transport is the process being upscaled. A closer look at the two fields shown in fig. 2 – fields 27 and 49 – can help explain this phenomenon. The connectivity measures for the two fields are listed in table 1.

Table 1. Connectivity measures for fields 27 and 49.

	Original		Modified		Relative Change	
Field #	27	49	27	49	27	49
CF_1	-0.016	-0.078	0.001	0.087	1.08	2.11
CF_2	0.83	0.46	1.02	2.49	0.22	4.44
CF_3	0.48	0.45	0.48	1.41	0.00	2.16
T_5	0.169	1.216	0.156	1.043	-0.08	-0.14
T_{AVE}	4.37E+4	4.21E+4	6.43E+5	2.08E+4	13.71	-0.51
CT_1	2.58E+5	3.47E+4	4.13E+6	1.99E+4	14.99	-0.43
CT_2	4.06	5.27	4.50	5.54	0.11	0.05

For field 27 the effective conductivity and thus CF_1 changes only insignificantly in response to the rearrangement of K-values. The flow lines depicted in fig. 2 show that the addition of connected features results in more concentrated flow along the main fast path through the domain. However, the fast path passes a low-K zone which probably is why the effective conductivity changes only slightly. On the other hand, CT_1 increases significantly, mainly because of an increase in the average arrival time T_{AVE} - probably due to the same low-K zone which constitutes a flow bottleneck. In other words, the presence of a low-K zone acting as a bottleneck along the fast path through the domain may affect flow connectivity more than transport connectivity. For field 49, the rearrangement caused CF_1 to increase. Apparently the two zones between the three connected high-K features do not significantly block the flow. T_{AVE} decreases whereas T_5 remains constant and as a consequence CT_1 decreases slightly. Contrary to field 27, the rearrangement of field 49 affected flow connectivity more than transport connectivity.

5. DISCUSSION AND CONCLUSIONS

Despite the apparent importance of connectivity, it has not yet been defined as a quantifiable parameter which is useful for hydrogeological research. We present several measures of flow and transport connectivity. The measures are analyzed with respect to their ability to explain the difference in flow and transport in multilog-Gaussian and non-multilog-Gaussian media. It is shown that for a given field, flow and transport connectivity, as measured by the two measures CF_1 and CT_1, respond differently to the same changes in the distribution of hydraulic conductivities. A measure of flow connectivity can increase while the measure of transport connectivity can decrease and vice-versa. This is partly due to fact that CF_1 and CT_1 contain different information on connectivity.

However, we believe that it also expresses a qualitative difference between flow and transport connectivity.

REFERENCES

1. Ambegaokar, V., Halperin, B.I., and Langer, L.S., 1971: Hopping Conductivity in Disordered Systems. Phys. Rev. B 4(8), 2612-2620
2. Davis, S.N., 1969: Porosity and Permeability of Natural Materials. In: R.J.M. de Wiest (Ed.), Flow Through Porous Media. Academic, New York, pp. 54-89
3. Fogg, G.E., 1985: Automatic Generation of Flow Nets with Conventional Ground-Water Modeling Algorithms. Ground Water 23(3), 336-344
4. Fogg, G.E., 1986: Groundwater Flow and Sand Body Interconnectedness in a Thick, Multiple-Aquifer System. Water Resour. Res. 22(5), 679-694
5. Fogg, G.E., Carle, S.F., Green, C., 2000: Connected-network paradigm for the alluvial aquifer system. In: Zhang, D. and Winter, C.L., eds., Theory, Modeling, and Field Investigation in Hydrogeology: A Special Volume in Honor of S. P. Neuman's 60th Birthday: Boulder Colorado, Geological Society of America, Special Paper 348, 25-42
6. Friedman, S.P. and Seaton, N.A., 1998: Critical path analysis of the relationship between permeability and electrical conductivity of three-dimensional pore networks. Water Resour. Res. 34(7), 1703-1710
7. Gomez-Hernandez, J.J. and Journel, A. G., 1993: Joint Sequential Simulation of Multi-Gaussian Fields. In: Troia '92, Amilcar Soares (Ed.), vol. 1, pp. 85-94, Kluwer.
8. Gomez-Hernandez and Wen, 1998: To be or not to be multi-Gaussian? A reflection on stochastic hydrogeology. Adv. Water. Res. 21, 47-61
9. Harbaugh, A.W., Banta, E.R., Hill, M.C., and McDonald, M.G., 2000: MODFLOW-2000, The U.S. Geological Survey Modular Ground-Water Model -- User Guide to Modularization Concepts and The Ground-Water Flow Process. USGS Open-File Report 00-92, 121 p.
10. Journel, A.G. and Deutsch, C., 1993: Entropy and spatial disorder. Math. Geol. 25(3), 329-355
11. Journel, A.G., 1983: Nonparametric estimation of spatial distributions. Math. Geol. 1(1), 79-96
12. Law, J., 1944: A Statistical Approach to the Interstitial Heterogeneity of Sand Reservoirs. Trans. Am. Inst. Mech. Eng. 155, 202-222
13. Matheron, G., 1967: Eléments pour une Théorie des Milieux Poruex. Masson, Paris.
14. Meier, P.M., Carrera, J., and Sanchez-Vila, X., 1998: An evaluation of Jacob's method for the interpretation of pumping tests in heterogeneous formations. Water Resour. Res. 34(5), 1011-1025
15. Renard, P. and de Marsily, G., 1997: Calculating Equivalent Permeability: A review. Adv. in Water Resour. 20(5-6), 253-278
16. Sánchez-Vila, X., Carrera, J. and Girardi, J.P., 1996: Scale Effects in Transmissivity. J. Hydrol. 183, 1-22
17. Scheibe, T.D. and Yabusaki, S., 1998: Scaling of flow and transport behavior in heterogeneous groundwater systems. Adv. Water Resour. 22(3), 223-238
18. Silliman, S.E. and Wright, A.L., 1988: Stochastic analysis of high hydraulic conductivity in porous media. Water Resour. Res. 24(11), 1901-1910
19. Stauffer, D. and Aharony, A., 1991: Introduction to Percolation Theory, 181 pp. Taylor and Francis, Philadelphia, PA

20. Webb, E.K. and Anderson, M.P., 1996: Simulation of preferential flow in three-dimensional heterogeneous conductivity fields with realistic internal architecture. Water Resour. Res. 32(3), 533-545
21. Western, A.W., Blöschl, G., and Grayson, R.B., 2001: Toward capturing hydrologically significant connectivity in spatial patterns. Water Resour. Res. 37(1), 83-97

MODELING OF REACTIVE CONTAMINANT TRANSPORT IN HYDRAULICALLY AND HYDROGEOCHEMICALLY HETEROGENEOUS AQUIFERS USING A GEOSTATISTICAL FACIES APPROACH

T. Ptak and R. Liedl
University of Tübingen, Center for Applied Geoscience, Sigwartstrasse 10, D-72076 Tübingen, Germany

Abstract: It is well known that aquifer structural properties and the resulting heterogeneous distribution of hydraulic conductivity and porosity significantly control groundwater flow and spreading of solutes. In addition to this, physico-chemical aquifer heterogeneity, i.e. different intra-particle sorption and diffusion properties for different source rocks of the aquifer material (lithological components) grouped in different grain size fractions, influence the interaction of reactive solutes with the aquifer material. To be able to consider both types of heterogeneity, a new 3D finite-difference reactive solute transport modeling approach was developed, being an essential component of a methodology allowing for the upscaling of small-scale laboratory measurements and for the assessment of parameter uncertainty. Sorption and desorption are introduced at grain scale through the simulation of a retarded intra-particle diffusion process in the heterogeneous aquifer material for each lithological component and each grain size fraction in every model cell. For a practical application of the code the data needed may be introduced into each model cell following a facies-based geostatistical approach. First modeling results emphasize the strong impact of the lithological aquifer material composition and confirm the need for a geostatistical process-based reactive transport modeling approach with spatially variable hydraulic and hydrogeochemical aquifer parameters.

X. Sanchez-Vila et al. (eds.), geoENV IV – Geostatistics for Environmental Applications, 247-258.

1. INTRODUCTION

Many organic contaminants in groundwater are not only subject to advection, dispersion and degradation, but also to sorption and desorption resulting in contaminant spreading which is retarded as compared to the transport of non-reactive compounds. It has been found that sorption and desorption processes in general cannot be adequately described by invoking the local equilibrium assumption (Sardin et al., 1991). Rather, sorption and desorption of organic compounds may exhibit a strong kinetic behavior, associated with an effective retardation factor increasing with time (e.g. Ball & Roberts, 1991a, b). Diffusive processes in intra-particle pores are mainly responsible for the sorption kinetics (e.g. Pignatello & Xing, 1996, Grathwohl, 1997).

Dealing with reactive transport modeling at field scale, both the hydraulic and the physico-chemical (hydrogeochemical) aquifer heterogeneities have to be considered. It is well known that aquifer structural properties, i.e. size, position and amount of clay lenses, sand and gravel layers, and the resulting heterogeneous distribution of hydraulic conductivity and porosity, significantly control groundwater flow and spreading of solutes (e.g. Dagan, 1989). In addition to this, hydrogeochemical aquifer heterogeneity, i.e. different intra-particle sorption and diffusion properties for different source rocks of the aquifer material (lithocomponents) grouped in different grain size fractions, influence the interaction of reactive solutes with the aquifer material (e.g. Kleineidam et al., 1999), and may tend to enhance tailing of reactive solutes, compared to non-reactive ones (e.g. Burr et al., 1994).

To be able to consider both the physical (hydraulic conductivity, porosity) and the hydrogeochemical aquifer heterogeneities (different intra-particle sorption and diffusion parameters for different lithological components of the aquifer material and the different grain size fractions), a 3D finite-difference solute transport modeling approach was developed. This approach is based on a sedimentological facies characterization using categorical variables and allows for upscaling of hydraulic and hydrogeochemical parameters, measured at laboratory scale, to field-scale scenarios.

2. THE 3D REACTIVE TRANSPORT MODELING APPROACH

The concept and the basic steps of the reactive transport simulation approach are summarized in Figure 1. The individual steps of the approach are described more detailed in the sections below.

2.1 Facies-based characterization of hydraulic and hydrogeochemical aquifer properties

For an application of the modeling approach, the aquifer is represented by a 3D finite-difference model grid. Due to the hydraulic and hydrogeochemical aquifer heterogeneity, the lithological composition and the grain size distribution may differ from one model cell to another. The data needed are introduced into each model cell following a facies-based categorical variable approach (Figure 1).

Since it is known that aquifer hydraulic properties are closely linked to the sedimentary lithofacies (for example well sorted sand, gravel with fine grain matrix etc., e.g. Kleineidam et al., 1999), the aquifer body is at first classified at the model cell scale (order of tens of cm) into typical lithofacies types (Ptak, 1997), which may be interpreted as aquifer material categories.

Herfort (2000) has shown that grain size distribution curves of aquifer material samples (usually sections of about 10 to 20 cm length of drill cores with 10 cm diameter) may be used for this classification, employing for example the K-means multivariate clustering algorithm (McQueen, 1967). In addition, categories or lithofacies types may also be attributed as soft information by an expert geologist (sedimentologist) through a visual inspection of drill core material. In this way, a large number of aquifer material category estimations, together with their position within the aquifer, can be obtained at affordable costs.

Then, for each lithofacies (aquifer material category, cluster of grain size distribution curves), characteristic sediment samples are collected, and a sediment material decomposition and analysis / batch experiment procedure is applied (Figure 1) to obtain the lithofacies-specific hydraulic parameters, mass fraction of (j,k)-grains (here j = index denoting a lithological component and k = index denoting a grain size class) and the lithocomponent-specific hydrogeochemical parameters, which are described below in Chapter 2.3. Of course, depending on the aquifer genesis, the facies properties may be site specific. The lithological composition of two facies from an experimental site in the Neckar Valley (South Germany) is shown as an example in Figure 2. The aquifer is composed of shallow Quaternary gravels with locally embedded sand, silt and loamy clay. Based on pumping tests, the average hydraulic conductivity was estimated as $2.5 \cdot 10^{-3}$ ms^{-1} (Herfort, 2000). The thickness of the aquifer varies between 0 m and 5.5 m, due to a structured base and a partial replacement by anthropogenic fills at the top.

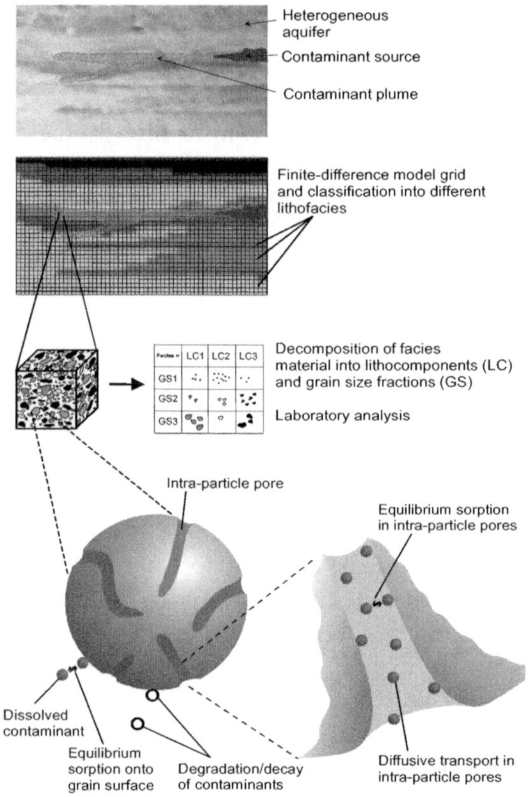

Figure 1. Concept and basic steps of the 3D reactive transport modeling approach.

2.2 Generation of 3D facies and fields of hydraulic and hydrogeochemical aquifer parameters

In the next step, a facies type respective aquifer material category (such as sand, gravel etc.) has to be assigned to each model cell. The 3D conditional sequential indicator simulation method (SIS) for categorical variables (Deutsch and Journel, 1992), which represent the different facies types, is applied to generate conditioned equiprobable 3D realizations of the facies fields. Figure 3 shows a realization of a site in the Neckar Valley. The parameters of the experimental and theoretical variograms shown in Table 1 are based on a geostatistical site characterization including 1420 data points from drilling logs and 120 sieve analyses of aquifer material (Peter, 2002).

With this approach, 3D hydraulic and hydrogeochemical parameter distributions are obtained simultaneously and can be used for reactive transport simulations at field scale, employing the numerical code described below. In this way, an upscaling of laboratory measurements for numerical field-scale simulations is achieved, without the need to define field-scale effective parameter values.

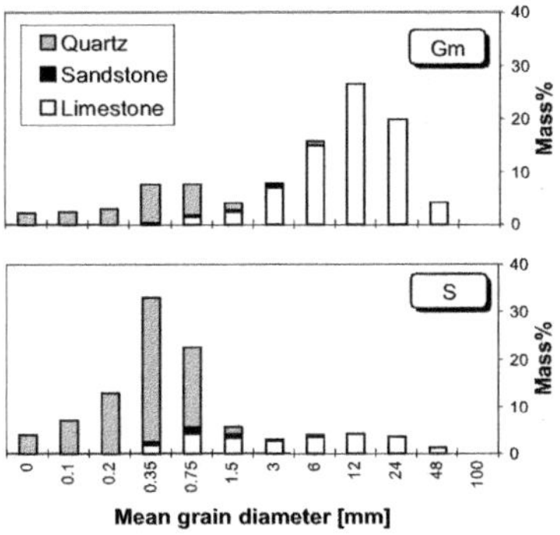

Figure 2. Lithological composition of two lithofacies (Gm = gravel with fine grain matrix, S = well sorted sand) from the Neckar Valley experimental site (Herfort, 2000).

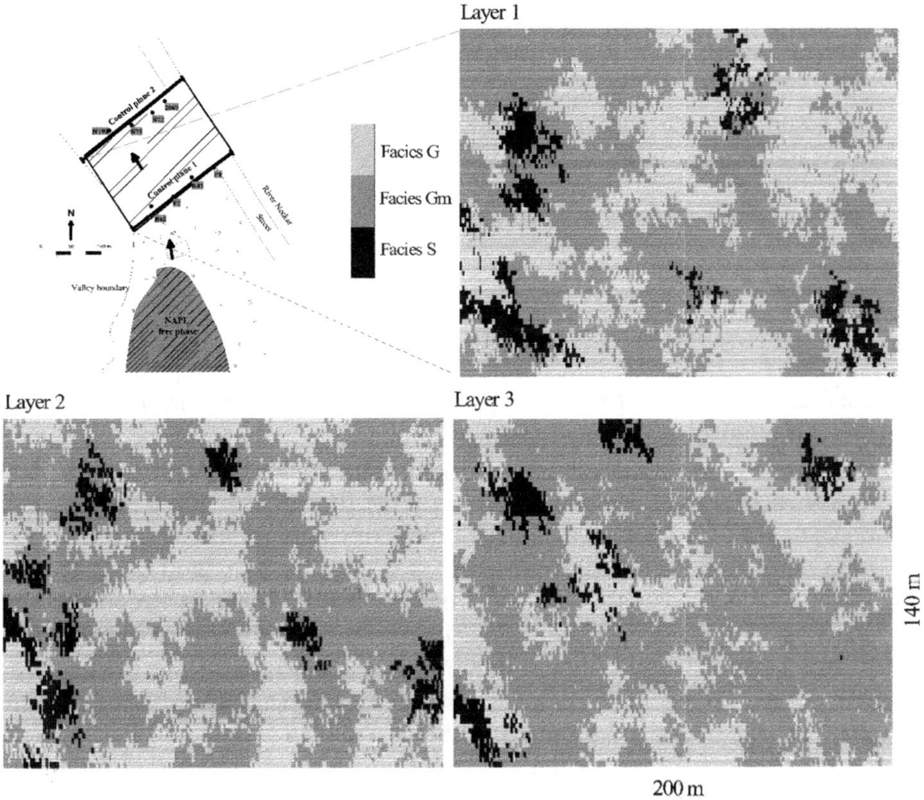

Figure 3. Typical facies distribution in the realizations at the Neckar Valley experimental site. Layer 1 denotes the top and Layer 3 the bottom of the aquifer (after Bockelmann, 2002).

Table 1. Parameters of the experimental and spherical theoretical variograms for each lithofacies. (ldis: lag distance; ltol: lag tolerance; n: nugget; s: sill; r: range) (Peter, 2002).

Direction[1]	Lithofacies G					Lithofacies Gm					Lithofacies S				
	Exper. Vario.		Sph. Theor. Vario.			Exper. Vario.		Sph. Theor. Vario.			Exper. Vario.		Sph. Theor. Vario.		
	ldis	*ltol*	*n*	*s*	*r*	*ldis*	*ltol*	*n*	*s*	*r*	*ldis*	*ltol*	*N*	*s*	*r*
1	25	12.5	0.06	0.22	58	20	10	0.05	0.26	45	20	10	0.006	0.046	60
2	25	12.5	"	"	48	20	10	"	"	38	20	10	"	"	40
3	0.2	0.1	"	"	5	0.2	0.1	"	"	5	0.2	0.1	"	"	2

[1] Investigated directions are (1) 112.5° east from north, (2) 22.5° east from north and (3) vertical direction

2.3 Modeling of flow and reactive transport

Sorption / desorption of reactive solutes is modeled by a diffusion-based formulation at grain scale, instead of employing transfer rate models involving empirical parameters. In each model cell, the retarded intra-particle diffusion process in the heterogeneous aquifer material is simulated for each lithological component and each grain size fraction. As the grains are assumed to be spherically symmetrical this equation can be written as (e.g. Grathwohl, 1997):

$$\frac{\partial}{\partial t}\left[\varepsilon_j c_{jk} + (1-\varepsilon_j)\ \rho_j\ \sigma_{jk}\right] = \frac{D_{aq}}{\tau_j}\frac{1}{r^2}\frac{\partial}{\partial r}\left[r^2\frac{\partial}{\partial r}\left(\varepsilon_j\ c_{jk}\right)\right] \tag{1}$$

with t = time [T], r = radial coordinate [L], D_{aq} = aqueous diffusion coefficient of the reactive solute [L^2T^{-1}], τ_j = tortuosity of intra-particle pores of a lithological component j [-], ε_j = intra-particle porosity of lithological component j [-], ρ_j = dry solid density of lithological component j [ML^{-3}], c_{jk} = concentration of chemical dissolved in the fluid phase within intra-particle pores of the (j,k)-grains [ML^{-3}], σ_{jk} = mass of chemical sorbed on surfaces of intra-particle pores of the (j,k)-grains per unit mass of the (j,k)-grains [-]. Equation (1) needs data obtained from the aquifer material decomposition, analysis and batch experiments (Figure 1). Diffusion is assumed to be retarded due to equilibrium sorption onto the surfaces of the intra-particle pores. This process can be quantified by $\sigma_{jk} = G_{jk}(c_{jk})$ where G_{jk} represents the type of sorption / desorption isotherm obtained from an aquifer material analysis (Figure 1) and batch experiments.

From the intra-particle diffusion equation (1) the mass of a chemical within the (j,k)-grains can be given per unit volume by an integral of equation (1) over the volume of the sphere:

$$m_{jk} = \frac{\rho f_{jk}}{\frac{4\pi}{3} R_k^3 \left(1-\varepsilon_j\right)\rho_j} \cdot \int_0^{R_k} 4\pi r^2 \left[\varepsilon_j c_{jk} + \left(1-\varepsilon_j\right)\rho_j \sigma_{jk}\right] dr \tag{2}$$

where R_k = grain radius of grain size class k [L] and f_{jk} = mass fraction of (j,k)-grains [-] according to the lithological and grain size decomposition (Figure 1). This mass is required for solving the reactive transport equation, extended with terms to consider (j,k)-grain-specific degradation and sorption / desorption of the chemical:

$$\frac{\partial}{\partial t}\left[nc + \sum_{j,k}\left(f_{jk}\rho s_{jk} + m_{jk}\right)\right] + \mathrm{div}\left(v_f c - nD\,\mathrm{grad}\,c\right) = Nc' - \lambda_d nc - \rho \sum_{j,k} f_{jk}\lambda_{d,jk} s_{jk} \tag{3}$$

with n = effective inter-particle porosity [-], ρ = bulk density [ML^{-3}], c = solute concentration in inter-particle pore space [ML^{-3}], D = local dispersion tensor [L^{2}T^{-1}], λ_d = degradation rate for the chemical dissolved in inter-particle pore space [T^{-1}], s_{jk} = mass of the chemical sorbed onto surfaces of the (j,k)-grains per unit mass of the (j,k)-grains [-], $\lambda_{d,jk}$ = degradation rate for the chemical sorbed onto surfaces of (j,k)-grains [T^{-1}], m_{jk} = mass of the chemical in intra-particle pores of the (j,k)-grains per unit volume [ML^{-3}], and c' = concentration of injected or withdrawn solute [ML^{-3}] (for withdrawal $c' = c$). Sorption of chemicals onto the outer grain surfaces is assumed to occur under equilibrium conditions so that isotherms $s_{jk} = F_{jk}(c)$ can be employed for each lithological component j and each grain size class k. F_{jk} may denote any type of isotherm such as linear, Freundlich, or others, obtained from batch experiments (Figure 1).

Equations (1) and (3) are coupled by $c_{jk}\left(\underline{x}, r = R_k, t\right) = c(\underline{x}, t)$ for any point $\underline{x}$ in a 1D, 2D or 3D model domain, i.e. solute concentration is assumed to vary continuously at the "interface" between inter- and intra-particle pore space ($r = R_k$). Additional initial and boundary conditions have to be specified for inter-particle transport according to the scenario to be modeled.

The mathematical model presented in this section has to be solved numerically due to the complex interaction of large (field) scale transport and local (grain) scale diffusion as well as linear or non-linear equilibrium sorption / desorption processes. For this purpose, the well known MT3D code (Zheng, 1990) and a new finite-difference code IPD (Jaeger & Liedl, 2000) for the intra-particle diffusion have been combined. The added IPD module simulates within each model cell the retarded intra-particle diffusion process in heterogeneous aquifer material by solving equation (1) for each lithological component j and each grain size class k. The extended version of MT3D is called MT3D-IPD.

Using MT3D-IPD, flow and transport simulations are finally performed for the reactive solutes within the generated hydraulic and hydrogeochemical parameter fields, with an aquifer geometry as well as initial and boundary conditions according to the field scale scenario. In addition, an ensemble of equiprobable realizations of the aquifer parameter fields may yield an assessment of parameter uncertainty in a Monte-Carlo-type stochastic framework.

3. EXAMPLES OF APPLICATION

3.1 Laboratory column

The first example refers to 1D solute transport in a hydraulically homogeneous column focusing on the impact of hydrogeochemical heterogeneity. The flow in the column is steady-state without sources or sinks ($N = 0$). Hydraulic conductivity and effective inter-particle porosity are set equal to $K = 10^{-4}$ ms^{-1} and $n = 0.4$, respectively. The length of the column is $L = 0.2$ m, and a constant head difference is maintained such that linear velocity $v_f n^{-1}$ equals 1 md^{-1} = $1.16 \cdot 10^{-5}$ ms^{-1}. Longitudinal dispersivity is assumed to be $\alpha_L = 4 \cdot 10^{-4}$ m. All modeling exercises refer to a continuous injection of Phenanthrene with an input concentration $c_{in} = c(x=0,t) = 40$ μgl^{-1} into an initially uncontaminated column. Phenanthrene was chosen here as a representative compound, since it belongs to the US-EPA priority pollutant list. Its aqueous diffusion coefficient equals $5.72 \cdot 10^{-6}$ cm^{2}s^{-1}. As the modeling studies are focused on the investigation of intra-particle diffusion, decay and sorption onto outer grain surfaces are neglected, i.e. $\lambda_d = \lambda_{d,jk} = 0$ and $s_{jk} = 0$, respectively. Sorption onto the walls of the intra-particle pores is modeled by Freundlich isotherms, i.e. $\sigma_{jk} = G_{jk}(c_{jk}) = K_{Fr,jk}\, c_{jk}^{n_{Fr,jk}}$ with $K_{Fr,jk}$ = Freundlich coefficient $[(M^{-1}L^3)^{nFr,jk}]$ and $n_{Fr,jk}$ = Freundlich exponent [-] of the (j,k)-grains.

The intra-particle diffusion of Phenanthrene is studied for three lithological components (or facies) with different sorption / desorption properties: sandstone, light-colored limestone and dark-colored limestone. Parameters of these lithological components resemble data published by Kleineidam et al. (1999) and are summarized in Table 2.

Table 2. Hydrogeochemical parameters of lithocomponents respective facies (model input data).

Parameter \ Lithocomponent	Sandstone	Light-colored limestone	Dark-colored limestone
Intra-particle porosity ε_j [-]	0.0195	0.0054	0.0035
Tortuosity τ_j [-]	12	20	590
Dry solid density ρ_j [gcm^{-3}]	2.69	2.73	2.74
Freundlich coefficient $K_{Fr,j}$ [(ml/µg)^{nFr,j}]	$2.9 \cdot 10^{-6}$	$5.5 \cdot 10^{-6}$	$3.6 \cdot 10^{-5}$
Freundlich exponent $n_{Fr,j}$ [-]	0.66	0.67	0.33

In order to maintain a constant surface-to-volume ratio all grains are assumed to have the same radius $R_k = 0.015$ cm. As an example of the results, Figure 4 shows breakthrough curves (BTCs) at the column outlet obtained from test runs simulating sorption. The runs can be distinguished by the mass fractions f_j for each lithological component.

Table 3. Mass fractions [%] of lithocomponents respective facies for the different model runs.

	Sandstone	Light-colored limestone	Dark-colored limestone
Run 1	33.33	33.33	33.33
Run 2	100	0	0
Run 3	0	100	0
Run 4	0	0	100

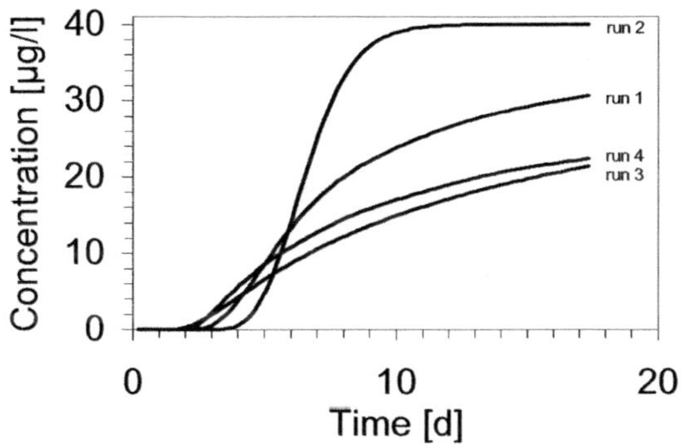

Figure 4. Breakthrough curves for sorption model runs(run 1: $f_j = 1/3$ for sandstone, light-colored and dark-colored limestone; run 2: sandstone only; run 3: light-colored limestone only, run 4: dark-colored limestone only).

It can be seen from Figure 4 that the first breakthrough for the sandstone column (run 2) occurs at later times than for the limestone columns. This is due to the higher apparent intra-particle diffusion coefficient of the sandstone grains The long-term behavior is explained by the sorption capacity factors providing a qualitative measure for the amount of contaminant which can be stored per solid mass in the intra-particle pores of each lithological component for a certain solute concentration in the inter-particle pores. The low capacity factor of the sandstone explains why the BTC belonging to model run 2 approaches the input concentration much more rapidly than the BTCs for the limestone columns. For comparison, Figure 4 also shows the BTC for a column filled with sandstone, light-colored limestone and dark-colored limestone at identical mass fractions (run 1). Of course, this BTC cannot be obtained by simple arithmetic averaging of the BTCs representing the lithologically homogeneous cases. This is mainly due to the temporally changing ratios of contaminant uptake by the three lithological components and the non-linear isotherms.

3.2 Reactive transport modeling at field scale

In the second example, Monte-Carlo type MT3D-IPD reactive transport simulations are conducted for a field-scale scenario using geostatistically generated facies-based aquifer realizations (Chapter 2.2). Figure 5 shows corresponding normalized breakthrough curves of Acenaphthene which can produce plumes of significant lengths and concentrations at gaswork sites. For the simulations, a cutout of a calibrated larger scale flow model was used, with constant heads at the two control planes (Figure 3) and no-flow boundaries elsewhere. A constant mass flow was applied at control plane 1. Integral (i.e. representative of a control plane as a whole) breakthrough curves of Acenaphthene mass flow were recorded at control plane 2.

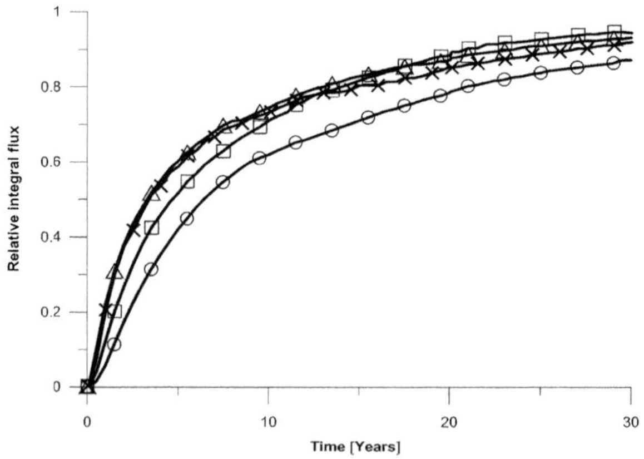

Figure 5. Examples of modeled relative integral mass flow of Acenaphthene (normalized with the input mass flow) at control plane 2 situated 140 m downgradient from the modeled source (control plane 1). Simulation started with an uncontaminated aquifer and accounted for intra-particle diffusion and non-linear intra-particle sorption of Acenaphthene. The differences between the realizations F1 (circles), F4 (triangles), F18 (squares), and F20 (crosses) decline with time as the partitioning of Acenaphthene approaches equilibrium (Bockelmann, 2002).

In all realizations, the diffusion-limited sorption of acenaphthene is a key attenuation process during the first 10 to 20 years of the contamination, assuming a time-invariant contaminant input at control plane 1 and a decoupling of diffusion-limited sorption and degradation. The latter was not included in the simulation in order to estimate the minimum time to achieve sorption equilibrium. Biodegradation of the contaminants might lead to an extension of the timeframe in which sorption is an important attenuation process. It can be seen from Figure 5 that even after 30 years of release the aquifer system is not at equilibrium. Using equilibrium retardation factors for reactive transport predictions, as it is very often done in practice, would strongly overpredict the retardation of the contaminant. Using MT3D-IPD the retardation can be modeled as an outcome of a diffusion-controlled sorption process in a hydraulically and hydrogeochemically heterogeneous aquifer, allowing to make physically correct predictions of contaminant spreading and plume development, without relying on, for example, simple fitted (first-order) rate models.

4. CONCLUSIONS AND FUTURE WORK

The modelling examples emphasize a strong impact of the lithological aquifer material composition on reactive solute transport predictions. Therefore, the joint simulation of sorption and desorption at small-scale and groundwater flow and transport at large-scale is regarded as an essential prerequisite for simulating field-scale scenarios of reactive solute spreading. The introduced generation of spatially variable facies distributions offers an effective possibility to consider heterogeneous hydraulic and hydrogeochemical aquifer parameter fields. The method allows for an upscaling of hydraulic and hydrogeochemical laboratory measurements to field-scale scenarios without the need to introduce empirical and / or scale-dependent effective parameter values, or some a priori correlation functions of hydraulic conductivities and distribution coefficients. It offers a broad field of applications, e.g. for the assessment of plume spreading and groundwater contamination risk at polluted sites, for the evaluation of the natural attenuation potential, or for the planning of active remediation activities.

REFERENCES

1. Ball, W. P. and Roberts, P. V. (1991a): Long-term sorption of halogenated organic chemicals by aquifer material – 1. Equilibrium. *Environ. Sci. Technol.* 25(7): 1223-1237.

2. Ball, W. P. and Roberts, P. V. (1991b): Long-term sorption of halogenated organic chemicals by aquifer material – 2. Intraparticle diffusion. *Environ. Sci. Technol.* 25(7): 1237-1249.
3. Bockelmann, A. (2002): Natural attenuation of organic contaminants: Integral mass flux estimation and reactive transport modeling in heterogeneous porous media. Ph.D. Thesis, University of Tübingen, Tübingen.
4. Burr, D.T., Sudicky, E.A. and Naff, R.L. (1994): Nonreactive and reactive solute transport in three-dimensional heterogeneous porous media: Mean displacement, plume spreading, and uncertainty, Water Resour. Res., 30(3), 791-815.
5. Dagan, G. (1989): *Flow and transport in porous formations.* Berlin: Springer.
6. Deutsch, C.V. and Journel, A.G. (1992): *GSLIB Geostatistical software library and user's guide.* Oxford University Press, New York, 340 pp.
7. Grathwohl, P. (1997): *Diffusion in natural porous media - Contaminant transport, sorption/desorption and dissolution kinetics.* Dordrecht: Kluwer.
8. Herfort, M. (2000): Reactive transport of organic compounds within a heterogeneous porous aquifer. Ph. D. Thesis, C54, University of Tübingen, Center for Applied Geoscience, Tübingen, 59 pp.
9. Jaeger, R. and Liedl, R. (2000): Prognose der Sorptionskinetik organischer Schadstoffe in heterogenem Aquifermaterial (Predicting sorption kinetics of organic contaminants in heterogeneous aquifer material). *Grundwasser* 5(2): 57-66.
10. Kleineidam, S., Rügner, H. and Grathwohl, P. (1999): Impact of grain scale heterogeneity on slow sorption kinetics. *Environ. Toxic. Chem.* 18(8): 1673-1678.
11. McQueen, J. (1967): Some methods for classification and analysis of multivariate observations. *5th Berkeley Symposium on Mathematics, Statistics and Probability*, 1, 281-298.
12. Peter, A. (2002): Assessing natural attenuation at field scale by stochastic reactive transport modeling. Ph.D. Thesis, University of Tübingen, Tübingen.
13. Pignatello, J. J. and Xing, B. (1996): Mechanisms of slow sorption of organic chemicals to natural particles. *Environ. Sci. Technol.* 30(1): 1-11.
14. Ptak, T. (1997): Evaluation of reactive transport processes in a heterogeneous porous aquifer within a non-parametric numerical stochastic transport modelling framework based on sequential indicator simulation of categorical variables. In A. Soares et al. (eds.), *geoENV I - Geostatistics for Environmental Applications*, Kluwer: 153-164.
15. Sardin, M., Schweich, D., Leij, F. J. and van Genuchten, M. T. (1991): Modeling the nonequilibrium transport of linearly interacting solutes in porous media: A review. *Water Resour. Res.* 27(9): 2287-2307.
16. Zheng, C. (1990): *MT3D – A modular three-dimensional transport model for simulation of advection, dispersion and chemical reactions of contaminants in groundwater systems.* S. S. Papadopoulos and Associates, Inc.

EFFECT OF HETEROGENEITY ON AQUIFER RECLAMATION TIME

M. Riva[1], X. Sanchez-Vila[2], M. De Simoni[1], A. Guadagnini[1], M. Willmann[2]
[1]D.I.I.A.R, Politecnico di Milano, Piazza L. Da Vinci, 32, 20133 Milano, Italy; [2]Departament of Geotechnical Engineering and Geosciences, Technical University of Catalonia, Gran Capità S/N, 08034 Barcelona, Spain

Abstract: We consider the effect of heterogeneity on estimation of the time that is necessary to reclaim an aquifer by means of a constant rate pumping well. We derive the predictor of *resident time* (rendered by its mean) together with the associated prediction error (rendered by its variance) for non reactive solute particles under mean radial flow conditions in a randomly, spatially correlated, heterogeneous aquifer. The solutions are obtained numerically following a Monte Carlo procedure, and compared with a newly developed first-order analytical approach. Agreement between analytical and numerical results is very good. One of the main results is that the mean travel time is always larger than the deterministic value obtained assuming a homogeneous media. Our analysis can be used in planning water resources protection strategies, since it would be possible to obtain an estimate of the maximum clean-up time, associated with a design probability, which would substitute the use of a single (underestimated) deterministic value.

1. INTRODUCTION

To design and implement properly clean-up of a contaminant plume in groundwater it is decisive to obtain a good estimation of the contaminant travel time which renders the aquifer reclamation time. Prediction of contaminant travel time in aquifers is commonly accomplished by means of models based on the assumption of a deterministic knowledge of the medium hydrogeological properties. However, hydraulic conductivity has been found to vary orders of magnitude at relatively close locations even in apparently fairly homogeneous aquifers.

X. Sanchez-Vila et al. (eds.), geoENV IV – Geostatistics for Environmental Applications, 259-270.

As our knowledge of the medium will never be complete, we should recognize the need to cast the equations that govern groundwater flow and contaminant transport within a stochastic framework. The latter is oriented towards rendering ensemble moments of quantities such as flux and travel time of solutes. Although of high relevance in practical applications, problems associated to contaminant transport in the vicinity of extraction wells in heterogeneous media have been tackled only recently (e.g. *Guadagnini and Franzetti* [1999], *Riva et al.* [1999], *Dagan and Indelman* [1999], *van Leeuwen et al.* [2000], *Feyen et al.* [2001]).

Here we consider the effect of random heterogeneity of the natural logarithm, Y, of transmissivity, T, upon the estimation of the time that is necessary to reclaim an aquifer by means of a constant rate pumping well, creating a mean radial flow within the polluted area. We derive the predictor of resident time (rendered by its mean) together with the associated prediction error (rendered by its variance) for non reactive solute particles injected at various distances from the well. The solutions are obtained numerically, by means of Monte Carlo simulations (detailed in Section 3.1) and compared to the analytical results recently developed by *Guadagnini et al.* [2001] (outlined in Section 3.2).

2. MATHEMATICAL STATEMENT OF THE PROBLEM

We consider incompressible groundwater steady state flow to a well located at the center of a polluted area (Figure 1). The modelling area is a circle of radius L. Inside this area there is another circle of radius r_0 ($< L$) corresponding to the limit of the polluted area. The radius of the well is small compared to all other relevant distances such as external radius or integral distance, and therefore we assume a zero radius. The well pumps at a constant deterministic rate, Q; head at the outer circular boundary remains at a constant deterministic value, H_L. We aim at evaluating the pumping time that is necessary to reclaim the aquifer for various levels of heterogeneity of the log-transmissivity field, $Y = ln\ T$, and relative extension of the polluted area.

We consider only the advective component of transport and disregard local dispersion. The two-dimensional scheme presented is useful for relatively thin aquifers in which vertical heterogeneity tends to be of minor concern relative to that in the horizontal plane.

In order to study reclamation time we concentrate on the pollutant particles that are initially located further from the well and study their travel time from their initial (at time $t = t_0 = 0$) location (point $r = r_0$, polar coordinates used throughout the text) to the well. The time for this particle to reach the well (residence time) is equal to [*Guadagnini et al.*, 2001]:

$$t(\mathbf{r}_0) = \int_{r_0}^{0} \frac{dr}{V_r\left(r, \varphi(r, \mathbf{r}_0)\right)} \tag{1}$$

where V_r is the radial component of the Lagrangian velocity vector and $\varphi(r, \mathbf{r}_0)$ is the trajectory of the particle initially at location $\mathbf{r}_0$ (i.e. is the angular displacement of the particle when reaching radial distance r from the well).

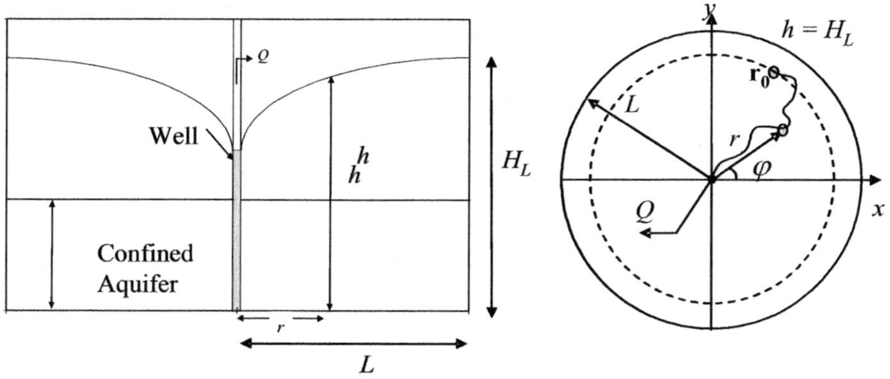

Figure1. Sketch of the domain. The pumping well is placed at the center of the domain (radius L). The dashed line indicates the border of the polluted area.

The randomness of transmissivity causes *residence time* to be also random and we aim to evaluate its statistical moments (in terms of ensemble mean and variance).

3. ENSEMBLE MOMENT OF SOLUTE RESIDENCE TIME

3.1 Numerical Monte Carlo simulations

We performed an extensive suite of numerical Monte Carlo simulations (MC) using an *ad hoc* code for steady state flow and transport in a square domain with 100 rows and 100 columns of uniform size ($\Delta x = \Delta y = \Delta$). A circular boundary of radius L = 50 Δ was defined about the well by designating all cells outside it as inactive.

We model the log hydraulic transmissivity, $Y(\boldsymbol{r}) = \ln\ T(\boldsymbol{r})$, as a statistically homogeneous and isotropic random function of space with covariance between two points $\mathbf{r}_{\mathrm{I}} \equiv (r_{\mathrm{I}}, \theta_{\mathrm{I}})$ and $\mathbf{r}_{\mathrm{II}} \equiv (r_{\mathrm{II}}, \theta_{\mathrm{II}})$ given, in dimensionless coordinate, $\xi_i = r_i/L$ (with i = I, II), by

$$C_Y(\xi_{\mathrm{I}}, \xi_{\mathrm{II}}, \theta_{\mathrm{I}} - \theta_{\mathrm{II}}) = \sigma_Y^2\ exp\left[-\omega^2\ d^2\right] \qquad (2)$$

where σ_Y^2 is the variance of Y, $\omega = \sqrt{\pi}\ L/(2\lambda)$, λ is the correlation length and d is the Euclidean distance ($d = \sqrt{\xi_{\mathrm{I}}^2 + \xi_{\mathrm{II}}^2 - 2\xi_{\mathrm{I}}\ \xi_{\mathrm{II}}\ cos(\theta_{\mathrm{I}} - \theta_{\mathrm{II}})}$).

The hydraulic head, H_L, along the circular boundary was set equal to 80 (in consistent units). A pumping well at a constant rate Q = 100 was placed at the central node of the grid. Gaussian sequential simulation (both codes SGSIM [*Deutsch and Journel*, 1998] and GCOSIM3D [*Gómez-Hernández*, 1991]) was used to generate random realizations of log Y on the above defined grid. Each realization constituted a sample from a multivariate Gaussian, statistically homogeneous field, $Y = \ln T$, with an isotropic Gaussian covariance (Eq. 2), with mean $\langle Y \rangle = 0$, variance σ_Y^2 ranging from 0.1 to 1.0 and spatial horizontal correlation length $\lambda = 0.1\ L$ or L. The effective porosity, n, is taken as a constant ($n = 0.3$).

Flow was solved by Galerkin finite elements using bilinear shape functions. An additional series of runs were performed with MODFLOW [*McDonald and Harbaugh*, 1988]. The results from our code were similar in terms of heads and fluxes, but convergence was faster. A number of Monte Carlo simulations ranging from 1000 (for the smaller σ_Y^2 and λ) to 4000 (for the larger σ_Y^2 and λ) was performed.

Solute transport in each realization is modeled by Particle Tracking, using an ad-hoc computer code validated with MODPATH [*McDonald and Harbaugh*, 1988]. To compute moments of aquifer reclamation time, an ideal tracer particle is located at grid nodes of radial distances from the well $\xi_0 = r_0/L$ = 0,1; 0.5; 0.9 and various angular positions, θ_0. Tracking was stopped when the particles reached one of the cells sharing the well node. To get the total travel time we added the time to get from the last trajectory end-point to the well (separated by a distance r), which is computed by means of the well known equation for the steady state radial flow in a homogeneous and isotropic field: $t = n\pi r^2 / Q$.

Due to the radial symmetry of the problem (domain, flow and boundary conditions) when the convergence is attained, the statistical moments of t are independent of the angular position of the solute starting points, θ_0.

3.2 Analytical solution

The analytical solution is based on exact nonlocal equations and their recursive approximations for (ensemble) moments of multidimensional steady state flow in bounded, randomly heterogeneous porous media developed by *Neuman and Orr* [1993], *Neuman et al.* [1996], and *Guadagnini and Neuman* [1999a-b] and makes full use of the analytical solutions developed by *Riva et al.* [2001] for mean radial flow taking place in the type of domain represented in Figure 1. In the following we report only the main results, while additional details can be found in *De Simoni* [2001] and *Guadagnini et al.* [2001].

We present our solution for mean residence time as an asymptotic expansion in the log-transmissivity variance, σ_Y^2, truncated up to first order

$$\langle t^{[1]} \rangle = \langle t^{(0)} \rangle + \langle t^{(1)} \rangle \tag{3}$$

where superscripts in angular bracket define the order of expansion and those in parentheses designate the order of its individual components. The zero order solution coincides with the travel time to a well in a homogeneous aquifer

$$\langle t^{(0)} \rangle = n \frac{\pi L^2}{Q} \xi_0^2 \tag{4}$$

The first order (in σ_Y^2) component of the mean travel time, for a Gaussian autocorrelation function of Y (Eq. 2), is given by

$$\langle t^{(1)} \rangle = n \frac{\pi L^2}{Q} \frac{\sigma_Y^2}{2} \frac{L^2}{\lambda^2} \left\{ \Im_1 + \frac{\Im_2}{2\pi} \right\} \tag{5}$$

where $\Im_1$ and $\Im_2$ are multidimensional integrals of cross-products between partial derivatives of the correlation function of Y and the Green's function of the zero-order flow problem (for details the reader is referred to *Guadagnini et al.*, 2001). We observe that $\langle t^{(1)} \rangle$ vanishes when the domain is very small with respect to the correlation scale of Y ($L/\lambda \to 0$) and the mean residence time (at least at first order in σ_Y^2) coincides with that obtained in a homogeneous aquifer (Eq. 4). This situation corresponds to the case where transmissivity is a random constant, so that the travel time to the well is given, for each realization, by the zero order solution.

To evaluate the three- ($\Im_1$) and five- ($\Im_2$) dimensional integrals in Eq. (5) we used Gaussian quadratures. In all the cases analyzed we obtained the convergence of the results with less than 80 Gauss points. Employing a 800 MHz Pentium III processor (RAM being immaterial) the computations took about 12 hours for each value of ξ_0 and ratio L/λ.

The first (and thus lowest) order approximation of residence time variance (for a Gaussian autocorrelation function of Y) is given by

$$\sigma_t^{2(1)}(\xi_0) = 4\pi^2 n^2 \frac{L^4}{Q^2}\sigma_Y^2 \left\{ \frac{2}{3\pi}\frac{\lambda^2}{L^2}\left(2\exp\left[-\omega^2\xi_0^2\right]\left(\xi_0^2 - \frac{1}{2\omega^2}\right) + \frac{1}{\omega^2} - 3\xi_0^2 \right) + \frac{2}{3}\frac{\lambda}{L}\xi_0^3 Erf\left[\omega\,\xi_0\right] - \frac{\xi_0^4}{4} + \frac{1}{\pi}\left(\frac{1}{8}\frac{L^2}{\lambda^2}\Im_3 + \Im_4 \right) \right\} \tag{6}$$

where $\Im_3$ and $\Im_4$ are multidimensional integrals of cross-products between partial derivatives of the correlation function of Y and the Green's function of the zero-order flow problem (for details the reader is referred to *Guadagnini et al.*, 2001). Consistently with our discussion about mean travel time, we observe that when the domain is very small with regard to the correlation scale of Y ($L/\lambda \rightarrow 0$), σ_t^2 vanishes independently of the size of the contaminated area, ξ_0.

To evaluate the two- ($\Im_4$) and four- ($\Im_3$) dimensional integrals in Eq. (6) we used Gaussian quadratures, obtaining convergence of the results with less than 80 Gauss points, taking a maximum of 15 minutes on a 800 MHz Pentium III.

4. RESULTS AND DISCUSSION

Figure 2 depicts the dependence on σ_Y^2 of dimensionless mean residence time, ⟨t⟩/C (where $C = n\pi L^2 / Q$), computed by our numerical Monte Carlo simulations and the first order analytical solution (3) – (5), when $L/\lambda = 1$ (a), and 10 (b) for three ξ_0 values. For reference, we also report the zero-order solution (Eq. 4). We judge the agreement between MC and our analytical solution as excellent, the largest discrepancy being only a few units percent for the larger value of heterogeneity considered (σ_Y^2=1). The most significant outcome is that while the homogeneous solution (zero-order solution) can be a good estimate of the mean aquifer reclamation time for weakly heterogeneity ($\sigma_Y^2 \leq 0.1$), it significantly underestimates the solute mean residence time for mildly and highly heterogeneous aquifers. The observed discrepancies are up to 20 % between zero- and first-order solutions. Therefore, ignoring the fact that heterogeneity causes an increase of our estimate of solute residence time would lead to inadequate aquifer reclamation designs.

Figure 3 depicts the dependence of dimensionless residence time variance, $\sigma_t^2/4C^2$ on σ_Y^2 for the same situations examined in Figure 2. Residence time variance increases monotonically with the distance between the release point and the well. Again, while the agreement between the MC results and our analytical solution is quite good, it deteriorates as σ_Y^2 increases, consistently with limitations inherent of perturbation results.

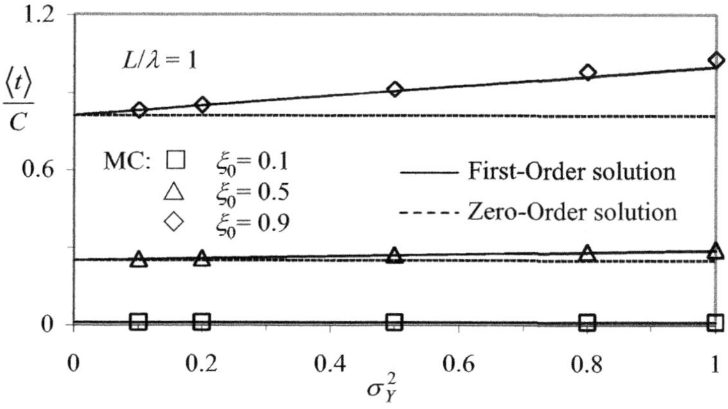

Figure 2a. Dimensionless mean residence time versus σ_Y^2 when L/λ = 1 and ξ_0 = 0.1; 0.5; 0.9.

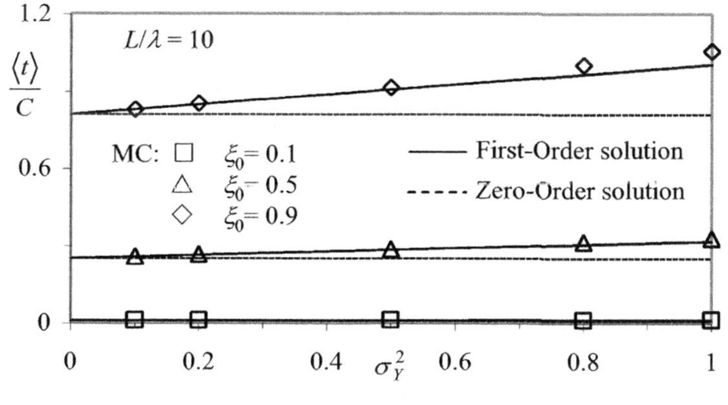

Figure 2b. Dimensionless mean residence time versus σ_Y^2 when L/λ = 10 and ξ_0 =0.1;0.5; 0.9.

In order to quantify reliability of predictions based only on the first two (statistical) moments of the state variable of interest, we tested the Gaussianity of the natural logarithm of residence time, $\tau = \ln(t/C)$, obtained from the Monte Carlo simulations by performing the χ^2 test. For all the cases considered τ passes the χ^2 test with a significance level of 5%. This result is in agreement with the findings of Riva et al. (1999) for the same type of flow. Thus, assuming t is log-normal, it is possible to calculate the pumping period that, with a given probability, is needed to claim the aquifer as a function of the plume size.

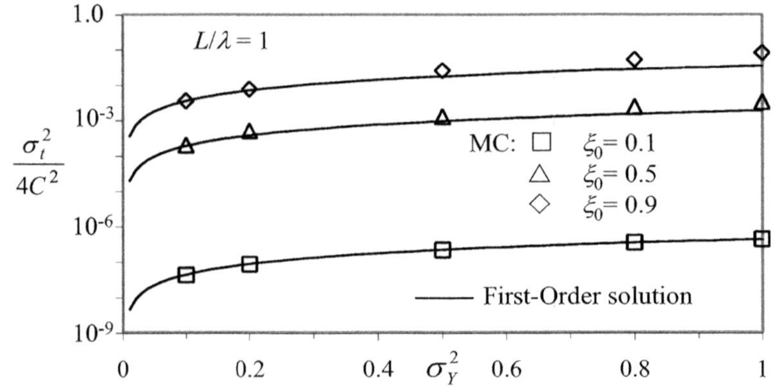

Figure 3*a*. Dimensionless residence time variance versus σ_Y^2 when $L/\lambda = 1$ and $\xi_0 = 0.1; 0.5; 0.9$.

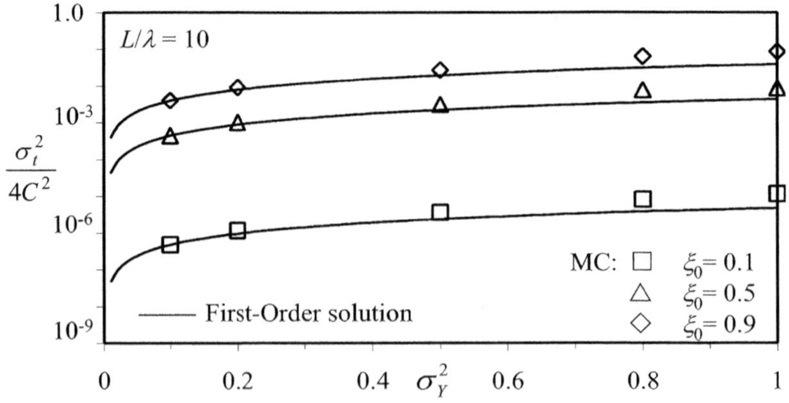

Figure 3*b*. Dimensionless residence time variance versus σ_Y^2 when $L/\lambda = 10$ and $\xi_0 = 0.1; 0.5; 0.9$.

Figure 4 depicts the first order approximation (in σ_Y^2) of the mean residence time, $\langle t^{[1]}\rangle$ versus σ_Y^2 for L/λ = 1 and plume sizes ξ_0 = 0.1, 0.5, 0.9. It is also reported the 95% confidence intervals obtained, assuming τ to be normal, as exp($\langle\tau^{[1]}\rangle \pm 1.96\,\sigma_\tau^{[1]}$) and the corresponding results computed by MC simulations.

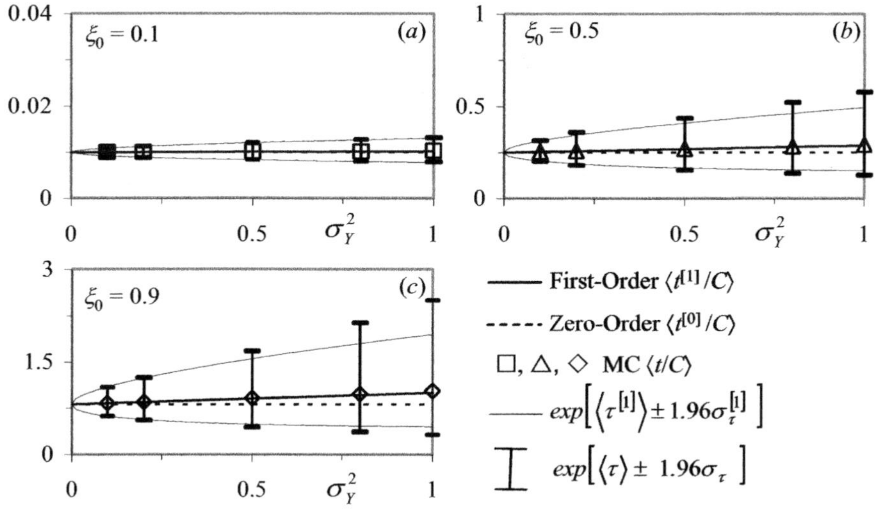

Figure 4. Dimensionless mean logarithm of the residence time and its standard deviation versus σ_Y^2 when L/λ = 1 and ξ_0 = 0.1 (*a*); 0.5 (*b*); 0.9 (*c*).

Figure 5 depicts the equivalent results for L/λ = 10. From these figures we can evaluate the pumping period (t_P) needed with a given probability P = 2.5% and 97.5% for cleaning an aquifer, as a function of aquifer heterogeneity and plume size.

The reclamation time, relative to a probability of 97.5%, $t_{0.975}$, is always larger than that predicted by a homogeneous approximation, $\langle t^{(0)}\rangle$ (Eq.4), on the other end of the spectrum there is the reclamation time associated to the 2.5% probability, $t_{0.025}$, that it is smaller than $\langle t^{(0)}\rangle$, as summarized in Table 1. For instance, if the pollution is released at ξ_0= 0.9, the time needed to clean the aquifer (with a probability of 97.5%) is more than twice what predicted under homogeneous assumptions if σ_Y^2=0.5, and increases up to more than three times $\langle t^{(0)}\rangle$ if σ_Y^2=1. Furthermore, a pollutant can reach the pumping well, with a probability of 2.25%, within approximately half the time it takes for a homogeneous aquifer if σ_Y^2 = 0.5 and even faster for larger degrees of heterogeneity.

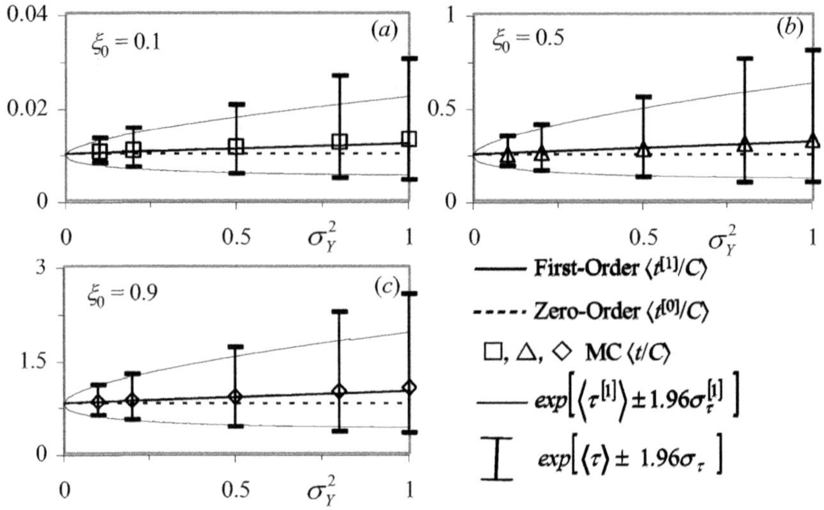

Figure 5. Dimensionless mean logarithm of the residence time and its standard deviation versus σ_Y^2 when $L/\lambda = 10$ and $\xi_0 = 0.1$ (*a*); 0.5 (*b*); 0.9 (*c*).

5. CONCLUSIONS

We consider the effect of heterogeneity on estimation of aquifer reclamation time by means of a single pumping well, located at the center of a contaminated area. We study the effects of the size of the polluted area, the correlation scale and the degree of heterogeneity characterized by σ_Y^2.

Our analysis leads to the following major conclusions:

1. the mean travel time needed to reclaim an aquifer increases with domain heterogeneity; ignoring this effect would lead to an inadequate reclamation design. In our simulations we observed discrepancies up to 20 % between first-order solutions and estimates based on a homogeneous approximation of the aquifer.
2. we identified the duration of pumping that, with a probability of 97.5%, is needed to clean an aquifer as a function of the field heterogeneity and starting location of the pollution. This operational time increases with σ_Y^2 and in the scenarios studied can be more than three times larger than that predicted by models relying on aquifer homogeneity.

Table 1. Comparison between the aquifer reclamation time needed with a given probability (P = 97.5% and 2.5%) for cleaning an aquifer and the homogeneous solution for different values of σ_Y^2, L/λ, and ξ_0.

ξ_0	σ_Y^2	L/λ	$t_{0.975}/\langle t^{(0)}\rangle$	$t_{0.025}/\langle t^{(0)}\rangle$
0.1	0.1	1	1.08	0.92
		10	1.34	0.80
	0.5	1	1.21	0.84
		10	2.05	0.58
	1	1	1.32	0.80
		10	3.03	0.45
0.5	0.1	1	1.26	0.81
		10	1.39	0.74
	0.5	1	1.74	0.63
		10	2.22	0.51
	1	1	2.29	0.51
		10	3.23	0.40
0.9	0.1	1	1.34	0.76
		10	1.36	0.75
	0.5	1	2.07	0.55
		10	2.11	0.53
	1	1	3.09	0.40
		10	3.16	0.41

ACKNOWLEDGEMENTS

This work was supported by the European Commission under Contract No. EVK1-CT-1999-00041 W-SAHARA.

REFERENCES

1. Dagan, G. and Indelman, P., 1999. Reactive solute transport in flow between a recharging and a pumping well in a heterogeneous aquifer. Water Resources Research, v.35, no.12, pp. 3639-3647.
2. De Simoni, M., Traiettoria e tempo di percorrenza di soluti in acquiferi eterogenei, Graduation Thesis (in Italian), Politecnico di Milano, 2001.
3. Deutsch, C. V. and Journel, A.G.: GSLIB Geostatistical Software Library and User's Guide. Oxford University Press, 1998.

4. Feyen, L., Beven, K.J., De Smedt, F., and Freer, J., 2001. Stochastic capture zone delineation within the generalized likelihood uncertainty estimation methodology: Conditioning on head observations. Water Resources Research, v.37, no.3, pp.625-638.
5. Guadagnini, A. and Franzetti, S., 1999. Time-related Capture Zones for Contaminants in Randomly Heterogeneous Formations. Ground Water, v.37, no.2, pp.253-260.
6. Gómez-Hernández J. J., A Stochastic Approach to the Simulation of Block Conductivity Fields Conditioned Upon Data Measured at a Smaller Scale, Ph. D. Dissertation, Stanford Univ., Stanford, Calif., 1991.
7. Guadagnini, A. and Neuman, S.P., 1999-a. Nonlocal and localized analyses of conditional mean steady state flow in bounded, randomly non uniform domains 1. Theory and computational approach. Water Resources Research, v.35, no.10, pp.2999-3018.
8. Guadagnini, A. and Neuman, S.P., 1999-b. Nonlocal and localized analyses of conditional mean steady state flow in bounded ,randomly non uniform domains 2. Computational examples. Water Resources Research, v.35, no.10, pp.3019-3039.
9. Guadagnini, A., Riva, M., and De Simoni, M., Travel time and trajectories of solutes in randomly heterogeneous aquifers, Deliverable D8 of Project "Stochastic Analysis of Well-Head Protection and Risk Assessment – W-SAHaRA", EU Contract EVK1-CT-1999-00041, Milano, 2001.
10. McDonald, M. G. and Harbaugh, A. W.: 1988, A modular three-dimensional finite-difference groundwater flow model. Manual 83-875. U.S. Geological Survey.
11. Neuman, S.P. and Orr, S., 1993. Prediction of Steady State Flow in Nonuniform Geologic Media by Conditional Moments: Exact Nonlocal Formalism, Effective Conductivities, and Weak Approximation. Water Resources Research, v.29, no.2, pp.341-364.
12. Neuman, S.P., Tartakovsky, D., Wallstrom, T.C. and Winter, C.L., 1996. Prediction of steady state flow in non uniform geologic media by conditional moments: Exact non local formalism, effective conductivities and weak approximation. Water Resources Research, v.29, no.2, pp.341-364.
13. Riva, M., Guadagnini, A. and Ballio, F., 1999. Time related capture zones for radial flow in two-dimensional randomly heterogeneous media. Stochastic Environmental Research and Risk Assessment, v.13, no.3, pp.217-230.
14. Riva, M., Guadagnini, A., and Neuman, S.P. and Franzetti, S., 2001. Radial flow in a bounded randomly heterogeneous aquifer. Transport in Porous Media, 45, 139–193.
15. van Leeuwen, M., Te Stroet, C.B.M., Butler, A.P., and Tompkins, J.A., 2000. Stochastic determination of well capture zones conditioned on regular grids of transmissivity measurements, Water Resources Research, v.36, no.4, pp.949-957.

SPATIAL PREDICTION OF CATEGORICAL VARIABLES: THE BME APPROACH

P. Bogaert

UCL/AGRO/MILA/ENGE. Place Croix du Sud, 2 box 16. 1348 Louvain-la-Neuve, Belgium

Abstract: Categorical variables often comes naturally and play an important role in environmental studies. Traditionally, they are processed in the geostatistical spatial estimation context using the indicator formalism. However, the indicator approach induces several and serious theoretical and practical problems. Among others, let us mention the inconsistencies and limitations of the linear model of coregionalisation, heavy computational load for taking simultaneously into account several categories, the limited pertinence of a linear predictor, and the incoherence of the predicted probabilities (negative probabilities, probabilities that do not sum up to one, etc.). This paper proposes a nonlinear approach that can be viewed as an extension of the Bayesian Maximum Entropy (BME) methods in the framework of categorical variables. The method is based on a maximum entropy reconstruction of high dimensional probabilities tables that are conditioned on their two-dimensional margins, followed by a conditioning of the table. The superiority of the BME approach over the indicator formalism is investigated both from the theoretical and practical point of views using an example.

1. INTRODUCTION

A new and powerful epistemic approach of random field estimation (mapping) accross space and time which combined Bayesian conditionalization of physical knowledge with stochastic information (entropy) maximization was proposed by Christakos (1990; 1991; 2000) and

X. Sanchez-Vila et al. (eds.), geoENV IV – Geostatistics for Environmental Applications, 271-282.

has been used with great success in a variety of scientific and engineering applications (see, e.g., Christakos & Li, 1998 ; D'Or *et al.*, 2001 ; Christakos *et al.*, 2002). In this work we propose an extension of the BME formalism in the case of discrete-valued (categorical) random field, which is a very powerful method for spatial prediction and mapping (Bogaert, 2002). Currently, the most widely used methods for mapping categorical variables rely on the geostatistical indicator formalism (Journel, 1983; Goovaerts, 1997), that offers the advantage of being simple to understand and easy to implement, though it suffers from many limitations that are both theoretical and methodological. This papers aims at emphasizing some of these limitations, as well as to show how BME can handle the prediction of categorical variables in a much more satisfactory way.

2. CATEGORICAL RANDOM FIELD

Consider a discrete or categorical random variable C, that can be nominal or ordinal, having $\Omega_C = \{c_1,\ldots,c_m\}$ as its finite set of possible outcomes. We will be interested in the case where the c_i's are forming a complete system of events, i.e. $c_i \neq \emptyset$, $c_i \cap c_j = \emptyset\ \forall i \neq j$, so that $\sum_i P(C = c_i) = 1$. Consider a continuous spatial domain D and an arbitrary location vector $\mathbf{x}$ such that $F_C = \{C(\mathbf{x}_\alpha),\ \mathbf{x}_\alpha \in D \subseteq \Re^n\}$ defines a discrete-valued random field with a continuous support over a n-dimensional space (see, e.g., Chiles & Delfiner (1999) for few examples of categorical random fields). Let

$$\pi_\alpha(i_\alpha) \equiv P(C(\mathbf{x}_\alpha) = c_{i_\alpha}) \quad (1)$$

be the univariate probability that the field is taking the value $c_{i\alpha}$ ($i_\alpha = 1,\ldots,m$) at location $\mathbf{x}_\alpha$. Similarly, let

$$\pi_{\alpha,\beta(i_\alpha,i_\beta)} \equiv P(C(\mathbf{x}_\alpha) = c_{i_\alpha} \cap C(\mathbf{x}_\beta) = c_{i_\beta}) \quad (2)$$

be the bivariate probability that the field is taking the modality $c_{i\alpha}$ and $c_{i\beta}$ at locations $\mathbf{x}_\alpha$ and $\mathbf{x}_\beta$, respectively. We will assume that by letting $\mathbf{h}_{\alpha\beta} = \mathbf{x}_\alpha - \mathbf{x}_\beta$ and $\mathbf{h}_{\chi\delta} = \mathbf{x}_\chi - \mathbf{x}_\delta$, we have $\pi_{\alpha,\beta} = \pi_{\chi,\delta(i,j)}$ if $\mathbf{h}_{\alpha\beta} = \mathbf{h}_{\chi\delta}$, i.e., bivariate probabilities are invariant under translation (an even stronger hypothesis would be $||\mathbf{h}_{\alpha\beta}|| = ||\mathbf{h}_{\chi\delta}||$, i.e. bivariate invariance under translation and rotation). It is clear that assuming this automatically implies that $\pi_{\alpha,\alpha(i,j)} = \pi_{\chi,\chi(i,j)}$, yielding $\pi_{\alpha(i)} = \pi_{\chi(i)} = \pi_i\ \forall \mathbf{x}_\alpha = \mathrm{x}_\chi$, i.e. invariance under translation of the univariate probability distribution.

The previous definition of a categorical random field is generic. It can be applied to a discretization of a continous-valued random field $F_Z = \{Z(\mathbf{x}) \in \Re^1,\ \mathbf{x}_\alpha \in D \subseteq \Re^n\}$ into a complete set of classes $c_i(\mathbf{x}) \equiv a_i < z(\mathbf{x}) \leq a_{i+1}$

where the a_i's are ordered quantiles, or to a real categorical random field that can be ordinal or nominal.

3. SECOND-ORDER PROPERTIES

Using the invariance under translation hypothesis, one can define the bivariate probability functions

$$\pi_{i,j}(\mathbf{h}) \equiv P(C(\mathbf{x}) = c_i \cap C(\mathbf{x}+\mathbf{h}) = c_j) \quad \forall i,j \tag{3}$$

such that $\pi_{i,j}(\mathbf{0}) = \pi_i \ \forall i = j$ and $\pi_{i,j}(\mathbf{0}) = 0 \ \forall i \neq j$. Assuming that dependence vanishes as $||\mathbf{h}|| \to \infty$, we also have $\pi_{i,j}(\infty) = \pi_i\pi_j$. Note that $\pi_{i,j}(\mathbf{h}) = \pi_{j,i}(-\mathbf{h})$, but this does not necessarily imply that $\pi_{i,j}(\mathbf{h}) = \pi_{i,j}(-\mathbf{h})$ or that $\pi_{i,j}(\mathbf{h}) = \pi_{j,i}(\mathbf{h})$.

Though bivariate probability function are seldom used in geostatistics (Carle & Fogg, 1996), they are directly linked to the indicator covariance functions and variograms, that are widely used. Using the Kronecker delta operator

$$\delta_i(\mathbf{x}) = \begin{cases} 1 & if \ C(\mathbf{x}) = c_i \\ 0 & \text{otherwise} \end{cases} \tag{4}$$

one can define

$$\begin{aligned} C_{\delta_i,\delta_j}(\mathbf{h}) &\equiv Cov(\delta_i(\mathbf{x}), \delta_j(\mathbf{x}+\mathbf{h})) \quad \forall i,j \\ &= \pi_{i,j}(\mathbf{h}) - \pi_i\pi_j \end{aligned} \tag{5}$$

which is referred to as the class indicator (cross-)covariance function. One can also define

$$\begin{aligned} \gamma_{\delta_i,\delta_j}(\mathbf{h}) &\equiv \frac{1}{2} Cov(\delta_i(\mathbf{x}) - \delta_i(\mathbf{x}+\mathbf{h}), \delta_j(\mathbf{x}) - \delta_j(\mathbf{x}+\mathbf{h}) \quad \forall i,j \\ &= \delta_{(i=j)}\pi_i - \frac{1}{2}(\pi_{i,j}(\mathbf{h}) + \pi_{i,j}(-\mathbf{h})) \end{aligned} \tag{6}$$

which is referred to as the class indicator (cross-)variogram function. Though (5) and (6) are popular because they are a straighforward extension of the traditional (cross-)covariance functions and (cross-)variograms in a categorical context, they do not however carry any extra information compared to (3), which is much simpler and much more intuitive in terms of probabilities. At this point, one canalso remark that, from (5) and (6), one can build the positive semi-definite matrix $\{C_{\delta_i\delta_j}(\mathbf{0})\} = \{\gamma_{\delta_i\delta_j}(\infty)\}$, such that

$$\left\{C_{\delta_i,\delta_j}(\boldsymbol{0})\right\} = \begin{pmatrix} C_{\delta_i,\delta_j}(\boldsymbol{0}) & \cdots & C_{\delta_1,\delta_m}(\boldsymbol{0}) \\ & \ddots & \vdots \\ (sym) & & C_{\delta_m,\delta_m}(\boldsymbol{0}) \end{pmatrix} = \boldsymbol{I}\pi - \pi'\pi \tag{7}$$

where $\pi' = (\pi_1 \ldots \pi_m)$, so that (7) has rank equal to m-1, with diagonal elements in]0,0.25[and negative off-diagonal elements in]-1,0[subject to the constraints $\pi_i > 0\ \forall i$ and $\sum_i \pi_i = 1$.

4. THE INDICATOR (CO)KRIGING APPROACH

In a spatial prediction context, what is sought is a predictor $p_{i0|\{i\alpha\}}$ for $\pi_{i0|\{i\alpha\}} \equiv P(C(\mathbf{x}_0 = c_{i0}| \{C(\mathbf{x}_\alpha = c_{i\alpha}\})$, i.e., for the conditional probabilities of the c_{i0}'s (i_0 = 1,…,m) at location $\mathbf{x}_0$ given the observed modalities at surrounding locations $\mathbf{x}_\alpha$ (α = 1,…,k). For all subsequent notations, the symbol $p(.)$ will denote an estimate of the corresponding $\pi(.)$ theoretical probability. Classically, eq. (5) and (6) are at the basis of the various indicator (co)kriging algorithms, that are obtained as a straighforward modification of the classical kriging algorithms for continous-valued random fields. Using the indicator coding (4) with $\delta_{i0}(\mathbf{x}_\alpha) = 1$ when $C(\mathbf{x}_\alpha) = c_{i0}$ and 0 otherwise, the indicator kriging (IK) predictor is

$$p_{i_o|\{i_\alpha\}} = \sum_\alpha \lambda_{i_0,\alpha} \delta_{i_0}(\mathbf{x}_\alpha) \quad \forall_{i_0} = 1,\ldots,m \tag{8}$$

where the weights $\lambda_{i0,\alpha}$ are obtained by solving a linear system of equations built from a valid choice for (5) or (6) (i.e., the $C_{\delta i0,\delta i0}(\mathbf{h})$'s are positive definite (p.d.) and the $\gamma_{\delta i0,\delta i0}(\mathbf{h})$'s are conditionally negative definite). This calls for several remarks. First, even if (8) makes use of an indicator (nonlinear) data coding, it remains a linear combination of these nonlinear (indicator) functionals, whereas $\pi_{i0|\{i\alpha\}}$ is clearly a nonlinear functional of the $C(\mathbf{x}_\alpha)$'s. Second, (8) is non-convex, and one easily end-up with predicted values outside the [0,1] interval. Last, as the predicted $p_{i0|\{i\alpha\}}$'s (i_0 = 1,…,m) are obtained separately, $\sum_{i0} p_{i0|\{i\alpha\}} \neq 1$ in general.

In order to simultaneously take into account the information for all classes, it has been suggested that indicator cokriging (ICK) would be more appropriate (e.g., Lajaunie, 1990). Experience however seems to suggest that results obtained from IK and ICK are quite similar (Goovaerts, 1997), an observation that would also suggest that ICK is making a poor use of the extra information that was not used by IK. Moreover, this entails new problems. The joint information is incorporated by using (5) or (6) in a Linear Model of Coregionalization (LMC, Journel and Huijbregts, 1978). E.g., in terms of covariance functions, this LMC is written as

$$\left\{C_{\delta_i,\delta_j}(\boldsymbol{h})\right\} = \begin{pmatrix} C_{\delta_1,\delta_1}(\boldsymbol{h}) & \cdots & C_{\delta_1,\delta_m}(\boldsymbol{h}) \\ & \ddots & \vdots \\ (sym) & & C_{\delta_m,\delta_m}(\boldsymbol{h}) \end{pmatrix} = \sum_k \boldsymbol{B}_k C_k(\boldsymbol{h}) \tag{9}$$

where the $\mathbf{B}_k$'s are p.d. matrices classically obtained through an iterative algorithm (Goulard & Voltz, 1992) and the $C_k(\mathbf{h})$'s are p.d. covariance models with $C_k(0) = 1$. First, according to (7), putting $\mathbf{h} = \mathbf{0}$ in (9) yields (7), so that $\mathbf{I}\pi - \pi'\pi = \sum_k \mathbf{B}_k$, which is of course impossible as the right-hand side is merely p.d. whereas the left-hand side is positive semidefinite and subject to numerous constraints. Second, notwithstanding the fact that the LMC is an invalid model, using it will let the user face considerable modeling difficulties. This is better illustrate with a simple example, where a categorical random field F_C has been built from a second-order stationary continuous-valued random field F_Z having a zero-mean unit variance multivariate Gaussian distribution, with $C(\mathbf{h})$ an exponential model with practical range equal to 0.5. Assume we define four classes as in Figure 1 such that $c_i(\mathbf{x}) \equiv a_i \leq z(\mathbf{x}) < a_{i+1}$, where the a_i's are the 0, 0.2, 0.5, 0.8 and 1 quantiles of the Gaussian distribution. It is then easy to compute (3) as well as, e.g., (5) by integration over bivariate distributions (Figure 2). Clearly, this set of covariance functions exhibits complex shapes that are unlikely to be captured by the use of (9); indeed, it can be proved (Bogaert, 2002) that the only possible valid model for (9) is the intrinsic coregionalization model. Finally, as ICK is a straighforward generalization of IK, all the limitations previously emphasized for IK still apply for ICK.

Indicator kriging and cokriging are somewhat abusively referred to as nonlinear methods (the only nonlinear part is the indicator coding of the data), whereas the conditional probability is a highly nonlinear function of the data. According to the numerous limitations of I(C)K, it would be more efficient to seek directly for this conditional probability, that can be obtained provided that the joint distribution $\pi_{0,\ldots,k(i0,\ldots,ik)} \equiv P(\cap_j C(\mathbf{x}_{ij}))$, $j = 0,\ldots,k$, $i_j = 1,\ldots,m$ can be estimated. This is precisely what BME is proposing to do.

5. THE BAYESIAN MAXIMUM ENTROPY APPROACH

The general process for BME prediction can be viewed as a three-stages procedure (Christakos, 2000); (i) at the first stage (the prior stage), one aims at finding a joint probability distribution that has a maximum entropy and respects some general constraints (e.g., a set of probabilities that are supposed to be known and are imposed) ; (ii) at the second stage (the meta-prior stage), the specific information about the data set under study are collected and translated into useable mathematical relations; (iii) the final stage (integration stage) is the computation of the posterior conditional distribution with respect to the maximum entropy distribution obtained at stage (i) and the information collected at stage (ii). Only stages (i) and (iii) are detailed hereafter.

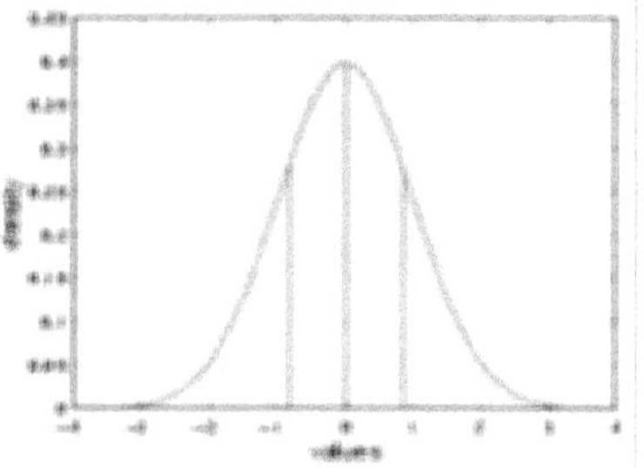

Figure 1. Four-classes partition of a $N(0,1)$ random variable.

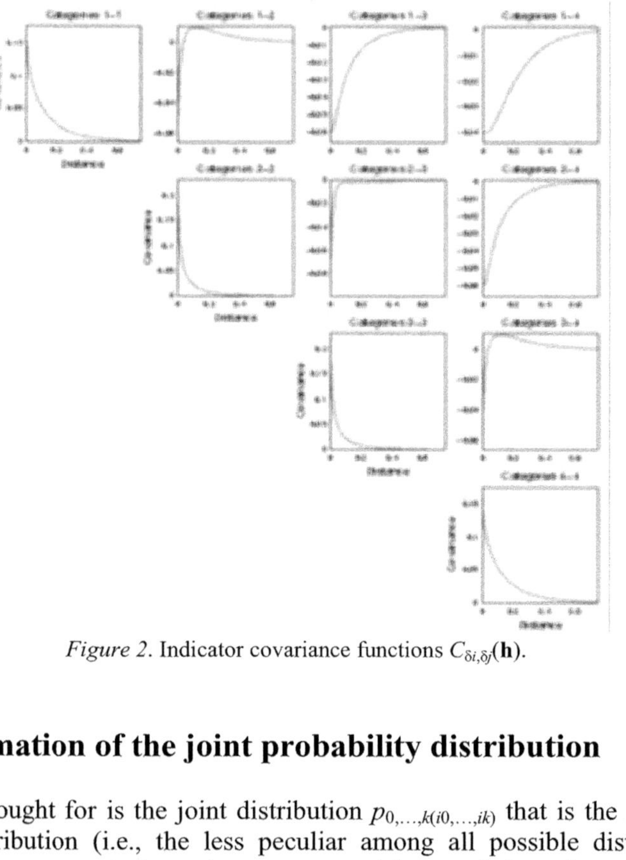

Figure 2. Indicator covariance functions $C_{\delta i,\delta j}(\mathbf{h})$.

5.1 Estimation of the joint probability distribution

What is sought for is the joint distribution $p_{0,\ldots,k(i0,\ldots,ik)}$ that is the maximum entropy distribution (i.e., the less peculiar among all possible distributions) subject to some constraints. Assume a partial knowledge about the field, represented by a set K_G that specify some of its properties. E.g., $K_G = \{\pi_{\alpha,\beta(i\alpha,i\beta)};\ \alpha,\beta = 0,\ldots,k\}$ if the sets of functions (3), (5) or (6) are known. If one think of $\pi_{0,\ldots,k(i0,\ldots,ik)}$ as a $m \times \ldots \times m$ (k times) hypersquare probability table having m^k cells, the $\pi_{\alpha,\beta(i\alpha,i\beta)}$'s are the margins or order 2 that are forming k^2 square $m \times m$ probability table. We will consider that these probability tables are the constraints for the maximum entropy estimation of $\pi_{0,\ldots,k(i0,\ldots,ik)}$.

The entropy of a distribution $p_{0,\ldots,k(i0,\ldots,ik)}$ is given by

$$H_{0,\ldots,k} = -\sum_{i_0,\ldots,i_k} p_{0,\ldots,k(i_0,\ldots,i_k)} \ln p_{0,\ldots,k(i_0,\ldots,i_k)} \tag{10}$$

where $\sum_{i0,\ldots,ik}$ denotes summation over all possible values for all indexes. We want to respect the contraints

$$p_{\alpha,\beta(i_\alpha,i_\beta)} = \pi_{\alpha,\beta(i_\alpha,i_\beta)} \; \forall \alpha \neq \beta, \forall i_\alpha, i_\beta \tag{11}$$

where $p_{\alpha,\beta(i\alpha,i\beta)} = \sum_{\{ij;j\neq\alpha,\beta\}} p_{0,\ldots,k(i0,\ldots,ik)}$.Using the Lagrangian formalism, maximizing (10) under (11) is equivalent to maximize

$$L_{0,\ldots,k} = H_{0,\ldots,k} + \sum_{\alpha,\beta;\alpha\neq\beta} \sum_{i_\alpha,i_\beta} \mu_{\alpha\beta(i_\alpha,i_\beta)} \left(p_{\alpha,\beta(i_\alpha,i_\beta)} - \pi_{\alpha,\beta(i_\alpha,i_\beta)} \right) \tag{12}$$

Setting the partial derivatives equal to zero with respect to $p_{0,\ldots,k(i0,\ldots,ik)}$ and $\mu_{\alpha\beta(i\alpha i\beta)}$ yields

$$\begin{cases} \dfrac{\partial L_{0,\ldots,k}}{\partial p_{0,\ldots,k(i_0,\ldots,i_k)}} = 1 - \ln p_{0,\ldots,k(i_0,\ldots,i_k)} + \sum_{\alpha,\beta;\alpha\neq\beta} \sum_{i_\alpha,i_\beta} \mu_{\alpha\beta(i_\alpha i_\beta)} = 0 \\ \dfrac{\partial L_{0,\ldots,k}}{\partial \mu_{\alpha\beta(i_\alpha i_\beta)}} = p_{\alpha,\beta(i_\alpha,i_\beta)} - \pi_{\alpha,\beta(i_\alpha,i_\beta)} = 0 \end{cases} \tag{13}$$

which must be solved with respect to the coefficients $\mu_{\alpha\beta(i\alpha i\beta)}$. The first part of (13) is the definition of a non-staturated log-linear model involving first-order interaction effects. The equivalence between maximum entropy probability distribution functions that satisfies marginal constraints and non-saturated log-linear models is well known (Good, 1963). Estimating the $\mu_{\alpha\beta(i\alpha i\beta)}$ is equivalent to fitting this non-saturated log-linear model, and classical algorithms like the iterative scaling procedure (Deming and Stephan, 1940) can be used.

5.2 Computation of the conditional probabilities

What is sought for at this step is an estimate of the conditional probabilities at an unsampled location $\mathbf{x}_0$ given some specific knowledge K_S, i.e.,

$$\begin{aligned} \pi_{i_0|K_s} &= P(C(\mathbf{x}_0 = c_{i_0})|K_s) \\ &= P(C(\mathbf{x}_0 = c_{i_0}) \cap K_s) / P(K_s) \qquad \forall_{i_0} \end{aligned} \tag{14}$$

with $P(K_S) = \sum_{i0} P(C(\mathbf{x}_0 = c_{i0}) \cap K_S)$. We will consider the cases where $P(C(\mathbf{x}_0 = c_{i0}) \cap K_S)$ can be computed univoquely from the joint distribution $\pi_{0,\ldots,k(i0,\ldots,ik)}$ or its estimate $p_{0,\ldots,k(i0,\ldots,ik)}$ obtained from the maximum entropy paradigm, so that $p_{i0|KS}$ is an estimate for $\pi_{i0|KS}$. A classical choice for K_S would be

$$K_s \equiv \bigcap_{j=1}^{k} C(\mathbf{x}_j) = c_{i_j} \tag{15}$$

(the category is known at each location $\mathbf{x}_1,\ldots,\mathbf{x}_k$), so that

$$P(C(\mathbf{x}_0 = c_{i_0}) \cap K_s) = \pi_{0,\ldots,k(i_0,\ldots,i_k)} \tag{16}$$

A more elaborate example would be

$$K_s \equiv \bigcap_{j=1}^{k} C(\mathbf{x}_j) \in E_j \tag{17}$$

where $E_j \subset \Omega_C\ \forall_j$ (the subset of possible categories is known at each location $\mathbf{x}_1,\ldots,\mathbf{x}_k$), so that

$$P(C(\mathbf{x}_0 = c_{i_0}) \cap K_s = \sum_{i_1 \in E_1,\ldots,i_k \in E_k} \pi_{0,\ldots,k(i_0,\ldots,i_k)} \tag{18}$$

In some instances, there could be a specific probability information that is made available, so that

$$K_s \equiv f_{1,\ldots,k(i_1,\ldots,i_k)} \tag{19}$$

where $f_{1,\ldots,k(i1,\ldots,ik)}$ is a joint probability distribution for the categories at locations $\mathbf{x}_1,\ldots,\mathbf{x}_k$, obtained independently from $\pi_{0,\ldots,k(i0,\ldots,ik)}$, so that

$$P(C(\mathbf{x}_0 = c_{i_0}) \cap K_s) = \sum_{i_1,\ldots,i_k} \pi_{0,\ldots,k(i_0,\ldots,i_k)} f_{1,\ldots,k(i_1,\ldots,i_k)} \tag{20}$$

Using the same reasoning, there is of course no problem for combining (15), (17) and (19), for including specific information that refers to $C(\mathbf{x}_0)$ itself (e.g., $C(\mathbf{x}_0) \in E_0$), or even for obtaining any multivariate conditional distribution.

5.3 Superiority of the BME approach

As seen from (14), the conditional probability estimates $p_{i0|KS}$ is a real nonlinear functional and not merely a linear combinations of indicator variables as it is the case for IK and ICK. It does not rely on the use of (5),(6) and (9), but instead it makes use of (3) as constraints in bivariate probability tables. This entails that the simple constraint that (3) must fulfill is that, for any distance $\mathbf{h}$, one have $\pi_{i,j}(\mathbf{h}) \geq 0\ \forall i,j$ and $\Sigma_{i,j}\pi_{i,j}(\mathbf{h}) = 1$. These conditions are considerably less restrictive than, e.g., the choice of a p.d. model for $C(\mathbf{h})$.

As all the conditional probabilities are computed from a valid joint distribution estimate $p_{0,\ldots,k(i0,\ldots,ik)}$, they automatically lead to valid conditional distributions, with $p_{i0|KS} \geq 0\ \forall i_0$ and $\Sigma_{i0}p_{i0|KS} = 1$.

The maximum entropy estimation of $\pi_{0,\ldots,k(i0,\ldots,ik)}$ that has been described has been conducted using as contraints the complete set of bivariate probabilities $\pi_{\alpha,\beta(i\alpha,i\beta)}$, but the methodology still holds if some of these $\pi_{\alpha,\beta(i\alpha,i\beta)}$ are omitted (e.g., because there are too few data for estimating them in a reliable way), or if higher order probabilities (e.g., trivariate probabilities $\pi_{\alpha,\beta,\chi(i\alpha,i\beta,i\chi)}$ are considered.

All these remarks emphasize the considerable generality and the power of the BME formalism. Various kind of knowledge are easily incoporated in a sound way and lead to valid nonlinear conditional probability estimates. All the theoretical restrictions and validity problems linked to the use of IK and ICK do not appear when using BME. The practical superiority of the BME approach over I(C)K is also emphasized in the next section.

6. BME AND INDICATOR (CO)KRIGING IN ACTION

As an example, assume that there are 100 locations that have been randomly sampled over a square of unit size. The prediction is conducted over a 100 by 100 grid covering the square. Continuous values z_j are jointly simulated at these 100 sampling locations (Figure 3a) and the 10000 prediction nodes using a sequential simulation method and an exponential model $C(\mathbf{h})$ with range and sill equal to 0.5 and 1, respectively, so that the distribution $f_z(z_0),\ldots,z_k)$ is multivariate Gaussian. The simulated values are then replaced by the interval $c(\mathbf{x}_j) \equiv I_j =]a_{ij},a_{ij+1}]$ to which they belong according to Figure 1 (Figure 3b). Due to the Gaussian hypothesis, knowing the mean and $C(\mathbf{h})$ is sufficient for computing any conditional probability. Assuming K_S as in (15), the conditional distribution $f_z(z_0| K_S)$ is given by

$$f_{\mathbf{z}}(z_0|K_s)=\frac{\int_{I_1}\ldots\int_{I_k} f_{\mathbf{z}}(z_0,\ldots,z_k)dz_k\ldots dz_1}{\int_{\mathbb{R}}\int_{I_1}\ldots\int_{I_k} f_{\mathbf{z}}(z_0,\ldots,z_k)dz_k\ldots dz_0} \tag{21}$$

where K_S is based on the set of 100 sampled locations. From (21), the conditional probabilities $\pi_{i0|KS}, i_0 = 1,\ldots,4$ are then obtained as

$$\pi_{i_0|K_s} = \int_{a_{i_0}}^{a_{i_0}+1} f_{\mathbf{z}}(z_0|K_s)dz_0 \tag{22}$$

These $\pi_{i0|KS}$'s (Figure 3c) are the reference values to which the IK, ICK and BME estimates will be compared, as they are the best probability estimates for the categories than one can get from the available information K_S. For comparison purposes, the theoretical functions (5) (see Figure 2) and (3) are used for all methods. Using them instead of estimating them allows us to obtain a fair and objective comparison of the performances for the different methods without taking into account complex inferential problems and methodological considerations. For all methods, a same neighborhood size consisting of the 5 closest sampled locations has been considered for each node of the prediction grid. As a first result, due to the inherent limitations of IK and ICK, the basic requirements for obtaining valid distributions are not met. For IK, 30% of the probabilities are summing out of the [0.95,1.05] interval (Figure 4a), whereas for ICK 19% of the probabilities are negative (Figure 4b). None of these problems are encountered with BME.

As a second result, a comparison of the conditional probability estimates obtained using BME shows that they are in very good agreement with the true conditional probabilities (Figures 3 and 5) In spite of the fact that BME does not explicitly use the information that the categories are strictly ordered, there is little information that has been lost. There is a very good agreement between BME and the true conditional distributions, as measured by the high correlation coefficient between probability estimates and the high frequency (95%) of identical maximum probability categories on the maps (Figure 3). The situation is much less favorable for IK and ICK. Conditional probabilities are quite different from the true ones, with plenty of values equal to 0 or 1, with a frequency of identical categories of only 75%. Note also the very marginal improvement when using ICK instead of IK, meaning that extra information carried by cross-covariance functions has been poorly used by ICK. Moreover, the spatial variations of the map are quite different. IK and ICK maps tend to show patchy areas where the expected progressive transitions from a category to an adjacent category is weekly apparent. The BME map is much more statisfactory, as there is a clear progressive transition between categories, that accounts correctly for the fact that these categories were ordered, even if this information was not explicitly incorporated in the estimation.

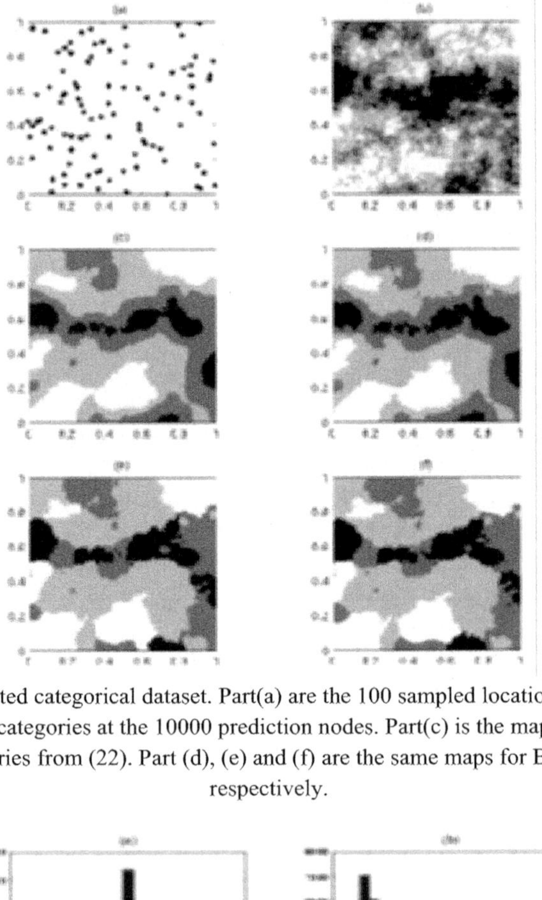

Figure 3. Simulated categorical dataset. Part(a) are the 100 sampled locations. Part(b) is the map of simulated categories at the 10000 prediction nodes. Part(c) is the map of the maximum probability categories from (22). Part (d), (e) and (f) are the same maps for BME, IK and ICK, respectively.

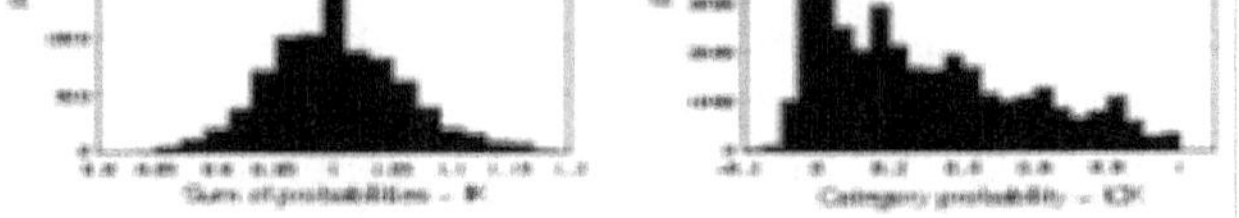

Figure 4.Non-validity of the (a) IK and (b) ICK conditional probabilities.

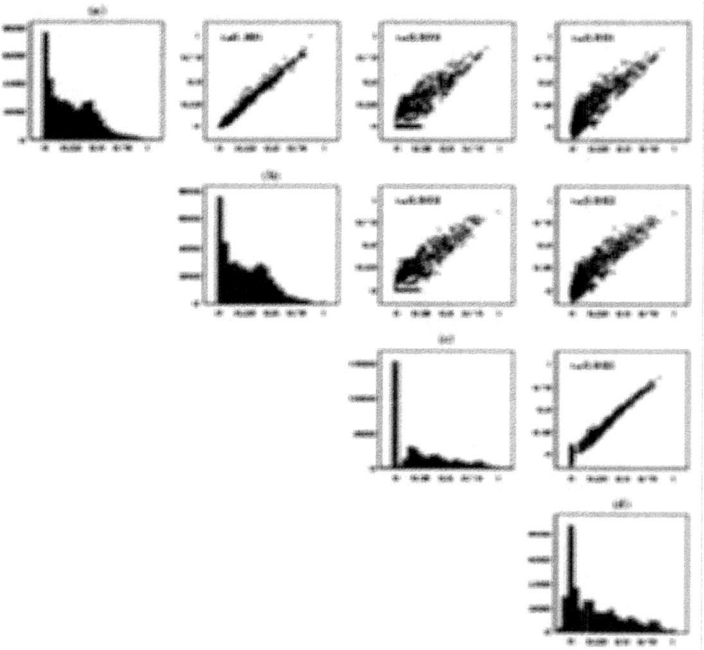

Figure 5. Comparison of the conditional probabilities obtained from (a) eq. (22) and from (b) BME, from (c) IK and from (d) ICK. For the sake of lisibility, only 10% of randomly chosen values have been plotted on the off-diagonal graphs.

7. CONCLUSIONS

As seen from theory and from the previous example, BME appears to be much more statisfactory than IK or ICK with respect to many points:

- It yields conditional probabilities that are automatically valid (no negative probabilities, probabilities that sum to one). These simple conditions cannot be enforced using IK and ICK;
- BME does not require the use of indicator (cross-)covariance or variogram models and does not rely on an invalid LMC, as it directly uses bivariate probability functions. Moreover, it is a real nonlinear method, whereas IK and ICK is a linear combination of indicator values;
- The method can be easily generalized, e.g., to obtain multivariate conditional probabilities, to account for multi-point probabilities (e.g., trivariate probabilities), to process incomplete information, etc. without any theoretical difficulties.

Although the methodology that has been presented here focused, for the sake of brievety, on a single spatial categorical variable, it can be generalized for dealing with space/time data, with several categorical variables at the same time, or even for combining both continuous and categorical variables. As a conclusion, BME can be considered as an extremely serious challenger for processing categorical data in a spatial estimation context.

REFERENCES

1. Bogaert, P. (2002). Spatial prediction of categorical variables: the Bayesian maximum entropy approach. Stochastic Environmental Research and Risk Assessment, 16: 425-448.
2. Carle, S.F., and Fogg, G.E. (1996). Transition probability-based indicator geostatistics. Mathematical Geology, 28: 453-476.
3. Chiles, J.-P., and Delfiner, P. (1999). Geostatistics - Modeling Spatial Uncertainty. Wiley, New York, 695pp.
4. Christakos G. (1990). A Bayesian/maximum entropy view to the spatial estimation problem. Mathematical Geology, 22: 763-776.
5. Christakos G. and Li, X. (1998). Bayesian maximum entropy analysis and mapping: a farewell to kriging estimators?. Mathematical Geology, 30: 435-462.
6. Christakos G. (1991). Some applications of the Bayesian maximum entropy concept in geostatistics. Fundamental Theories of Physics, Kluwer Acad. Publ., Boston, 215-229.
7. Christakos G. (2000). Modern Spatiotemporal Geostastistics. Oxford University Press, New York, 312 pp.
8. Christakos, G., Bogaert, P. and Serre, M. (2002). Temporal Geographical Information Systems: A Bayesian Maximum Entropy Primer for Natural and Epidemiological Sciences. Springer-Verlag, New York, 250 pp.
9. Deming, W.E. and Stephan, F.F. (1940). On a least squares adjustement of a sampled frequency table when the expected marginal totals are known. Ann. Math. Statist., 11: 427-444.
10. D'Or, D., Bogaert, P. and Christakos, G. (2001). Application of the BME approach to soil texture mapping. Stochastic Environmental Research and Risk Assessment: 15, 87-100.
11. Good, I.J. (1963). Maximum entropy for hypotheses formulation especially for mutidimensional contingency tables. Ann. Math. Statist., 34: 911-934.
12. Goulard, M. and Voltz, M. (1992). Linear coregionalization model : Tools for estimation and choice of cross-variogram matrix. Mathematical Geology, 24: 269-286.
13. Goovaerts, P. (1997). Geostatistics for Natural Resources Evaluation. Oxford University Press, New York, 496 pp.
14. Journel, A.G. (1983). Non-parametric estimation of spatial distributions. Mathematical Geology, 15: 445-468.
15. Journel, A.G., and Huijbregts, C.J. (1978). Mining Geostatistics. Academic Press, New York, 600 pp.
16. Lajaunie, C. (1990). Comparing some approximate methods for building local confidence intervals for predicting regionalized variables. Mathematical Geology, 22: 123-144.

OPTIMIZING SAMPLING FOR ACCEPTABLE ACCURACY LEVELS ON REMEDIATION VOLUME AND COST ESTIMATIONS

An iterative approach, based on geostatistical methods, illustrated on a former smelting works polluted with lead

H.Demougeot-Renard[1], C. de Fouquet[2] and M. Fritsch[3]

[1]*Eidgenössische Technische Hochschule Zürich, Institut für Raum und Landschaftsentwicklung, Hönggerberg, 8093 Zürich, Switzerland. At present, Université de Neuchâtel, Centre d'Hydrogéologie de Neuchâtel, 11 rue Emile Argand, 2007 Neuchâtel, Switzerland;* demougeot.renard@unine.ch [2] *Ecole Nationale Supérieure des Mines de Paris, Centre de Géostatistique, 35 rue Saint Honoré, 77305 Fontainebleau, France ;* fouquet@cg.ensmp.fr [3] *Environmental Management and Communication, Wildbachstrasse 46, CH-8008 Zürich ;* mfritsch@emac.ch

Abstract: The aim of the paper is to present an iterative approach, based on geostatistical methods, to optimize sampling according to financial and environmental criteria. At current stage j of sampling, if the accuracy on remediation volume and cost estimates is not considered as sufficient, we try to anticipate the number of samples that needs to be collected at stage j+1, to reach an acceptable accuracy level. Sampling of various numbers Nj+1 of additional data is modelled, based on one simulation of the pollutant concentrations generated at stage j, conditioned with the available experimental data, in the area where the probabilities of exceeding the remediation cutoff are too high. In the variogram model fitted at stage j, remediation volumes and costs are re-calculated with the various Nj+1 additional conditional data. If necessary, the process is repeated. The approach is illustrated on the site of a former smelting works presenting a lead pollution. Since the uncertainties on the remediation volume and cost estimates at the sixth real sampling stage are not satisfactory, a number of additional N7 data is chosen according to volume and cost forecasts calculated for various N7. The choice is non unique since various criteria, objectives, constraints and decision makers preferences can be taken into account. As an example, it is shown which number N7 will be chosen by a risk averse decision maker or by a risk prone decision maker, according to four common environmental and financial objectives.

X. Sanchez-Vila et al. (eds.), geoENV IV – Geostatistics for Environmental Applications, 283-294.

Key words: geostatistics, soil pollution, lead, smelting works, optimizing sampling, uncertainty, remediation, volume, cost, conditional simulation, decision making

1. INTRODUCTION

Evaluating the volume of soil that requires remediation, and its accuracy, is an important step of the process of restoring an industrial polluted site. A high uncertainty on the estimated volume implies an environmental risk: soils may remain while they require remediation; and a financial risk: during the remediation works, unexpected polluted soils may be discovered, or soils may be excavated for remediation although their pollutant concentrations are inferior to the remediation cutoff.

A sampling designed to provide an acceptable level of accuracy should help to reduce those risks. But in practice, data are collected above all to answer the questions raised by the risk assessment study. As a result, the accuracy level of the estimated volumes is not really chosen.

The following question is then raised: since the number of available samples has a major influence on the accuracy level, is it possible, for a pollution showing a spatial structure, to anticipate the number of samples required to provide an acceptable accuracy on the remediation volumes and costs? More precisely, for investigations made of different stages, is it possible, at stage j , to forecast the number of additional samples that needs to be collected at stage j+1 to reach an acceptable accuracy?

This question emphasises another difficulty: what is an "acceptable" accuracy level? The answer is non unique, since it depends on particular criteria, goals, constraints, and decision makers preferences.

Since the beginning of the 1990s[1], sampling designs have been proposed to improve the accuracy of geostatistical estimations on polluted soils, but their goals are different from those mentioned above. We propose a methodology, based on geostatistics that helps to optimize the sampling strategy in order to reach an "acceptable" uncertainty level on remediation volume and cost estimations. The methodology is applied to a real former smelting works, polluted with lead.

[1] The letter j indexes the investigation stages.

2. METHODOLOGY

The proposed methodology [3] is iterative, including two steps after each new sampling stage. Step 1: the remediation volume and cost, as well as their uncertainties, are evaluated, on the basis of the experimental data available at stage j. Step 2: if the volume and cost uncertainties are too high at stage j, the influence of additional data Nj+1 is examined, based on volume and cost forecasts calculated for various numbers Nj+1 of simulated data. The two steps are repeated until the number of actually collected samples at stage j+1 is sufficient to reach a correct accuracy on the volume and cost estimations.

2.1 Step 1: estimations at stage j

Suppose that investigations of an industrial site in j stages, as well as a risk assessment study, have shown that part of the soil requires remediation. It is further assumed that soils with pollutant concentrations superior to the remediation cutoff (S) will be treated either on site or ex situ, since nowadays, these remediation techniques are the most frequently applied. Consequently, the soils will be segregated and excavated before being sent to the treatment unit.

2.1.1 Volumes and uncertainty

Following the common practice, the remediation volumes are estimated in two steps. First, the volume of soil that needs to be extracted for remediation (notation:Vexc) is delineated on the basis of the available investigation data. Second, the volume requiring remediation, included in the volume delineated for excavation, is estimated; soils are segregated based on remediation data collected systematically, in blocks of a regular sampling grid applied to the volume delineated for excavation.

Non-linear geostatistics are necessary to estimate the volumes with concentrations superior to S [4]. Block conditional simulations are generated and used to calculate the probabilities that a block pollutant concentration is above S. Vexc is then defined as the set of blocks with probabilities superior to a maximal "acceptable" probability (b). The major difficulty is to define b, taking into account the risk assessment results, the possible re-use of the restored site, and the general context in which restoration takes place.

In Vexc, the volume where pollutant concentrations exceed S is calculated for each block conditional simulation. The resulting volume distribution provides an estimate of the volume requiring clean up (Vc), materialized by the mean or the median of the distribution, and an estimate of its accuracy, in the form of a variance or an inter-quantile interval.

The probability map allows for estimating two other important volumes. First, the environmental risk may be considered as non-significant for blocks where the probability is inferior to a threshold value, called a. These blocks can be considered as a residual volume (Vr), which can remain without any remediation or additional monitoring. Second, blocks where the probability is superior to a and inferior to b represent a volume for which environmental and financial risks are still significant: either blocks with pollutant concentrations above S may remain, or blocks with pollutant concentrations inferior to S may be excavated for remediation, although unnecessary. This so-called in-between volume (Vu) represents the uncertainty remaining on the site at stage j.

Furthermore, because the volumes depend on them [3], usual remediation conditions are modelled: (1) Vexc includes non-polluted blocks that have to be excavated to make the polluted blocks accessible; their global volume Vnp is the complementary of Vc in Vexc (Vexc = Vc + Vnp). (2) The support effect: the size of the investigation samples is smaller than the size of the remediation blocks. (3) The information effect: the true block pollutant concentration, always unknown, is estimated by the concentration measured on a composite of small samples collected in the block. (4) Real data are always affected by sampling errors.

2.1.2 Restoration cost and uncertainty

The global restoration cost (Ctotal) is defined as the sum of investigation cost (Ci), remediation cost (Cc) and uncertainty on remediation cost (Cu): Ctotal = Ci + Cc + Cu. Ci depends on the number of collected samples: it is calculated, using a specific investigation cost-function [3]. Cc depends on Vexc, Vc and Vnp while Cu depends on Vu: they are calculated with the same remediation cost-function [3]. The two cost-functions have been fully parameterized [3] so that restoration budgets can be calculated for various pollution scenarios and for various commercial and technical proposals.

2.2 Step 2: forecasts for stage j+1

If at stage j, the uncertainties remaining on the volume designed for excavation and on the remediation cost estimates are too high for the decision-maker(s), we propose to anticipate the volume and cost that could be estimated with additional samples collected at stage j+1 in the following way.

In order to calculate block simulations, conditional simulations of point pollutant concentrations are generated in the variogram model adjusted at stage j, based on a fine rectangular grid [3]: one of these point simulations is selected randomly. It is taken as the reference for the state of pollution of the industrial site. The consequence is that the reference is now supposed perfectly known. Various numbers of simulated point values are selected as samples from the reference, according to a grid, in Vi defined at stage j. The size of the rectangular grid depends on the number of additional data to consider. The selected values are taken as new conditional data, collected at stage j+1, and added to the data set available at stage j.

The volumes Vexc, Vc, Vnp, Vr and Vu are re-calculated with new block conditional simulations, in the model of variogram adjusted at stage j, for the various data sets modelled for stage j+1.

In the same way, the investigation cost-function is applied to the various data sets modelled for stage j+1, in order to forecast the sampling cost estimates at stage j+1. The remediation cost-function is applied to the volumes anticipated for stage j+1, in order to forecast the remediation cost and the financial risk estimates at stage j+1.

3. APPLICATION ON A FORMER SMELTING WORKS

3.1 Site description

The former smelting works covered a surface of 3 hectares. The investigation of the site and its neighbourhood (45 hectares) has provided 75 lead concentrations in 6 stages, with a homogeneous sampling support. The detailed risk assessment study has shown that soil whose [Pb] exceeds S = 300 ppm involves a risk for human health, due to dust inhalation. The value of 300 ppm was taken as the legal remediation cutoff, and an *on site* soil washing was applied. The zone requiring excavation for remediation, delineated at the end of the risk assessment study (without geostatistics) was segregated during the remediation works according to 212 estimated block [Pb], defined on a regular grid. One block was 10 m side length and 0.30 m height. The soil was excavated in 3 layers. The lead concentrations were measured on a composite sample, made of four small samples taken at the corners and one small sample taken at the centre of each block.

3.2 Step 1: estimations at stage 6

First, the remediation volumes and the restoration costs were estimated on the basis of the 75 real investigation data available at stage 6. The validity of the adopted models was checked, using the 212 real remediation data [3].

3.2.1 Volumes and uncertainty

The gaussian transformed investigation data show an anisotropic spatial structure, which has been modelled with a combination of a nugget effect and two spherical models: $\gamma(h) = 0.1 + 0.5\ \mathrm{Sph}(50\ m_{NE}, 70\ m_{NW}, 0.7\ m_{Vert}) + 0.65\ \mathrm{Sph}(140\ m_{NE}, 1000\ m_{NW}, 4\ m_{Vert})$ [2]. The anisotropy of the variogram is consistent with the principal directions of wind, which is responsible for lead dispersion (*Figure 1)*. The variogram model is chosen showing a stationarity outside the limits of the domain of study.

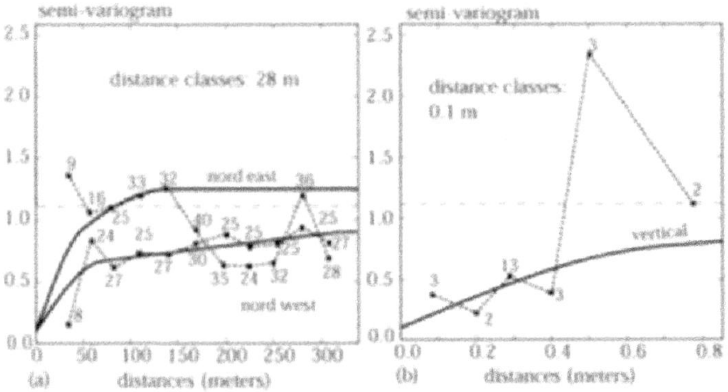

Figure 1 .Experimental variogram and fitted variogram model of the gaussian transformed investigation data

A total of 200 conditional simulations of point lead concentrations have been generated with the turning band method, on a fine grid, in the frame of a multigaussian model, in a unique neighbourhood. Every fine grid mesh was 4.30 m side length and 0.30 m height, oriented according to the anisotropy axis. These point simulations were used to calculate block simulations, accounting for:

- The support effect. The lead concentrations were simulated in blocks similar to those actually used for excavating the soil.
- The information effect. Simulated block [Pb] were considered as the mean of five point simulated values, by analogy with the real block concentrations measured during remediation.

[2] NE: Nord East, NW: Nord West, Vert: vertical directions, Sph : spherical model

- The sampling error. It has been shown that high errors have affected the real block [Pb] [3]. They are modelled by: $Z(x) + Z(x).\varepsilon(x)$, where x is the position of the block in the geometrical space, Z is the random variable figuring the block concentrations, and ε is a uniform distribution on the interval [-1;+1], whose variance is high, equal to 0.33.

For each block, the probability that block lead concentration exceed 300 ppm was calculated as the ratio of the number of simulated values exceeding 300 ppm and the total number of simulated values at that block.

If it is supposed that remediation is required for soils with a probability above $\beta = 0.6$, Vexc is evaluated to 10912 m^3. In addition, assuming that soils can remain without any remediation or additional monitoring if their probability is below $\alpha = 0.2$, Vu is evaluated to 23327 m^3.

3.2.2 Restoration costs and uncertainty

The investigation costs at stage 6, calculated with real unit prices, were estimated to Ci = 68600 Euros. The soil washing costs, calculated with actual market unit prices, were estimated to Cc = 984821 Euros. The financial risk was estimated to Cu = 1295817 Euros.

3.3 Step 2: forecasts for stage 7

Since Vu represents **214 %** of Vexc and since Cu corresponds to **132 %** of Cc, the environmental and financial risks remaining at stage 6 are considered as too high. Consequently, an additional sampling stage has to be designed.

Various numbers N_7 = 12, 25, 50, 100 and 200 additional lead concentrations were selected successively from the reference point conditional simulation, selected randomly among the 200 simulations generated at stage 6. The point values were chosen in the limits of Vu (*Figure 2*).

Using the variogram model fitted at stage 6 (see paragraph 3.2), point conditional simulations were calculated successively, with data sets including the 75 real investigation data of the 6 first stages, and the N_7 simulated values selected on the reference. The probabilities of exceeding 300 ppm (*Figure 2*) and the volumes forecasted for stage 7 (*Table 1* and *Figure 3*) were then calculated for each data set, for a block support, as explained in paragraph 3.2.

Using the same unit prices applied at stage 6, for each data set modelled for stage 7, the investigations costs, the *on site* soil washing costs and the financial risks (*Figure 4* and *Table 2)* were calculated as explained in paragraph 2.1.

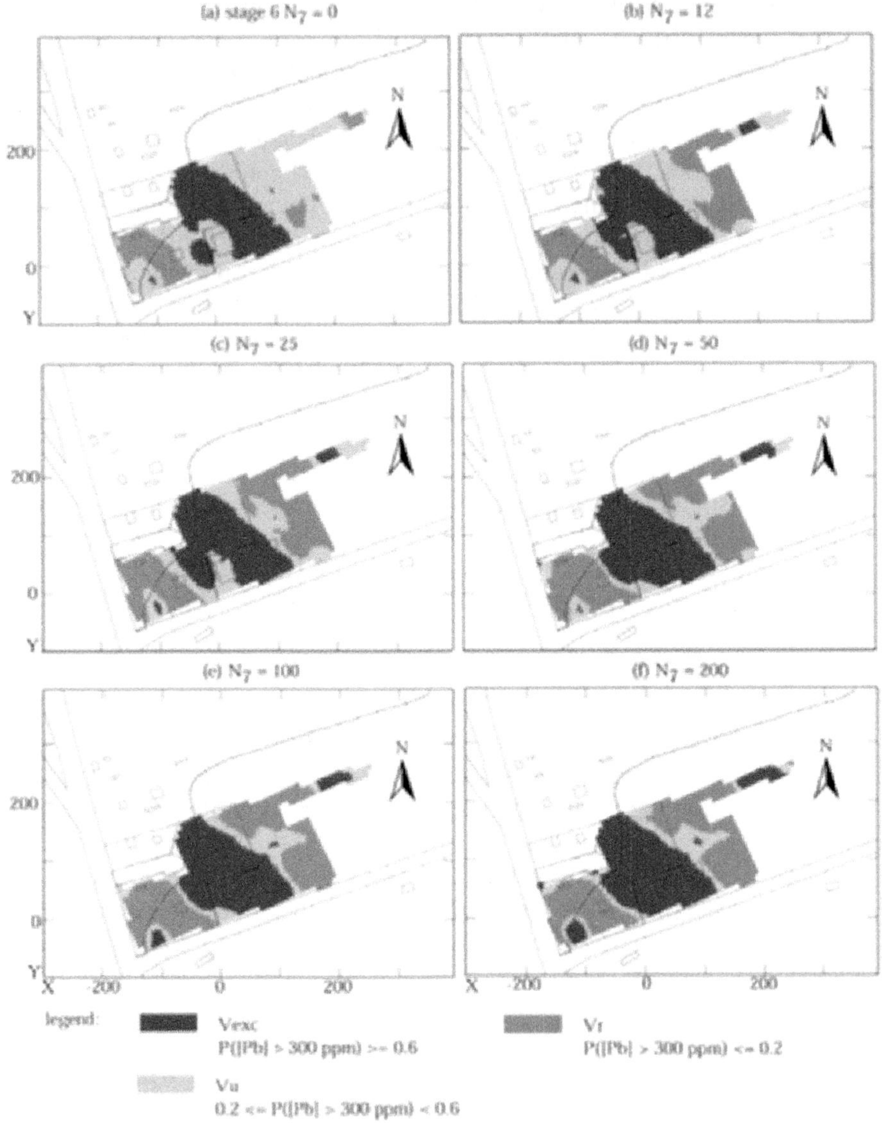

Figure 2. Simplified map of probabilities that block lead concentrations exceed 300 ppm, in the superficial layer of 0.30 m height, estimated at stage 6 with the 75 real investigation data, and forecasted at stage 7 with N_7 additional data whose sampling has been modelled. (a) stage 6, $N_7 = 0$ (b) stage 7, $N_7 = 12$ (c) stage 7, $N_7 = 25$ (d) stage 7, $N_7 = 50$ (e) stage 7, $N_7 = 100$ (f) stage 7, $N_7 = 200$

Table 1. Forecasted volumes Vexc requiring excavation for remediation, and volumes Vu materializing uncertainties on Vexc, according to the number N_7 of additional data whose sampling is modelled at stage 7 (unity: cubic meter)

N_7	0	12	25	50	100	200
Vexc	10912	11212	13287	13767	16743	17555
Vu	23327	18276	14910	12385	9078	7996
Vu / Vexc	214 %	163 %	112 %	90 %	54 %	46 %

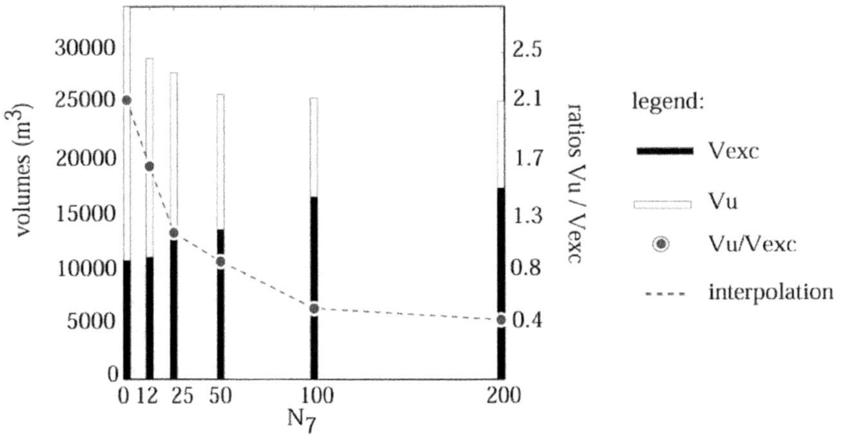

Figure 3. Volume forecasts graph: Volume Vexc requiring excavation for remediation, and volume Vu materializing volume uncertainties, as a function of the number N_7 of additional data whose sampling is modelled at stage 7

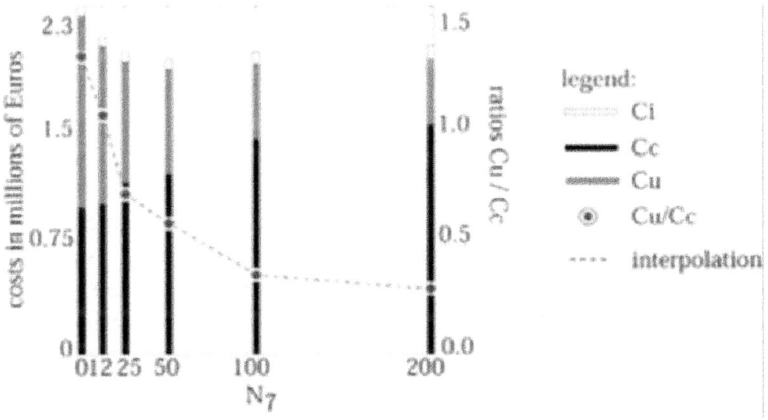

Figure 4. Cost forecasts graph: investigation costs Ci, remediation costs Cc and financial risk Cu, as a function of the number N_7 of additional data whose sampling is modelled at stage 7

Table 2. Forecasted investigation costs Ci, remediation costs Cc and financial risk Cu, according to the number N_7 of additional data whose sampling is modelled at stage 7 (unity: euro)

N7	0	12	25	50	100	200
Ci	68600	71651	73176	77749	83847	94518
Cc	984821	1007688	1105990	1215019	1445217	1542784
Cu	1295817	1064094	815602	702790	506131	445151
Cu / Cc	132 %	106 %	71 %	58 %	35 %	29 %
Ci + Cc + Cu	2347715	2141909	2041292	1995558	2035194	2082454
Ci / Cc	6.9 %	7.1 %	6.4 %	6.4 %	5.8 %	6.1 %

3.4 Decision making: a "best compromise" for stage 7

3.4.1 Working hypothesis

The aim of this paragraph is to show the usefulness of the forecasts graphs to plan the number of additional samples required to get an "acceptable" remediation volume and cost uncertainty level. It is important to underline that there is never an optimum for this number, but only a "best compromise", because the criteria, goals, constraints and preferences of the decision makers are various, depending on the context in which restoration takes place. The proposed graphs help to decide according to environmental criteria, in terms of remediation volumes and uncertainties; and according to financial criteria, in terms of investigation costs, remediation costs and uncertainties. These criteria allow for defining specific objectives and constraints. As an illustration, we selected two objectives on uncertainties:

1. Minimizing the ratio Vu / Vexc
2. Minimizing the ratio Cu / Cc

But we took also into account one common financial objective and one common financial constraint:

1. Minimizing the global restoration cost Ctotal
2. Ci / Cc $\leq$ given percentage

The weights applied on these objectives depend on the decision makers profiles. As an illustration, we consider a so-called risk averse decision maker, that is trying to avoid risks: he (she) will probably assign heavier weights to (1) and (2) than to (3) and (4). On the opposite, a so-called risk prone decision maker, that is ready to take risks, will probably assign heavier weights to (3) and (4) than to (1) and (2).

3.4.2 Discussion

The volume forecasts graph (*Figure 3*) shows that Vu / Vexc decreases as N_7 increases, and tends to stabilize for $N_7 \geq 100$. For $N_7 = 100$, Vu represents half of Vexc, instead of the 214 % estimated at stage 6. Similarly, the cost forecasts graph (*Figure 4*) shows that Cu / Cc decreases as N_7 increases, and tends to stabilize for $N_7 \geq 100$. For $N_7 = 100$, Cu is forecasted at 35 % of Cc, instead of the 132 % estimated at stage 6. When $N_7 > 100$, the gain in accuracy on the forecasted volumes and costs is negligible.

The global cost Ctotal = Ci + Cc + Cu is minimum for $N_7 = 50$. The ratio Ci / Cc slightly declines as N_7 goes up, until $N_7 = 100$, and grows for $N_7 > 100$. The minimum ratio, calculated for $N_7 = 100$, is forecasted to 5.8 %.

3.4.3 Decisions

Independently from other objectives or constraints than those defined on paragraph 3.4, we can reasonably think, according to these forecasts, that a risk averse decision maker would choose to collect $N_7 = 100$ additional samples at stage 7.

Similarly, we can think that a risk prone decision maker would prefer $N_7 = 50$, because, even though the corresponding Vu / Vexc and Cu / Cc ratios are higher than those forecasted with $N_7 = 100$, that number of additional data minimizes the global restoration cost, and maintains the ratio Ci / Cc to an average value (6.4 %).

4. CONCLUSIONS

The presented approach can only be applied if a spatial structure is visible. In case of a pure nugget effect, no additional sampling will improve the existing uncertainty levels: uncertainties are inevitable. It is quite often difficult to highlight the spatial structure of soil pollution, due to a high heterogeneity and data scarcity, but recent works show that it is possible with adapted variographic tools [2].

Another limitation of the methodology is linked to the conditional simulation chosen as a reference, since it is one possible realization of the random function representing the soil pollution phenomenon, but it is not reality. As a consequence, the forecasted volumes and costs may differ from the real ones. An improvement should consist in repeating volumetric and cost calculations for various simulations taken as references, and in comparing the results.

Lastly, the volumes and costs are anticipated in the variogram model fitted at stage j. The variogram model that will be fitted to the real data at stage j+1 may differ from that model, especially at the beginning of the investigations when few data are available, inducing a bad quality of variogram fitting.

A necessary step to finalize that work should consist in applying the methodology at the early sampling stages of real polluted sites, in order to (1) test its practical validity in various cases, (2) study its co-ordination with other steps of the restoration process, especially with the risk assessment sampling, and (3) assess its interest when it is coupled with quick *on site* chemical analysis.

ACKNOWLEDGEMENTS

The authors thank M^s Martine Louvrier and Mr Philippe Bégassat of the ADEME for the data and their technical support, and the ADEME and Gaz de France for their financial support.

REFERENCES

1. Englund, E. and Heravi, N., 1992. Conditional Simulation: Practical Application for Sampling Design Optimization, Geostatistics Troia 92, pp. 613-624
2. Jeannée, N., 2001, Caractérisation Géostatistique de Pollutions Industrielles de Sols. Cas Des Hydrocarbures Aromatiques Poycycliques sur D'anciens Sites de Cokeries., PhD. Thesis, Ecole des Mines de Paris (F)
3. Renard-Demougeot, H., 2002, De la reconnaissance à la réhabilitation des sols industriels pollués: estimations géostatistiques pour une optimisation multicritère, PhD. Thesis n°14615, ETHZ (CH)
4. Rivoirard, J., 1994. Introduction to Disjunctive Kriging and Non-Linear Geostatistics. Clarendon press, Oxford.

COMBINING CATEGORICAL INFORMATION WITH THE BAYESIAN MAXIMUM ENTROPY APPROACH

D. D'Or and P. Bogaert

Université catholique de Louvain, Dept. of Environmental Sciences and Landuse Planning – Environmetry, Place Croix du Sud, 2 bte 16 1348 Louvain-la-Neuve, Belgium. dor@enge.ucl.ac.be, bogaert@enge.ucl.ac.be

Abstract: The estimation of categorical variables is a recurrent problem in geostatistics. Beyond traditional and well-known methods like indicator kriging (IK) or classification, the Bayesian Maximum Entropy (BME) approach offers a new sound theoretical framework for modeling the spatial correlation and for computing estimates for categorical variables. In this paper, we show how the BME approach can be used for estimating a categorical variable by combining multiple sources of information. This methodology is illustrated with a practical example dealing with the estimation of soil drainage classes. Data involved consist in a set of punctual observations and a pre-existing exhaustive soil map. Estimates are obtained with BME using various combinations of the data, i.e., (i) the soil map only, (ii) the punctual observations only, and (iii) both of them. For the latter, the relation between the two data sets is taken into account by the way of a double entry probability table when obtaining the maximum entropy joint distribution. The strong advantages of BME over IK are explained at the light of the results that are obtained. The important case of conflicting informations is also discussed at the light of the way BME merges these information.

Key words: categorical data, spatial estimation, Bayesian maximum entropy, Kullback-Leibler

X. Sanchez-Vila et al. (eds.), geoENV IV – Geostatistics for Environmental Applications, 295-306.

1. INTRODUCTION

Handling qualitative information is very common in soil sciences: about 90% of the variables collected during soil surveys are either ordinal or nominal (Bregt *et al.*, 1992). However, spatial interpolation of such data is scarce in the literature. It is traditionally performed using the indicator kriging approach (Bierkens and Burrough, 1993a,b), but this approach suffers from severe and well-documented limitations (Goovaerts, 1997; Bogaert and D'Or, 2002). The main reason for these limitations is the lack of a strong theoretical justification for this approach, leading to many intern incoherences and often inconsistant results. Solutions traditionally proposed to overcome these imperfections consist mainly in tricks and are again not supported by any theoretical concepts. This is for example the case for the order relation problems, corrected by several algorithmic tips (see e.g., Goovaerts, 1997).

In a spatial mapping context, another concern is that several sources of information may be available at the same time. The objective when combining them is to get better estimates than what would have been obtained by using them in a separate way. However, a serious problem occurs when at the same location these various informations are in disagreement. E.g., at a given sampling location, some of these sources can give contradictory information about the category. Clearly, the occurrence frequency of this kind of problem is expected to increase with the number of sources that are taken into account and depends largely on the reliability of each of them.

In order to find solutions to the two problems mentioned here above, a new method should be developed with the following features: (i) having clear theoretical basis for each step of the information processing; (ii) allowing the user to give as input only fragmentary information (e.g, those that are reasonnably reliable; (iii) yielding results that clearly agree with the intuitive reasoning, especially when merging contradictory information. The Bayesian Maximum Entropy approach has recently proved to be a powerfull tool to process spatial data sets (Christakos, 2000, 2002; D'Or *et al.*, 2001). Based on sound information processing rules and classical probability laws, it was first developed for continuous variables but it can be extended for the processing of categorical data Bogaert (2002). The use of this approach will be illustrated here, based on a real data set.

2. THE DATA SET

The study zone is located in the sandy area of Flanders (*région sableuse*), around the city of Mechelen (Belgium) and occupies a 30 by 30 km^2 area extending from Vilvoorde (South) to Antwerpen (North) (Fig. 1). Most of the soils are classified as Spodosols. In the alluvial plains of the Grote Nete, Dijle and Zenne, soils are more clayey and classified as Fluvents. Average elevation is around 15 m and the topography is very flat.

The available information about drainage is coming from two sources. The first is the Aardewerk database (Van Orshoven *et al.*, 1998), from which we extracted the 347 soil profile descriptions that are available over the 30 by 30 km^2 area represented in Fig. 1a. The second source is a spatially exhaustive digitalized version of a pre-existing soil map that was manually contoured based on auger boring sampling campaigns (Fig. 1b). For both sources, three soil drainage classes have been defined by grouping the nine original classes used in the Belgian soil map. The three classes are c_1≡"excessive to good drainage", c_2≡"good to moderately bad drainage", and c_3≡"moderately bad to very bad drainage". The set of possible outcomes is thus $\Omega_C=\{c_1,c_2,c_3\}$. Let us define the digitalized map as $\{ M(\mathbf{x}_\alpha), \mathbf{x}_\alpha \in D\}$, where D is the area represented in Fig. 1b, and the Aardewerk database as $\{A(\mathbf{x}_1), \ldots, A(\mathbf{x}_\beta), \ldots, A(\mathbf{x}_k)\}$ (k=347), where the $M(\mathbf{x}_\alpha)$'s and $A(\mathbf{x}_\beta)$'s are categorical variables with Ω_C as possible outcomes.

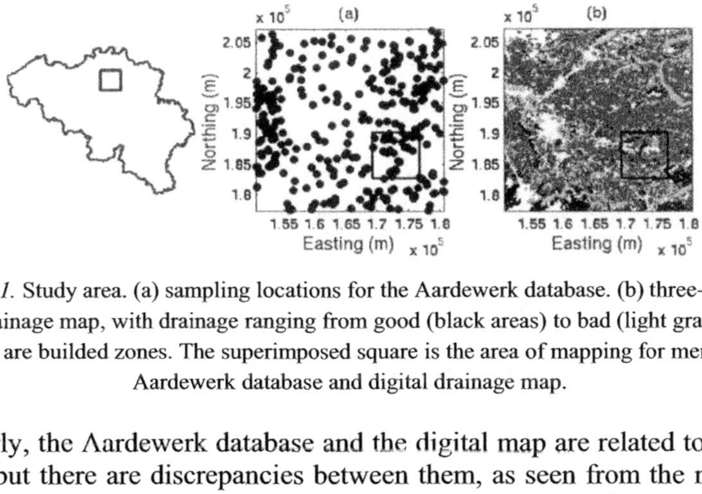

Figure 1. Study area. (a) sampling locations for the Aardewerk database. (b) three-classes digital drainage map, with drainage ranging from good (black areas) to bad (light gray). White areas are builded zones. The superimposed square is the area of mapping for merging Aardewerk database and digital drainage map.

Clearly, the Aardewerk database and the digital map are related to a large extent, but there are discrepancies between them, as seen from the non-null probabilities for the off-diagonal cells in the joint probability table (Table 1a). At a given sampling location, the Aardewerk database and the digitalized map can give contradictory information about the drainage class. These discrepancies are easily explained. The Aardewerk database consists of a set of detailed profile descriptions that can be assumed as error-free, whereas the drainage map is the digital version of a manually contoured pedological map, based on a set of regular auger boring samples. Due to the limited accuracy obtained when using auger borings as well as due to the interpolated nature of the map, it is expected that some conflict may occur. However, both informations have useful features: the limited Aardewerk database can be assumed as error-free, whereas the digital map is somewhat approximate but spatially exhaustive. Combining these two sources is thus more reasonable than discarding one of them for pure convenience reasons.

Table 1. Estimated probabilities for Aardewerk & digital Map classes.

(a) $\hat{P}(A(\mathbf{x}) = c_i \cap M(\mathbf{x}) = c_j)$

	j=1	j=2	j=3	Σ_j
i=1	0.133	0.042	0.007	0.182
i=2	0.044	0.415	0.054	0.513
i=3	0.003	0.124	0.178	0.306
Σ_i	0.180	0.581	0.239	1

(b) $\hat{P}(A(\mathbf{x}) = c_i | M(\mathbf{x}) = c_j)$

	j=1	j=2	j=3
i=1	0.737	0.072	0.029
i=2	0.246	0.714	0.225
i=3	0.017	0.214	0.746
Σ_i	1	1	1

As the local uncertainty about the digitalized map is unknown and is expected to vary from places to places, we focused on the Aardewerk database for estimating the spatial structure of drainage classes over the area. Using the 347 Aardewerk sampled locations, the bivariate probability functions for the three drainage classes can be estimated. The theoretical bivariate probability functions $\pi_{i,j}(\mathbf{h})$ are defined as

$$\pi_{i,j}(\mathbf{h}) \equiv P(A(\mathbf{x}) = c_i \cap A(\mathbf{x}+\mathbf{h}) = c_j) \quad \forall i, j = 1,...,3 \tag{1}$$

and can be estimated using

$$p_{i,j}(\mathbf{h}) \equiv \frac{1}{N(\mathbf{h})} \sum_{N(\mathbf{h})} \delta(A(\mathbf{x}) = c_i \cap A(\mathbf{x}+\mathbf{h}) = c_j) \tag{2}$$

where $N(\mathbf{h})$ refers to the number of pairs separated by a distance **h** and where $\delta(.)$ is the Kronecker delta, equal to 1 or 0 if the condition between brackets is verified or not, respectively (Fig. 2).

Remark that, for any given distance **h**, (2) fulfills the validity conditions $p_{i,j}(\mathbf{h}) \geq 0 \ \forall i,j$ and $\Sigma_{i,j} \, p_{i,j}(\mathbf{h}) = 1$. As these functions exhibit considerable variability and are computed for a limited set of distances, it is useful to get smoother estimates that can be obtained, e.g., through a Gaussian kernel smoothing procedure as shown on Fig. 2. There is thus no need for parametric assumption when modeling the $p_{i,j}(\mathbf{h})$'s.

From Fig. 2, one can see that there is little evidence of a clear spatial structure in the data for distances greater than 2 km.

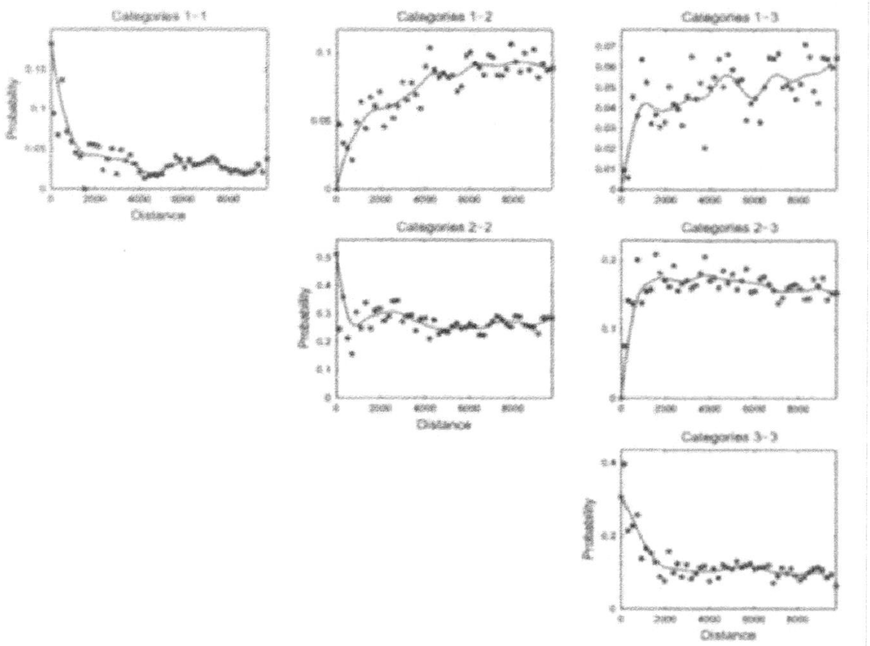

Figure 2. Estimated and kernel smoothed bivariate probability functions for the three drainage classes (1=good, 2=moderate, 3=bad).

3. CONDITIONAL DISTRIBUTIONS

What is sought for is a method of merging the Aardewerk and the digital map information, in order to get probability estimates for each drainage class at any arbitrary unsampled location $\mathbf{x}_0$. This method should account for several facts: (i) we want to use the knowledge of the spatial structure of the data brought by the $p_{i,j}(\mathbf{h})$'s, (ii) we want to account for the relation between the Aardewerk database and the digital map, brought by the $\hat{P}(A(\mathbf{x}) = c_1 \cap M(\mathbf{x}) = c_j)$'s, without having to explicitly specify the spatial structure of the digital map, and (iii) we want to solve the problem of possibly conflicting information, as the Aardewerk database and the digital map may indicate different categories for the same location (27% of the cases out of the 347 Aardewerk sampled locations).

At the unsampled location $\mathbf{x}_0$, we can define the *a priori* distribution $\pi_{i0} \equiv P(A(\mathbf{x}_0) = c_{i0})$, $i_0 = 1, \ldots, 3$ as well as the conditional distribution $\pi_{i0}| K_S \equiv P(A(\mathbf{x}_0) = c_i| K_S)$, where K_S refers to some specific knowledge. For our case study, different cases for K_S can been considered: (i) if we use only Table 1b along with the digital map, we have $K_{S,M} \equiv M(\mathbf{x}_0) = c_{j0}$; (ii) if we use only the

Aardewerk database, we have $K_{S,A} \cap_\beta A(\mathbf{x}_\beta) = c_{i\beta}$; (ii) if we use both the Aardewerk database and the digital map, we have $K_{S,AM} \equiv K_{S,A} \cap K_{S,M}$.

In a spatial context, the prediction of categorical variables is traditionally based on indicator (co-)kriging (Journel, 1983). The application of IK or ICK is straightforward when based on $K_{S,A}$, but then, the information provided by $K_{S,M}$ is neglected. If one wants to incorporate $K_{S,M}$ also, a "soft" indicator formalism could be used (Journel, 1986; Goovaerts, 1997), specifying at each location $\mathbf{x}_0$ a "soft" indicator coding, that corresponds to the probabilities coming from Table 1b. However, as $K_{S,M}$ is spatially exhaustive, and due to the well-known exactitude property of I(C)K, this would return the soft information as estimate of the conditional distribution, which is of course a useless result. So using (soft) I(C)K, one is faced with an impossible choice: (i) neglecting the spatially exhaustive map information when computing $p_{i0}| K_{S,A}$, or (ii) neglecting the Aardewerk information when computing $p_{i0}| K_{S,AM}$, as by property $p_{i0}| K_{S,AM} = p_{i0}| K_{S,,M}$ which is one of the columns of Table 1b.

It is in order to overcome this kind of paradoxes that the BME approach for categorical variables was designed (Bogaert, 2002), as an extension of the BME principle for continuous variable (Christakos, 2000; Christakos *et al.*, 2002).

Instead of relying on a (soft) indicator coding of the data in a linear kriging system, BME will directly incorporate all the available information in order to built a joint probability table, that can be used afterward for deriving any kind of conditional distributions.

4. THE BME APPROACH FOR CATEGORICAL VARIABLES

Denote K_G as the general knowledge that we have about the variables under study. For our case, we can define $K_{G,A} \equiv \{p_{i,j}(\mathbf{h}), \text{i,j} = 1, \ldots,3\}$ as the set of bivariate probability functions and $K_{G,M} \equiv \{\hat{P}\ (A(\mathbf{x}) = c_i \cap M(\mathbf{x}) = c_j)\}$as Table 1a, so that $K_{G,AM} = K_{G,A} \cap K_{G,M}$. What is sought for is an estimate $p_{0,\ldots,k}(i_0,\ldots,i_k,j_0)$ for the joint distribution $\pi_{0,\ldots,k}(i_0,\ldots,i_k,j_0)$, where

$$p_{0,\ldots,k}(i_0,\ldots,i_k,j_0) \equiv \hat{P}((\bigcap_{\beta=0}^{k} A(x_\beta) = c_{i_\beta}) \cap M(x_0) = c_{j_0}) \quad (3)$$

The maximum entropy estimate of (3) is obtained by maximizing

$$H_{0,\ldots,k} = -\sum_{i_0,\ldots i_k,j_0} p_{0,\ldots,k}(i_0,\ldots i_k,j_0) \ln p_{0,\ldots,k}(i_0,\ldots i_k,j_0) \quad (4)$$

under the constraints provided by the general knowledge $K_{G,AM}$, so that

$$\begin{cases} p_0(i_0,j_0) = \hat{P}(A(x_0) = c_{i_0} \cap M(\boldsymbol{x}_0) = c_{j_0}) \\ p_{\beta,\beta'}(i_\beta,j_{\beta'}) = p_{i_\beta,i_{\beta'}}(\boldsymbol{h}_{\beta\beta'}) \end{cases} \quad \begin{matrix} \forall i_0, j_0 = 1,2,3 \\ \forall \beta,\beta' = 1,\ldots,k \\ \forall i_\beta, i_{\beta'} = 1,2,3 \end{matrix} \tag{5}$$

where the $p_0(i_0,j_0)$'s and the $p_{\beta,\beta'}(i_\beta,i_{\beta'})$'s are marginal distributions obtained by summation over $p_{0,\ldots,k}(i_0,\ldots,i_k,j_0)$, with

$$\begin{aligned} p_0(i_0,j_0) &= \sum_{i_1,\ldots,i_k} p_{0,\ldots,k}(i_0,\ldots i_k,j_0) \\ p_{\beta,\beta'}(i_\beta,i_{\beta'}) &= \sum_{\{i_n;n\neq\beta,\beta'\}} p_{0,\ldots,k}(i_0,\ldots i_k,j_0) \end{aligned} \tag{6}$$

The maximization of (4) can be accomplished using an iterative scaling algorithm, where the constraints (5) appears as imposed bivariate probability tables that are margins of the probability table $p_{0,\ldots,k}(i_0,\ldots,i_k,j_0)$.

It is worth noting that even if we did not specify any of the probabilities $\hat{P}(A(\mathbf{x}_\beta) = c_{i\beta} \cap M(\mathbf{x}_0) = c_{j0})$ (with $\beta \neq 0$), the maximum entropy algorithm will provide $p_{0,\beta}(i_\beta,j_0)$ as their estimates, where

$$p_{0,\beta}(i_\beta,j_0) = \sum_{\{i_j;j\neq\beta\}} p_{0,\ldots,k}(i_0,\ldots i_k,j_0) \tag{7}$$

In other words, even if a fragmentary information is provided about the relation between the two variables using Table 1a, the maximum entropy algorithm will give back an estimate for the missing values. The same property would have applied if, e.g., some of the bivariate probabilities $p_{i\alpha},i_\beta(h_{\alpha\beta})$ would have been left unspecified due to a lack of reliable information. This is a very nice feature, as it offers considerable flexibility to the user.

After $p_{0,\ldots,k}(i_0,\ldots,i_k,j_0)$ has been estimated, the conditional distributions can of course be easily obtained from it, with

$$p_{i_0|K_{S,AM}} = \frac{p_{0,\ldots,k}(i_0,\ldots i_k,j_0)}{\sum_{i_0} p_{0,\ldots,k}(i_0,\ldots i_k,j_0)} \qquad i_0 = 1,2,3 \tag{8}$$

where the index values $(i_1,\ldots,i_k,j_0)$ are known from the specific knowledge $K_{S,AM}$.

5. RESULTS

For mapping purposes, a smaller 7.5 by 7.5 km^2 area around Tremelo (see superimposed square on Fig. 1) has been considered in order to illustrate the use of the method. The conditional probabilities for each drainage class have been estimated at the nodes of a 101 by 101 square grid (the grid spacing is thus 75 m), based on the neigbouring Aardewerk sampled locations (Fig. 3a) and the digital map (Fig. 3b), so that maps of the maximum probability drainage classes can be obtained. Three cases can be considered:

- if only $K_{G,M}$ is used, the $p_{i0}| K_{S,M}$'s are those given by the corresponding columns in Table 1b, and as seen from this table the maximum probability occurs when $i_0 = j_0$, so that the map of the maximum $p_{i0}| K_{S,M}$'s corresponds to the digital map given in Fig. 3b;
- if only $K_{S,A}$ is used (Fig. 3c), the map of the maximum $p_{i0}| K_{S,A}$'s is very smooth with few details, according to the limited number of sampled Aardewerk locations over the area;
- by using $K_{S,AM}$ (Fig. 3d), it is easy to see that the map of the maximum $p_{i0}| K_{S,AM}$'s is close to the digital map when one is far from the Aardewerk sampled locations; main differences between the two maps appear in the areas close to these sampled locations.

The BME algorithm is thus attaching importance to the Aardewerk information when the location $\mathbf{x}_0$ is close from a sampled location $\mathbf{x}_\beta$, whereas it neglects this information when $\mathbf{x}_0$ is far from it, as there is no more correlation between the drainage class at locations $\mathbf{x}_0$ and $\mathbf{x}_\beta$.

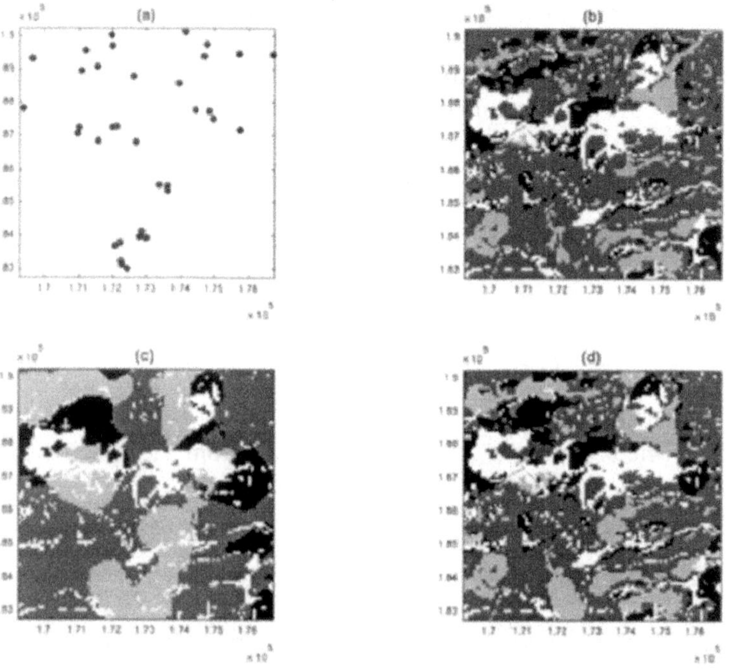

Figure 3. Maps of the maximum probability drainage classes, with (a) the Aardewerk sampled locations, (b) digital drainage map, (c) BME map using only Aardewerk data, and (d) BME map using both Aardewerk data and digital map.

6. MERGING VARIOUS INFORMATION SOURCES WITH BME

According to the previous results, we can focus on the way BME is doing when merging both sources of information for computing $p_{i0}| K_{S,AM}$. This can be done using the Kullback-Leibler (KL) distance or relative entropy (Kullback and Leibler, 1951).

The KL distance $KL(\pi_a \| \pi_b)$ between two arbitrary distributions $\pi_a = (\pi_a,1,\ldots, \pi_{a,m})$ and $\pi_b = (\pi_b,1,\ldots, \pi_{b,m})$.

$$KI(\pi_b \| \pi_b) = \sum_i \pi_{a,i} \ln \frac{\pi_{a,i}}{\pi_{b.i}} \quad (9)$$

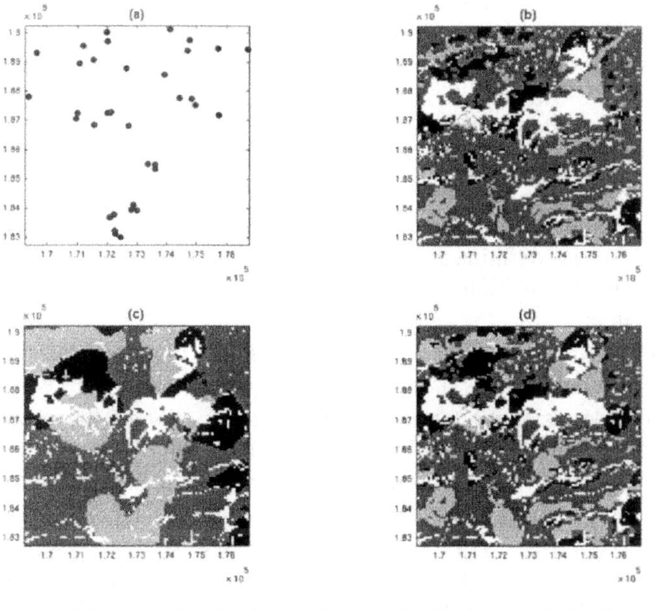

Figure 4. maps of the entropies fo rthe conditional distributions, with (a) the Aardewerk sampled locations, (b) the digital drainage entropy map, (c) the BME entropy map using only Aardewerk data, and (d) the BME entropy map using both Aardewerk data and digital map. Values are ranging from 0 (black) to ln(3) (white).

where $KL(\pi_a || \pi_b) \geq 0$ and is null if and only if $\pi_a = \pi_b$. It is thus a measure of the "distance" between π_a and π_b. This measure is very useful in our context for assessing the information that has been brought by the use of a spatial information for the mapping. Assume that our reference distribution is the *a priori* distribution π_{i0}, $i_0 = 1,2,3$. What is sought for is a measure of the additional information content in the conditional distributions $\pi_{i0}| K_{S,A}$, $\pi_{i0}| K_{S,M}$ and $\pi_{i0}| K_{S,AM}$ compared to π_{i0}. We can thus compute the KL distance between each of these conditional distributions and the *a priori* distribution, and see how BME is operating when merging possibly conflicting specific knowledge $K_{S,A}$ and $K_{S,M}$ in order to get the conditional distribution $\pi_{i0}| K_{S,AM}$. For each possible location $\mathbf{x}_0$, we can compute these various conditional distributions, as well as their entropy (H) and the KL distance from the {\it a priori} distribution. Additionally, we can also compute the KL distance $KL(A||M)$between $\pi_{i0}| K_{S,A}$ and $\pi_{i0}| K_{S,M}$, which is a measure of the disagreement between these two distributions. Let us examine the posterior distributions at three specific locations selected to illustrate three characteristic situation:

- In Table 2a, it appears that KL is low for $\pi_{i0}| K_{S,A}$, showing that $\pi_{i0}| K_{S,A}$ is close to π_{i0} and thus the Aardewerk database is weakly informative (it did not modify substantially the *a priori* probabilities). On the opposite, KL is higher for $\pi_{i0}| K_{S,M}$, showing that the digital map is more informative than the Aardewerk database. Note also that there is a disagreement between $\pi_{i0}| K_{S,A}$ and $\pi_{i0}| K_{S,M}$, as the first one favors category 3 whereas the second one favors category 2. As a consequence, $\pi_{i0}| K_{S,AM}$ is close to $\pi_{i0}| K_{S,M}$ as $K_{S,M}$ is considered as more relevant than $K_{S,A}$;
- In Table 2b, KL is high both for $\pi_{i0}| K_{S,A}$ and $\pi_{i0}| K_{S,M}$, so that both sources are informative. Moreover, $KL(A||M)$ is low, showing that both sources are in agreement for giving preference to category 1. As a consequence, $\pi_{i0}| K_{S,AM}$ is definitively heading for a preference for category 1, as translated by its high conditional probability, the low H value and the high KL value;
- In Table 2c, the reverse situation occurs. KL is high too both for $\pi_{i0}| K_{S,A}$ and $\pi_{i0}| K_{S,M}$, but there is a strong disagreement between the two distributions ($KL(A||M)$ is high), as one favors category 1 and the other one favors category 3. As both information are valuable but contradictory, they tend to annihilate each other. The H value for $\pi_{i0}| K_{S,AM}$ is higher than for $\pi_{i0}| K_{S,A}$ or $\pi_{i0}| K_{S,M}$, with a less clear-cut choice between categories 1 and 3.

It is worth noting that the entropy H is an absolute measure of the uncertainty associated with these conditional distributions, whereas the KL distance is a relative measure by comparison with the *a priori* distribution. As a consequence, it is KL and not H that should be used if what is sought for is a quantification of the accomplishment made by the method, compared to what was known prior to the use of it. E.g., for an imaginary situation summarized in Table 3, all the H values for the conditional distributions $\pi_{i0}| K_{S,AM}$ are equal, whereas the KL distances are quite different and reflect correctly the gain of information that has been obtained using the method, as measured by the divergence between $\pi_{i0}| K_{S,M}$ and π_{i0}.

As a summary, BME is correctly processing the various information sources according to their relative information content as well as according to their agreement or disagreement about this content. This is a very important feature of the method, as it makes sure that logical rules are automatically translated in sound mathematical results, that reflect these rules in the approriate way.

7. CONCLUSIONS

The BME approach proves to be a useful and flexible way for processing categorical data in a spatial context. Being a nonlinear method, it does not rely on the traditional linear paradigm used by IK or ICK, which suffer from serious theoretical and practical problems when dealing with this kind of variables

Table 2. Conditional distributions at various locations $\mathbf{x}_0$.

(a) weak Aardewerk & strong Map information ($KL(A \parallel M)$=0.40)

	$i_0 = 1$	$i_0 = 2$	$i_0 = 3$	H	KL
π_{i_0}	0.182	0.513	0.306	1.01	
$\pi_{i_0 \mid K_{S,M}}$	0.029	0.225	0.746	0.66	0.48
$\pi_{i_0 \mid K_{S,A}}$	0.074	0.604	0.322	0.86	0.06
$\pi_{i_0 \mid K_{S,AM}}$	0.016	0.256	0.728	0.64	0.39

(b) strong Aardewerk & Map information ($KL(A \parallel M)$=0.17)

	$i_0 = 1$	$i_0 = 2$	$i_0 = 3$	H	KL
π_{i_0}	0.182	0.513	0.306	1.01	
$\pi_{i_0 \mid K_{S,M}}$	0.737	0.246	0.017	0.64	1.01
$\pi_{i_0 \mid K_{S,A}}$	0.709	0.162	0.129	0.80	0.61
$\pi_{i_0 \mid K_{S,AM}}$	0.979	0.020	0.002	0.11	2.94

(c) strong Aardewerk & Map information ($KL(A \parallel M)$=2.05)

	$i_0 = 1$	$i_0 = 2$	$i_0 = 3$	H	KL
π_{i_0}	0.182	0.513	0.306	1.01	
$\pi_{i_0 \mid K_{S,M}}$	0.029	0.225	0.746	0.66	0.48
$\pi_{i_0 \mid K_{S,A}}$	0.728	0.144	0.129	0.77	0.66
$\pi_{i_0 \mid K_{S,AM}}$	0.303	0.120	0.578	0.93	0.46

(see e.g. Bogaert, 2002). The method also does not rely on any parametric hypothesis. Based on a maximum entropy algorithm, it provides also the most general estimates for the joint distribution, that respects constraints specified by the user. In this study, the constraints are a complete set of bivariate probabilities for the Aardewerk variable and a partial knowledge about the relation between this variable and a digital map, showing that partial knowledge about a variable can be easily processed too. Finally, BME is able to deal with possible conflicting sources of information, as it translates logical rules into mathematically sound results. All these advantages make BME a very promising method, opening e.g. new possibilities for updating old maps with recently collected samples.

Table 3. Kullback-Leibler distance as a measure of the gain.

	$i_0 = 1$	$i_0 = 2$	$i_0 = 3$	H	KL
π_{i_0}	0.182	0.513	0.306	1.01	
	0.190	0.560	0.250	0.99	0.01
$\pi_{i_0 \mid K_{S,AM}}$	0.190	0.250	0.560	0.99	0.17
	0.560	0.190	0.250	0.99	0.39

REFERENCES

1. Bierkens, M.F.P. and P.A. Burrough (1993a). The indicator approach to categorical soil data. 1. theory. *Journal of Soil Science*, **44**(2):361-368.
2. Bierkens, M.F.P. and P.A. Burrough (1993b). The indicator approach to categorical soil data. 2. application to mapping and land-use suitability analysis. *Journal of Soil Science*, **44**(2):369-381.
3. Bogaert, P. (2002). Spatial prediction of categorical variables: the Bayesian Maximum Entropy approach. To appear in *Stoch. Env. Res. Risk A.*
4. Bogaert, P. and D. D'Or (2002). Spatial prediction of categorical variables: the BME approach. GeoEnv IV.
5. Bregt, A.K., J.J. Stoorvogel, J. Bouma, and A. Stein (1992). Mapping ordinal data in soil survey - a costa-rican example. *Soil Sci. Soc. Am. J.*, **56**(2):525-531, 1992.
6. Christakos, G. (2000). *Modern Spatiotemporal Geostatistics*. Oxford University Press, New York.
7. Christakos, G., P. Bogaert, and M.L. Serre (2002). *Temporal GIS*. Springer-Verlag, New York.
8. D'Or D., Bogaert P. and G. Christakos (2001). Application of the BME Approach to Soil Texture Mapping. *Stoch. Env. Res. Risk A*. **15**(1):87-100.
9. Goovaerts, P. (1997). *Geostatistics for Natural Resources Evaluation*. Oxford University Press, New York.
10. Journel, A.G. (1983). Nonparametric estimation of spatial distributions. *Mathematical Geology*, **15:**445-468.
11. Journel, A.G. (1986). Constrained interpolation and qualitative information – the soft kriging approach. *Mathematical Geology*, **18**:269-86.
12. Kullback, S. and R.A. Leibler (1951). On information and sufficiency. *Annals of Mathematical Statistics*, **22**:76-86.
13. Van Orshoven, J., J. Maes, H. Vereecken, J. Feyen, and R. Dudal (1988). A structured database of Belgian soil profile data. *Pedologie*. **38**:191-206.

SEQUENTIAL UPDATING SIMULATION

R. Froidevaux

FSS Consultants SA, Geneva, Switzerland.

Abstract: This paper presents a new implementation of the sequential simulation principle, within a multi-Gaussian framework. In this approach, the local conditional distribution functions, from which simulated values are drawn by Monte-Carlo, are updated iteratively rather than re-estimated at each step. This new implementation offers several significant advantages: the local distribution functions, from which simulated values are drawn, are conditional to all hard and previously simulated data, rather than to data within a search neighbourhood only; there is no need to assign existing hard data to the nearest grid nodes; the local means and variances are estimated from the available data at their exact locations; and the updating process does not involve any longer the solving of a linear system of equations. This, in turns, relaxes the constrains on the spatial correlation models which can be used. This new approach is illustrated by a case study in soil contamination.

1. INTRODUCTION

Sequential simulation is a wide class of simulation algorithms, all based on a recursive implementation of the Bayes axiom whereby the modelling of the multivariate distribution function, which fully describes a random function Z at any location $\mathbf{u}$, is replaced by the product of a set of univariate conditional cdfs:

X. Sanchez-Vila et al. (eds.), geoENV IV – Geostatistics for Environmental Applications, 307-318.

$$F(\mathbf{u}_1,...,\mathbf{u}_N;z_1,...,z_N\,|(N_0)) = F\left(\mathbf{u}_N;z_N\,|(N_0+N-1)\right) \cdot F\left(\mathbf{u}_{N-1};z_{N-1}\,|(N_0+N-2)\right)\cdot ... \cdot F\left(\mathbf{u}_1;z1|(N_0)\right)$$

with

$$F(\mathbf{u}_1,...,\mathbf{u}_N;z_1,...,z_N\,|(N_0)) = \text{Prob}\left\{Z(\mathbf{u}_1)\le z_1,...,Z(\mathbf{u}_N)\le z_N\,|(N_0)\right\}$$

and N_0 denotes the number of original data values.

Discussions on various implementations of sequential simulation can be found in Verly, 1986, Journel and Alabert, 1988, Gómez-Hernandez and Journel, 1993, Xu and Journel, 1994, and Soares, 2001.

Sequential Gaussian simulation is an implementation of the sequential simulation paradigm under the multiGaussian random function model which is used to simulate continuous variables. In its traditional implementation, sequential Gaussian simulation proceeds as follows (see Gooverts, 1997, p.380).

1. The set of data values $\{z_1,\ldots,z_N\}$ is transformed into a corresponding set of normal scores $\{y_1,\ldots,y_N\}$ using an appropriate transform $Z = \mathbf{\Phi}^{-1}(Y)$, and a multiGaussian hypothesis is assumed
2. The normal scores values $\{y_1,\ldots,y_N\}$ are assigned to nearest node of the grid to be simulated
3. A random path, visiting each node of the grid, is defined
4. At each grid node the local mean and variance of the local Gaussian ccdf is estimated by simple kriging. A simulated value $y(\mathbf{u})$ is drawn from this local ccdf and added to the data set
5. Once all the nodes have been visited the simulated normal scores are back-transformed into simulated values of the original variable using the inverse of the Gaussian transform used to calculate the original normal scores.

The new approach proposed in this paper differs from this classical implementation in that the local ccdfs are not estimated at each grid node before drawing a simulated value. Rather, the local ccdfs are initialized before performing any conditioning or simulation and then are updated sequentially after each drawing of a simulated value.

2. THE SEQUENTIAL UPDATING APPROACH

Consider the N grid $\mathbf{u}_i$, i = 1,…,N nodes discretizing the domain D to be simulated and denote by *k* the iteration index for visiting all the nodes.

At the initial stage (*k* = 0), before any conditioning or simulation, the standard multivariate Gaussian random function $Y(\mathbf{u})$ is fully defined by the following stationary moments:

$$\text{expected value}: m_0(\mathbf{u}) = E\{Y(\mathbf{u})\} = m = 0; \forall \mathbf{u} \subset D$$

$$\begin{aligned}\text{covariance}: C_0(\mathbf{u},\mathbf{u}+\mathbf{h}) &= E\{Y(\mathbf{u})\cdot Y(\mathbf{u}+\mathbf{h})\} - E\{Y(\mathbf{u})\}\cdot E\{Y(\mathbf{u}+\mathbf{h})\} \\ &= C_0(\mathbf{h}); \forall \mathbf{u} \subset D\end{aligned}$$

$$\text{variance}: \sigma_0^2(\mathbf{u}) = C_0(0) = 1; \forall \mathbf{u} \subset D$$

$$\text{correlogram}: \rho_0(\mathbf{u},\mathbf{u}+\mathbf{h}) = C_0(\mathbf{h}) / \sigma_0^2; \forall \mathbf{u} \subset D$$

Once a value $y(\mathbf{u}_\alpha)$ is drawn by Monte-Carlo, the posterior local ccdfs $Y(\mathbf{u}|(1))$, conditional to this value, become non-stationary. Hence the multi-Gaussian model becomes location dependent and it can be shown (Anderson, 1984, p 41) that, at iteration *k*, its parameters are given by:

$$\rho_k(\mathbf{u},\mathbf{v}) = \frac{\rho_{k-1}(\mathbf{u},\mathbf{v}) - \rho_{k-1}(\mathbf{u},\mathbf{u}_\alpha)\cdot\rho_{k-1}(\mathbf{v},\mathbf{u}_\alpha)}{\sqrt{1-\rho_{k-1}^2(\mathbf{u},\mathbf{u}_\alpha)}\cdot\sqrt{1-\rho_{k-1}^2(\mathbf{v},\mathbf{u}_\alpha)}} \tag{1}$$

$$\sigma_k^2(\mathbf{u}) = \sigma_{k-1}^2(\mathbf{u})\cdot\left(1-\rho_{k-1}^2(\mathbf{u}-\mathbf{u}_\alpha)\right) \tag{2}$$

and:

$$m_k(\mathbf{u}) = m_{k-1}(\mathbf{u}) + \rho_{k-1}(\mathbf{u}-\mathbf{u}_\alpha)\cdot\frac{\sigma_{k-1}(\mathbf{u})}{\sigma_{k-1}(\mathbf{u}_\alpha)}\cdot\left(y(\mathbf{u}_\alpha) - m_{k-1}(\mathbf{u}_\alpha)\right) \tag{3}$$

Thus, the key idea of sequential updating simulation is to visit randomly all grid nodes, to draw by Monte-Carlo a simulated value at each location, and to condition the moments of the local gaussian ccdfs to this newly simulated value before moving to the next location.

Because of the iterative way in which the local ccdfs are updated, the equations (1), (2) and (3) allow to generate a correlated gaussian field only if the correlogram is defined by a single structure with no nugget effect.

Indeed equation (2) results in a gradual reduction of variance. In the case of a multi-structure variogram (for instance a 50% nugget effect and a 50% large range variogram model), this reduction will affect the overall variance and lead, eventually, to a complete obliteration of the short scale variability.

Hence, in order to reproduce a multi-structure correlogram model, the local cdf $Y(\mathbf{u})$ needs to be interpreted as a linear combination of $Ns+1$ (structure 0 is the nugget effect) $Y^l(\mathbf{u})$ independent random functions with parameters:

$m_o^l(\mathbf{u}) = 0; \quad \forall l; \forall \mathbf{u}$

$\sigma_0^{2l}(\mathbf{u}) = C_o^l(0) =$ increment of the *lth* structure of the variogram.

Thus, at each location $\mathbf{u}_\alpha$, a set of simulated values $\{y^l(\mathbf{u}), l=0,..., Ns\}$ is drawn from the corresponding set of ccdfs and recombined into a single simulated value:

$$y(\mathbf{u}_\alpha) = \sum_{l=0}^{N_S} y^l(\mathbf{u}_\alpha)$$

and the updating of local parameters is performed independently for each structure using equations (1) to (3).

Remarks:

1. In the Sequential Updating approach, the locals ccdfs, from which simulated values are drawn, are conditioned to all hard data and previously simulated values: there is no neighbourhood search nor a maximum number of data to be considered. In practice, however, the updating of local ccdfs is performed only within correlation distance of the location $\mathbf{u}_\alpha$.
2. In order to ensure that the conditional correlogram values remains between -1 and +1, the numerator of equation (1) must verify the inequality:

$$\rho_k^2(\mathbf{u},\mathbf{v}) - \rho_k(\mathbf{u},\mathbf{u}_\alpha)\cdot\rho_k(\mathbf{v},\mathbf{u}_\alpha) \geq 0$$

3. CONDITIONING TO DATA

In the Sequential Updating approach, conditioning to existing hard data is achieved as an initial updating of the local cdfs $Y(\mathbf{u})$:

- The data set $\{z(\mathbf{u}_j), j=1,...,n\}$ is first transformed into a corresponding set of normal scores $\{y(\mathbf{u}_j), j=1,...,n\}$
- Then, each normal score $y(\mathbf{u}_j)$ value is split into its structural components: $y^l(\mathbf{u}_j) = y(\mathbf{u}_j) \cdot C^l(0)$
- Finally, each set of values $\{y^l(\mathbf{u}_j), l=0,...,Ns\}$ is used, successively, to update the mean, variance and correlogram of the local gaussian ccdfs using equations (1) to (3).

Remarks

1. Unlike the traditional implementation of Sequential Gaussian simulation, the conditioning data are not re-assigned to nearest grid node: the exact location of each datum is used for performing the updating.
2. The order in which the n conditioning data are used for updating varies randomly from one realization to the next.

4. SIMPLE NON-CONDITIONAL EXAMPLE

Let's consider first a non conditional simulation of an attribute over a 4000 metres by 4000 metres grid. The a priori distribution function is Gaussian with a mean equal to 0 and a variance equal to 1. A spherical, anisotropic variogram model is assumed with a short range of 250 metres a large range of 750 metres and a direction of maximum continuity of 135°.

Three realizations are generated using the proposed Sequential Updating approach and three other realizations are generated using the traditional Sequential Gaussian Simulation algorithm.

Figure 1 presents the two sets of realizations and Figure 2 shows the average histograms and average variograms, calculated over the three realizations, for the Sequential Updating results and for the traditional Sequential Gaussian Simulation results.

As can be seen the results are very close and confirm that the two approaches are equivalent.

5. SOIL CONTAMINATION EXAMPLE

In this example the objective is to simulate the polycyclic aromatic hydrocarbon (PAH) concentrations in view of delineating potentially hazardous zones requiring clean-up (Colin *et al.*, 1996). The available data consisted of chemical measurements of PAH concentrations from 51 boreholes located on the disaffected industrial site. In addition, an electrical resistivity survey was available, which led to the definition of two types of soil:

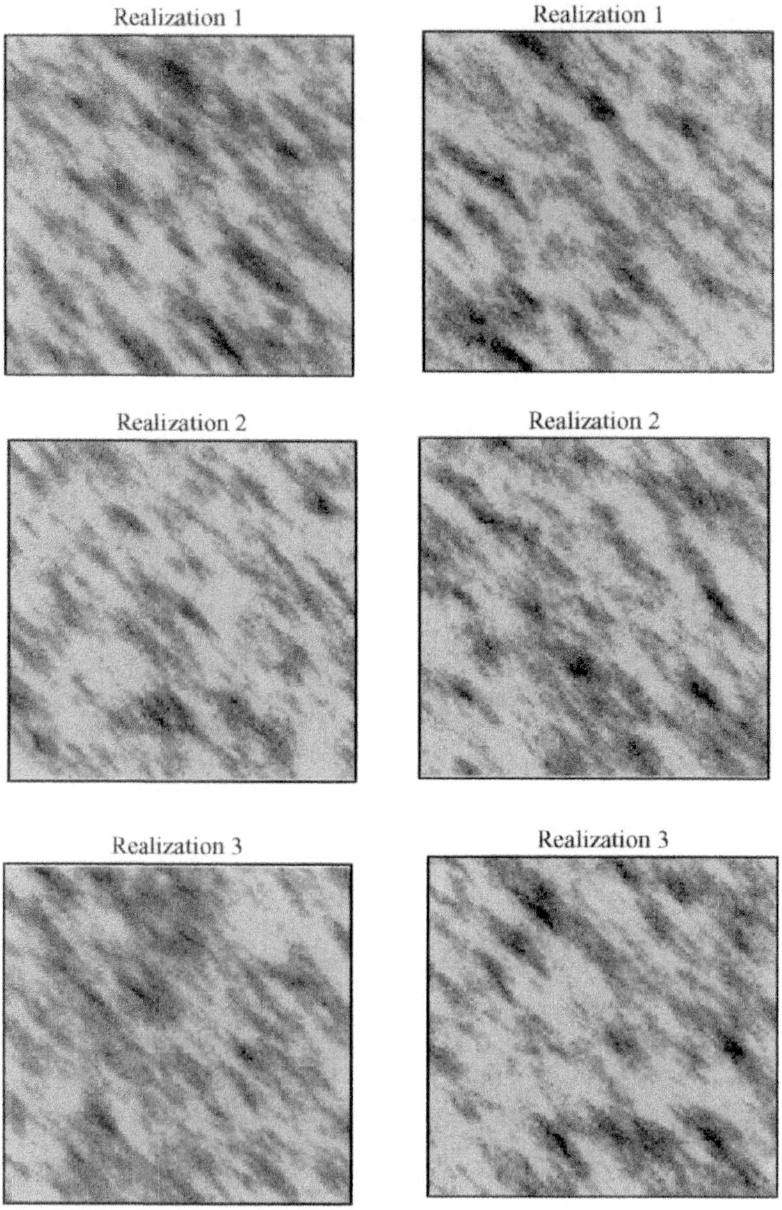

Figure 1. Non conditional simulation. Comparison of Sequential Updating simulation (left column) and Sequential Gaussian Simulations (right column).

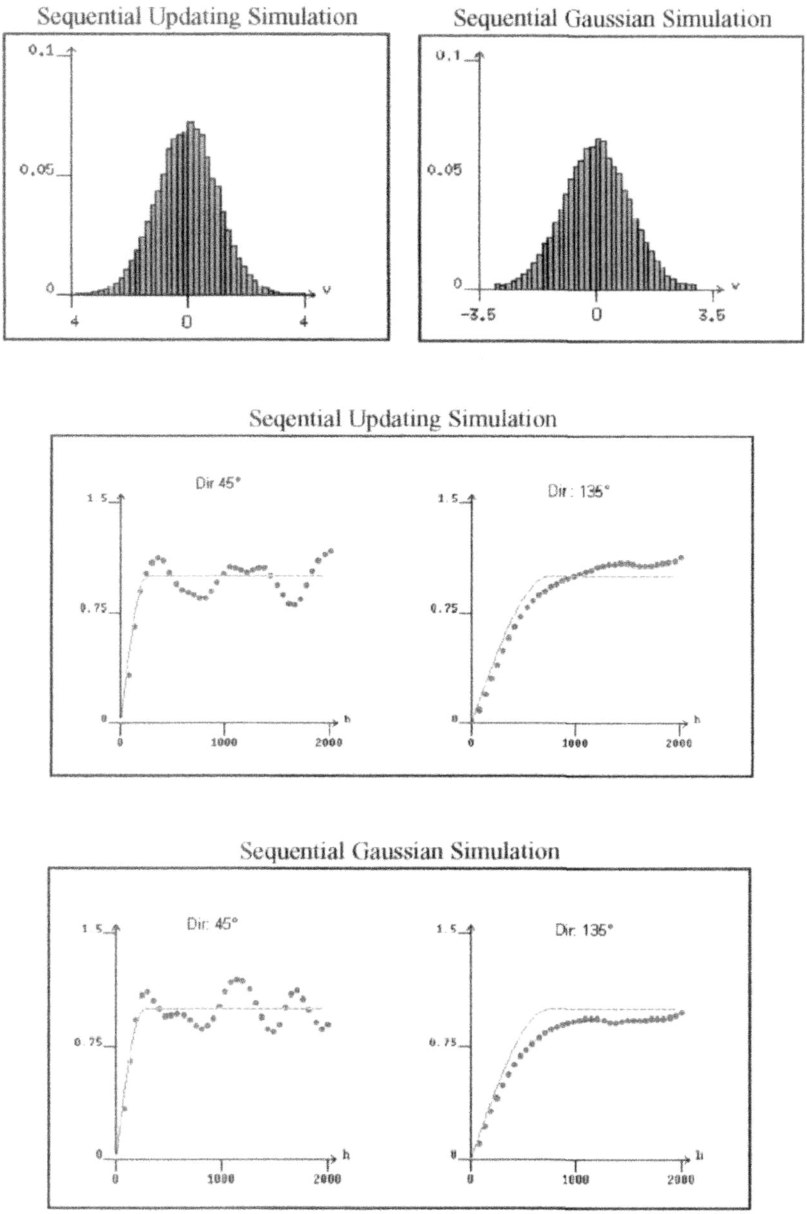

Figure 2. Average histograms and average variograms. Comparison between Sequential Updating Simulation and traditional Sequential Gaussian Simulation results.

- A first zone with rather low electrical resistivity values and where the PAH concentrations are above 35ppm.
- A second zone with higher resistivity values and lower PAH concentrations.

Figure 3 shows the location map of the boreholes, the electrical resistivity map and the zone map. Base on a preliminary exploratory data analysis of the available data, the following distribution and variogram models were selected:

	Distribution model	Variogram model
Zone 1	Lognormal	Spherical
	mean . 12 ppm standard deviation: 10 ppm	Isotropic, range 50 metres
Zone 2	Non parametric	Exponential
	Range: 0 to 500 ppm	Isotropic, range 75 metres

Equi-probable images of PAH concentration were generated using the Sequential Updating approach. Figure 4 shows three realizations and Figure 5 the comparison between the prior models and the posterior statistics. As can be seen, the simulation results are consistent with the specified prior models.

6. DISCUSSION

The Sequential Updating approach, proposed in this paper, offers an alternative to the classical implementation of the sequential gaussian simulation with the following attractive features:

1. By construction, all local ccdfs are fully conditioned to all simulated values. This is not the case in the classical implementation since the local ccdfs are estimated on the basis of a limited number of data within a search ellipse.
2. It does not require the solving of any system of linear equations. As a result Sequential Updating is generally faster than the classical implementation and is not prone to the sometimes annoying numerical problems found in kriging.

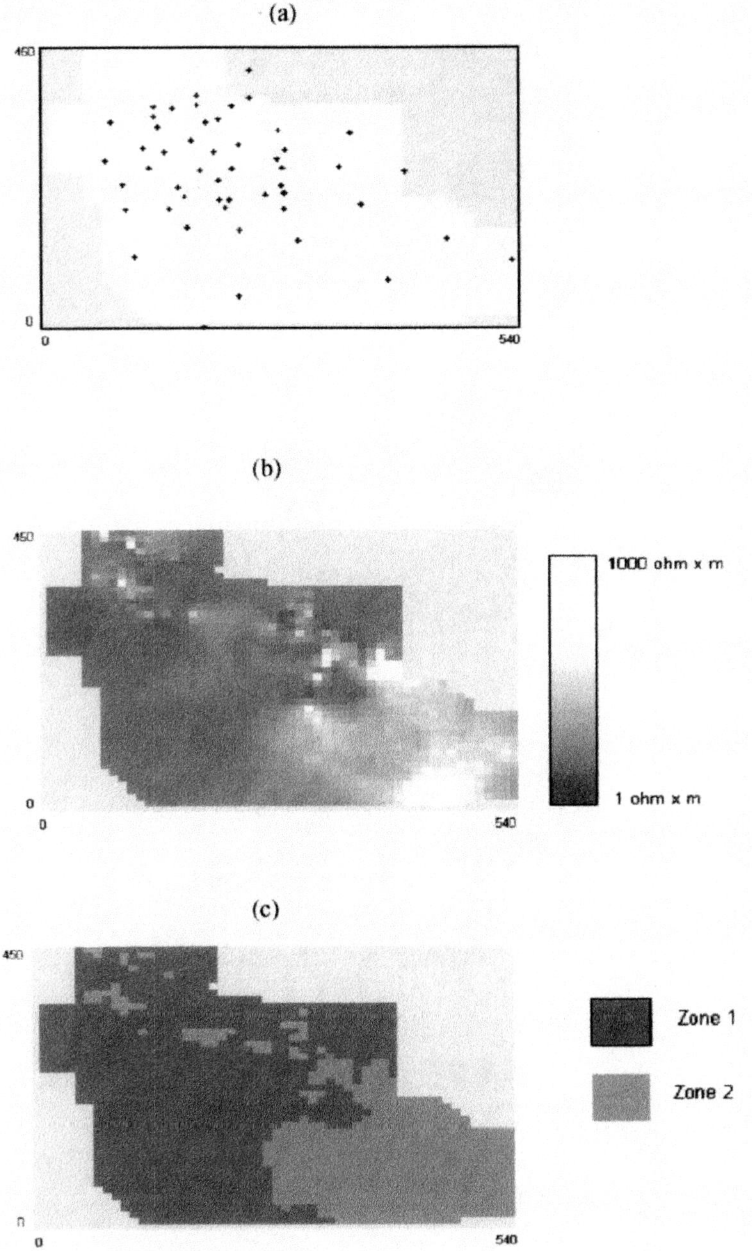

Figure 3. Boreholes location map (a), electrical resistivity map (b) and soil type map (c).

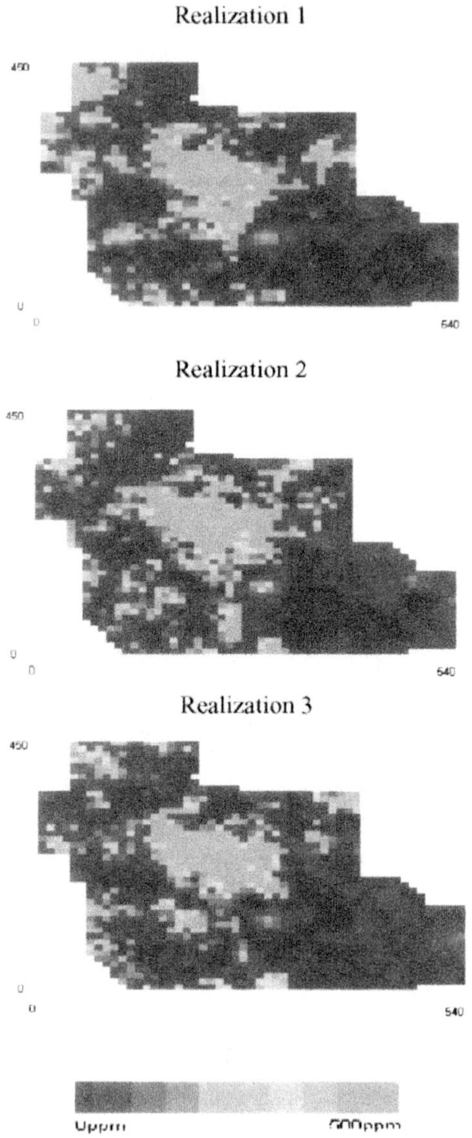

Figure 4. Three realizations of PAH concentration.

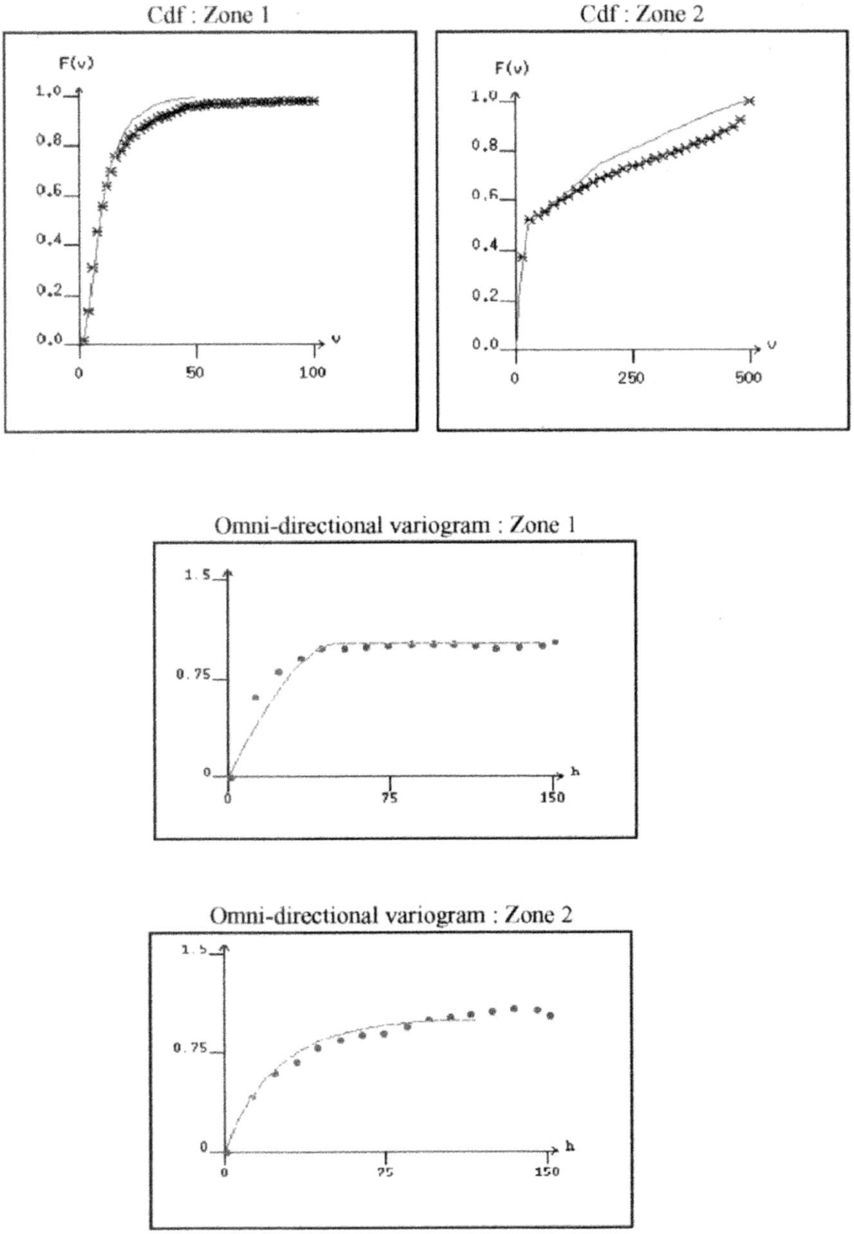

Figure 5. Prior models and posterior statistics.

3. Since no kriging system needs to be solved, spatial continuity for the attribute to be simulated can be specified with less restrictions than before.
4. The conditional hard data are not assigned to the nearest grid node, but are used at their exact locations. Although this data re-allocation has never been a serious concern in 2D, it has represented a problem in 3D if the vertical grid node spacing is larger than the vertical data spacing. In this situation several data may share the same grid node and a decision must be taken on which one takes precedence.

7. REFERENCES

1. Anderson, T.W (1984). An introduction to multivariate statistical analysis. John Wiley & Sons, 675p.
2. Colin, P., Froidevaux, R., Garcia, M. and Nicoletis, S. (1996). Integrating Geophysical Data for Mapping the Contamination of Industrial Sites by Polycyclic Aromatic Hydrocarbons: a Geostatistical Approach. In R.M. Srivastava, S. Rouhani, M.V. Cromer, A.I. Johnson, editors. Geostatistics for Environmental and Geotechnical Applications, ASTM STP 1238.
3. Gómez-Hernández, J. and Journel, A. (1993). Joint sequential simulation of multiGaussian fields. In A. Soares, editor, Geostatistics Troia'92, **1**:85-94.
4. Gooverts, P. (1997). Geostatistics for Natural Resources Evaluation. Oxford University Press, 483 p.
5. Journel, A. and Alabert, F. (1988). Focusing on spatial connectivity of extreme valued attributes: stochastic indicator models of reservoir heterogeneities. SPE paper # 18324.
6. Soares, A. (2001). Direct sequential co-simulation. In Monestiez, P., Allard. D. and Froidevaux, R., editors, geoENV III - Geostatistics for Environmental Applications. Kluwer Academic Publishers.
7. Verly, G. (1986). MultiGaussian kriging - A complete case study. In Ramani R., editor, Proceedings of the 19th International APCOM Symposium, pp. 283-298. Society of Mining Engineers.
8. Xu, W., Tran, T., Srivastava, R.M. and Journel, A. (1992). Integrating seismic data in reservoir modeling: The collocated cokriging alternative. SPE paper # 24742.

VARIANCE-COVARIANCE MODELING AND ESTIMATION FOR MULTI-RESOLUTION SPATIAL MODELS

G. Johannesson and N. Cressie

Department of Statistics. The Ohio State University. Columbus, OH 43210. USA. gardar@stat.ohio-state.edu, ncressie@stat.ohio-state.edu

Abstract: The tree-structured multi-resolution spatial models (MRSMs) yield optimal and computationally feasible spatial smoothers of massive spatial data with nonstationary behavior. The nonstationary spatial correlation structure of MRSMs is the result of inhomogeneous stochastic parent-child relationships at adjacent resolutions. Likelihood-based methods are presented for the estimation and modeling of variance-covariance parameters associated with the parent-child relationships, resulting in data-adaptive, nonstationary covariance structure. An application of the MRSMs is given to total column ozone (TCO) data obtained from a polar-orbiting satellite.

Key words: Nonstationarity, RESL estimation, RESREL estimation, total column ozone, tree-structured models, covariance-parameter estimation

1. INTRODUCTION

As a consequence of new remote-sensing technology, spatio-temporal environmental data have become more massive in their raw form. Provided with such rich datasets, scientists eye new opportunities, but at the same time they are faced with new challenges. The massiveness of the data is in most cases due to both *fine-resolution* sampling and a *large* spatial domain. An example is Total Column Ozone (TCO), sampled remotely by satellites over the entire globe on a daily basis. Due to the large size of the spatial domain,

X. Sanchez-Vila et al. (eds.), geoENV IV – Geostatistics for Environmental Applications, 319-330.

stationarity assumptions about the process of interest do not typically hold. Hence, computationally tractable spatial models for massive data, with nonstationary spatial dependence, are in great demand.

Tree-structured multi-resolution spatial models (MRSMs) (see e.g., Huang et al., 2002) are able to handle massive spatial data with nonstationary spatial correlation structure. In Section 2, we shall review the MRSM and the associated fast, change-of-resolution Kalman-filter algorithm for optimal spatial prediction. At the core of the MRSM is the specification of the spatial covariance structure through a coarse-to-fine-resolution process model. Section 3 considers such models and proposes a parameterization that allows one to capture smooth changes in (nonstationary) spatial covariance structure. We also show in Section 3 how to estimate the model parameters using resolution-specific likelihood-based methods. An application to a day's worth of TCO satellite data is presented in Section 4.

2. MULTI-RESOLUTION SPATIAL MODELS

In this section, we review briefly the multi-resolution spatial model (MRSM) as given in Huang et al. (2002). Let D be the spatial domain of interest. The domain D is partitioned into n_0 grid cells, which make up the coarsest resolution (resolution-0). Each grid cell at resolution $r = 0,\ldots,R-1$, is then successively partitioned into m_r smaller grid cells. Thus, we obtain a nested partition of D at $(R+1)$ resolutions. At the r-th resolution, there are $n_r = n_0\, m_0 \ldots m_{r-1}$ grid cells given by $\{D(i,r)\}_{i=1}^{nr}$. We call $(i^*,r+1)$ a child of (i,r) if $D(i^*,r+1) \subset D(i,r)$, and we denote the set of the children of (i,r) by $ch(i,r) \equiv \{ch(i,r)_1,\ldots, ch(i,r)_{mr}\}$. Then

$$D(i,r) = \cup_{j=1}^{m_r} \mathbf{D}(ch(i,r)_j); \quad (i,r) \in N_{\mathbf{R}-1,}$$

where $N_u \equiv \{(i,r)\colon i = 1,\ldots, n_r,\ r = 0,\ldots, u\}$. Figure 1 shows an example of a multi-resolution partition at resolutions r=0,1,2.

Let $\{Y(\mathbf{s})\colon \mathbf{s} \in D\}$ be a Gaussian spatial process ofinterest defined on D, and define the multi-resolution aggregated Y-process as

$$Y(i,r) \equiv \frac{1}{v(i,r)} \int_{D(i,r)} Y(\mathbf{s})\mathbf{ds}; \quad (i,r) \in N_{\mathbf{R}},$$

where $v(I,r) \equiv |D(i,r)|$ denotes the area (volume) of $D(i,r)$. The aggregated Y-process is not observed directly, but indirectly through the additive-measurement-error model,

$$Z(i,r) = Y(i,r) + \nu(i,r); \quad (i,r) \in N_R, \tag{1}$$

where $\{Z(i,r)\}$ are (potentially) observed data, and the measurement errors $\nu(i,r) \sim \text{Gau}(0, \sigma^2 V(i,r))$ are independent with $\{V(i,r)\}$ known. Henceforth, we refer to (1) as the data model. It should be noted that observations are not needed at all resolutions and can be missing for some cells within a resolution. For example, in the ozone example considered in Section 4, the Y-process is taken to be the underlying TCO process at different resolutions and the data are noisy satellite observations of TCO, reported (incompletely) at the finest resolution, resolution-R.

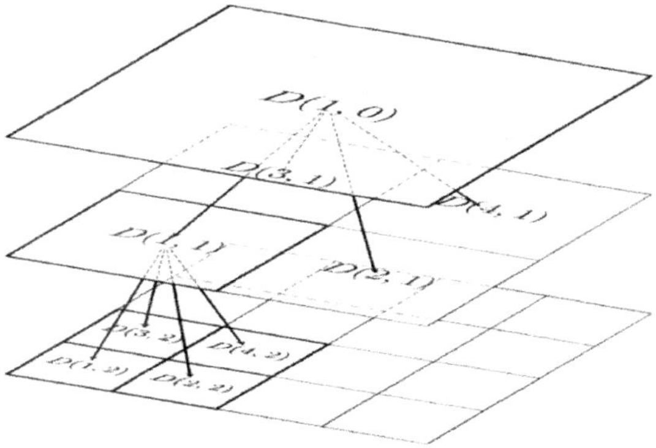

Figure 1. An example of a spatial multi-resolution tree-structure partition.

The spatial variance-covariance structure associated with the Y-process is specified indirectly through the following coarse-to-fine-resolution model:

$$\mathbf{Y}(i,r) = \mathbf{1}Y(i,r) + \boldsymbol{\omega}(i,r); \quad (i,r) \in N_{\mathbf{R}-1}, \tag{2}$$

where $\mathbf{Y}(i,r) \equiv (Y(ch(i,r)_1),\ldots, Y(ch(i,r)_{mr}))'$ and $\omega(i,r) \sim \text{Gau}(\mathbf{0}, \sigma^2\mathbf{W}(i,r))$, independently. Henceforth, we refer to (2) as the process model. Hence, the Y-process at the children cells is just taken to be equal the the Y-process at the parent cell plus an error term. The process model is completed by specifying the distribution of the Y-process at the coarsest resolution; here we simply assume that $(Y(1,0),\ldots, Y(n_r,0))' \sim \text{Gau}(\mathbf{a}(0), \sigma^2\mathbf{R}(0))$.

To match the notation style used for the process model in (2), it will be more convenient to write the data model in (1) as

$$\mathbf{Z}(i,r) = \mathbf{Y}(i,r) + \mathbf{v}(i,r); \quad (i,r) \in N_{\mathbf{R}-1}, \tag{3}$$

where $\mathbf{Z}(i,r) \equiv (Z(ch(i,r)_1),\ldots, Z(ch(i,r)_{mr}))'$ and $\nu(i,r) \sim \text{Gau}(\mathbf{0}, \sigma^2\mathbf{V}(i,r))$, independently, with $\mathbf{V}(i,r) \equiv \text{diag}(V(ch(i,r)_1),\ldots, V(ch(i,r)_{mr}))$.

2.1 Constrained *Y*-Process

Note that the process model in (2) does not have a one-to-one mapping between $\mathbf{Y}(i,r)$ and $\{Y(i,r), \omega(i,r)\}$; $\mathbf{Y}(i,r)$ is a vector of length m_r, but $\{Y(i,r), \omega(i,r)\}$ has a total of $(m_r + 1)$ elements. Consequently, different configurations of $\{Y(i,r), \omega(i,r)\}$ can yield the same $\mathbf{Y}(i,r)$. However, by placing a single linear constraint on the error term $\omega(i,r)$, a one-to-one mapping is achieved. That is, we constrain

$$\mathbf{q}(i,r)'\boldsymbol{\omega}(i,r) = 0; \quad (i,r) \in N_{\mathbf{R}-1}, \tag{4}$$

for some chosen constraining vectors $\{\mathbf{q}(i,r)\}$. To satisfy (4), let $\mathbf{Q}(i,r)$ be any m_r x (m_{r-1}) orthonormal matrix with columns that span the space orthogonal to $\mathbf{q}(i,r)$ (i.e., $\mathbf{q}(i,r)'\ \mathbf{Q}(i,r) = \mathbf{0}$ and $\mathbf{Q}(i,r)'\mathbf{Q}(i,r) = \mathbf{I}$). Then any $\omega(i,r)$ satisfying (4) can be written as

$$\boldsymbol{\omega}(i,r) = \mathbf{Q}(i,r)\boldsymbol{\omega}^*(i,r); \quad (i,r) \in N_{\mathbf{R}-1},$$

for some unconstrained $\omega(i,r) \in \mathrm{R}^{mr-1}$. The constrained *Y*-process can therefore be written as:

$$\mathbf{Y}(i,r) = \mathbf{1}Y(i,r) + \mathbf{Q}(i,r)\boldsymbol{\omega}^*(i,r); \quad (i,r) \in N_{\mathbf{R}-1}, \tag{5}$$

where $\omega^*(i,r) \sim \mathrm{Gau}(0, \sigma^2\mathbf{W}^*(i,r))$, independently. In terms of the process model in (2), we have constrained $\mathbf{W}(i,r)$ to be of the form:

$$\mathbf{W}(i,r) = \mathbf{Q}(i,r)\mathbf{W}^*(i,r)\mathbf{Q}(i,r)'; \quad (i,r) \in N_{\mathbf{R}-1}. \tag{6}$$

Huang et al. (2002) proposed choosing $\mathbf{q}(i,r) = \mathbf{v}(i,r)$, where $\mathbf{v}(i,r) \equiv (v(ch(i,r)_1),\ldots,v(ch(i,r)_{mr}))'$. This choice results in a physically *mass-balanced* process model, since it follows that

$$\mathbf{v}(i,r)\mathbf{Y}(i,r) = \sum_{j=1}^{m_r} v\left(ch(i,r)_j\right)Y\left(ch(i,r)_j\right); \quad (i,r) \in N_{\mathbf{R}-1}. \tag{7}$$

2.2 Posterior Inference

Given all the variance-covariance parameters associated with the data model in (1) and the process model in (2), our goal is to predict the hidden process $\{Y(i,r)\}$ from noisy and incomplete data $\{Z(i,r)\}$. Optimal prediction is obtained from the *posterior distribution* of $\{Y(i,r)\}$, which can be calculated rapidly using the change-of-resolution Kalman-filter algorithm (Chou et al., 1994; Huang and Cressie, 2001). The algorithm consists of two major steps, namely the leaves-to-root step and the root-to-leaves step. The leaves-to-root step consists of recursively deriving the distribution of $Y(i,r)$ conditional on all data observed at all descendents of (i,r) and at (i,r) itself. At the end of the leaves-to-root recursion, we obtain the distribution of $\{Y(\mathrm{i},0)\}$ conditional on all the data (i.e., the posterior distribution of $\{Y(\mathrm{i},0)\}$). The root-to-leaves step starts at the root node, and then traces down the tree, recursively computing the posterior distribution of $Y(\mathrm{i},0)$ at every node in the tree. The algorithm is fast; it requires computations only proportional to the number of nodes in the tree, with a small computational overhead at each node. Computation times are discussed in Section 4.

3. VARIANCE-COVARIANCE MODELING AND ESTIMATION

In Section 2, the scalars $\{V(i,\mathrm{r})\}$ associated with the measurement errors in (1), and the parameters $\{\mathbf{W}^*(i,r)\}$, $\mathbf{a}(0)$, and $\mathbf{R}(0)$ associated with the process model in (5), were assumed known. This assumption is realistic for the $\{V(i,\mathrm{r})\}$, since they reflect the relative accuracy (weight) of each observation. On the other hand, the matrices $\{\mathbf{W}^*(i,r)\}$ and $\mathbf{R}(0)$ determine the variance-covariance structure of the hidden process $\{Y(i,r)\}$, *a priori*. A common approach in spatial statistics is to use the data to assist in specifying the variance-covariance structure of the Y-process, which can be thought of as an empirical Bayes approach. For example, when doing kriging (e.g., Cressie, 1993, Chapter 3), the data are typically used to estimate variance-covariance parameters using, for example maximum likelihood (ML) or restricted maximum likelihood (REML) estimation (e.g., Cressie, 2002). We follow a similar approach here by parameterizing the $\{\mathbf{W}^*(i,r)\}$ matrices and then estimating any unknown parameters using ML- and REML-based methods. Estimation of $\mathbf{a}(0)$ and $\mathbf{R}(0)$ is discussed in Section 4.

3.1 Variance-Covariance Modeling

The (m_{r-1}) x (m_{r-1}) matrix $\mathbf{W}^*(i,r)$ has at most $(m_{r-1})m_r/2$ unknown parameters associated with it that need to be estimated. Denote by $\theta(i,r)$ the unknown parameter vector associated with $\mathbf{W}^*(i,r)$, and write

$$\mathbf{W}^*(i,r) = \mathbf{W}_r^*\left(\boldsymbol{\theta}(i,r)\right); \quad (i,r) \in N_{\mathbf{R}-1}. \tag{8}$$

An example of a $\mathbf{W}^*$-model is the single-parameter-per-scale (SPPS) model:

$$\mathbf{W}^*(i,r) = \boldsymbol{\theta}(i,r)\mathbf{C}_0(i,r); \quad (i,r) \in N_{\mathbf{R}-1}, \tag{9}$$

where $\{\mathbf{C}_0(i,r)\}$ are known positive-definite matrices and $\{\theta(i,r)\}$ are unknown, positive, scaling parameters.

As presented above, the different $\{\boldsymbol{\theta}(i,r)\}$ in (8) are not related in any way. However, one could expect that cells within the same resolution that are nearby (in space) will have similar $\boldsymbol{\theta}$-parameters. Let $\{\mathbf{s}(i,r)\}$ be a set of representative point locations for $\{D(i,r)\}$ (e.g., using the centroids of each cell). Then, in the case of the SPPS model (9), for example, one could assume

$$\log \boldsymbol{\theta}(i,r) = \sum_{j=1}^{p_r} \psi_j\left(\mathbf{s}(i,r)\right)\beta(r)_j, \tag{10}$$

within each resolution r, where $\psi_1(\cdot),\ldots,\psi_{pr}(\cdot)$ are known, smooth basis-functions of spatial locations, $\boldsymbol{\beta}(r) \equiv (\beta(r)_1,\ldots,\beta(r)_{pr})'$ are unknown parameters to be estimated, and $p_r \in \{1,2,\ldots\}$. We now present likelihood-based methods for estimating $\{\beta(r)\}$.

3.2 Likelihood-based Parameter Estimation

Denote by

$$p\left(\mathbf{Z}(i,r)\middle|\mathbf{Y}(i,r)\right) \quad \text{and} \quad p\left(\mathbf{Y}(i,r)\middle|Y(i,r);\boldsymbol{\theta}(i,r)\right), \tag{11}$$

the conditional Gaussian probability densities associated with the data model (3) and the process model (5), respectively, and assume for the moment that σ^2, the variance-scaling parameter in (3) and (5) is known. With very little loss of generality, assume further that the data are only observed at the finest resolution, resolution-R. Due to the conditional structure of the MRSM, the joint density of $\{Z(i,R)\}$ and $\{Y(i,r)\}$ is given simply by a product of conditional densities. That is,

$$p\left(\{Z(i,R)\},\{Y(i,r)\};\{\boldsymbol{\theta}(i,r)\}\right) = \left(\prod_{i=1}^{n_{R-1}} p\left(\mathbf{Z}(i,R-1)\middle|\mathbf{Y}(i,R-1)\right)\right)$$
$$\times\left(\prod_{r=1}^{R-1}\prod_{i=1}^{n_r} p\left(\mathbf{Y}(i,r)\middle|Y(i,r);\boldsymbol{\theta}(i,r)\right)\right)p\left(\{Y(i,0)\}\right), \tag{12}$$

where $\{\mathbf{Z}(i,R\text{-}1)\}$ is equivalent to $\{Z(i,R)\}$ and recall that the last factor is the density of the multivariate Gau($\mathbf{a}(0),\sigma^2\mathbf{R}(0)$). However, for maximum-likelihood inference, the marginal distribution of the data $\{Z(i,R)\}$ is needed, which is the integral the joint distribution above with respect to $\{Y(i,r)\}$. This integration is not at all straightforward, and it leaves us with a likelihood that has to be simultaneously maximized with respect to all variance-covariance parameters. However, as we shall see, it is possible to extract information from the data that is relevant to each resolution separately, leading to fast, resolution-specific likelihood inference. One such approach, given by Kolaczyk and Huang (2001), is to combine a recursive integration of (12) with recursive aggregation and transformation of the data. The resulting marginal distribution of the transformed data factors into resolution-specific likelihood (RESL) components, with each component being only informative for the variance-covariance parameters associated with that particular resolution. Another such approach, which mirrors REML estimation in mixed-effects models (e.g., McCulloch and Searle, 2001), is to form contrasts among the data such that the distribution of the contrasted data only depends on the variance-covariance parameters associated with a single resolution. The resolution-restricted likelihood-based (RESREL-based) estimates derived using this latter approach are not in general the same as the RESL-based estimates obtained from the first approach. However, when estimating variance-covariance parameters in Gaussian mixed-effect models and in Gaussian spatial models, REML estimators are in many cases preferred (see, e.g., McCulloch and Searle, 2001, Section 6.10; Cressie, 2002). We now present briefly both estimation approaches; see Johannesson (2003) for full details.

The RESL is derived by effectively integrating (12), resolution-by-resolution, with the help of a recursive decomposition of the data. Let $r = R$-1. Integrating (12) with respect to $\{\mathbf{Y}(i,r)\}_{i=1}^{nr}$, results in most terms coming outside the integral, leaving behind

$$\prod_{i=1}^{n_r} \int_{\mathbf{Y}(i,r)} p\big(\mathbf{Z}(i,r)\big|\mathbf{Y}(i,r)\big)\, p\big(\mathbf{Y}(i,r)\big|Y(i,r);\boldsymbol{\theta}(i,r)\big)\, d\mathbf{Y}(i,r), \tag{13}$$

for $r = R$-1. The i-th integral in (13) is easily seen to be $p(\mathbf{Z}(i,r) \mid Y(i,r); \boldsymbol{\theta}(i,r))$, which can be obtained from the additive model,

$$\mathbf{Z}(i,r) = \mathbf{1}Y(i,r) + \mathbf{Q}(i,r)\boldsymbol{\omega}^*(i,r) + \boldsymbol{\upsilon}(i,r); \quad i = 1,\ldots,n_r. \tag{14}$$

Instead of proceeding to next resolution and taking a second integral of (12), now with respect to $\{\mathbf{Y}(i,r\text{-}1)\}_{i=1}^{nr-1}$, we decompose the $\{\mathbf{Z}(i,r)\}$ into aggregated global components $\{\mathbf{Z}(i,r)\}$ and detail local components $\{\mathbf{d}(i,r)\}$. Define

$$\begin{bmatrix} Z(i,r) \\ \mathbf{d}(i,r) \end{bmatrix} \equiv \begin{bmatrix} \tilde{\mathbf{q}}(i,r)' \\ \mathbf{P}(i,r)' \end{bmatrix} \mathbf{Z}(i,r); \quad i = 1,\ldots,n_r,\ r = R-1, \tag{15}$$

where $\mathbf{q}(i,r)$ is given in (4), $\tilde{\mathbf{q}}(i,r) \equiv \mathbf{q}(i,r)(\mathbf{1}'\mathbf{q}(i,r))^{-1}$, assuming that $\mathbf{1}'\mathbf{q}(i,r) \neq 0$; and $\mathbf{P}(i,r)$ is any $m_r \times (m_r\text{-}1)$ matrix satisfying $\mathbf{P}(i,r)'(\mathbf{1} - \mathbf{k}(i,r)) = \mathbf{0}$, $\mathbf{k}(i,r) \equiv \mathbf{V}(i,r)\mathbf{q}(i,r)V(i,r)^{-1}$, and $V(i,r) = \tilde{\mathbf{q}}(i,r)'\mathbf{V}(i,r)\mathbf{q}(i,r)$. Given that the transformation in (15) is one-to-one and does not depend on $\boldsymbol{\theta}(i,r)$, the joint density of $\{Z(i,r), \mathbf{d}(i,r)\}$ provides identical likelihood inference for $\boldsymbol{\theta}(i,r)$, conditional on $Y(i,r)$. Its advantage over using the conditional density of $\mathbf{Z}(i,r)$ given $Y(i,r)$, follows from the fact that

$$p\big(Z(i,r),\mathbf{d}(i,r)\big|Y(i,r);\boldsymbol{\theta}(i,r)\big) = p\big(\mathbf{d}(i,r)\big|Z(i,r);\boldsymbol{\theta}(i,r)\big)\, p\big(Z(i,r)\big|Y(i,r)\big),$$

where $p(Z(i,r) \mid Y(i,r))$ is a Gaussian density with mean $Y(i,r)$ and variance $\sigma^2 V(i,r)$, and $p(\mathbf{d}(i,r) \mid Z(i,r);\, \boldsymbol{\theta}(i,r))$ is a multivariate Gaussian density with mean $\mathbf{P}(i,r)'\mathbf{k}(i,r)Z(i,r)$ and variance-covariance matrix

$$\mathbf{P}(i,r)'\left(\mathbf{Q}(i,r)\mathbf{W}_r^*\left(\theta(i,r)\right)\mathbf{Q}(i,r)'+\mathbf{V}(i,r)-V(i,r)\mathbf{k}(i,r)\mathbf{k}(i,r)'\right)\mathbf{P}(i,r).$$

That is, (15) factorizes the information content of $\mathbf{Z}(i,r)$ into what is relevant to $\boldsymbol{\theta}(i,r)$, through $\mathbf{d}(i,r)$, and what is relevant to all coarser-resolution $\boldsymbol{\theta}$-parameters, through $Z(i,r)$; $r = R$-1. Note that $\{Y(i,R\text{-}1)\}$ is equivalent to $\{Y(i,R\text{-}2)\}$ and hence the second integration of (12) yields a term equivalent to (13) with $r = R$-2. By repeating the integration-factorization process outlined above, until the final integration with respect to $\{Y(i,0)\}_{i=1}{}^{n0}\}$, we obtain the likelihood of $\{\boldsymbol{\theta}(i,r)\}$ to be proportional to

$$\prod_{r=0}^{R-1}\prod_{i=1}^{n_r} p\left(\mathbf{d}(i,r)\middle|Z(i,r);\theta(i,r)\right), \tag{16}$$

where $\mathbf{d}(i,r)$ and $Z(i,r)$ are obtained from (15), generalized for all $r = R$-1,…,0. The estimation of $\boldsymbol{\theta}(r) \equiv \{\boldsymbol{\theta}(i,r){:}\ i = 1,\ldots,n_r\}$ (or equivalently $\beta(r)$) is then carried out using the resolution-specific likelihood (RESL),

$$L_r^{(d)}(\boldsymbol{\theta}(r)) \equiv \prod_{i=1}^{n_r} p\left(\mathbf{d}(i,r)\middle|Z(i,r);\theta(i,r)\right);\quad r = R-1,\ldots,0, \tag{17}$$

resulting in a fast, resolution-specific estimation procedure.

RESL-based estimates of $\{\boldsymbol{\theta}(i,r)\}$ are identical to maximum-likelihood estimates if the transformation in (15) is one-to-one. Kolaczyk and Huang (2001) point out that a necessary and sufficient condition for this is $\boldsymbol{q}(i,r) = (V(ch(i,r)_1)^{-1},\ldots, (V(ch(i,r)_{\mathrm{mr}})^{-1})'$. Generally, this is different from the mass-balance constraint $\boldsymbol{q}(i,r) = \mathbf{v}(i,r)$, but is the same when the measurement-error variance is inversely proportional to the area of the cell. However, if the transformation in (15) is not one-to-one, the likelihood decompositon in (16) is not exact, and hence the RESL estimates derived using (17) are only approximately ML estimates. We therefore consider an alternative likelihood-type quantity to maximize, namely the resolution-specific restricted likelihood (RESREL).

In place of maximum likelihood estimation of the $\{\boldsymbol{\theta}(i,r)\}$, the fine-resolution data $\{Z(i,\mathrm{R})\}$ and the aggregated data $\{Z(i,\mathrm{R})\}$; $r = R$-1,…,0, can be used to construct a sequence of resolution-specific restricted likelihoods (RESRELs), such that the r-th likelihood is used to estimate $\boldsymbol{\theta}(r)$; $r = R$-1,…,0. Just as for REML, let $\mathbf{E}(r)$ be any $m_r \times (m_r\text{-}1)$ matrix such that $\mathbf{E}(r)'\mathbf{1} = \mathbf{0}$, and define the contrasts,

$$\mathbf{e}(i,r) \equiv \mathbf{E}(r)'\mathbf{Z}(i,r);\quad (i,r) \in N_{R-1}. \tag{18}$$

Then, using (14),

$$\mathbf{e}(i,r) = \mathbf{Q}_e(i,r)\boldsymbol{\omega}^*(i,r) + \boldsymbol{\upsilon}_e(i,r); \quad (i,r) \in N_{R-1},$$

where $\mathbf{Q}e(i,r) \equiv \mathbf{E}(r)'\mathbf{Q}(i,r)$ and $\boldsymbol{\upsilon}_e(i,r) \equiv \mathbf{E}(r)'\boldsymbol{\upsilon}_e(i,r)$. That is,

$$\mathbf{e}(i,r) \sim \text{Gau}\left(\mathbf{0}, \sigma^2\left(\mathbf{Q}_e(i,r)\mathbf{W}_r^*(\boldsymbol{\theta}(i,r)), \mathbf{Q}_e(i,r)' + \mathbf{V}_e(i,r)\right)\right),$$

where $\mathbf{V}_e(i,r) \equiv \mathbf{Q}_e(i,r)\mathbf{V}(i,r)\mathbf{Q}(i,r)'$. Note that within each resolution r, the $\{\mathbf{e}(i,r)\}$ are independent. One can then use the resolution-specific restricted likelihood (RESREL),

$$L_r^{(e)}(\boldsymbol{\theta}(r)) \equiv \prod_{i=1}^{n_r} p\left(\mathbf{e}(i,r); \boldsymbol{\theta}(i,r)\right), \tag{19}$$

for inference on $\boldsymbol{\theta}(r)$, where $p(\mathbf{e}(i,r); \boldsymbol{\theta}(i,r))$ is the Gaussian density associated with $\mathbf{e}(i,r)$; $i = 1,\ldots, n_r$, $r = R\text{-}1,\ldots,0$.

Note that the RESREL is not tied to any particular set of constraining vectors $\{\mathbf{q}(i,r)\}$, as is the case for RESL. However, the choice of $\{\mathbf{q}(i,r)\}$ does determine how the fine-resolution data $\{Z(i,R)\}$ will be aggregated.

Hitherto, we have assumed that σ^2 is known and there is no missing data in $\{Z(i,R)\}$. In the more realistic situation where σ^2 is unknown, one can estimate σ^2 at a fixed resolution, say the finest-resolution (using either RESL or RESREL), and use the resulting σ^2 estimate when estimating $\{\boldsymbol{\theta}(i,r)\}$.

If some of the elements of $\mathbf{Z}(i,R\text{-}1)$ are missing (unobserved), it is not possible to decompose $\mathbf{Z}(i,R\text{-}1)$ into the two components, $Z(i,R\text{-}1)$ and $\mathbf{d}(i,R\text{-}1)$ with the right factorization properties needed for RESL. One solution is to ignore those i for which $\mathbf{Z}(i,R\text{-}1)$ has any missing elements. A similar strategy can be taken for the RESREL approach.

4. APPLICATION: TOTAL COLUMN OZONE (TCO)

Our data consist of spatially and temporally irregular TCO observations sampled on October 2, 1988, by the total ozone mapping spectrometer (TOMS) instrument on the Nimbus-7 satellite. In a single day, the satellite is able to achieve approximately global coverage, with a slight overlap in consecutive orbits. Under perfect conditions, this generates about 200,000 TCO observations within a single day. In practice, a number of observations are missing and others are removed by a quality-control procedure, resulting in 162,265 valid observations for October 2, 1988. In our analysis of the TCO data, we shall use five spatial resolutions, as in Huang et al. (2002):

Resolution:	R–1	R–2	R–3	R–4	R–5
Cell size (lon x lat):	45°x36°	15°x12°	5°x4°	2.5°x2°	1.25°x1°
Number of cells:	40	360	3,240	12,960	51,840

The TCO data are initially aggregated to the finest resolution, R-5, yielding the (potential) data $\{Z(i,5), V(i,5)\colon i = 1,\ldots, 51{,}840\}$, where $Z(i,5)$ is

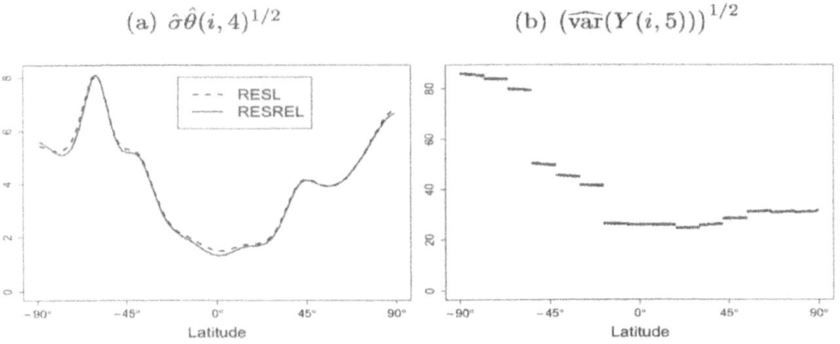

Figure 2. (a) The RESL and RESREL estimates of $\sigma\theta(i,4)^{1/2}$ as a function of latitude. (b) The estimated standard deviation of $\{Y(i,5)\}$, as a function of latitude, based on RESREL estimation of σ^2 and $\{\theta(i,r)\}$.

defined as the average of all observations within $D(i,5)$, and $V(i,5)$ is taken to be the reciprocal of the number of observations within $D(i,5)$; $i = 1,\ldots,$ 51,840. In our case, 7,382 R-5 cells do not contain any observations, resulting in 7,382 missing observations in the R-5 dataset (Figure 3, top).

To apply the spatial multi-resolution model of Section 2 to the TCO data, the matrices $\{\mathbf{W}(i,r)\}$ and σ^2 need to be estimated. Although the optimal predictor does not depend on σ^2, we need it for prediction variances. We assume that the TCO process follows the mass-balanced, coarse-to-fine-resolution process model (5), with $\{\mathbf{W}^*(i,r)\}$ given by the SPPS in (9) and $\mathbf{C}_0(i,r) = \mathbf{I}$. An exploratory data analysis indicates that most of the between-resolution variation is latitudinal. Based on this, $\{\log \theta(i,r)\}$ is modeled as a smooth function of latitude only, within each resolution r, using a linear combination of B-spline basis functions, as in (10), with 4, 7, 10, and 14 knots at resolutions 1-4, respectively. The B-splines were constrained to have zero derivative at the poles, resulting in a smooth surface on the sphere. Estimation of the parameter vectors $\{\beta(r)\}$ was carried out using both the RESL in (17) and the RESREL in (19), with σ^2 estimated at the finest resolution in each case. Only aggregated data $\{\mathbf{Z}(i,r)\}$ with no missing elements were used in the estimation process. At the coarsest resolution, R-1, recall that $(Y(1,1),\ldots,Y(40,1))' \sim \mathrm{Gau}(\mathbf{a}(0), \sigma^2\mathbf{R}(0))$. We assume that the trend $\mathbf{a}(0)$ is a linear combination of 25 spherical harmonics (i.e., $\mathbf{a}(0) = \mathbf{X}\beta(0)$) and $\mathbf{R}(0)$ is given by an exponential covariance function. Unknown parameters of this model were estimated from the coarsest-resolution aggregated data $\{Z(i,0)\}_{i=1}^{40}$ using REML. An alternative approach would be to detrend the original, massive TCO data, as in Cressie (2003).

Figure 2(a) shows both the RESL and the RESREL estimates of $\sigma\theta(i,4)^{1/2}$, plotted versus latitude. We note first that the two estimates are basically identical, both showing that the difference between the aggregated Y-process at R-4 and R-5 has least variability around the equator. Figure 2(b) shows the marginal variance of $\{Y(i,5)\}$, based on the RESREL estimates of σ^2 and $\{\theta(i,r)\}$. The stepwise appearance in Figure 2(b) is due to the change-of-resolution nature of the MRSM. Finally, Figure 3 shows the TCO data $\{Z(i,5)\}$, and the posterior mean and standard deviation given by the MRSM after substituting in RESREL estimates of σ^2 and $\{\theta(i,r)\}$.

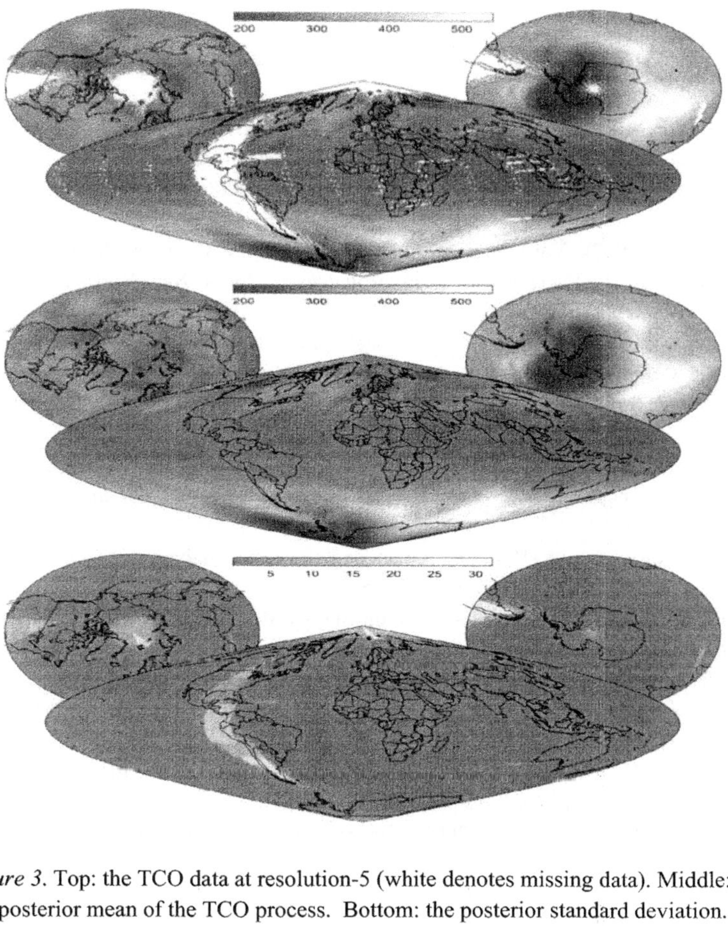

Figure 3. Top: the TCO data at resolution-5 (white denotes missing data). Middle: the posterior mean of the TCO process. Bottom: the posterior standard deviation.

The MRSM has enormous advantages, computationally. The program used for the analysis in this paper was written using the statistical programming language R (Ihaka and Gentleman, 1996). The whole execution time of the program, from creating the spatial tree-structure, through to computing the estimates used in Figures 2 and 3, took about 3 minutes on a linux computer with an Atholon MP 1800 processor.

ACKNOWLEDGEMENTS

This research was supported by the U.S. Environmental Protection Agency under Assistance Agreement R827257-01-0. The authors would like to thank Hsin-Cheng Huang for his helpful comments.

REFERENCES

1. Chou, K. C., A.S. Willsky, R. Nikoukhah (1994). Multiscale systems, Kalman filters, and Riccati equations, *IEEE Transactions on Automatic Control*, **39**:479-492.
2. Cressie, N. (1993). *Statistics for Spatial Data (Revised Edition)*, New York: Wiley, pp. 1-900.
3. Cressie, N. (2002). Variogram estimation. In: A.H. El-Shaarawi, Abdel H. and W.W. Piegorsch (eds.): *Encyclopedia of Environmetrics*, Vol. 4. New York: Wiley, pp. 2316-2321.
4. Huang, H.-C., N. Cressie (2001). Multiscale graphical modeling in space: application to command and control. In: M. Moore (eds.): *Spatial Statistics: Methodological Aspects and Some Applications*, Vol. 159 of Springer lecture Notes in Statistics. New York: Springer, pp. 83-113.
5. Huang, H.-C., N. Cressie, J. Gabrosek (2002). Fast, resolution-consistent spatial prediction of global processes from satellite data. *Journal of Computational and Graphical Statistics*. **11**:63-88.
6. Ihaka, R., R. Gentleman (1996). R: A Language for Data Analysis and Graphics. *Journal of Computational and Graphical Statistics*. **5**:299-314.
7. Johannesson, G. (2003) Multi-resolution statistical modeling in space and time with application to remote sensing of the environment. Ph.D. thesis, The Ohio State University.
8. Johannesson, G., N. Cressie (2003). Finding large-scale spatial trends in massive, global, environmental datasets. Technical report, Department of Statistics, The Ohio State University.
9. Kolaczyk, E.D., H. Huang (2001). Multiscale statistical models for hierarchical spatial aggregation. *Geographical Analysis*. **33**:95-118.
10. McCulloch, C.E., S.R. Searle (2001). Generalized, Linear, and Mixed Models. New York: Wiley, pp. 1-325.

GEOSTATISTICAL INTERPOLATION AND SIMULATION IN THE PRESENCE OF BARRIERS

K. Krivoruchko and A. Gribov
Environmental Systems Research Institute, 380 New York Street, Redlands, CA 92373, kkrivoruchko@esri.com, agribov@esri.com

Abstract: Statistical correlation between spatial variables depends on the distance between locations and the direction of travel from one to the other. Geostatistical interpolation most often uses the Euclidean distance between observations. But since most surfaces in nature are convoluted, with edges and breaks, anything that travels along them is thereby constrained. Smog, for instance, is blocked by hills and mountains. Animals migrate around lakes, mountains, and settlements. Contaminants in water follow the coastline. This paper proposes using cost weighted distance, a common raster function in GIS (Geographical Information Systems) that calculates the cost of travel from one cell of a grid to the next, making it the natural choice of the distance metric for spatial interpolation. Determining cost value at each location is discussed, as is calculation of distances between sampled locations and unsampled ones. Also covered is how to choose a valid covariance model with barriers defined by cost surface. We illustrate the approach using publicly available ozone data in California, where mountains are the natural barriers for smog propagation, and nutrients data in the Chesapeake Bay, where the coastline forms non-transparent barrier for chemical propagation.

1. INTRODUCTION

Spatial interpolation assumes that locations close together are more similar than locations that are far apart. Most interpolators use Euclidian distances to calculate the weight of neighboring data, which they use to predict the value of unsampled locations. In geostatistics, weights are calculated according to the value of covariance or of semivariogram, the statistical variant of distances

X. Sanchez-Vila et al. (eds.), geoENV IV – Geostatistics for Environmental Applications, 331-342.
Published by Kluwer Academic Publishers.

between locations. But even though points farther away from where a value needs to be predicted are not necessarily weighted less than points that are closer, both semivariogram and covariance are still functions of distance, traditionally Euclidean distance.

Predicting values for unknown locations becomes difficult in the presence of barriers, as illustrated in figure 1a. The straight-line distance between B and C is shorter over land than around that spit. But a chemical spilled in the water at B would travel to C by sea, not land. And though C is closer to B than A, over land, we can see that A is much likelier to be contaminated than C is. Using the length of the shortest path in the water between two locations (as the fish swims), A is closer to B than C is. This example illustrates the need of spatial correlation model that is consistent with the physical process under study.

Barriers are rarely considered in geostatistics because one of the assumptions in traditional geostatistics is that predictions can be based on just one realization of the random function of the *unrestricted spatial process*, see Gandin, 1963.

Similar to water contamination is contamination of the air. In figure 1b, the arrow shows the direction smog will take from east of Los Angeles. Mountains on the right side of figure block the smog, and from about 600 meters above sea level the air is typically cleaner than in the valley. Air quality is also much worse along freeways so the physical model of the pollution distribution dictates that the statistical distance between locations along roads should be different from the distance across the roads.

In California, at least three variables known for each location influence the level of pollutants in the air: elevation, distance from the ocean, and distance from the road. Figure 2a presents a 3D view of distances from the freeways and distance from the ocean. These surfaces, together with the surface of California elevation, in figure 2b, can be used to improve geostatistical models of air pollutant prediction.

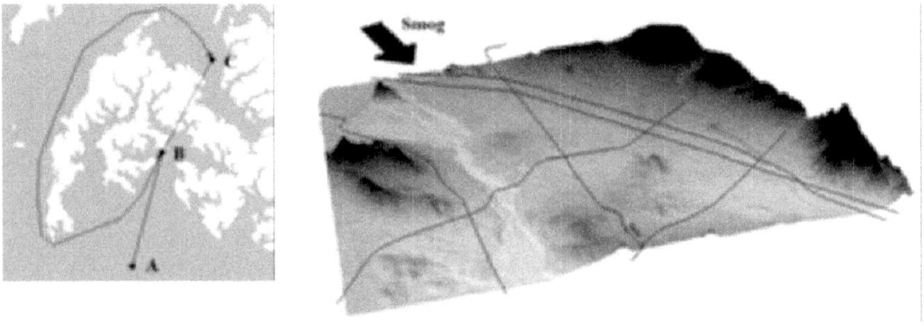

(a) (b)

Figure 1. a) Modeling water contamination needs a non-Euclidean metric. b) Freeways and mountains affect how smog is produced and how it spreads. Geography of Chesapeake Bay and Southern California is used.

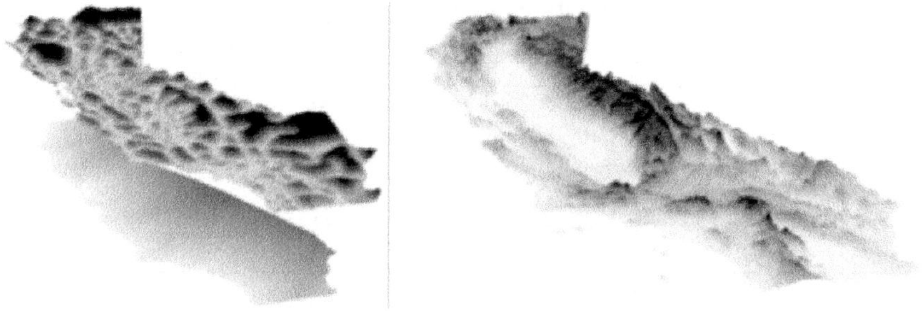

(a) (b)

Figure 2. a) Distance from major roads (top) and from the ocean (bottom). b) Elevation enlarged by a factor of 15. Geography of Southern California is used.

The rest of the article discusses interpolation and simulation using non-transparent and semi-transparent barriers. We calculate distance between locations using a cost surface grid. We illustrate the approach using publicly available ozone data in California and nutrient data in the Chesapeake Bay. There is no intention to present a complete analysis of these data, however.

2. INTERPOLATION WHEN DETAILED INFORMATION ON SECONDARY VARIABLES IS AVAILABLE

For data interpolation of air pollution in California, we have a limited number of data measurements and detailed information on secondary variables. Among the possible approaches to model such data are the followings:

- Universal kriging with external trend, see Ver Hoef, 1993;
- Cokriging, see Gandin, 1963;
- Changing the definition of the covariance model, see Carroll and Cressie, 1996.

Universal kriging with external trend assumes that the mean of the primary variable changes locally and can be estimated as a function of the secondary variable. This assumption is often appropriate for aggregated polygonal data and rarely works well for continuous ones.

Figure 3a presents the result of an ozone prediction using a cokriging model, with ozone as the primary variable and a grid of distances from major California roads, see figure 2a, as the secondary variable. Major roads are displayed as the top layer of the map. Table 1 presents cross-validation statistics for ordinary kriging and ordinary cokriging, with the second variable as distance to a major road.

Table 1. Comparison of cross-validation statistics.

Cross-validation statistics	Ordinary kriging	Ordinary cokriging, with second variable as distance to a major road
Mean error	0.00026	0.00038
Root-mean-square error	0.01268	0.01206
Average standard error	0.01628	0.01476
Mean standardized error	0.0112	0.01928
Root-mean-square standardized error	0.7845	0.8484

The best model is the one that has the smallest root-mean-squared prediction and average standard errors, and the standardized root-mean-squared prediction error nearest to one, see Cressie, 1993. Thus, using distance from a road as a secondary variable improves the prediction of ozone pollution. One problem with cokriging is how to model cross-correlation between variables. Figure 3b shows the cross-covariance cloud and the exponential model used to create the map in figure 3a. The largest correlation occurs at the non-zero distance between the monitoring stations and the data on the grid. Cross-correlation is anisotropical and shifted, so it is difficult to find the optimal cross-covariance model in this situation.

Carroll and Cressie, 1996, added information on elevation, slope, and aspect to the definition of the covariance model. Such distance metric is a particular case of the city-block distance metric, which is valid for an exponential covariance model, see the section "Interpolation and simulation using non-Euclidian distances" below.

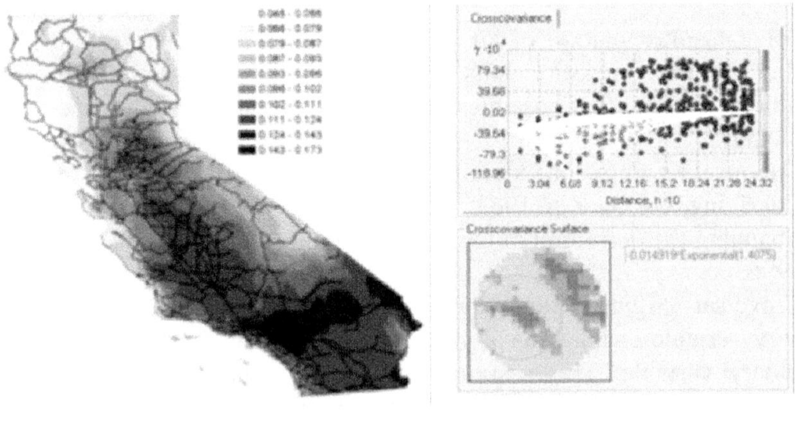

(a) (b)

Figure 3. a) Ordinary cokriging prediction of the maximum one-hour annual value of ozone in California in 1999 using distance to the road from the monitoring stations as the secondary variable. b) Cross-covariance between ozone and distance from the major roads.

3. DISTANCE BASED ON COST SURFACE

We propose a new approach to the problem of data interpolation and simulation with non-Euclidean distances, one based on cost weighted distance. Cost weighted distance is a common raster function in GIS that calculates the cost of travel from one cell of a grid to the next, making it the natural choice of the distance metric for spatial interpolation. Typical examples of cost surfaces are travel time, dollars, and alternate routes.

The value of each cell in the cost surface represents the resistance of passing through the cell and may be expressed in units of cost, risk, or travel time. Figures 4a illustrates a cost surface usage for interpolation purposes using a side view of elevation. The x-axis shows cell locations and the y-axis shows cost value assigned to grid cells. We want to penalize moving up and down, because a car, for example, uses more gas to go up hill and has more brake wear going down. On a flat surface the distance between points is calculated without penalties: moving from cell 3 to cell 4 is not penalized. Going uphill, from cell 4 to cell 5, we add distance to the path because of the difference between cost surface values in the neighboring cells, using either

(average cost value in the neighboring cells) (distance between cell centers)*

or

(difference between cost values in the neighboring cells) + (distance between cell centers)

formula. Cell locations where distance is changed are highlighted.

The templates in figure 4b show four ways to calculate distance between centers of neighboring cells. The more directions used, the closer the distance between points will be to optimal trajectory. However, the more directions used, the more time calculation takes.

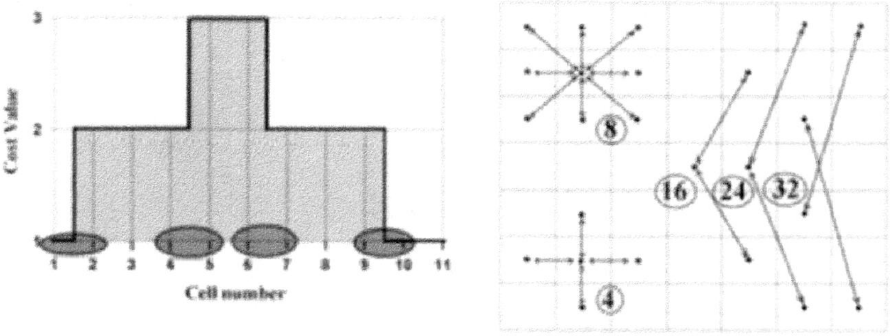

(a) (b)

Figure 4. a) Cost surface usage. b) Distance calculation using 4, 8, 16, and 24 directional templates.

Figure 5a shows the variable range of data correlation found using moving window covariance modeling, when analyzing nonstationary phosphorus data in a farm field in Illinois, Krivoruchko and Gribov, 2002.

This parameter of the geostatistical model can be used in the calculation of the cost surface grid: *cost value=(maximum range)/(range in the cell)*. Then the size of the moving kernel will change according to change of the range of data correlation (see discussion on kernel approach below).

To create a raster cost surface using detailed information on California elevation, data were reclassified according to our observations on smog propagation in summer time in Southern California shown in table 2.

Table 2. Relationship between California elevation and cost surface values.

Elevation, meters	*<100*	*100-200*	*200-300*	*300-450*	*450-600*	*600-750*	*750-900*	*900-1200*	*1200-1500*	*>1500*
Cost value	***1.0***	***1.1***	***1.2***	***1.3***	***1.5***	***2.0***	***3.0***	***5.0***	***10.0***	***100.0***

We used Dijkstra's source-sink shortest path algorithm, see Sedgewick, 2002: given a start vertex A and a finish vertex B, find the shortest path in the graph from A to B given weights equal to a cost value in each grid cell. If there is no path from A to B, infinite weight is assigned to the vertices.

4. INTERPOLATION AND SIMULATION USING NON-EUCLIDIAN DISTANCES

Figure 5b presents an example of a naïve approach to interpolation in the presence of barriers, which is implemented in some spatial data analysis programs. In this approach, if the straight line between two locations intersects a line or a polygonal barrier, then the points do not "see" each other and are excluded from the searching neighborhood and, in the case of the kriging model, from the list of empirical semivariogram pairs of points.

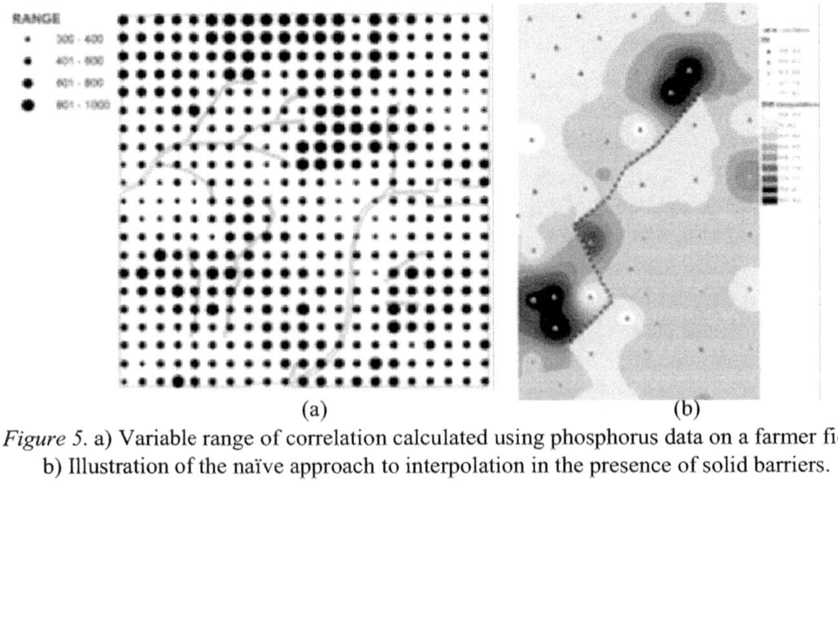

(a) (b)

Figure 5. a) Variable range of correlation calculated using phosphorus data on a farmer field. b) Illustration of the naïve approach to interpolation in the presence of solid barriers.

One of the problems with this method is that prediction abruptly changes near the line barrier (or corners of the polygonal barrier) without any physical reason for it.

Little et al, 1997, Curriero, 1997 and 2003, Rathbun, 1998, and Higdon, 1998, discussed using non-Euclidean distances in geostatistics. They considered a distance metric that must satisfy the following geometric properties:

$$d(s_1, s_2) = d(s_2, s_1);$$
$$d(s_1, s_2) \geq 0, \text{ with equality if and only if } s_1 = s_2;$$
$$d(s_1, s_3) \leq d(s_1, s_2) + d(s_2, s_3),$$

where s_1, s_2, s_3. are coordinates of the data locations, and d(•) is a distance between two locations.

Curriero, 2003, pointed out that conditions of a metric are not sufficient proof of the validity of distance to yield positive definite functions. Such distances cannot be used without proof in covariance and semivariogram models. Covariance $\mathrm{cov}(d(s_i, s_j))$ calculated using metric $d(s_i, s_j)$ must satisfy the non-negative definiteness property:

$$\sum_i \sum_j b_i b_j \,\mathrm{cov}(d(s_i, s_j)) \geq 0$$

An important result of previous research is that most traditional parametric covariance models, including spherical one, are not valid for non-Euclidean distances. One exception is an exponential covariance model, $\mathrm{cov}(d) = e^{-d}$, which is valid for the city-block distance metric,

$$d_{cb}(s_1, s_2) = | x_1 - x_2 | + | y_1 - y_2 | .$$

The city-block distance metric corresponds to the template with four possible directions, see bottom left of figure 4b. A geostatistical process with a city-block distance metric and an exponential covariance model would be constructed as follows:

Consider n independent random processes with the exponential covariance model in one dimension $r_i(s), i = \overline{1,n}$:

$E\{r_i(s)\} = 0$ и $\mathrm{cov}\{r_i(s), r_i(t)\} = e^{-\alpha|t-s|} \forall i = \overline{1,n}$, where α is a constant inversely proportional to the range of data correlation.

Construct a process, $r(s) = \prod_{i=1}^{n} r_i(s_i)$. The expected value of this process is zero, $E\{r(s)\} = 0$, and covariance represents statistical distance using the city-block metric:

$$\mathrm{cov}\{r(s),r(t)\} = \mathrm{cov}\left\{\prod_{i=1}^{n} r_i(s_i), \prod_{i=1}^{n} r_i(t_i)\right\} = E\left\{\prod_{i=1}^{n} r_i(s_i) r_i(t_i)\right\} =$$

$$= \prod_{i=1}^{n} E\{r_i(s_i) r_i(t_i)\} = \prod_{i=1}^{n} e^{-\alpha|s_i - t_i|} = e^{-\alpha\sum_{i=1}^{n}|s_i - t_i|}$$

It is possible that the other templates presented in figure 4b are also valid for calculating distance in the exponential covariance using a cost surface with the same values in each grid cell. Unfortunately, exponential covariance is not valid when distance may change according to cost values in the neighboring cells.

We do not know how to find a valid parametric covariance model in the presence of semitransparent barriers, but it is possible to examine a selected model for non-negative definiteness. We simulated distances between points with a valid space metric and calculated the exponential covariance matrix to check its non-negative definiteness property. In about one case out of three thousand, the resulting covariance matrix was negatively definite.

Because commonly used theoretical covariance models are not valid in the case of distances modified by cost surface values, we propose using a moving average approach for covariance model estimation. Notable references on such flexible covariance modeling are Barry and Ver Hoef, 1996; Higdon, 1998; Yao and Journel, 1998, and Ver Hoef, et al., 2001.

A modeling process based on distances defined by cost surface using a moving average can be constructed as follows:

- In each grid cell, model the independent random variable $\xi_{t,s}$ with zero mean and variance σ^2.
- Based on the cost value in each grid cell, find the distance $\rho_{(i,j),(t,s)}$ to points in the specified neighborhood, where pairs (i,j) and (t,s) refer to grid rows and columns.
- Define kernel function $f(\rho_{(i,j),(t,s)})$.

Then process is defined as $n_{i,j} = \dfrac{\sum_{t,s} f(\rho_{(i,j),(t,s)}) \cdot \xi_{t,s}}{\sqrt{\sum_{t,s} f^2(\rho_{(i,j),(t,s)})}}$ with covariance

equals $\mathrm{cov}(n_{i,j}, n_{i',j'}) = \sigma^2 \cdot \dfrac{\sum_{t,s} f(\rho_{(i,j),(t,s)}) \cdot f(\rho_{(i',j'),(t,s)})}{\sqrt{\sum_{t,s} f^2(\rho_{(i,j),(t,s)})} \cdot \sqrt{\sum_{t,s} f^2(\rho_{(i',j'),(t,s)})}}$

The important step is the choice of the appropriate kernel. Two different strategies can be used here: defining the kernel itself or calibrating a spatial moving average using well-known spatial covariance functions. The former approach can be based on a known or an estimated range of data correlation to define size of the kernel, data variance to define the height of the kernel and the underlying physical process to define shape of the kernel. The later approach is discussed in detail in Oliver, 1995, and Cressie and Pavlicova, 2002. In both situations, the kernel should correspond to the covariance with the dependence disappearing after a fixed distance.

The situation here is more complicated than those described in Ver Hoef et al, 2001, because the kernels are not symmetrical near barriers.

There is a choice between prediction and simulation.

Covariance $\mathrm{cov}\left(n_{i,j}, n_{i',j'}\right)$ is known for all pairs of grid locations and can be used to solve simple or ordinary kriging equations. Because of data uncertainties, such as inaccuracy in the measurement device, rounding off, and local integration errors, and because of the locational errors introduced when data locations are moved to the center of the nearby grid cell, filtered versions of kriging are preferred.

Alternatively, given unconditional simulations in the grid cells $\boldsymbol{n_{i,j}}$, conditioning to the observations can be made. Conditioning to the data should take into account the measurement error component in the kriging model. A geostatistical conditional simulation model using simple and ordinary filtered kriging can be found in Aldworth, 1998.

To show the influence of cost surface on simulations, we used Chesapeake Bay geography and a cost surface defined so that the water surface received the value of one and land received a very large value, making it a non-transparent barrier for chemical propagation. Figure 6a shows unconditional simulation using a moving cylindrical kernel with radius displayed in the top left corner and height corresponding to the unity variance over simulated Gaussian white noise. The white contour indicates the border between land and water. However, the difference between water and land was ignored. The same kernel was used to smooth out the same noise in the map in figure 6b, but the process was estimated on water only, with land as non-transparent barrier. The kernel is circular if it does not touch the barrier (land). Near the barrier, the kernel changes its shape as in the top left corner of figure 6b.

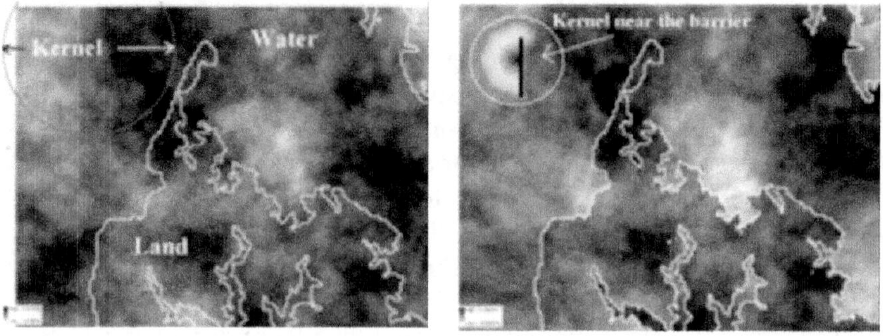

(a) (b)

Figure 6. a) Unconditional simulation on a flat surface. b) Unconditional simulation using a cost surface with non-transparent barriers.

The difference between maps is significant (see for example the central part of the figures under the label "Water") because the kernel constrained by land surface does not use significantly different information on the land, but searches for the data along the river surface only.

Figure 7 shows two simple kriging interpolations of ozone in Southern California, one using Euclidean distance, the other using non-Euclidean. For prediction, the difference between models is not as significant as for prediction standard errors. Prediction standard error mapping is of greater importance in environmental applications because it can indicate areas where predictions are unreliable. The kind of maps often used in decision-making, quantile and probability maps, are essentially based on predicted standard errors; see for example Krivoruchko, 2001, and Krivoruchko and Gribov, 2002. Figure 7a is the map of simple kriging standard error based on Euclidean distances. Figure 7b shows simple kriging standard error when distances are calculated using a 200 by 200 cost surface grid based on the relationship between elevation and cost values and superimposed over the California territory in table 2, bottom row.

Interpolation in the presence of semitransparent barriers, figure 7b, shows the uncertainty of prediction based both on density of observations and on elevation. This makes sense because smog is blocked by mountains but can travel through gorges. The interpolation based on straight line distances ignores the mountains, so prediction uncertainty is based only on data density. As a result, prediction errors are underestimated when predicting smog in the hills and mountains using measurements in the valley.

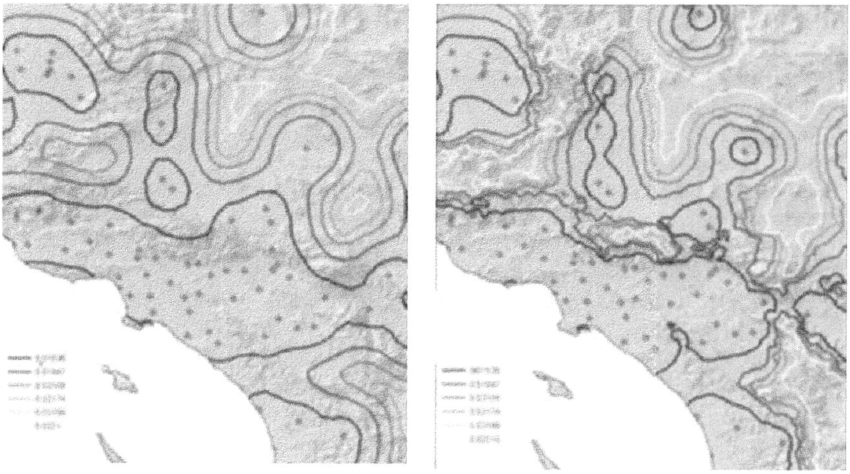

(a) (b)

Figure 7. Simple kriging prediction error using a straight line distance (a) and using the least accumulative cost distance (b).

5. CONCLUSION

Euclidean distance is the default distance in nearly all geostatistical applications. However, it may not be the best distance metric when the process being modeled is affected by natural factors, such as elevation, geological faults, and coastlines. Thus, statistical distances that account for these natural factors can be defined by introducing non-transparent and semitransparent barriers for movement from one location to another. In this paper, we proposed that interpolation in the presence of such barriers be based on cost surface, a common GIS modeling option. The cost value at each location can be a function of several variables and all the cells in the grid. Recently Dubois, 2001, used a cost surface for calculation of the semivariograms used with kriging interpolators. However, theoretical covariance models are not valid when distances change randomly between neighboring cells. Thus, we proposed a solution that uses the moving window kernel approach for calculation of the spatial correlation. The important step is the choice of the appropriate kernel. Two different strategies can be used here: defining the kernel itself or calibrating spatial moving averages using well-known spatial covariance functions. In both situations, the kernel should correspond to the covariance, with dependence vanishing after a fixed distance. Using an appropriate cost surface with geostatistical models produces more reliable prediction and prediction standard errors, as we demonstrated using air quality data in California.

REFERENCES

1. Aldworth, J. (1998). Spatial Prediction, Spatial Sampling, and Measurement Error. Ph.D. thesis. Iowa State University.
2. Barry, R. P. and Ver Hoef, J. (1996). Blackbox kriging: spatial prediction without specifying variogram models, *J. Ag. Bio. and Eco. Statist.* **3**: 1–25.
3. Carroll, S.S. and Cressie, N. 1996. A comparison of geostatistical methodologies used to estimate snow water equivalent. Journal of the American Water Resources association, Vol. 32, No. 2, 267-278.
4. Cressie, N. 1993. Statistics for spatial data, Wiley, New York.
5. Cressie , N. and Pavlicova M. 2002. Calibrated spatial moving average simulations. Statistical Modeling, 2, 1-13.
6. Curriero, F. C. 1997. The use of non-Euclidean distances in geostatistics, PhD thesis.
7. Curriero, F. C. 2003. Norm dependent isotropic covariogram and variogram models. In print.
8. Dubois, G. 2001. Intégration de Systèmes d'Informations Géographiques (SIG) et de méthodes géostatistiques. University of Lausanne, Ph. D. Thesis (In French). 260 pp.
9. Gandin, L.S. 1963. Objective Analysis of Meteorological Fields. Gidrometeorologicheskoe Izdatel'stvo (GIMIZ), Leningrad (translated by Israel Program for Scientific Translations, Jerusalem, 1965).
10. Higdon, D. 1998. A process-convolution approach to modeling temperatures in the North Atlantic Ocean, Journal of Environmental and Ecological Statistics 5, 173-190.
12. Krivoruchko K., 2001. Using linear and non-linear kriging interpolators to produce probability maps. Available from ESRI online at http://www.esri.com/software/arcgis/ arcgisxtensions/geostatistical/ research_papers.html
13. Krivoruchko K. and Gribov A. 2002. Working on Nonstationarity Problems in Geostatistics Using Detrending and Transformation Techniques: An Agricultural Case Study. Available from ESRI online at http://www.esri.com/software/arcgis/arcgisxtensions/geostatistical/research_papers.html
14. Little LS, Edwards D, Porter D.E. 1997. Kriging in estuaries: as the crow flies or as the fish swims? Journal of experimental marine biology and ecology, 213, pp. 1-11.
15. Oliver, D.S. 1995. Moving averages for Gaussian simulation in two and three dimensions. Mathematical Geology, 27, 8, 939-960.
16. Rathbun, S. 1998. Spatial modeling in irregularly shaped regions: kriging estuaries. Environmetrics, 9, 109-129.
17. Sedgewick, R. 2002.Algorithms in C++. Graph algorithms. Addison-Wesley, 496 p.
18. Ver Hoef, J.M. 1993. Universal kriging for ecological data. Pages 447-453 in Ver Hoef, J. and Barry, R. P. (1998). Constructing and fitting models for cokriging and multivariable spatial prediction. J. Statist. Planning and Inference 69, 275-294.
19. Ver Hoef, J. M., Cressie, N., and Barry, R. P. (2001). Flexible spatial models for kriging and cokriging using moving averages and the Fast Fourier Transform (FFT). Department of Statistics Preprint No. 675, The Ohio State University.
20. Yao, T. and Journel, A.G. 1998. Automatic modeling of (cross)covariance tables using fast Fourier transform. Mathematical Geology 30, 589-615.

A SPECTRAL TEST OF NONSTATIONARITY FOR SPATIAL PROCESSES

J. Mateu and P. Juan

Universitat Jaume I, Department of Mathematics. Campus Riu Sec, E-12071, Castellón, Spain. Email: mateu@mat.uji.es. Fax: +34.964.728429.

Abstract: We present a test for the detection of nonstationary spatial processes using spectral methods. The spatial field is represented locally as a stationary isotropic random field, but only the parameters of the stationary random field that describe the behaviour of the process at high frequencies are allowed to vary across in space, reflecting the lack of stationarity of the process.

Key words: Geostatistics, Nonstationarity, Spatial statistics, Spectral density.

1. INTRODUCTION

Spectral analysis of stationary processes is particularly advantageous in the analysis of large data sets and in studying properties of multivariate processes. Geostatistical data are usually collected over a large region, and handling large data sets is often problematic for the commonly used techniques: inversion of a large covariance matrix to compute the likelihood function may not be possible or may require a long time in computation. The use of a Fast Fourier transform (FFT) algorithm for spectral densities can be a good solution for these problems. However, FFT can be applied only to regularly gridded data, though this disadvantage is not that important as there are theoretical connections between the estimators of the spectral densities in both the regular lattice and irregular spaced data (Renshaw, 2002). The periodogram, a nonparametric estimate of the spectral density, is

X. Sanchez-Vila et al. (eds.), geoENV IV – Geostatistics for Environmental Applications, 343-354.

a poweful tool for studying the properties of stationary processes observed on a two-dimensional lattice (Stein, 1999).

Spatial processes in environmental sciences are generally nonstationary, in the sense that the spatial structure depends on location. Therefore, standard methods of spatial interpolation are inadequate. Thus, it is clear that many environmental problems have to deal with nonstationarity of the underlying spatial process. The decision to treat the problem at hands as stationary or nonstationary is often not based on a mere statistical scrutiny of data values, but on considering the physics of the problem. However, at least two reasons reinforce the necessity of statistical tools for detecting nonstationarity: (a) the researcher does not always have access to all the physical data, and can not evaluate an appropriate underlying physical model, and (b) as a nonstationarity validation procedure, even knowing a priori the data behaves as such. In recent years, probably the most extensively studied method for nonstationary spatial processes is the deformation approach due to Sampson and Guttorp (1992). Recently, several spectral methods for analysis and interpolation of environmental nonstationary processes have been presented (Fuentes, 2001, 2002; Fuentes and Smith, 2002).

In this paper we focus our attention on the above methodology by Fuentes, for spatial interpolation of nonstationary processes using spectral methods, to propose a simple diagnostic test of nonstationarity of a spatial process. In Section 2 we introduce the spectral representation of spatial processes. Section 3 is devoted to modeling the spatial structure in terms of the periodogram, as an estimate of the spectral density. A test of nonstationarity is presented and evaluated in Section 4.

2. NONSTATIONARY SPATIAL PROCESSES. SPECTRAL METHODS

Let Z be a nonstationary process observed on a region D. Suppose D is covered by well-defined subregions $S_1,...,S_k$, and consequently, Z can be written as a weighted average of orthogonal local stationary processes Z_i for i=1,...,k, with $cov(Z_i(\mathbf{x}),Z_j(\mathbf{y}))=0$ for $i \neq j$. We have

$$Z(\mathbf{x}) = \sum_{i=1}^{k} Z_i(\mathbf{x})K_i(\mathbf{x}) \qquad (1)$$

where Zi is a local stationary process in the subregion S_i, $K_i(\mathbf{x})$ is a positive kernel function centered at the centroid of S_i. The weighted average (1) is the discrete representation of the process Z, but we could write this average as an integral to obtain a continous representation.

The nonstationary covariance of Z is defined in terms of the local stationary covariances of the processes Z_i for i=1,…,k,

$$cov(Z(\mathbf{x}),Z(\mathbf{y})) = \sum_{i=1}^{k} K_i(\mathbf{x})K_i(\mathbf{y})cov(Z_i(\mathbf{x}),Z_i(\mathbf{y})) \tag{2}$$

where $cov(Z_i(\mathbf{x}),Z_i(\mathbf{y})) = C_{\theta i}(\mathbf{x}-\mathbf{y})$, the covariance parameter θ_i varies with the subregion measuring the lack of stationarity of Z.

The two-dimensional random field Z_i, with i=1,... k can then be represented in the form of the following Fourier-Stieltjes integral (Cressie, 1993)

$$Z_i(\mathbf{x}) = \int_{\Re^2} \exp(i\omega^T \mathbf{x}) dY_i(\omega) \tag{3}$$

where Y_i are random functions with *uncorrelated increments* and ω are the frecuencies. The representation (3) is called *spectral representation* of Z_i. The spectral representation describes the harmonic analysis of a general stationary process, i.e. its representation in a form of a superposition of harmonic oscillations.

Let the function F_i be a positive finite spectral measure for Z_i, defined by $E|Y_i(\omega)|^2=F_i(\omega)$. If F_i has a density with respect to the Lebesgue measure, this density is the *spectral density* f_i, defined as the Fourier transform of the autocovariance function C_i,

$$f_i(\omega) = \frac{1}{(2\pi)^2} \int_{\Re^2} \exp(-i\omega^T \mathbf{x}) C_i(\mathbf{x}) d\mathbf{x} \tag{4}$$

By Bochner's theorem, the function C_i is an autocovariance if and only if can be represented as in (4), where F_i is a positive, finite measure. Thus, the spatial structure of Z_i could be analyzed with a spectral approach or equivalently by estimating the autocovariance function.

Focussing now on the nonstationary process Z, defined as a mixture of the stationary processes $Z_1,\ldots,Z_k$ as in (1), the spectral representation of Z is $Z(\mathbf{x}) = \int_{\Re^2} (i\omega^T \mathbf{x}) dY(\boldsymbol{\omega})$ with $Y(\boldsymbol{\omega}) = \Sigma^k_{i-1} \bar{K}_i * Y_i(\boldsymbol{\omega})$ for $\bar{K}_i$ the Fourier transform of K_i, and * denotes the convolution. The covariance of Z can be defined in terms of the covariance of the orthogonal local stationary processes Z_i, as in (2), defining a valid nonstationary covariance. The corresponding spectral density is given by $f(\boldsymbol{\omega}_1,\boldsymbol{\omega}_2) = \Sigma_{i=1}^{k} \hat{f}_i * [\bar{K}_i(\boldsymbol{\omega}_1)\, \bar{K}_i(\boldsymbol{\omega}_2)]$, where $\bar{K}$ is the FT of K.

3. MODELING THE SPATIAL STRUCTURE

3.1 Spectral domain: tapered periodogram

The spatial periodogram is a nonparametric estimate of the spectral density, and a powerful tool for studying the properties of random fields observed on a lattice. It is the modulus-squared of a finite Fourier transform for the observed region of the process, introduced to search for hidden periodicities of processes. The periodogram itself is not a consistent estimator of the spectral density, but consistency can be achieved by applying linear smoothing filters to the periodogram. Smoothing the periodogram, as is frequently done in time series does not remove large edge-effects in two or more dimensions. The sidelobes (subsidiary peaks) occurring on smoothing filters cause unnecessary large values of the periodogram ordinates for high frequencies and result in substantial bias. This phenomenon is called leakage. Instead of smoothing biased periodogram estimates, direct filtering of the data with a data taper before computing the periodogram can also provide a consistent estimate of the spectral density. The information lost through powerful frequencies by smoothing the periodogram can be better recovered by data tapers (Fuentes, 2001, 2002).

Data tapers put relatively less weight on the boundary points and is highly effective in removing edge-effects, even in higher dimensional problems. Moreover, in the fixed-domain aymptotics where the number of observations in a fixed study area increases, it has been shown that using the periodogram of the raw data without any data tapers applied can yield highly misleading results (Stein, 1999).

Consider a spatial stationary process $Z(\cdot)$ with covariance parameter θ which is assumed here to be known. We observe the process at N equally spaced locations in a regular grid D $(n_1 \times n_2)$, where $N=n_1 n_2$.

We define $I_N(\omega)$ to be the periodogram at a frequency ω,

$$I_N(\omega) = (2\pi)^{-2}(n_1 n_2)^{-1}\left|\sum_{x_1=1}^{n_1}\sum_{x_2=1}^{n_2} Z(\mathbf{x})\exp\left\{-i\mathbf{x}^T\omega\right\}\right|^2. \tag{5}$$

In practice, the periodogram estimate for ω is computed in the set of Fourier frequecies $2\pi f/n$ where $f/n=(f_1/n_1, f_2/n_2)$, and $f \in J_N$, for

$$J_N = \left\{\lfloor -(n_1-1)/2 \rfloor, \ldots, n_1 - \lfloor n_1/2 \rfloor\right\} \times \left\{\lfloor -(n_2-1)/2 \rfloor, \ldots, n_2 - \lfloor n_2/2 \rfloor\right\} \tag{6}$$

where $\lfloor u \rfloor$ denotes the largest integer less or equal that u.

The expected value of the spatial periodogram is not $f(\omega)$, but a weighted integral of $f(\omega)$. In terms of an increasing density asymptotics, it is asymptotically unbiased, its asymptotic variance is $f^2(\omega)$, and the periodogram values $I_N(\omega)$ and $I_N(\omega')$ for $\omega \neq \omega'$, are asymptotically uncorrelated (Fuentes, 2002). The asymptotic independence of the periodogram estimates is one of the big advantages of the spectral analysis.

We then use a data taper to prevent the leakage from far away frequencies that could have quite a lot of power, though every time we do tapering we lose information. Thus, we form the product $h(\mathbf{x})Z(\mathbf{x})$ for each value of $\mathbf{x} = x_1,x_2$, where $\{h(\mathbf{x})\}$ is a suitable sequence of real-values constants called a data taper. The traditional tapers used for two-dimensional data are the tensor product of two one-dimensional data tapers, $h_M(\mathrm{j})=h_1(j_1)h_2(j_2)$, where $j=(j_1,j_2)$, $1 \leq j_1 \leq n_1$ and $1 \leq j_2 \leq n_2$. For instance, $h_1(\cdot)$ and $h_2(\cdot)$ could be a m-cosine taper, where $1 \leq \mathrm{m} < \frac{n_1}{2}$ (Fuentes, 2002; Fuentes and Smith, 2002). $h_M(\cdot)$ is usually called the *multiplicative data taper* for two-dimensional data. In this paper, and following Fuentes (2002) and Fuentes and Smith (2002), we focus on a *rounded taper*, as seems to show better behaviour than the multiplicative one. This kind of taper is defined in terms of two parameters controling the rounded region.

3.2 Models for spectral densities

Consider the decomposition (1) of the nonstationary two-dimensional process Z into the local stationary processes Z_i for k subregions covering the region D. A class of practical variograms and autocovariance functions for a process Z_i can be obtained from the Matérn class of spectral densities

$$f_i(\omega) = \phi_i\left(\alpha_i^2 + |\omega|^2\right)^{(-\upsilon_i-1)} \tag{7}$$

with parameters $v_i > 0$, $\alpha_i > 0$ and $\phi_i > 0$. Here, the vector of covariance parameters is $\theta_i = (v_i,\alpha_i,\phi_i)$. The parameter α_i^{-1} can be interpreted as the autocorrelation range. The parameter v_i measures the degree of smoothness of the process Z_i, the higher the value of v_i the smoother Z_i would be, ϕ_i is the ratio of the variance σ_i^2 and the range (σ_i^{-1}) to the $2v_i^{\text{th}}$ power, $\phi_i=\sigma_i^2\alpha_i^{2v}$.

The corresponding covariance for the Matérn class is

$$C_{\theta_i}(x) - \frac{\pi\phi_i}{2^{\upsilon_i-1}\Gamma(\upsilon_i+1)\alpha_i^{2\upsilon_i}}(\alpha_i|x|)^{\upsilon_i}k_{\upsilon_i}(\alpha_i|x|) \tag{8}$$

where κ_{vi} is a modified Bessel function of order v_i. For instance, when $v_i=(1/2)$ we get the exponential covariance function, $C_{\theta i}(x)=\pi\phi_i\alpha_i^{-1}\exp(-\alpha_i|x|)$.

After accounting for the degree of smoothing, the validity of standard local interpolation procedures depends only on the parameterϕ_i. Even if the range parameter varies with location, local interpolation procedures (kriging) can be shown to be asymptotically optimal as long as v_i and ϕ_i are constant over the domain. This is a consequence of the fact that low frequency behavior of the spectrum (small values of $|\omega|$) should have little effect on interpolation. It is possible to provide a qualitative theory supporting this point of view (Stein, 1999).

Thus, if the goal is spatial interpolation, it is preferable to work on the spectral domain and focus on high frequency values (Fuentes, 2002). An approximate expression for the spectral density of the Matén class for high frequency values is obtained from expression (7) by letting $|\omega|$ go to ∞. As a consequence, the degree of smoothness, v_i and ϕ_i are the critical parameters (and not the range α_i^{-1}).

4. A FORMAL TEST OF NONSTATIONARITY FOR A SPATIAL PROCESS

Suppose we wish to test whether a two-dimensional spatial process Z is nonstationary, with the aim of further spatial interpolation. Further, suppose Z is measured at $N = n_1 \times n_2$ regularly spaced data. At this point, it is worth noting that: (a) In practice, we can have missing data at several locations in the lattice; (b) The source data could be sampled over an irregular grid. In this case, and taking into account the result in Renshaw (2002), we could define an appropriate regular grid to approximate the irregular data locations, and proceed normally with our proposed test. Then, we decompose Z into a sum of local stationary processes Z_i for k subregions covering the region of interest, say D. The number of regions (k) can be found using an AIC criterium (Fuentes, 2001), or by experimentation depending, for example, on the a priori knowledge of the physical characteristics of the region D. However, the number k is restricted by the number of original sampled locations N. Note that each subregion S_i will only have a subset of the N data, and as will be shown later, the power of the proposed statistical test depends very much on the number of data locations in each subregion. Furthermore, each subregion need not be the same size compared to the others. However, defining all k subregions under equal sizes will be useful for further test comparisons.

Suppose the spectral density of the process Z belongs to the Matérn class, given by (7), with vector of positive parameters given by $\theta_i=(\phi_i v_i \alpha_i)$. Recall that parameter ϕ_i measures the spatial variability. Focusing on high frequency values, an approximate expression of (7) is given by $f_i(\omega) = \phi_i(|\omega|^2)^{(-vi-1)}$, with v_i and ϕ_i as critical parameters. Working now in the log scale, we can fit the following linear model $log(f_i(\omega))=\beta_{oi}+\beta_{oi}log(|\omega|)$, where $\beta_{oi}=log(\phi_i)$ and $\beta_{oi}=-2(v_i+1)$.

In practice, $f_i(\omega)$ is estimated by the corresponding tapered periodogram $I^i_N(\omega)$. Taking into account that the periodogram values are asymptotic independent, we can use regression techniques to estimate the values of intercept and slope. A diagnostic to detect nonstationarity of Z should be able to detect differences among the values of β_{0i} and β_{1i} for different subregions (different values of i). Various alternatives can be considered. A simple map of these values in the region of interest could be a first step. However, we need a formal test to detect possible significant differences. The following test for the equality of regression equations is considered (see Rao, 1973 p.281). Let $y_i = log(I^i_N(\omega))$ and $x_i = log(|\omega|)$, to form the regression equation $y_i = \beta_{oi} + \beta_{1i}x_i$. Consider another equation given by $y_j = \beta_{oj} + \beta_{1j}x_i$. Suppose n_i and n_j stand for the sample size. Here we adapt a known test in the regression context, to check the null hypothesis H_o: $\beta_{oi} = \beta_{oj}$ and $\beta_{1i} = \beta_{1j}$. This procedure can then be easily generalized to comparison of any regression equations. Define the corrected sums of products for the second series as: $S^j_{xy} = \Sigma(x_j - \bar{x}_j)(y_j - \bar{y}_j)$ and $S^j_y = \Sigma(y_j - \bar{y}_j)^2$. These quantities are sufficient to determine the regression function $y_i = \beta_{oj}' + \beta_{1j}'x_i$. Then, the residual sum of squares for the separate regressions case is calculated by $R^2_0 = S^j_y\ \beta_{1j}'S^j_{xy} + S^i_y\ \beta_{1i}'S^i_{xy}$ and has $n_i + n_j$-4 degrees of freedom. We further consider the samples in the different subregions all together and consider them as a single sample. Proceeding in the same way as above, we calculate R^2_1, the residual sum of squares for the common regression case, which is associated to $n_i + n_j - 2$ degrees of freedom. We finally set up the analysis of variance to test the equality of the regressions (Rao, 1973).

A step-by-step guideline for practical implementation is the following. Given a spatial process sampled at regularly spaced data (see comments above, in case of irregular locations): (a) Select a number of subregions with equal sizes, if possible, using an AIC criterium (Fuentes, 2001); (b) For each subregion, estimate the spectral density over a range of Fourier frequencies by means of the periodogram. Select a data taper, for example, a rounded-type; (c) For each subregion, estimate the intercept and slope of the corresponding regression equation. Thus, calculate the parameter estimates of the spatial covariance function; (d) Evaluate our statistical test to assess spatial differences, i.e. different local behaviour of the spatial process.

To know more about the behaviour of our methodology, we performed a simulation study to: (a) evaluate the estimation procedure of the parameters ϕ_i and ν_i based on $log(f_i(\omega)) = \beta_{oi} + \beta_{1i}\ log(|\omega|)$; (b) assess spatial differences through graphical tools; (c) analyze both type I error and power of the test. We thus considered the following procedure. Estimate, using the regression technique, the corresponding parameter vector θ_i, for any given subregion. Choose, among the fitted values, one parameter value, say θ_0, which will be kept fixed for all the subregions. Then, use a Monte Carlo test based on simulations of the underlying spatial process with θ_0 to perform a formal hypothesis test, where the condition of *no difference between two parameter values* defines the null hypothesis. Note that the observed parameter value of *ith* subregion, θ_i, is compared with a number of simulations (400 in our case) under θ_0. If θ_I ranks $4 \times$ p*th* largest, the attained significance level of the test is p*th*.

4.1 Simulation study

We used a simulated case study to quantify and compare the proposed methodology for testing nonstationarity under different experimental circumstances. In this simulation report, we kept fixed the number of subregions ($k = 4$), and parameter $\nu = 1$, to cover the case when the process is mean square differentiable. We then considered several scenarios: (a) several values for the grid size at which the process in each subregion is observed, $N = 20 \times 20$, 10×10, 5×5; (b) several combinations for the sill and range parameters, focusing on those cases for which there are both small and big differences between the ranges. The aim is to see how good the test is in detecting differences. The parameter combinations (sill,range) we looked at were: (2.9,300), (2.9,10), (2.0,10), (2.9,166), (2.9,200). We used 400 Fourier frequencies for the rounded tapered periodogram evaluation.

Table 1. Simulation results for the following setup: $\nu = 1$, $N = 20 \times 20$ and the corresponding combinations of (sill,range). Means and standard deviations of estimated ϕ and ν parameters based on 1000 simulations.

sill	range	real (ϕ)	mean (ϕ)	s.d. (ϕ)	mean (ν)	s.d. (ν)
2.9	300	3.22e-05	4.75e-05	3.444e-05	0.830	0.173
2.9	10	0.02900	0.01118	0.00371	0.688	0.135
2.0	10	0.02000	0.00723	0.00284	0.667	0.122
2.9	166	0.00011	0.00014	0.00012	0.786	0.165
2.9	200	3.27e-05	9.97e-05	8.513e-05	0.796	0.172

A first analysis consisted of the evaluation of the estimation of the spatial process parameters (ν,ϕ) through the simple linear regression defined over the logarithm of the periodogram. We simulated a spatial process with a Matérn covariance function with any parameter combination considered above. Then, we estimated the parameters, and this procedure was repeated 1000 times. The results of these simulations are reported in Tables 1-2. The regression procedure provided the best estimates for grid sizes 20×20 and for the bigger ranges. However, the procedure provided misleading results, when the number of evaluation points was 5×5 (and we decided not to show them).

Table 2. Simulation results for the following setup: $\nu = 1$, $N = 10 \times 10$ and the corresponding combinations of (sill,range). Means and standard deviations of estimated ϕ and ν parameters based on 1000 simulations.

sill	range	real (ϕ)	mean (ϕ)	s.d. (ϕ)	mean (ν)	s.d. (ν)
2.9	300	3.22e-05	4.29e-05	2.85e-05	0.612	0.131
2.9	10	0.02900	0.01177	0.00683	0.516	0.152
2.0	10	0.02000	0.00826	0.00518	0.518	0.156
2.9	166	0.00011	0.00012	0.00011	0.587	0.125
2.9	200	3.27e-05	8.64e-05	7.83e-05	0.592	0.125

A further evaluation of this technique was to assess spatial differences, and this was done through a graphical procedure applied over the Monte Carlo p-values. Suppose in each region the parameter vector θ_i has been estimated. We wish to compare this value with a fixed parameter value, θ_0, for the four subregions, where θ_0 may or may not be the same as θ_i. We set now 400 simulations of the θ_0 condition and calculated a set of p-values for each θ_i, obtained by the ranking procedure explained earlier in this paper. Under the null hypothesis of $\theta_0 = \theta_i$, the p-values should behave as Uniform on the range (0,1). And when a real difference exists, the p-values should be far from Uniform. This is based on the concept of Expected Significance Level (ESL), which was first introduced by Dempster and Schatzoff (1965). The results for $N = 20 \times 20$ and $N = 10 \times 10$ are shown in Figures 1-2. Note that for the case $N = 5 \times 5$, the procedure could not detect a real difference (for example, the case range=200, 166), though found the difference when this is big enough (range=166, 10). The corresponding line for the null hypothesis of no difference is given by circles and behaved at all times as Uniform. Lines marked by crosses indicated the p-values when a real difference existed. If they were compatible to the Uniform distribution, the test could not identify spatial differences. In general, the procedure worked well for any case, particularly when using bigger grid sizes.

The above set of simulations were also considered to evaluate the proposed test based on the equality of the regression coefficients, following Rao (1973). To analyze the behaviour under the null hypothesis of *no difference among parameters for the* $k = 4$ *subregions*, i.e. to evaluate the type I error, we considered the same parameter combinations in the four subregions and performed 1000 simulations, under the same conditions as above. The analysis of variance was then derived. The rejection rates at a significance level of 0.05, are shown in Table 3. To analyze the behaviour under the hypothesis of *differences among parameters for the 4 subregions*, i.e. to evaluate the power of the test, we considered the following subset of parameters: (2.9,300), (2.9,200), (2.9,10). In this case, two of the subregions were defined with one combination and the other two with other combination different from the first one. The procedure was repeated again 1000 times. The estimated powers are shown in Table 4. Looking at Tables 3 and 4, we can see that this test can be aimed to detect spatial differences when they are present and to detect stationarity when there are no spatial differences, when the grid size at which the process is observed is at least 10 × 10. The test gave misleading results for smaller sizes.

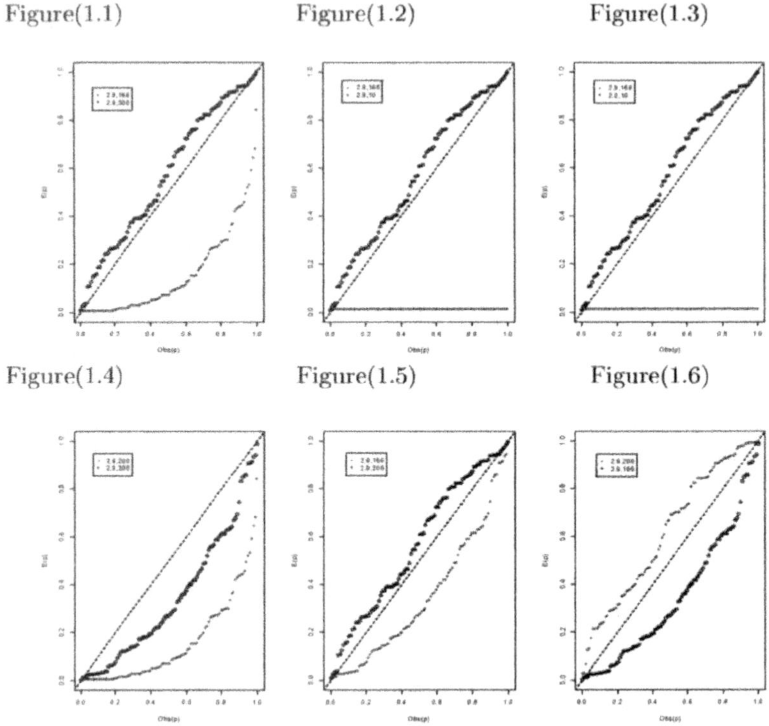

Figure 1. Observed versus Expected p-values under H_o (no differences) and H_a (real differences) in the case where the data are observed in a regular grid of $N = 20 \times 20$. (1.1) (2.9,166) vs (2.9,166) (Circles) and (2.9,300) (Crosses); (1.2) (2.9,166) vs (2.9,166) (Circles) and (2.9,10) (Crosses); (1.3) (2.9,166) vs (2.9,166) (Circles) and (2.0,10) (Crosses) (1.4) (2.9,200) vs (2.9,200) (Circles) and (2.9,300) (Crosses) (1.5) (2.9,166) vs (2.9,166) (Circles) and (2.9,200) (Crosses) (1.6) (2.9,200) vs (2.9,200) (Circles) and (2.9,166) (Crosses).

Table 3. Rejection rates (in percent) at 0.05 (Type I error) under the null hypothesis for the test based on equality of regression coefficients.

sill	range	$N = 20 \times 20$	$N = 10 \times 10$	$N = 5 \times 5$
2.9	300	1.8	2.9	12.3
2.9	10	2.8	4.1	19.1
2.0	10	3.0	4.7	18.3
2.9	166	1.4	3.1	20.6
2.9	200	1.6	2.9	16.7

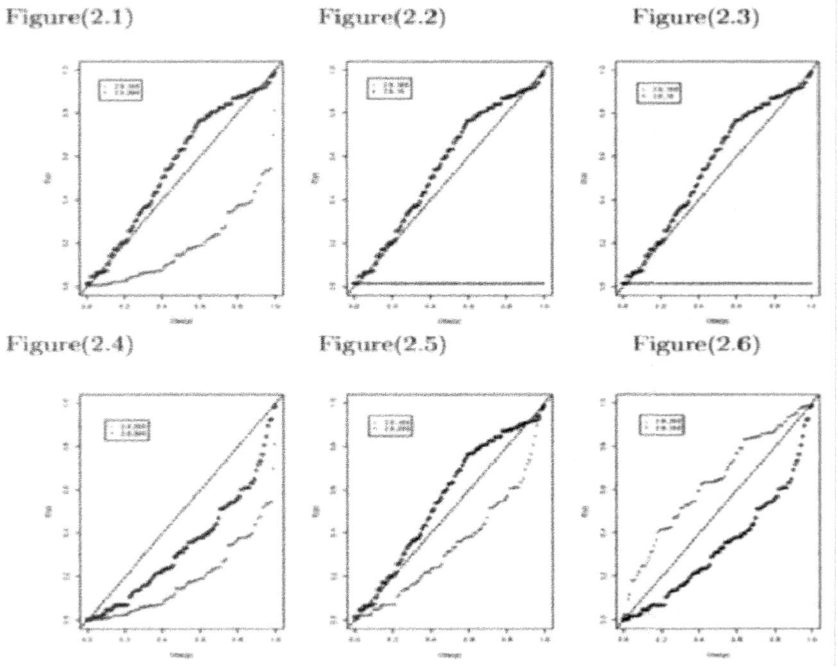

Figure 2. Observed versus Expected p-values under H_o (no differences) and H_a (real differences) in the case where the data are observed in a regular grid of N = 10 × 10. See Figure 1 for the label of each subfigure.

Table 4. Estimated powers (in percent) at 0.05 under the alternative hypothesis for the test based on equality of regression coefficients.

(sill, range)	$N = 20 \times 20$	$N = 10 \times 10$	$N = 5 \times 5$
(2.9,300) vs (2.9,200)	98.3	88.6	37.7
(2.9,300) vs (2.9,10)	99.2	92.2	40.2
(2.9,200) vs (2.9,10)	99.1	89.6	39.1

5. CONCLUSIONS AND DISCUSSION

In this paper we have shown a procedure to deal with nonstationarity based on the spectral representation of a nonstationary process and on a particular property of the Matérn family of spectral densities. The test has been shown to detect differences, when they really exist and to detect stationarity when it should. However, in real-life problems, the nonstationarity might be smooth, regions might not be well known and the date might be irregularly spaced. Some of these issues have been considered in the paper, but we have only analyzed a limited number of possible scenarios, and ideally we should take into account much more possibilities.

ACKNOWLEDGEMENTS

This research has been partially supported by grant BFM2001-3286. The referees are acknowledged for their helpful comments and suggestions.

REFERENCES

1. Cressie, N.A.C. (1993). *Statistics for Spatial Data*, Wiley, Revised version.
2. Dempster, A.P. and Schatzoff, M. (1965). Expected Significance Level as a sensitivity index for test statistics. *Journal of the American Statistical Association*, **60**:420-436.
3. Fuentes, M. (2001). A high frequency kriging approach for nonstationary environmental processes. *Environmetrics*, **12**:469-483.
4. Fuentes, M. (2002). Spectral methods for nonstationary spatial processes. *Biometrika*, **89**. To appear.
5. Fuentes, M. and Smith, R. (2002). A new class of models for nonstationary processes. Submitted.
6. Rao, C.R. (1973). *Linear Statistical Inference and its Applications*. 2nd Edition, Wiley, New York.
7. Renshaw, E. (2002). Two-dimensional spectral analysis for marked point processes. *Biometrical Journal*, **6**:718-745.
8. Sampson, P. and Guttorp, P. (1992). Nonparametric estimation of nonstationary spatial covariance structure. *Journal of American Statistical Association*, **87**:108-119.
9. Stein, M. (1999). *Interpolation of Spatial Data: some theory for kriging*. Springer-Verlag, New York.

OPTIMIZATION OF AN ESTUARINE MONITORING PROGRAM: SELECTING THE BEST SPATIAL DISTRIBUTION

S. Caeiro[1], L. Nunes[2], P. Goovaerts[3], H. Costa[4], M.C. Cunha[5], M. Painho[6], L. Ribeiro[7]

[1]IMAR, Depart. of Exact and Technological Sciences of the Portuguese Distant Learning University, .R. Escola Politecnica, 147, 1200 Lisbon, Portugal. scaeiro@univ-ab.pt; [2]CVRM, Faculty of Marine and Environmental Sciences, University of Algarve, Portugal, Campus de Gambelas, 8000 Faro, Portugal, lnunes@ualg.pt; [3] Biomedware, Inc. 516 North State Street, Ann Arbor MI 48104, USA, goovaerts@biomedware.com; [4]IMAR, Faculty of Science and Technology of the New University of Lisbon, Quinta da Torre, 2829-516 Caparica, Portugal, mhcosta@fct.unl.pt; [5]Department of Civil Engineering, University of Coimbra, Pinhal de Marrocos, 3030 Coimbra, mccunha@dec.uc.pt; [6]ISEGI/CEGI, Institute for Statistics and Information Management of the New University of Lisbon, Campus de Campolide, 1070 – 312, Lisboa, Portugal, painho@isegi.unl.pt; [7]CVRM, Lisbon Technological Institute, Technical University of Lisbon, R, Rovisco Pais,1096 Lisboa Codex, nlrib@alfa.ist.utl.pt

Abstract: Monitoring estuarine programs are fundamental to evaluate pollution abatement actions, fulfillment of environmental quality standards and compliance with permit conditions. Their sampling designs should provide statistically unbiased estimates of the status and trends with quantitative confidence limits on spatial scale. The aim of this work is to select a subset of monitoring sampling stations based on locations from an extensive sediment campaign (153 sites) in the Sado estuary (Portugal). In each location three sediment parameters were determined with the objective of defining spatially homogenous environmental areas. The new monitoring program is based on fewer and on the most representative monitoring stations inside each homogeneous environmental area for their future contaminant assessment. Simulated annealing was used to iteratively improve on the mean square error of estimation, by removing one station at a time and estimating it by indicator kriging using the remaining stations in the sub-set, within a controlled non-exhaustive looping scheme. Different sub-set cardinalities were tested in order to determine the optimal cost-benefit relationship between the number of stations and monitoring costs. The model results indicate a 60 station design to be optimal, but 17 additional stations were added based on expert criteria of proximity to point sources and characterization of all homogenous areas.

X. Sanchez-Vila et al. (eds.), geoENV IV – Geostatistics for Environmental Applications, 355-366.

Key words: Optimization, monitoring sampling, indicator kriging, estuarine sediments.

1. INTRODUCTION

Estuaries are coastal transitional water bodies with natural resources of high preservation values, providing important habitats for different species of organisms. The uses inside the estuary and around it have impacts on the water and sediment quality that may put at risk the equilibrium of the ecosystem. Environmental management of these ecosystems cannot be conducted effectively without reliable information on changes in the environment and on the causes of those changes. Ecological monitoring programs can represent an important source of that information. However many of the existing programs are not effective. To assure effectiveness, monitoring programs should be well designed, to enable the statistical analysis and interpretation needed to relate cause and effects (Olsen *et al.*, 1999 and Vos *et al.*, 2000).

The reliability of the sampling design depends on such a large degree on the sampling spatial distribution and size that their importance should not be underestimated (Haining, 1990). One or more of the following principles could govern the size of the sample (Cochran 1977; Clark and Hosking 1986; Strobel *et al.*, 2000):i) the required sampling size can be found if we have reasonable estimates of the population variance measured through a preliminary pilot survey; ii) certain statistical tests require a reasonable sample size; although no fixed minimum can be stated, a sample size of at least 30 is usually employed; iii) too large sample implies a waste of resources, and too small diminishes the utility of the results; iv) finance and time may dictate a certain maximum sample size.

In ecosystems like estuaries the spatial variability of key ecological indicators could be a measure to determine the appropriate monitoring sampling design (Strobel *et al.*, 2000).

The kriging interpolation is very useful to minimise the estimation variance for any fixed sampling design. The plot of the maximum value of the minimised estimation variance against sampling interval, or sample size, can be used to select sample size to achieve a required level of precision (Haining, 1990). For operational, economic or political reasons sometimes sampling sites for monitoring must be reduced and resource allocation optimized (Cochran, 1977). Optimal sampling scheme can then be designed by deleting sites from a current network so as to minimize the variance of estimation error, which means deleting the site that can be predicted best from the remaining sites (Cressie, 1993). Clever search algorithms like simulated annealing can then help designing the best sampling scheme. Difficulties usually arise in finding an optimal sampling plan and optimal kriging weights. Sampling plans can be important factors when looking for optimal spatial designs. Using the mean-squared prediction error of predictors, the rate of convergence to zero is faster for stratified random sampling than random or systematic random sampling designs (Cressie, 1993).

The sampling optimality criteria should not only be statistical but also cost related or economical (Cochran, 1977, Cressie, 1993, Vos *et al*, 2000). Sampling and parameters measurement costs are very important limitations and should be taken into account in the optimization procedure.

The aim of this work is to select, due to budget constrains, a subset of monitoring sampling stations from an extensive stratified random campaign of estuarine sediments. This subset will be used to assess Sado Estuary sediment contamination in management areas previously delineated. Spatial simulated annealing was used to optimize the sample locations. These data will be further integrated in an environmental management system for Sado Estuary.

2. CASE STUDY

The Sado Estuary, located in the West Coast of Portugal, is the second largest in Portugal with an area of approximately 24,000 ha. The estuary comprises the Northern and the Southern Channels, partially separated by intertidal sandbanks. Most of the water exchange is made through the southern Channel. The estuary is linked to the ocean by a narrow and deep channel that makes a major contribution to the general pattern of the estuarine circulation (Neves, 1986). Most of the estuary is classified as a Nature Reserve. There are many industries mainly on the northern margin of the estuary. Furthermore the harbour associated activities and the city of Setúbal along with the mines on the Sado watershed also releases contaminants into the estuary. In other areas around the estuary, intensive farming, mostly rice fields, is the main land use together with traditional saltpans and increasingly intensive fish farms. Most of these activities have negative impacts on water, sediment and biotic communities namely because they discharge to the estuary contaminants like heavy metals, or organic compounds (Caeiro *et al.*, 2002b).

3. METHODS

3.1 Sediment Homogenous Areas Delineation

In a first extensive campaign 153 sediment locations were sampled for analysis of properties of general characterisation: fine fraction (FF), organic matter (OM), and redox potential (Eh). These key ecological parameters explain main variations in the type and behaviour of benthic organisms as well as contaminant mobility/accumulation (Rodrigues and Quintino, 1993). One method of determining sample size for multiple parameters assessment, is to specify margin error for the items that are regarded as most vital to the survey (Cochran, 1977). A systematic unaligned sampling design with a grid size equal to 0.365 km^2 was used based on prior information on the spatial variation of sediment granulometry (Figure 1) (Caeiro *et al.*, 2002a).

This extensive campaign was intended to help defining homogeneous areas (future management areas) for Sado Estuary within which contamination would be monitored using smaller sample sets.

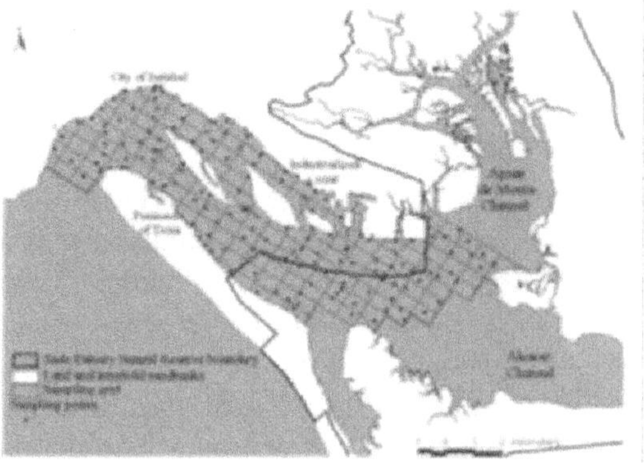

Figure 1. Sado Estuary sediment sampling design (Adapted from Caeiro *et al.*, 2002a).

These homogenous areas were delineated in 5 steps based on grouping individual sampling sites that have similar physicochemical properties while being geographically close (Caeiro *et al.*, submitted): 1)Principal component (PC) extraction of the 3 sediment properties variability (FF, OM and Eh); 2)Variogram fitting of a spherical model to 1st PC factor scores; 3)Dissimilarity matrix determination; 4)Cluster analysis using the complete linkage rule on the dissimilarity. matrix to estimate the probability of occurrence of four selected clusters at sampled stations; 5)Indicator kriging to interpolate these probabilities at unsampled stations; 6)Maximum likelihood classification of these unsampled stations.

The dissimilarity between any two sampling sites *i* and *j* (step 3) was computed following Oliver and Webster (1989) equation with spherical model adjustment (Goovaerts, 1997) to take into account the form of spatial variation. Step 5, started with an indicator coding of classification results (x_α) at each sampled station x_α:

$$i(x_\alpha; z_l) = \begin{cases} 1 \text{ if } z(x_\alpha) = z_l \\ 0 \text{ otherwise} \end{cases} \quad l=1,\ldots, L \tag{1}$$

where *L* is the number of clusters (four selected). For each cluster z_l, experimental indicator variograms are then computed and modelled:

$$\gamma(\mathrm{h}; z_l) = \frac{1}{2N(\mathrm{h})} \times \sum_{\alpha=1}^{N(\mathrm{h})} \left[i(x_\alpha; z_l) - i(x_\alpha + \mathrm{h}; z_l) \right]^2 \tag{2}$$

The probability of occurrence of the *l*-th cluster at the unsampled station *x* is estimated as a linear combination of indicator data:

$$\hat{p}(x; z_l \mid B) = \sum_{\alpha=1}^{n_c} \lambda(x_\alpha; z_l) \times i(x_\alpha; z_l) \tag{3}$$

where B is the set of n_c surrounding data $\{z(u_\alpha), \alpha=1,\ldots, n_c\}$. The weights $\lambda(x_\alpha; z_l)$ are solutions of an indicator kriging system and account for data configuration and spatial continuity of clusters as modelled by indicator variograms. In theory, indicator cokriging estimator is better than the indicator kriging estimator because it accounts for additional information available across categories. However, indicator cokriging improves little over indicator kriging according to Goovaerts (1994).

3.2 Optimization model

The stations that produce the lowest estimation error variance, estimated using cross-validation technique (Deutsch and Journel, 1998), result in a spatial distribution with the highest accuracy. The objective function considers a set, S, of all the original stations, with cardinality Ω, and take a subset, S', with cardinality ω, such that $\omega < \Omega$.

Minimize

$$s^2_{fp} = \frac{1}{\omega}\sum_{\alpha=1}^{\omega}\left[i(x_\alpha;z_l)-i^*(x_\alpha;z_l)\right]^2, \omega \in S', S' \subset S \qquad (4)$$

Subject to:

$$\Psi_{S'}(A;z_l) \approx \Psi_S(A;z_l) \qquad (5)$$

s^2_{fp} is the mean squared error of estimation and equal to the variance of the estimation error if zero mean estimation errors are considered (i.e. no bias). $i^*(x_\omega z_l)$ is the indicator kriging estimated value, $\psi_S(A,z_l)$ and $\psi_{S'}(A,z_l)$ are the marginal probabilities of finding stations with values in $]z_{l-1}, z_l]$ in the original data set and in the candidate solution, respectively.

The new design S' must reflect the sediment physical and chemical variability detected with the prior sampling campaign. Therefore we imposed the constraint that the proportions of monitoring stations in each of the identified homogeneous areas are similar to the proportions in the original sampling campaign (Table 1). Van Groenigen *et al.* (2000) also successfully used sampling constraints in spatial annealing to optimise sampling scheme. The condition is not equality because, for practical computation, floating-point variables equality is machine dependent and varies with the precision. Instead, $\Psi_{S'}(A,z_l)$ may be bounded, and the constraint becomes:

$$\Psi_S(A;z_l)(1-\delta) \leq \Psi_{S'}(A;z_l) \leq \Psi_S(A;z_l)(1+\delta) \qquad (6)$$

A conditioning on the objective function with $\delta = 0.3$ was imposed. This condition is necessary to correct the bias introduced by variogram models fitting errors (when adjusting the theoretical models to the experimental variogram). If no conditioning is used increasing the number of stations will result in higher estimation error variances. This is due to the fact that at very low ω only stations with low estimation error in the optimal solution are included; as ω increases higher estimation error stations are included (Nunes *et al.*, unpublished).

Simulated Annealing (SA) algorithm with the Metropolis iterative improvement procedure (Metropolis *et al.*, 1953) was then used to solve the optimisation model. This procedure generalises by incorporating controlled uphill steps (to worse solutions). The procedure states the following: consider one small random change in the system at a certain temperature (the control parameters t is usually termed temperature); the change in the objective function is ΔOF; if $\Delta OF \leq 0$, then the change in the system is accepted and the new configuration is used as the starting point in the next step; if ΔOF > 0 then the probability that the change is accepted is determined by $P(\Delta OF) = exp(-\Delta OF/t)$; a random number uniformly distributed in the interval (0,1) is taken and compared with the former probability; if this number is lower than $P(\Delta OF)$ then the change is accepted. The SA algorithm runs in the following way: i) the system is *melted* at a high temperature (initial temperature, t_I); ii) the temperature is decreased gradually until the system *freezes* (no further OF change occurs); iii) at each iteration the Metropolis procedure is applied; iv) if any of the stopping criteria is reached the algorithm is stopped and the best solution found is presented. The generic SA algorithm for a minimisation, considering a neighbourhood structure N, a current solution X, a best solution found so far X_{best}, a solution space χ, a α temperature decrease control parameter and an objective function OF has the following pseudo-code.

```
Select an initial solution X_best;
Select an initial temperature t_I>0;
Select a temperature reduction factor;
Repeat
  Repeat
    Randomly select X∈ N(X_best);
     ΔOF = OF(X) – OF(X_best);
     IF ΔOF<0 then
    X_best = X
       else
   generate random z uniformly in (0,1);
         if z < exp(-ΔOF/t) then X_best = X;
  Until iterations = max_iterations
  Set t = αt;
Until stopping condition = true;
X_best is the optimal solution found.
```

In order to speed-up the process several improvements have been proposed, namely by limiting the number of iterations at each temperature, i.e., defining the number *max_iterations*. The dimension of the Markov chain has been proposed to be a function of the dimension of the problem (Kirkpatrick *et al.*, 1983): temperature is maintained until 100Ω solutions (iterations), or 10Ω successful solutions have been tested, whichever comes first. Ω stands for the number of variables (stations) in a problem.

A specific computer code in FORTRAN that incorporates both the estimation error variance and the SA algorithm was developed by (Nunes *et al.*, unpublished) to optimise location problems and adapted to this specific problem. Runs were made on PC Intel 2000 MHz machines.

Fourteen different monitoring network dimensions (cardinality of S': ω) were tested, {25,30,35,40,45,50,60,70,80,90,100,110,120,130} according to the following scheme: i) impose a number of monitoring stations (ω) to be included in the new design; ii) find the optimal allocation solution with SA; iii) increase ω and return to i). SA solutions are considered optimal when more than 70% out of 20 consecutive runs with the same objective function conditions (ω, δ) and SA parameters have the lowest and equal s_{fp}^2 value.

A complementary analysis comparing the loss in accuracy versus reduction in exploration costs as stations are removed was also performed. For that purpose a cost per sampling was computed based on the previous sampling campaign and laboratory analysis costs (official costs of the laboratory where the analysis are going to be made - ControLab, lda.): i) linear distance between n sampling point: n/study area (56 km^2); ii) boat velocity: 12,8 km^2; iii) hour of work per day:7 h/day; iv) time for sampling: 20 min; v) Boat cost per day: 250 Euros; vi) Cost per total contaminant analysis: 500 Euros (discount: 25 % from 20 to 50 stations, 30 % from 55 a 100 stations and 40 % from 105 to 135 stations).

4. RESULTS AND DISCUSSION

Table 1 lists four different physical and chemical homogeneous areas (clusters) based on the sampling campaign data and results from hierarchical classification (step 4), and their frequencies in the study area.

For each cluster, the indicator variogram was computed along four directions and a geometric anisotropic model was fitted (Figure 2).

Table 1.Physical and chemical parameters of each cluster and their frequency.

	Clusters (s)			
Sediment Parameter	High organic load (z_1)	Medium high organic load (z_2)	Medium organic load (z_3)	Low organic load (z_4)
OM (%)	8.6 ± 2.4	4.2 ± 1.4	1.9 ± 0.7	0.9 ± 0.3
FF (%)	60.4 ± 27	21.7 ± 11.8	9.1 ± 7.8	1.5 ± 1.3
Eh (mV)	-278.9 ± 68.6	-178.8 ± 72.6	-137.4 ± 50.9	74.4 ± 49
Freq. (%)	11.76	37.91	23.53	26.80

Figure 3 shows the spatial accuracy plotted versus the monitoring network dimension. Beyond 60, each new added station had little effect on the monitoring spatial accuracy (s_{fp}^2). Sixty is therefore considered as the optimal ω value. The resulting network was overlaid on the sediment homogenous areas within the estuary coast line (Caeiro *et al.*, 2002a) using Arcview/arcinfo 3.2 GIS software (Figure 4a). In cluster one and two (z_1 and z_2) the estimation errors are higher, therefore leading the optimisation algorithm to select preferentially the two remaining clusters with lower estimation errors. These clusters are therefore more densely sampled than in the original data set, as a way to compensate for the bias introduced. Also when high or low values of a cluster are grouped in small areas scattered in the study area, their relative frequencies are low or data values is too random, the variogram fitting becomes difficult and prone to error. The result is the fitting of theoretical variograms that only roughly approximate the real variability and large estimation errors. This does not hinder the geostatistical method, but justifies the need to impose reproduction of the original proportions (Nunes *et al.*, unpublished).

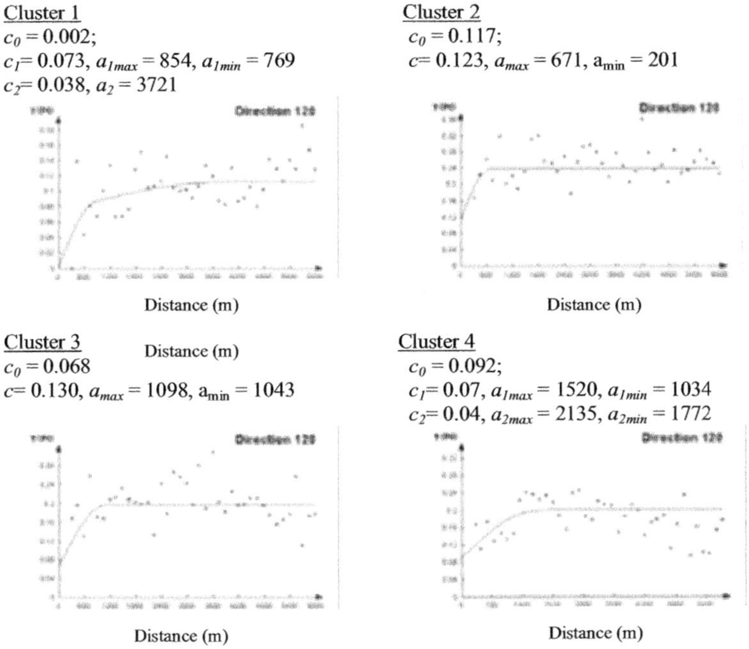

Figure 2. Cluster experimental directional variograms and spherical model fitted for 120°, the major direction of anisotropy. Other directions (not shown) included 30°, 75° and 165°.

Figure 4a) indicates that not all the homogenous areas are sampled in the optimal scheme solution, in particular areas belonging to clusters with high organic load (1 and 2), for the reasons explained earlier. Most of these cluster 1 and 2 areas are near contaminant point sources, mainly in the North Channel. Thus 17 stations were added to the optimal ω value according to expert knowledge aiming to characterize the impact of those point sources and homogenous areas not included in the optimised network (Fig. 4b).

The number of stations to evaluate contamination in the study area (77 stations/56 km^2, corresponding to 1.38 stations/km^2) is within the average of sediment sample size of Environment Monitoring Assessment Program (EMAP) of United States Environmental Protection Agency (USEPA) for small estuaries. The sample sizes for the different estuaries of EMAP vary from 0.11 to 4.16 stations/km^2 (Strobel *et al.*, 2000). Such a wide interval might be related to the spatial variability of sediment parameters in each coastal zone, which is caused by differences related to geomorphological, biological and human pressures.

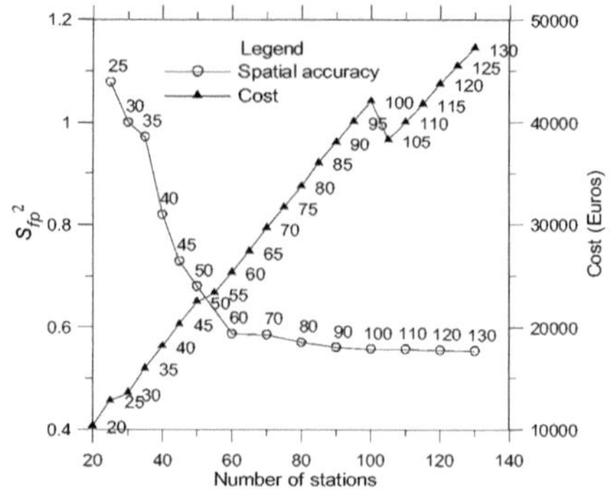

Figure 3. Estimation error variance and cost versus number of monitoring stations.

The exploration costs analysis (Figure 3) showed that costs are always increasing and only for large number of stations (from 110 to 115) does the cost decrease. Indeed the cost of contamination concentration analyses has a high weight in the total cost and only for 105 laboratory analysis does the laboratory discount significantly affect the total cost.

Although seventy-seven stations still represent a high cost (about 60 % of stations total number cost), this budget figure is considered necessary at the present time for a contamination assessment. For any future long-term monitoring program to assess estuary ecological condition, a reduced number of sampling sites should be chosen. Thirty sampling stations should represent a good number for a monitoring program since: i) each of the 19 management areas could be sampled at least at one location or two in case of larger areas, ii) it is a statistical minimum required; iii) the cost is not too high (and similar to 25 stations – see Figure 3). Nevertheless, 30 stations will represent a 40 % loss in spatial accuracy (see Figure 3).

In the future developments for a monitoring program of the environmental management system of the estuary, the model should take into account two strata in the study area. One in the North Channel near pollution sources and the other in the South where the hydrodynamics is highest and the pollution sources are non-point. Vos *et al.*, (2000) discuss that the identification of relevant subsystems or strata for monitoring purpose, is very important to maximise diagnostic of ecological changes. In these strata changes in the anthropogenic inputs or "controlled variables" are expected. Also, once contaminants have been measured at the 77 sampling points a new optimisation criterion could be developed to sample preferentially areas with high priority (e.g. high concentrations). Van

Groenigen *et al.* (2000) used a spatial weight function in spatial simulated annealing that allows distinguishing between areas with different contamination priorities. This could be achieved through Weighted Mean of Shortest Distance; i.e. the fitness is extended with a location-dependent weighing function, or/and using probability maps of contamination and indicator kriging. In particular in our case the weight function should take into account small areas and distance to contaminant sources.

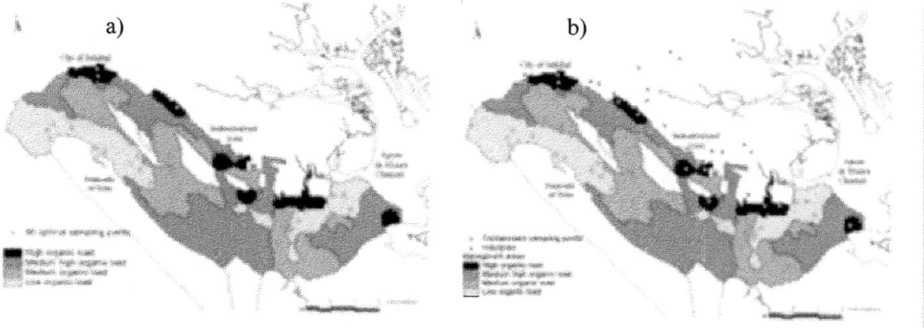

Figure 4. Monitoring networks a) for ω value = 60 stations; b) with 60 optimal stations and additional expertise criteria (17) (Location of industries from Araujo *et al.* (2002).

5. CONCLUSIONS

Monitoring programs should be planned in order to provide quantitative and scientific assessments of pollutants' complex effects on these systems. Optimal sampling designs for ecological condition assessment should take into account not only statistical criteria but also historical knowledge about the study area. In particular estuaries have always areas with different priorities (e.g. human pressures or more sensitive areas). From an extensive campaign including 153 sampling points, a sampling design with 77 stations was selected for sediment contaminant assessment in Sado estuary. This selection was based on minimization of indicator kriging mean square error estimation and expertise knowledge. For a future long–term monitoring program of the estuary condition assessment a reduced subset of 30 stations should be chosen based on definition of contaminant priority areas.

ACKNOWLEDGEMENTS

The two first authors had a PRODEP scholarship. The research was partially financed by the Portuguese Science and Technology Foundation (Research Project BSE/35137/99-00).

REFERENCES

1. Araujo, R., Vasconcelos, L. and Painho, M. SADIND - Sistema de Visualização Interpretativa para a Gestão Ambiental, *Biologia* (In Press), 2002.
2. Caeiro, S., Goovaerts, P., Painho, M., Costa, M. H. and Sousa, S. Optimal spatial sampling design for mapping estuarine sediment management areas. Ruiz, M., Gould, M., and Ramon, J. (Ed.) 5th AGILE Conference on Geographic Information Science; Palma, Spain. Universitat de les Illes Balears, 389 – 396, 2002a.
3. Caeiro, S., Painho M., Costa, M. H. and Ramos, T. B. Sado Estuary Ecosystem: a management methodology. Fernando Pessoa University. Duarte, P. (Ed.) Proceedings of International Conference on Sustainable Management of Coastal Ecosystems; Porto, Portugal. Fernando Pessoa University, 2002b.
4. Clark, W. and Hosking, P. *Statistical Methods for Geographers*. John Wiley & Sons, 1986.
5. Cochran, W. *Sampling Techniques*. 3rd edition ed. New York: John Wiley & Sons, 1977.
6. Cressie, N. *Statistics for Spatial Data*. Revised Edition. John Wiley & Sons, 1993.
7. Deutsch, C. and Journel, A. G. GSLIB. *Geostatistical Software Library and Users's Guide*. 2nd edition. Oxford University Press, 1998.
8. Goovaerts, P. Comparative performance of indicator algorithms for modeling conditional probability distribution functions. *Mathematical Geology*, 1994; 26(3):389-411.
9. Goovaerts, P. *Geostatistics for Natural Resources Evaluation*: Oxford University Press, 1997.
10. Haining, R. *Spatial Data Analysis in the Social and Environmental Sciences*. Cambridge University Press, 1990.
11. Kirkpatrick, S., Gellat, Jr. C. D. and Vecchi, M. P. Optimization by simulated annealing: Science, 1983; 220:671-680.
12. Metropolis, N., Rosenbluth, A. W., Rosenbluth, M. N., Teller, A. H. and Teller, E. Equation of state calculations by fast computing machines: *The Journal of Chemical Physics*, 1953; 21: 1087-1092.
13. Neves, R. J. J. Hidrodynamical Modelling as a Toll in Waste Disposal Selection. A Case Study on Sado Estuary. Kullenberg (ed.). *The Role of the Oceans as a Waste Disposal Option*, 1986, pp. 563-576.
14. Oliver, M. A. and Webster, R. A Geostatistical Basis for Spatial Weighting in Multivariate Classification. *Mathematical Geology*, 1989; 21(1):15-35.
15. Olsen, A. R., Sedransk, J., Edwards, D., Gotway, C. A., Liggett, W., Rathbun, S., Reckhow, K., Young, L. Statistical issues for monitoring ecological and natural resources in the United States. *Environmental Monitoring and Assessment*, 1999; 54:1 – 45.
16. Rodrigues, A. and Quintino, V. Horizontal Biosedimentary Gradients Across the Sado Estuary, W. Portugal. *Netherlands Journal of Aquatic Ecology*, 1993; 27(2-4):449-464.
17. Strobel, C., Paul, J., Hughes, M., Buffum, H., Brown, B. and Summers, K. Using Information on Spatial Variability of Small Estuaries in Designing Large-scale Estuarine Monitoring Programs. *Environmental Monitoring and Assessment*, 2000; 63:223-236.
18. Van Groenigen, J. W., Pieters, G. and Stein, A. Optimizing spatial sampling for multivariate contamination in urban areas. *Environmetrics*, 2000; 11:227-244.
19. Vos, P., Meelis, E. and Keurs, W. J. A framework for the design of ecological monitoring programs as a tool for environmental and nature management. *Environmental Monitoring and Assessment*, 2000; 61:317-344.

GEOSTATISTICAL ANALYSIS OF THREE DIMENSIONAL CURRENT PATTERNS IN COASTAL OCEANOGRAPHY: APPLICATION TO THE GULF OF LIONS (NW MEDITERRANEAN SEA)

P. Monestiez[1], A. Petrenko[2], Y. Leredde[2] and B. Ongari[2]

[1]*Unité de Biométrie, INRA, Domaine St Paul, Site Agroparc, 84914 Avignon Cedex 9, France.*
[2]*Centre d'Océanologie de Marseille, LOB, Campus de Luminy, Case 901, 13288 Marseille Cedex 9, France*

Abstract: Two geostatistical methods are used to map hydrodynamic patterns in the Gulf of Lions (Mediterranean Sea). The aims are both methodological – mapping vectorial data raises some difficulties – and applied – sampling schemes from boat cruise are not convenient to get maps or to compare with model output. From a large data set that was obtained from a shipboard ADCP (Acoustic Doppler Current Profiler), stationary isotropic geostatistical models were fitted for several horizontal layers. Vectors of current are characterized by two components or by intensity and direction. A linear model of coregionalization was used on vector components and compared to a second approach that considers vectors as elements of the complex plane ¢. Then interpolated maps were computed by ordinary cokriging and by ordinary kriging in the complex plane for two different depths. Although some difficulties remain unsolved due to the effect of time in the sampling scheme or to some constrains (physical equations and limit conditions) that currents must satisfy, the first results are already satisfactory and allow a better understanding of spatial patterns than the simple plots of original data. The same data set were also used in parallel for hydrological modelling using a physical circulation model. Then the complex kriging approach was used to address the spatial analysis of the residuals, i.e. difference between predicted and observed current vectors. Residuals were highly structured in space.

X. Sanchez-Vila et al. (eds.), geoENV IV – Geostatistics for Environmental Applications, 367-378.

Geostatistical methods confirmed their potential as complementary tools in physical circulation model validation and error reduction for current pattern predictions.

1. INTRODUCTION

The hydrodynamics of coastal areas is a central issue to assess and to understand spatio-temporal distribution of biological and chemical parameters related to resources, global fluxes or pollution assessment. The Gulf of Lions system is mainly forced by strong physical and meteorological influences as the Rhone river plume, very strong winds and mesoscale current patterns (Millot, 1990). Preliminary studies at one permanent monitoring site (SOFI monitoring station) and on the whole gulf during cruises (MOOGLI cruises) were not able to deal with spatial pattern descriptions but showed their importance and the necessity to get accurate measurement of currents for the whole area in a short time interval (Petrenko et al., 2002). This was done during the ten SARHYGOL cruises in 2000 and 2001 to identify main patterns and seasonal effects. In this study we focused on one cruise (June 14-15, 2000) which was used for a physical modeling study, that allowed us to compare data interpolation and circulation model output.

It is not frequent in geostatistics to deal with directional or vectorial data. Lajaunie and Béjaoui (1991) proposed a kriging in the complex plane and developed some theory on the spatial covariance models. It is probably the first example of covariance modelling in the complex plane. They applied it to a case study on tidal data which was very similar to our problem. In fact, their data derived from a physical model solved by a finite element method and then sampled to test the geostatistical approach. However, they did not restraint the model to a variogram structure in the complex plane which would have been a less rich model. Some other cases can be found in Chilès and Delfiner (1999) with the interpolation of directional derivatives of a variable. They used a model of coregionalisation and then the cokriging. Grzebyk (1993) in a different way developed some theory in the field, but did not work on vectorial structures. His main results were an extention of the classical linear model of coregionalization for two or more variables to complex framework in order to model asymmetrical crosscovariances. A synthesis of all theses first approaches is given in Wackernagel (1995) completed by some considerations and analysis on the available models and methods.

2. DATA, MODELS AND METHODS

2.1 Data set

Current data are measured by shipboard ADCP (Acoustic Doppler Current Profiler) following a broken line route all over the Gulf of Lions (Figure 1) in less than 48 hours. The sampling frequency is dense along the line (2773 points) and in depth, every 4 meters from 8 meters to 240 meters deep, with a limit of a few meters over the bottom. For several reasons and after checking the experimental variograms on the total data set of 2773 locations, we regularized the original data by pooling 8 successive points. In place of one measurement every minute and every 250 m, we will work on one measurement every two kilometers. The first advantage is to reduce the 2773 measurement to 367 with very small changes on the experimental variograms (no nugget effect and linear behavior close to the origine for most of them). The second advantage is to reduce the number of redundant data that are very close along the trajectory in kriging and cokriging, and to give consequently the possibility to enlarge the neighborhood. The third reason comes from further comparison to physical circulation model that needs to smooth very local turbulence patterns and to compare data and model output on similar supports. Although patterns seems to be very homogeneous for the greater depths, no regularization was done along the vertical axis to keep a precise description of the variabilities close to the surface.

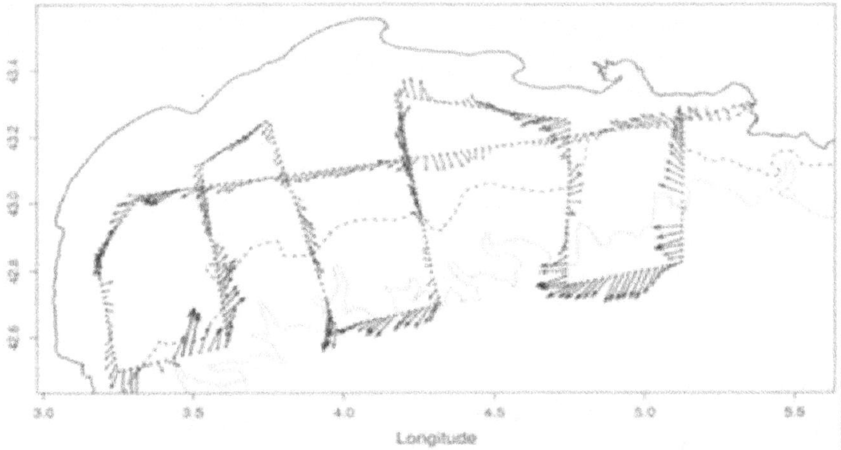

Figure 1. Trajectory of the cruise on June 14-15, 2000 from and to the port of Marseilles. 2773 localized points with 58 different depths were measured but only the 367 regularized points are plotted. The coast line is plotted as solid line. Dashed line is the 100m-deep isoline and dotted line the 500m-deep isoline. Current data at the depth of 8m are symbolized by vectors.

2.2 Models and methods

Let be $Z_E(x_\alpha)$and $Z_N(x_\alpha) \in R^2$ the two East and North components of current vector measured at site x_α.

2.2.1 Linear model of coregionalization and cokriging on isotopic data

Variograms and crossvariograms are defined by

$$\gamma_{ij}(h) = \frac{1}{2}\mathrm{E}\big(Z_i(x) - Z_i(x+h)\big)\big(Z_j(x) - Z_j(x+h)\big) \text{ where } i,j \in \{E,N\}\times\{E,N\}$$

They are modelled using a linear model of coregionalization and fitted using the least squares procedure described in Goulard and Voltz (1992).

$$\boldsymbol{\Gamma}(h) = \sum_{u=0}^{S} \mathbf{A}_u g_u(h)$$

where $\boldsymbol{\Gamma}(h)$ is the 2 X 2 matrix of $\gamma_{ij}(h)$, S an adequate number of nested models, $\mathbf{A}_u$ are positive semi-definite matrices and $g_u(h)$ are normalized univariate variograms.

Then two ordinary cokriging are solved at a site x_o to compute for both the East and North components

$$Z_E^*(x_o) = \sum_{\alpha=1}^{n} \omega_\alpha^E Z_E(x_\alpha) + \sum_{\alpha=1}^{n} \omega_\alpha^N Z_N(x_\alpha)$$

$$Z_N^*(x_o) = \sum_{\alpha=1}^{n} \omega'^N_\alpha Z_N(x_\alpha) + \sum_{\alpha=1}^{n} \omega'^E_\alpha Z_E(x_\alpha)$$

where n is the number of isotopic data. We have the following conditions

$$\sum_{\alpha=1}^{n} \omega_\alpha^E = \sum_{\alpha=1}^{n} \omega'^N_\alpha = 1 \quad \text{and} \quad \sum_{\alpha=1}^{n} \omega_\alpha^N = \sum_{\alpha=1}^{n} \omega'^E_\alpha = 0$$

Whether the component k that we are interpolating is E or N, the ordinary cokriging system is given by

$$\begin{cases} \displaystyle\sum_{j\in\{\mathrm{E,N}\}} \sum_{\beta=1}^{n} \omega_\beta^j \gamma_{ij}(x_\alpha - x_\beta) + \mu_i = \gamma_{ik}(x_\alpha - x_o) & \text{for} \quad i \in \{\mathrm{E,N}\}; \alpha = 1, n \\ \displaystyle\sum_{\beta=1}^{\mathrm{n}} \omega_\beta^i = \begin{cases} 0 & \text{if } i \neq k, \\ 1 & \text{if } i = k, \end{cases} & \text{for} \quad i \in \{\mathrm{E,N}\} \end{cases}$$

where the left terms remain constant for a given neighborhood $x_i,\ldots,x_n$ and the right terms depend on $k \in \{\mathrm{E,N}\}$ x_o.

2.2.2 Complex random field and ordinary complex kriging

Let $Z(x)=Z_E(x) + i\ Z_N(x)$ be a random field that is defined on R^2 and has values in the complex plane ¢. The covariance and the variogram are then defined by

$$C(h) = \mathbf{E}\left[Z(x+h)\overline{Z(x)} \right] \quad \text{where } \overline{Z(x)} \text{ is the conjugate of } Z(x)$$

$$\gamma(h) = \frac{1}{2}\mathbf{E}\left[\left(Z(x) - Z(x+h)\right) \ \overline{\left(Z(x) - Z(x+h)\right)} \right]$$

$$= \frac{1}{2}\mathbf{E}\left[\left\| \left(Z(x) - Z(x+h)\right) \right\|^2 \right]$$

Modelling $\gamma(h)$ does not raise any specific difficulties because it is always real, and any model classically used in R may be used in ¢. It is not the case for the covariance function that is richer than the variogram and allows a real and an imaginary part. More details are given in Wakernagel (1995). In this study we limit ourselves to model a variogram structure after computing the experimental variogram on the norms of vector differences.

Complex kriging is then defined by

$$Z^*(x_o) = \sum_{\alpha=1}^{n} \omega_\alpha Z(x_\alpha)$$

In theory, the weights ω_α that are solution of a classical ordinary kriging system belong to ¢, but because the variogram model is real, all imaginary parts vanish in the solution. The kriging becomes vector as a vectorial sum of data vectors, weighted by real coefficients. That would not be the case with a covariance model including imaginary terms, and the kriging system resolution would lead to complex weights.

2.2.3 Circulation hydrodynamic model

In parallel to this study, a circulation model, named "SYMPHONIE", has been implemented on a slightly larger area including the Gulf of Lions and a deeper area east of Marseilles. This model, which is not the object of this paper, is described in Estournel et al. (2002). The hydrodynamic equations are implemented using a finite difference method on a grid with a horizontal lag of three kilometers and varying vertical lag adjusted on the depth and denser close to the sea surface. It starts from general conditions, and it is forced during two weeks before measurements by meteorological conditions over the area (wind and temperature) and by data on the input fluxes of the Liguro Provencal Current (coming from the East) and of the Rhone river. The output of this model, i.e. the circulation patterns, were then compared to the shipboard measures at the same dates. The residuals were obtained by differences between computed and measured currents, on both components, East and North, after a linear interpolation of model output between the computation nodes.

3. RESULTS

3.1 Vertical variogram structures

Experimental variograms were computed along the vertical using all data points, i.e. 2773 boat locations. Each measurement boat location was considered as a replicate, so no hypothesis of stationarity on the vertical direction was necessary. It is possible to get an experimental value of the variogram for every pairs of depths. On Figure 2, the variograms are displayed respectively for the depths 8 m, 24 m, 48 m and 80 m with all the other measured depths every four meters. For example, the variogram computed for differences of currents between 24 m and other depths is null for 24 m and increases for depths that are higher or lower. We can notice a non stationary pattern, with a larger variation of current component for the same depth difference when the pair is closer to the surface. Variogram lines for 48 m and 80 m show a smaller increase for small differences and a lower sill on the right than those for 8 m or 24 meters.

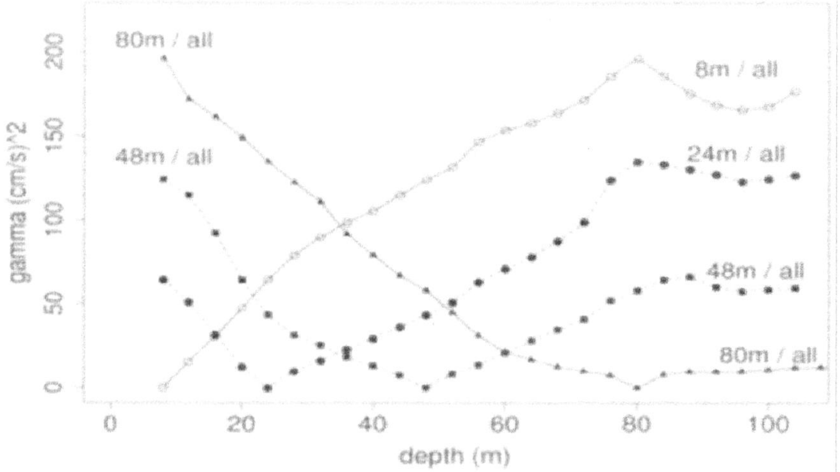

Figure 2. Experimental variograms on vertical direction. Semi-variance of the differences between the North components of current at 8~m and at all other depths (solid line and circles), and respectively at 24 m (dotted line filled circles), 48 m (dotted line squares), 80 m (solid line triangles).

It would be a difficult but possible task to model variograms including non-stationarity, in order to get a general 3D model of spatial variations. It was not done in this paper because the ADCP give systematically a dense sampling scheme along the vertical from the depth of 8 m to a depth close to the sea bottom. Consequently, spatial interpolations on the vertical direction do not pose difficulties.

3.2 Coregionalization and cokriging horizontal maps

A coregionalization model was fitted to horizontal spatial variation at several depths. Shown here are the results at two depths: 8 and 80 meters that are representative of behavior close to the sea surface and at large depths. Data at depths greater than 100 m were not processed because a too small part of the study area was concerned due to bottom profile.

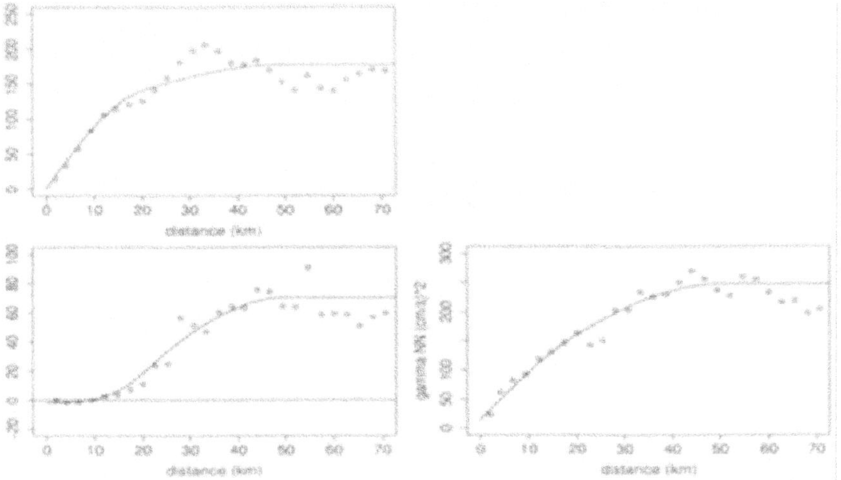

Figure 3. Experimental variograms and covariogram for North and East components of current at the depth of 8 meters. Distances are in km on the horizontal plane. The fitted model of linear coregionalization is plotted with solid lines.

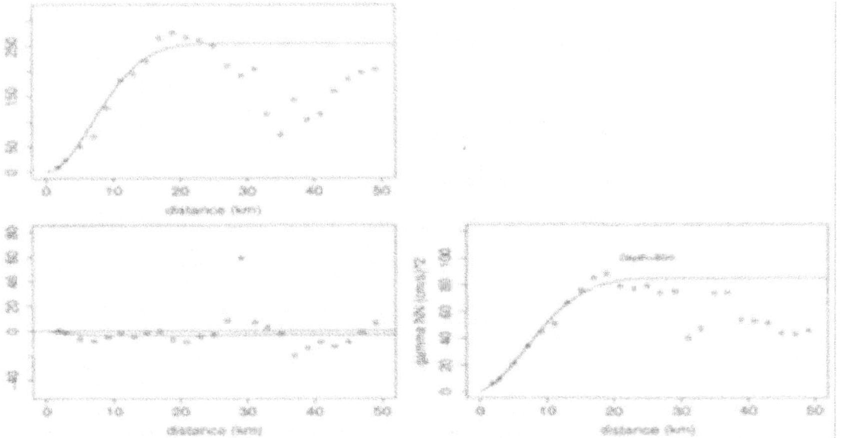

Figure 4. Experimental variograms and covariogram for North and East components of current at the depth of 80 meters. Distances are in km on the horizontal plane. The fitted model of linear coregionalization is plotted with solid lines.

The two variables are the East and the North components of the horizontal current. Experimental variograms and crossvariogram are shown in figure 3. We assumed isotropy. A model based on three elementary structures was fitted. The structures were a nugget effect and two nested spherical models of range 20 km and 50 km respectively. The nugget effect remains very small and differs slightly from zero for the variogram on North component. The cross variogram shows that spatial variations of North and East components are quite uncorrelated at short distance, but become correlated when distance increases, showing dependence patterns at the scale of 40 km or more.

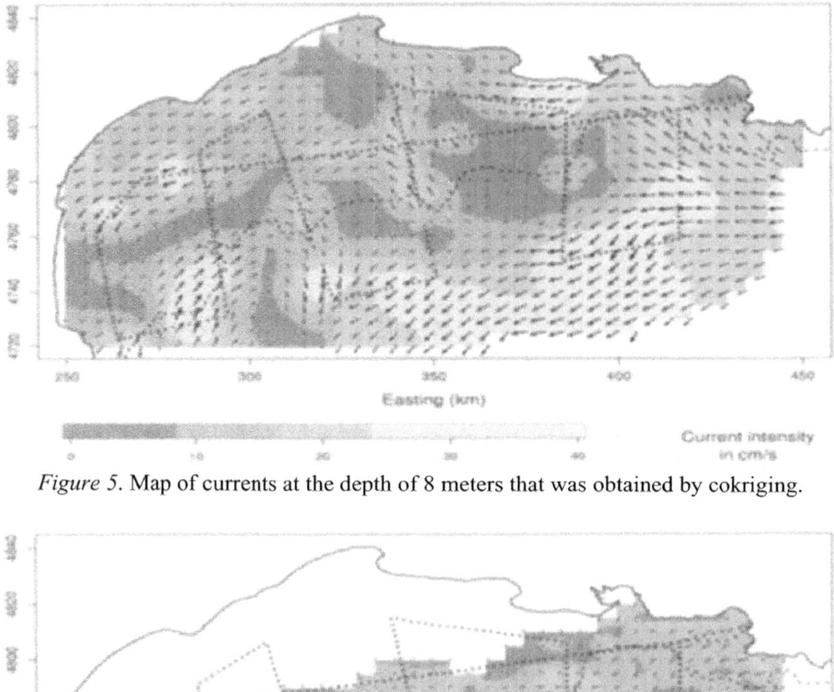

Figure 5. Map of currents at the depth of 8 meters that was obtained by cokriging.

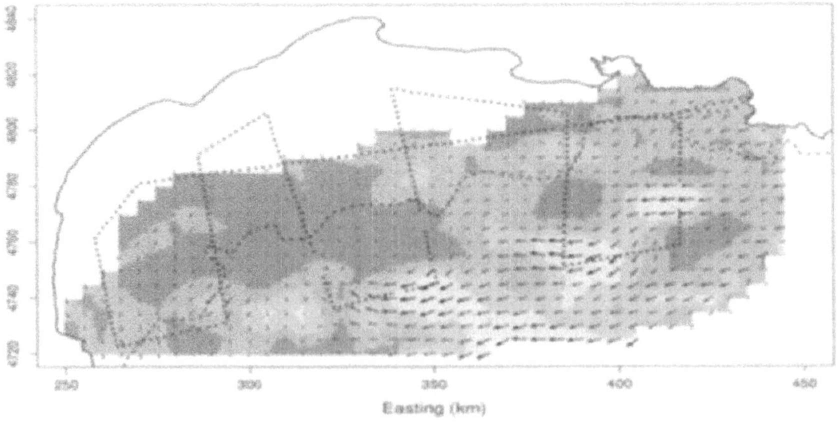

Figure 6. Map of currents at the depth of 80 meters that was obtained by cokriging. Image legend is the same than for figure 5.

At the depth of 80 meters, the cross structure vanish and the two variograms show more regularity close to the origin (Figure 4). The coregionalization was decomposed in a spherical model of range 25 km and a gaussian variogram model with a "range" parameter of 10 km. No nugget effect was needed.

In a second stage, cokriging of both N and E components were performed and the map of currents rebuilt from the two components (Figure 5 and 6). On a qualitative point of view the maps feature currents that are coherent with known circulation pattern in this area. Crude interpolation did not lead to obviously erroneous patterns as, for example, current pointing to the coast or currents converging to a single point.

3.3 Variogram and kriging in the complex plane

In the complex plane, experimental variograms corresponding to the variogram that was introduced in section 2.2.2 were computed and fitted, with a spherical model for the 8m depth, and with a nested model composed of a spherical and a gaussian variogram model for the 80 m depth (Figure 7).

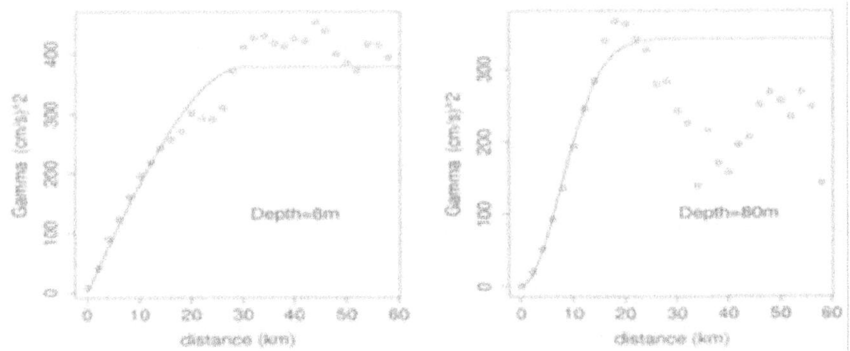

Figure 7. Experimental variograms on horizontal direction computed on the complex plane.

Complex kriging was then applied to the map of currents at the depth of 8m and results are shown in Figure 8. This method seems to work as well as the cokriging of the current components. In fact, in this study, differences between maps obtained by the two different krigings are smaller than differences we can observe when we choose other variogram models as nested or simple exponential, suppressing or not the nugget effect, in place of the two nested spherical models.

3.4 Comparison with circulation model and kriging of residuals

To compare kriging interpolation to results of the physical circulation model presented in section 2.2.3, we selected from the simulation output those which correspond to the data measured on the boat trajectory, and then we computed the residual vectors. A map of the data compared to the simulated current values along the boat trajectory is given by Figure 9.

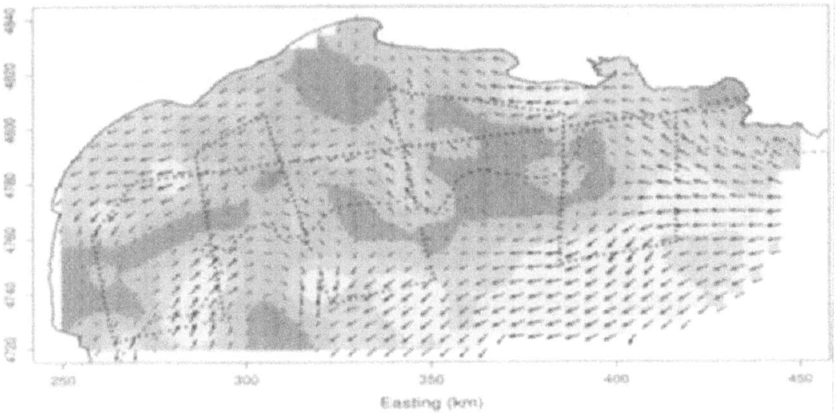

Figure 8. Map of currents at the depth of 8 meters that was obtained by complex kriging. Image legend is the same than for figure 5.

Using these residual vectors as data, we reapply the complex kriging method - after residual variogram fitting - in order to get a kriged map of residuals. Figure 10 shows the result for the depth of 8 meters. It looks like an over estimation of main stream current (or a bad positioning too close to the coast) by the model and a wrong rotation pattern in the western part of the Gulf of Lions.

4. CONCLUSION AND DISCUSSION

We proposed two different kriging methods, coregionalization and complex, for interpolating maps of vectors. The theory tells us that the coregionalization approach on the components should be the more effective. In fact, for our data set, the two approaches give very close results, the second one with complex kriging is a lot simpler to implement with the restriction to variogram and not with the richer class of complex covariance. Nevertheless, it is too early to generalize this result, and other patterns in data could make the coregionalization approach more relevant.

Another point is that we did quite crude interpolation ignoring that the vector field has to honor physical differential equations, or at least to be reasonably close to solutions of the physical circulation equations. In our case, results are quite good although the coast was only considered as a mask and not as a boundary. Chilès (2001) and in Chilès and Delphiner (1999) proposed some interpolations that honor known boundary conditions on flux. We could here impose to the current component that is orthogonal to the coast line to be null. More generally, specific models of covariance have been proposed to honour the physical framework of differential equations and get ad-hoc kriging or conditional simulations (Chilès and Delphiner, 1999). For that purpose, the physical circulation model is plugged in the simulation procedure and the covariance estimation. A counter part is a great loss in generality and simplicity when researchers in oceanography need simple and robust tools to help understanding differences between modeled circulation patterns and those derived from sparse data or boat cruises.

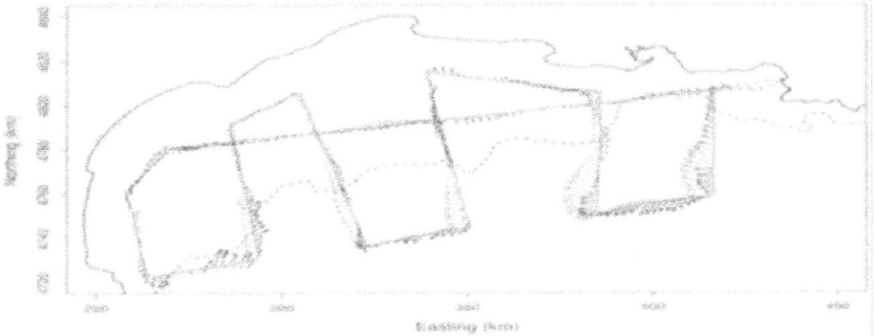

Figure 9. Measurement of currents (in black) along the trajectory compared with the output of the circulation model (in gray, same scale) at same time, same location and at the depth of 8 meters.

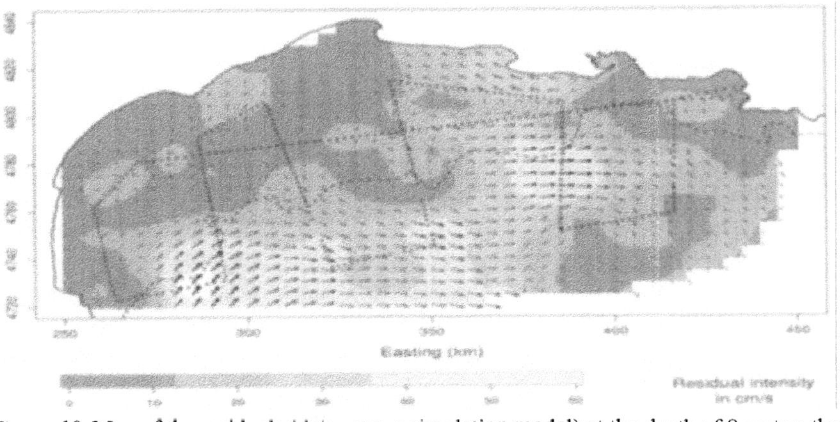

Figure 10. Map of the residuals (data versus circulation model) at the depth of 8 meters that was obtained by complex kriging.

A last point concerns the time. The boat cruise takes some time and the currents may change in between. This can be checked when the boat crosses it own trajectory. We tried to take into account the time but there was some time-space confounding effects in the measurement procedure itself. For most data, because of the constant boat speed, a pair of points at a given distance corresponds to a given time lag. So it is impossible to get points at different distances for a given time lag excepted if two boats are simultaneously measuring currents, and it was not possible to get regularly spaced time intervals for points at a given small distance for this kind of boat trajectories. Improvement could be easily done with a boat trajectory that often comes back on itself or stops to get a better and "orthogonal" sampling of time and space pairs. Modelling fully time and space components in the variogram model should improve comparisons between circulation models and data.

ACKNOWLEDGMENTS

This work was supported by the PNEC (National Program for Coastal Environment) of the CNRS (French National Institute for Scientific Research). During most part of this research, Pascal Monestiez was visiting the Centre d'Océanologie de Marseille (Université de la Méditerranée - Marseille II) which supported his stay.

REFERENCES

1. Chelton D. B. (1994). Physical oceanography: a brief overview for statisticians. *Statistical Science*, **9**:150-166
2. Chilès, J-P. (2001). Hydrogeology and geostatistics. In *geoENV III: Geostatistics for Environmental Applications*, Monestiez P., Allard D. and Froidevaux R. Eds, Kluwer Academic Publishers, Dordrecht, 1-16.
3. Chilès, J-P. and Delfiner, P. (1999).*Geostatistics: Modeling Spatial Uncertainty*. Wiley Series in Probability and Statistics, Wiley. 695 p.
4. Courault, D. and Monestiez, P. (1999). Spatial interpolation of air temperature according to atmospheric circulation patterns in Southeast France. *International Journal of Climatology*, **19**:365-378.
5. Cressie, N. (1993). *Statistics for Spatial Data, rev. edn.* Wiley: NY, 900 pp.
6. Estournel C., Durrieu de Madron, X., Marsaleix, P. Auclair, F., Julliand, C. and Vehil, R. (2002). Oberservation and modelisation of the winter coastal oceanic circulation in the Gulf of Lions under wind conditions influenced by the continental orography (FETCH experiment). *J. Geophys. Res.* (in press).
7. Goulard, M. and Voltz, M. (1992). Linear coregionalization model: tools for estimation and choice of multivariate variograms. *Mathematical Geology*, **24**:269-286.
8. Grzebyk, M. (1993). *Ajustement d'une Coré gionalisation Stationnaire*. Thesis, Ecoles des Mines de Paris, 154 pp.
9. Lajaunie Ch. and Béjaoui R. (1991). Sur le krigeage des fonctions complexes. Note interne N-23/91/G. Centre de Géostatistique, Ecole des Mines de Paris, Fontainebleau, 24pp.
10. Millot, C. (1990) The Gulf of Lion's hydrodynamics. *Continental shelf Research*, **10**:885-894.
11. Panel on statistics and oceanography (1994). Report on statistics and physical oceanography (with discussion) *Statistical Science*, **9**:167-221
12. Petrenko, A. (2002). Circulation features in the Gulf of Lions, NW Mediterranean sea; summer versus winter conditions. *Oceanol. Acta*, N° special PNEC-Golfe du Lion (in press).
13. Wackernagel, H. (1995). Multivariate Geostatistics, Springer-Verlag, Berlin, 256 pp.

INTERPOLATION OF RAINFALL AT SMALL SCALE IN A MEDITERRANEAN REGION

Relation between rainfall and altitude within a simple space-time model

D. Béal[1], G. Guillot[2], D. Courault[3] and C. Bruchou[4]

[1]*DESS SITN, Université Claude Bernard Lyon 1. Now at Unité Climat Sol e Environnement, INRA, Avignon. David.Beal@avignon.inra.fr*
[2]*Unité de Biométrie, INRA, Avignon. Now at Institut National Agronomique de Paris-Grignon. Gilles.Guillot@inapg.inra.fr*
[3]*Unité Climat Sol et Environnement, INRA, Avignon. Dominique. Courault@avignon.inra.fr*
[4]*Unité de Biométrie, INRA, Avignon. Claude.Bruchou@avignon.inra.fr*

Abstract: This paper describes a statistical space-time model for rain-fall in Mediterranean regions. The rain-fall is supposed to be the sum of a deterministic component and a random function enjoying spatial second-order stationary and without temporal correlation. Under these assumptions, we analyse the dependence of the trend upon time and compute the optimal linear predictor. The proposed methods are implemented and discussed on a two year data set of daily recordings.

1. INTRODUCTION

Despite the development of dense rain-gauge networks of typical densities of one station for a few hundreds of squared kilometers, the spatial prediction of rainfall at small scale remains an issue in the Mediterranean regions. Indeed, for a substantial part of it, annual rainfall in these regions is due to very localized events which affect very small areas only. In an agro-

X. Sanchez-Vila et al. (eds.), geoENV IV – Geostatistics for Environmental Applications, 379-389.

meteorological context, poor rainfall predictions cumulated over several cultural cycles may affect considerably prescribed technical practices.

Our study is part of a larger project concerned with global fruit production management (Habib, 2002) in French Mediterranean regions. It is aimed at prescribing better technical practices so as to reduce inputs such as water, nitrogen, pesticides. This paper suggests several rainfall predictors when a typical research rain-gauge networks is available. The scale considered is the 24 hours duration at point support. We examine linear predictors and describe how prediction can be improved by accounting for a deterministic trend related to local relief. The data are described in section 1, the model and predictors are presented in section 2, the results of an implementation are reported in section 3, while technical computations appear in the appendix.

2. DATA SET AND EXPLORATORY DATA ANALYSIS

We study a region of 100kmX200 km located in the South of France nearby the Mediterranean Sea (see figure 1). Rainfall measurements from a network of 32 stations are available from 01/01/2000 to 31/03/2002 at a daily time step. These rain-falls are related to elevations as available from a terrain model at a resolution of 75 meters.

The points of the network will be denoted by $\{s_\alpha\}_{\alpha=1,\ldots,ns}$, the dates of measurements by $\{t_i\}_{i=1,\ldots,nt}$, and a rainfall value at point x and date t by $R(x,t)$.

The point distribution is strongly non Gaussian with a lot of zero recordings and a marked assymetry. Therefore, the covariance is probably not the best tool to capture weak and non-linear space-time dependences among stations. However, we investigate the space-time covariance structure of rain-falls by means of the empirical space-time covariance function defined as

$$C^*(h,\tau) = \frac{1}{n_{h,\tau}} \sum_i \sum_{\|s_\alpha - s_\beta\| \approx h} R(s_\alpha, t_i) R(s_\beta, t_i + \tau) - \overline{R}^2 \tag{1}$$

where $\overline{R}$ is the overall mean and $n_{h,\tau}$ the number of pairs of data involved in the sum.

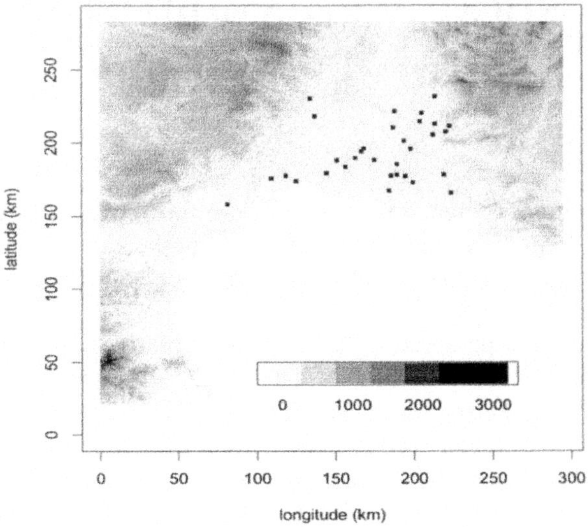

Figure 1. Elevation in the region under study and rain-gauge network.

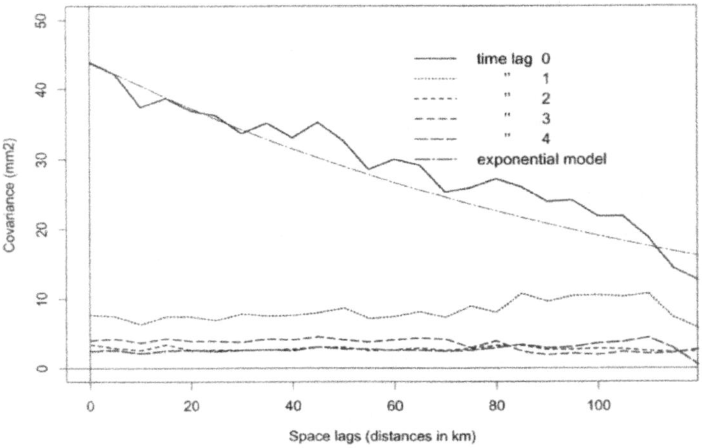

Figure 2. Empirical space-time covariances $C^*(h,\tau)$ of daily rainfall measurements at various lags (exponential fit at lag 0).

The spatial covariances at five time lags are displayed on figure 1. An obvious spatial structure appears at lag 0. At the other lags, there is also a marked structure: the covariances are not equal to zero, which corresponds to a spatial correlation accross different days. They are flat which might be interpreted as large scale (or low frequency features) of rainfall events. However the variance involvedd is at least five times smaller than the variance at lag 0, and even smaller than the covariance at lag 0 for distances of about 100 km.

3. STATISTICAL MODEL AND DEFINITION OF PREDICTORS

3.1 Hypotheses

From the previous elementary analysis we formulate a space-time second-order model defined through the following hypotheses:

1. $R(x,t)=m(x,t) + Y(x,t)$ where m is a deterministic trend and Y a random residual
2. $m(x,t)$ admits a parametric decomposition onto a family of known basis functions where the weights of the decomposition are allowed to vary in time, namely $m(x,t) = \Sigma^{L}_{l=1}\, a_l(t)\, f_l(x)$,
3. $Y(x,t)$ is a zero mean second-order random function uncorrelated in time i.e $Cov[Y(x,t),Y(x',t')] = 0$ for $t \neq t'$,
4. The covariance function of Y is stationary and isotropic in space i.e $Cov[Y(x,t),Y(x',t')] = C(x-x')$.

The covariance function C being unknown, we adopt a purely frequentist approach and estimate it off line considering that the number of independent replications is large and that we are close to the "true" model. A parametric fit is performed with an exponential model $C(h) = \sigma^2 \exp(-h/r)$, where σ^2 and r are taken to be respectively 45 mm^2 and 120 km so as to fit the empirical curve.

The decomposition of $m(x,t)$ involves weights $a_l(t)$ varying in time. This is a major difference as compared to the usual universal kriging model (Chilès and Delfiner, 1999), and a natural generalization in this context where a deterministic influence of variables like elevation, distance to the sea could be of varying amplitude along the year. However there is no way to check the hypotheses of time variations of the trend. Therefore we compute optimal linear estimation within this model, considering first the particular case where the trend is constant and then the general case.

3.2 Model with constant trend in time

We consider the problem of estimating $m(x_0,t_{i0})$ and $R(x_0,t_{i0})$ under the assumption that $m(x,t) = m(x) = \Sigma_l\, a_l\, f_l(x)$ and therefore introduce the space-time linear estimators:

$$\hat{m}(x_0) = \sum\nolimits_{\alpha,i} \lambda^{m}_{\alpha,i} R(x_\alpha, t_i) \qquad (2)$$

and

$$\widehat{R}(x_\alpha, t_{i_0}) = \sum\nolimits_{\alpha,i} \lambda^{R}_{\alpha,i} R(x_\alpha, t_i) \qquad (3)$$

Denoting by $\lambda^{m}_{.i0} = (\lambda_{1,i0},\ldots, \lambda_{ns,i0})^{t}$ the vector of kriging weights corresponding to date t_{i0} and by $\{\lambda^{m}_{.i}\}_{i=1,\ldots,nt}$ $i \neq i_0$ the vectors of weights for the other dates, the optimal weights are as follows :

The $\{\lambda^{m}_{.i}\}_{i=1,\ldots,nt}$ are all equal to $(1/n_t)\lambda^{m}$, where λ^{m} the solution of the fixed time universal kriging system of the trend, namely

$$\begin{pmatrix} \mathbf{C} & \mathbf{F} \\ \mathbf{F}^{t} & 0 \end{pmatrix} \begin{pmatrix} \lambda^{m} \\ \mu \end{pmatrix} = \begin{pmatrix} 0 \\ \mathbf{F}_0 \end{pmatrix} \tag{4}$$

where $\mathbf{F} = (f_l(x_\alpha))_{\alpha,l}$, the details of computation being given in appendix.

The vectors of weights involved in the estimation of R satisfy

$$\forall i \neq i_0 \quad \lambda^{R}_{.i} = \tilde{\lambda}^{R} = \frac{1}{n_t}\mathbf{F}_0^{t}(\mathbf{F}^{t}\mathbf{C}^{-1}\mathbf{F})^{-1}\mathbf{F}^{t}\mathbf{C}^{-1}$$

$$-\frac{1}{n_t}\mathbf{C}_0^{t}\mathbf{C}^{-1}\mathbf{F}(\mathbf{F}^{t}\mathbf{C}^{-1}\mathbf{F})^{-1}\mathbf{F}^{t}\mathbf{C}^{-1} \tag{5}$$

$$\text{and} \qquad \lambda^{R}_{.i_0} = \tilde{\lambda}^{R} + \mathbf{C}^{-1}\mathbf{C}_0 \tag{6}$$

These results have the following interpretation: despite the absence of temporal correlation, there is a transfer of information across time, each date taking in charge a fraction $1/n_t$ of the estimation of m. In the estimation of R, the symmetry among dates is broken and the date i_0 plays a special role.

In the fixed time case, the usual Universal Kriging predictor can be decomposed into the sum of the estimated mean plus the simple kriging of the estimated residual (see (Chilès and Delfiner, 1999), pp 182-183). After simple computations (see section A.1.3) it appears that this nice property also holds in the present framework: $\hat{R} = \hat{m} + (R-\hat{m})^{SK}$, where $(R-\hat{m})^{SK}$ is the simple kriging predictor of the estimated residuals.

3.3 Model with a trend varying in time

Under the assumption that $m(x,t) = \Sigma_l\, a_l(t)\, f_l(x)$ and keeping the same notations, we now obtain

$$\lambda^{m}_{.i_0} = \lambda^{m} \tag{7}$$

$$\lambda^{m}_{.i} = 0 \quad \forall i \neq i_0 \tag{8}$$

and

$$\lambda^{R}_{.i_0} = \lambda^{R} \tag{9}$$

$$\lambda^{R}_{.i} = 0 \quad \forall i \neq i_0 \tag{10}$$

where λ^{R} is solution of the UK system :

$$\begin{pmatrix} \mathbf{C} & \mathbf{F} \\ \mathbf{F}^t & 0 \end{pmatrix} \begin{pmatrix} \lambda^R \\ \mu \end{pmatrix} = \begin{pmatrix} \mathbf{C}_0 \\ \mathbf{F}_0 \end{pmatrix} \tag{11}$$

In this slightly less simple model, there is no transfer of information across time. The best strategy for estimating m or R at (x_0,t_{i0}) is to perform a fixed time UK, using measurements of date t_{i0} only.

3.4 Choice of basis functions

3.4.1 Polynomial functions

In the usual Universal Kriging framework, the basis functions are always taken as polynomial functions of increasing degrees. This family is conceptually simple, it might account for various patterns in the trend. It has also the nice property to provide estimations independent from the location of the origin of axis. On the other hand, one could imagine that in various situations the polynomial form of the trend is physically questionable and that one might capture more feature of the actual rain/space relation considering more physically rooted basis functions.

3.4.2 Weights derived of a principal components analysis of the terrain model

The latter idea was the aim of the so-called Aurelhy method developed at the French meteorological service (Benichou and Breton, 1986) and widely used in operational contexts. Although not originally written in term of Universal Kriging we can reformulate this method in our framework as follows:

Consider a regular grid of N points encompassing the region under study, and an N X n_n matrix $\mathbf{H}$ whose lines denoted by $\mathbf{h}(x_i)$ are filled (in a prescribed fixed order) with the heights of the n_n nearest neighbors of each x_i. Each line contains the information of the relative relief around x_i at a certain scale. This information can be summarized via a principal component analysis (PCA) whose principal components will be denoted by $p_1(x),...,p_{nn}(x)$. Note that these functions can be evaluated at any point x, (not necessarily those included in $\mathbf{H}$ provided that the local relief $\mathbf{h}(x)$ is known.

The first principal components among $(p_1(x),...,p_{nn}(x))$ provide a statistical summary of $\mathbf{h}(x)$ with a straightforward physical interpretation: the $p_i(x)$ are the weights of a decomposition onto elementary local landscapes (such as local minima or maxima of the height, constant slopes and passes along specific directions, etc), these elementary relieves being simply the eigen vectors of the PCA. They are therefore natural candidates to explain the deterministic variations of rain.

In the actual Aurelhy method, the trend is assumed to be variable in time, but the estimator described in (Benichou and Breton, 1986) is not the one proposed in section 2.3. It turns out that in Aurelhy, the implicit estimation of m is performed by an ordinary least square minimization, whereas the optimal weights must be obtained with an implicit estimation of m obtained through generalized least squares (UK of m).

4. IMPLEMENTATION

The various methods considered above have been implemented and are compared computing the Cross Validation Mean Squared Error defined as

$$EQM = 1/n_s n_t \sum_{\alpha,i} \left(R^*(x_\alpha,t_i) - R(x_\alpha,t_i) \right)^2 \tag{12}$$

where each estimation $R^*(x_\alpha,t_i)$ is performed on the data set deprived from $R(x_\alpha,t_i)$. As the optimal estimator with a varying trend consist in performing a classical fixed time single date UK, it will referred hereafter as fixed time UK. The results of Ordinary Kriging are also given as a reference.

All the methods lie within a small interval. UK with a fixed trend is better than under the varying trend hypothesis, whatever the choice of basis functions, the latter being worse than Ordinary Kriging. The space-time UK with fixed trend performs slightly better than Ordinary Kriging but the magnitude of the improvement does not sound to be significant.

These results are probably strongly dependent of the topography of the region under study and also of its climatology. The relief in our domain is not very marked. This could explain the poor results obtained when using basis functions derived from the PCA of the terrain model. It remains interesting to note that the fixed time trend model under which a space-time linear prediction is realized performs notably better than the fixed time UK.

5. CONCLUSION

A simple space-time model for the study of environmental variables has been proposed. It appears that even under the assumption of time independence, a space-time predictor should be used. This predictor has the nice feature to estimate implicitly the spatial trend using all the information available from the various dates. On our data set it also seems to perform at

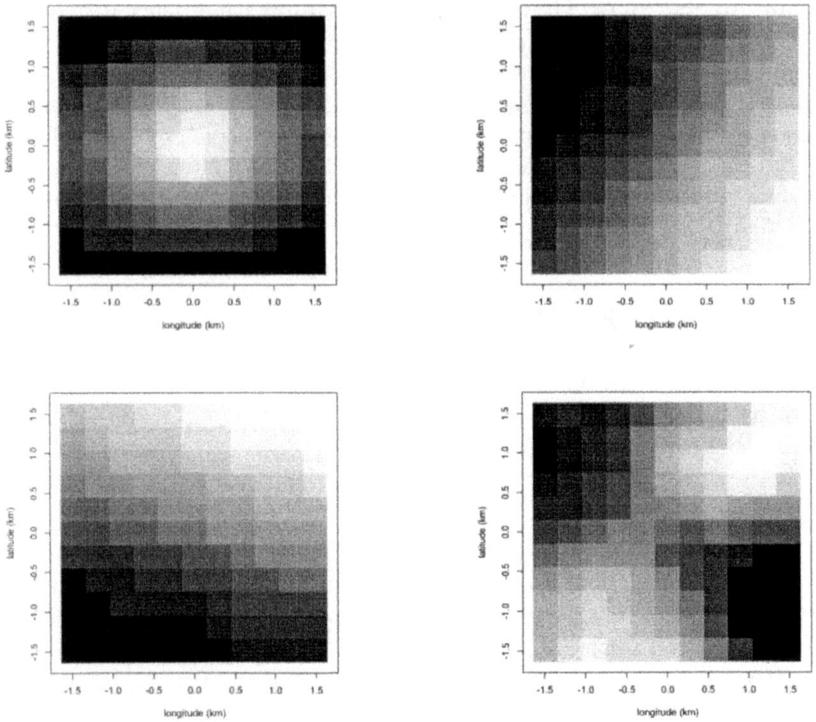

Figure 3. Eigen vectors displayed as images associated to the 1st, 2nd, 3rd and 14th eigen values (from top to bottom and from left to right) derived from the principal component analysis of the terrain model.

Table 1. Cross Validation mean squared error.

Method	MSE (mm^2)
Space-time Universal kriging with polynomial basis functions	11
Space-time Universal Kriging with principal components as basis functions	11.2
Ordinary kriging	11.2
Fixed time Universal Kriging with polynomial as basis functions	14.7
Fixed time Universal Kriging with principal components as basis functions	15.2

least as well than usual fixed time predictors. Concerning the decomposition of the trend, the functions derived from an analysis of the terrain model which sounded intuitively to be physically related to rain-fall do not help to improve the estimation and the classical polynomial basis function seem to remain the best tool in this framework. This empirical result might off course be different for other regions. This work was simply a modest contribution to define a framework suitable to analyze space-time non-stationary linear methods.

ACKNOWLEDGE

We acknowledge Thomas Nesme and Robert Habib from INRA-Avignon and Centre d'Information Régional Agro-Météorologique (CIRAME) for providing rain-fall data.

APPENDIX: COMPUTATION OF SPACE-TIME KRIGING WEIGHT

A.1 Model with constant trend in time

A.1.1 Estimation of *m*

The non bias condition is

$$\sum_{\alpha,i} \lambda_{\alpha,i} \mathbf{f}_l(x_\alpha) - \mathbf{f}_l(x_0) = 0 \quad \text{for } l = 1,\dots,L \tag{A.1}$$

The error variance is

$$\sum_i \sum_{\alpha,\beta} \lambda_{\alpha,i} \lambda_{\beta,i} \mathbf{C}_{\alpha,\beta} \tag{A.2}$$

Then after taking derivative of the objective function, the kriging weights appear to be solution of the block matrix system

$$\begin{pmatrix} \mathbf{C} & \mathbf{0} & \cdots & \mathbf{0} & \mathbf{F} \\ \mathbf{0} & \ddots & \ddots & \vdots & \vdots \\ \vdots & \ddots & \ddots & \mathbf{0} & \\ \mathbf{0} & \cdots & \mathbf{0} & \mathbf{C} & \mathbf{F} \\ \mathbf{F}^t & \cdots & \mathbf{F}^t & \mathbf{0} & \mathbf{0} \end{pmatrix} \begin{pmatrix} \lambda_{.1} \\ \vdots \\ \vdots \\ \lambda_{.n_t} \\ \mu \end{pmatrix} = \begin{pmatrix} \mathbf{0} \\ \vdots \\ \vdots \\ \mathbf{0} \\ \mathbf{F}_0 \end{pmatrix} \tag{A.3}$$

A.1.2 Estimation of *R*

The non bias condition remains the same whereas the error variance is:

$$\sum_i \sum_{\alpha,\beta} \lambda_{\alpha,i} \lambda_{\beta,i} \mathbf{C}_{\alpha,\beta} - 2\sum_{\alpha} \lambda_{\alpha,0} \mathbf{C}_{\alpha,0} + \mathbf{C}_{0,0} \tag{A.4}$$

Introducing a vector μ of Lagrange multipliers, we obtain the following block matrix system:

$$\begin{pmatrix} \mathbf{C} & \mathbf{0} & \cdots & \mathbf{0} & \mathbf{F} \\ \mathbf{0} & \ddots & \ddots & \vdots & \vdots \\ \vdots & \ddots & \ddots & \mathbf{0} & \\ \mathbf{0} & \cdots & \mathbf{0} & \mathbf{C} & \mathbf{F} \\ \mathbf{F}^t & \cdots & \mathbf{F}^t & \mathbf{0} & \mathbf{0} \end{pmatrix} \begin{pmatrix} \lambda_{.1} \\ \vdots \\ \vdots \\ \lambda_{.n_t} \\ \mu \end{pmatrix} = \begin{pmatrix} \mathrm{U} \\ \vdots \\ \mathbf{C}_0 \\ \vdots \\ \mathbf{0} \\ \mathbf{F}_0 \end{pmatrix} \tag{A.5}$$

A.1.3 Relation between m^* and R^*

Denoting $(\mathbf{F}^t\mathbf{C}^{-1}\mathbf{F}^{-1})^{-1}\mathbf{F}^t\mathbf{C}^{-1}$ by $\mathbf{B}$ and $(R(x_1,t_i),\ldots,R(x_{ns},t_i))^t$ by $\mathbf{R}_{.i}$ we have:

$$\begin{aligned} \widehat{R}(x_0,t_{i_0}) &= (\mathbf{C}^{-1}\mathbf{C}_0 + \widetilde{\lambda})\mathbf{R}_{.i_0} + \sum_{i\neq i_0} \lambda^R_{.i}\mathbf{R}.i \\ &= \mathbf{C}^{-1}\mathbf{C}_0\mathbf{R}_{.i_0} + \frac{1}{n_t}\mathbf{F}_0^t\mathbf{B}\mathbf{R}_{.i_0} - \frac{1}{n_t}\mathbf{C}_0\mathbf{C}^{-1}\mathbf{F}\mathbf{B}\mathbf{R}_{.i_0} \\ &\quad + \sum_{i\neq i_0}\frac{1}{n_t}\mathbf{F}_0\mathbf{B}\mathbf{R}_{.i} - \sum_{i\neq i_0}\frac{1}{n_t}\mathbf{C}_0^t\mathbf{C}^{-1}\mathbf{F}\mathbf{B}\mathbf{R}_{.i_0} \\ &= \mathbf{C}_0^t\mathbf{C}^{-1}(\mathbf{R}_{.i_0} - \sum_i \frac{1}{n_t}\mathbf{F}\mathbf{B}\mathbf{R}_{.i_0}) + \sum_i \frac{1}{n_t}\mathbf{F}_0\mathbf{B}\mathbf{R}_{.i} \\ &= (R-\widehat{m})^{SK}(x_o,t_{i_0}) + \widehat{m}(x_o,t_{i_0}) \end{aligned} \tag{A.6}$$

A.2 Model with a trend varying in time

A.2.1 Estimation of m

The bias takes the following form:

$$\sum_{\alpha,i}\sum_{l}\left(a_l(t_i)\mathbf{f}_l(x_\alpha) - \mathbf{f}_l(x_0)\right) \tag{A.7}$$

This term has to be zero whatever the $a_l(t_i)$ which requires that

$$\sum_\alpha \lambda_{\alpha,i}\mathbf{f}_l(x_\alpha) = 0 \quad \forall l = 1,\ldots,L \quad ,\forall i = 1,\ldots,n_t \quad i \neq i_0 \tag{A.8}$$

and $$\sum_\alpha \lambda_{\alpha,i_0}\mathbf{f}_l(x_\alpha) = \mathbf{f}_l(x_0) \qquad \forall l = 1,\ldots,L \tag{A.9}$$

We know have n_t sets of constraints, that is n_t vectors of Lagrange multipliers. The error variance being still given by expression (a.2), the optimal weights are solution of the block matrix system:

$$\begin{pmatrix} \mathbf{C} & \mathbf{0} & \cdots & \mathbf{0} & \mathbf{F} & \mathbf{0} & \cdots & \mathbf{0} \\ \mathbf{0} & \ddots & \ddots & \vdots & \mathbf{0} & \ddots & \ddots & \vdots \\ \vdots & \ddots & \ddots & \mathbf{0} & \vdots & \ddots & \ddots & \mathbf{0} \\ \mathbf{0} & \cdots & \mathbf{0} & \mathbf{C} & \mathbf{0} & \cdots & \mathbf{0} & \mathbf{F} \\ \mathbf{F}^t & \mathbf{0} & \cdots & \mathbf{0} & \mathbf{0} & \cdots & \cdots & \mathbf{0} \\ \mathbf{0} & \ddots & \ddots & \vdots & \vdots & \ddots & \ddots & \vdots \\ \vdots & \ddots & \ddots & \mathbf{0} & \vdots & \ddots & \ddots & \vdots \\ \mathbf{0} & \cdots & \mathbf{0} & \mathbf{F}^t & \mathbf{0} & \cdots & \cdots & \mathbf{0} \end{pmatrix} \begin{pmatrix} \lambda.1 \\ \vdots \\ \vdots \\ \lambda_{.n_t} \\ \mu_1 \\ \vdots \\ \vdots \\ \mu_{n_t} \end{pmatrix} = \begin{pmatrix} \mathbf{0} \\ \vdots \\ \vdots \\ \vdots \\ \mathbf{F}_0 \\ \mathbf{0} \\ \vdots \\ \mathbf{0} \end{pmatrix} \tag{A.10}$$

A.2.2 Estimation of *R*

Keeping the same non bias conditions and the error variance of expression (A4) we get the block matrix system:

$$\begin{pmatrix} \mathbf{C} & \mathbf{0} & \cdots & \mathbf{0} & \mathbf{F} & \mathbf{0} & \cdots & \mathbf{0} \\ \mathbf{0} & \ddots & \ddots & \vdots & \mathbf{0} & \ddots & \ddots & \vdots \\ \vdots & \ddots & \ddots & \mathbf{0} & \vdots & \ddots & \ddots & \mathbf{0} \\ \mathbf{0} & \cdots & \mathbf{0} & \mathbf{C} & \mathbf{0} & \cdots & \mathbf{0} & \mathbf{F} \\ \mathbf{F}^t & \mathbf{0} & \cdots & \mathbf{0} & \mathbf{0} & \cdots & \cdots & \mathbf{0} \\ \mathbf{0} & \ddots & \ddots & \vdots & \vdots & \ddots & \ddots & \vdots \\ \vdots & \ddots & \ddots & \mathbf{0} & \vdots & \ddots & \ddots & \vdots \\ \mathbf{0} & \cdots & \mathbf{0} & \mathbf{F}^t & \mathbf{0} & \cdots & \cdots & \mathbf{0} \end{pmatrix} \begin{pmatrix} \lambda_{.1} \\ \vdots \\ \vdots \\ \lambda_{.n_t} \\ \mu_1 \\ \vdots \\ \vdots \\ \mu_{n_t} \end{pmatrix} = \begin{pmatrix} \mathbf{0} \\ \mathbf{C}_0 \\ \mathbf{0} \\ \vdots \\ \vdots \\ \mathbf{0} \\ \mathbf{F}_0 \\ \vdots \\ \mathbf{0} \end{pmatrix} \qquad \text{(A.11)}$$

REFERENCES

1. Benichou, P. and Breton, L. (1986). Prise en compte de la topographie pour la cartographie des champs pluviométriques statistiques: la méthode Aurelhy. In *Agrométéorologie des régions de moyennes montagnes*, number 39, pages 51-68. INRA.
2. Chilès, J. and Delfiner, P. (1999). *Geostatistics*. Wiley.
3. Habib, R. (2002). Action transversale Production Fruitière Intégrée. Technical report, INRA - Avignon.

AUTOMATIC MODELING OF CROSS-COVARIANCES FOR RAINFAL ESTIMATION USING RAINGAGE AND RADAR DATA

E.F. Cassiraga, C. Guardiola-Albert and J.J. Gómez-Hernández

Departamento de Ingeniería Hidráulica y Medio Ambiente. Technical University of Valencia, 46071 Valencia, Spain. E-mail: efc@dihma.upv.es, guardiola@dihma.upv.es, jaime@dihma.upv.es

Abstract: The use of radar data is a powerful tool to improve rainfall spatio-temporal estimation. Geostatistical techniques are well suited to combine both raingage and radar measurements for this purpose. The main problem of this application, particularly in the context of real time estimation, is the definition of a positive definite model of cross-correlation between radar and rainfall. We propose the direct use of the experimental surface variogram, after filtering the spectra and cross-spectra in the frequency domain to ensure positive definiteness of the model. This technique, which has been proposed in the literature, is suitable for its introduction in a real time forecasting system in which fast estimation of the rainfall spatial distribution is needed. A case study shows its application with a real data set corresponding to the Barcelona radar and its watershed pluviographs.

1. INTRODUCTION

Accurate and reliable real time forecasting of areal rainfall, at the basin scale, has been one of the unresolved needs of hydrology. Flood timing and peak intensity in natural catchments are heavily influenced by the space-time variability of rainfall. Thus, good rainfall spatial distribution estimations are important for a better real-time flood forecasting. In this sense, radar sensors

X. Sanchez-Vila et al. (eds.), geoENV IV – Geostatistics for Environmental Applications, 391-399.

have turned out to be of wide applicability in the measurement of rainfall fields, mainly due to their ability to map the spatial characteristics of rainfall.

Good estimation of rainfall spatio-temporal distribution using fast and reliable techniques is required in flood forecasting systems for the success of hydrological alert systems, such as the Spanish SAIHs.

Several attempts have been made to utilize both raingage measurements and radar rainfall data in rainfall estimation, using geostatistical techniques (Krajewski, 1987; Creutin et al., 1988, Seo et al. 1990; Seo et al. 1990, Cassiraga et al. 1997, Goovaerts, 2000, Sun et al. 2000). The main problem of this application, in the context of real time estimation, is to fit, quickly, a positive definite model for the cross-correlation between rain and radar data. The model must be positive definite in order to ensure existence and uniqueness of the kriging system solution. The traditional modeling approach only considers positive linear combinations of basic models that are known to be positive definite, under very restrictive conditions; the so-called linear model of coregionalization (LMC). Not only LMC is too restrictive but also determining the linear combination that bests fits the experimental information is difficult and time consuming.

In this paper, we propose the direct use of the experimental surface variogram, after filtering the spectra and cross-spectra in the frequency domain. This technique (Yao et al., 1998) could be made automatic and introduced in a real time forecasting system in which fast estimation of the rainfall spatial distribution is needed.

2. AUTOMATIC COVARIANCE MODELING WITH FAST FOURIER TRANSFORM (FFT): A SHORT RECALL

The proposed algorithm capitalizes upon the fact that the positive definiteness constraints on the covariance are mapped into simpler constraints on its density spectrum. The main idea is to transform the experimental (cross-)covariance tables into density spectrum tables using FFT. These density spectrum tables are then smoothed under the constraints of positivity and unit sum. A back transform through inverse FFT yields permissible (jointly) positive definite (cross-)covariance tables. Through this method, permissible (cross-)covariance tables are obtained automatically without calling for any analytical model nor for the linear coregionalization model. (Notice that the difficult-to-verify-in-real-space positive definite condition, is straightforward in the frequency domain, the only condition is that the density function obtained by FFT of the correlogram must be positive and its integral be one).

The algorithm proceeds as follow:

1. First, calculate the experimental covariance maps from the sample data. These maps must have enough resolution, i.e., contain enough number of lag vectors **h**, for later use in kriging at unsampled locations. There may be large experimental fluctuations and many missing entries in these maps due to data sparsity.
2. Perform a preliminary smoothing of these experimental covariance maps to fill-in all missing entries of the covariance table and to filter the most severe fluctuations. These completed covariance maps are not positive definite, because each lag is calculated independently of the others using different data pairs, so its use may result in singular kriging matrices or negative estimation variances.
3. The previous gridded and completed covariance map is transformed into a gridded spectrum map by FFT. To ensure positivity of the density spectrum table and their unit sum, these experimental spectrum values are further smoothed under these two constraints. Such smoothing also removes the last unwanted sample fluctuations and the result is a licit spectral probability density function.
4. This licit and smooth probability density function is back-transformed by inverse FFT into a licit covariance look-up table for estimation in the spatial domain.

The interested reader could find the details of the algorithm in Yao (1998) o en Yao et al. (1998).

3. CASE STUDY

The case study corresponds to the estimation of rainfall in the Barcelona watershed making use of data from the Barcelona radar and of the pluviographs in the watershed. The data base consists of a set of radar images with the log$Z$ in dB (decimal logarithms of reflectivity in decibels) in intervals of ten minutes and the corresponding pluviographs measurements given intensity of precipitation in mm/h.

The area of study is a square of 140 km by 140 km, discretized into cells of 1 km by 1 km. The Barcelona city radar is located in the center of the area. In Figure 1 we can see the situation of the pluviographs and the corresponding radar image for one selected time step. The data set is integrated by 77 raingage measurements and the radar image has 19600 log$Z$ data.

We are going to calculate different spatial correlation measurements in order to make estimations using different kinds of kriging. We will perform all calculations after standardization of both variables, therefore we will use correlograms and cross-correlograms instead of covariances and cross-covariances throughout.

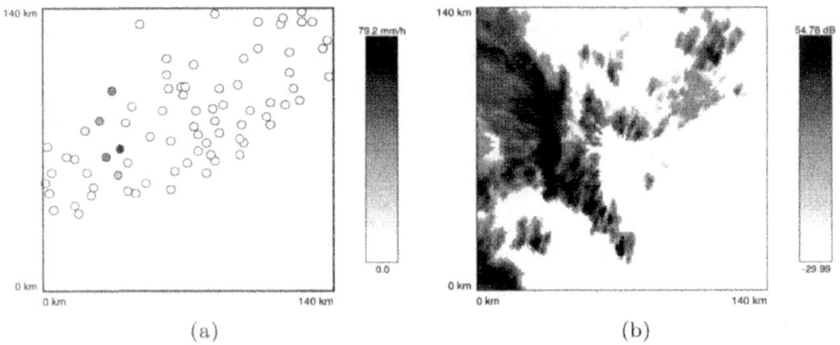

Figure 1. Data used in the case study. (a) Raingage locations: values of intensity of precipitation in mm/h. (b) Radar image: values of log$Z$ in dB (the white pixels correspond to log$Z$= –30 dB, that is associated with zero precipitation.

For the purpose of demonstrating the algorithm we will consider only a single time slice, therefore the possible temporal correlation that may exist is not accounted for.

As it has been mentioned before, different geostatistical methods can be used to estimate the rainfall field with radar data. We have applied the algorithm described for the computation of the necessary correlograms to be used by three estimating approaches: kriging, cokriging, and kriging with an external drift.

3.1 Kriging: auto covariance modeling

The simplest technique that we can use to obtain a rainfall field is to interpolate just the rainfall data without accounting for the radar information. For the interpolation of a single attribute we can use ordinary or simple kriging. In either case a model for the autocorrelogram is needed.

In recent years, in surface hydrology, it is becoming usual to extrapolate the spatial pattern from the radar image to rainfall. There are always more radar than rainfall data, and therefore it is easier to obtain a correlogram model for radar. In this case, the rainfall data only work as conditioning on the resulting map.

The rainfall and radar correlogram maps were calculated and are shown in Figure 2. The rainfall correlogram (Figure 2(a)) was calculated using only the 77 rainfall data. This map shows unwanted fluctuations and too little structure that arise because of the limited rainfall data. The radar correlogram map (Figure 2(b)) corresponds to the radar image. It displays a more structured phenomena with a clear anisotropy and lacks the unwanted fluctuations of the rainfall map.

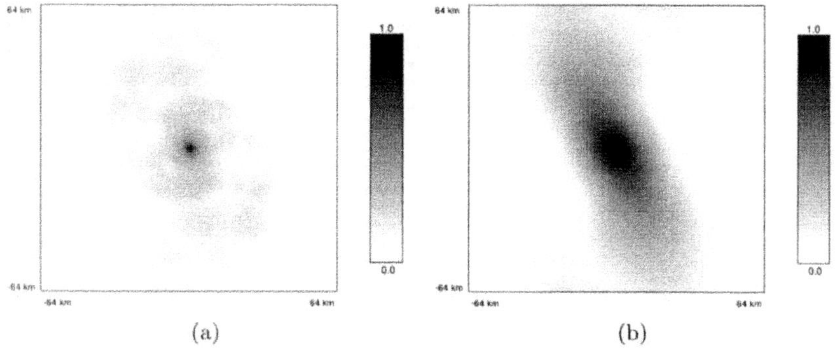

Figure 2. The correlogram maps of rainfall and radar data calculated independently. (a) Rainfall. (b) Radar.

3.2 Cokriging: joint cross-covariance modeling

In presence of multiple cross-correlated variables, the auto and cross-correlograms cannot be modeled independently. They must all be modeled simultaneously to ensure the positive definiteness of any cross-correlation matrix built from them. The linear model of coregionalization was introduced for this purpose. In practice, the model of coregionalization becomes unwieldy as the number of coregionalized variables increases.

This difficulty may make unfeasible the use of cokriging in a real time flood forecast system. As already mention, to avoid this difficulties and to provide an effective way of computing permissible coregionalization models we use the smoothing of the spectra in the frequency domain.

The resulting model of coregionalization obtained by the automatic modeling of (cross-)correlogram tables proposed is shown in Figure 3.

3.3 Kriging with an external drift: residual auto covariance modeling

Kriging with an external drift assumes that the rainfall data should be modeled as a drift term plus a residual, and that the drift term is an unknown linear function of the radar data. The application of kriging with an external drift requires modeling the covariance of the residuals. Assuming that we want to impose the spatial pattern of the radar image to the interpolated rainfall field and that the relation between rainfall and radar data is linear, we propose the follow methodology in order to obtain a valid correlogram map:

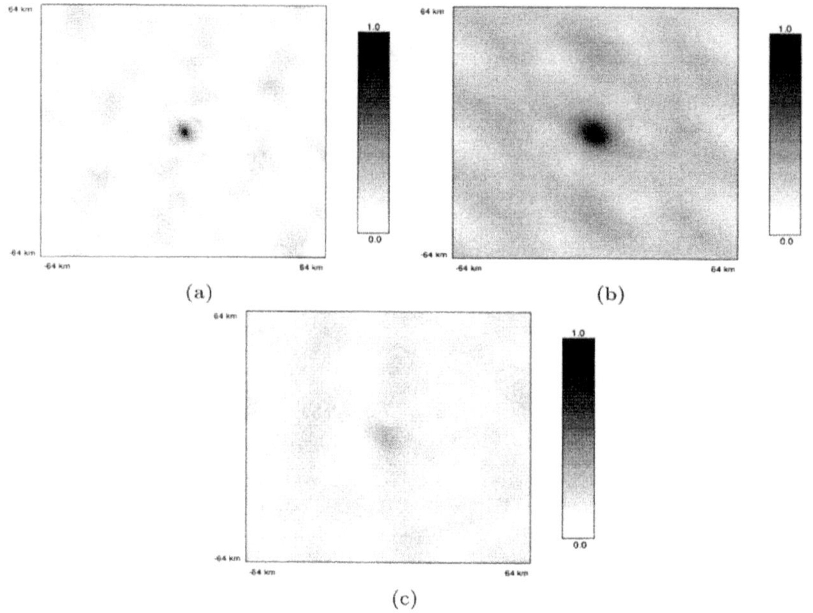

Figure 3. The (cross-)correlogram maps of rainfall and radar data. (a) Rainfall correlogram. (b) Radar correlogram. (c) Rainfall-radar cross-correlogram.

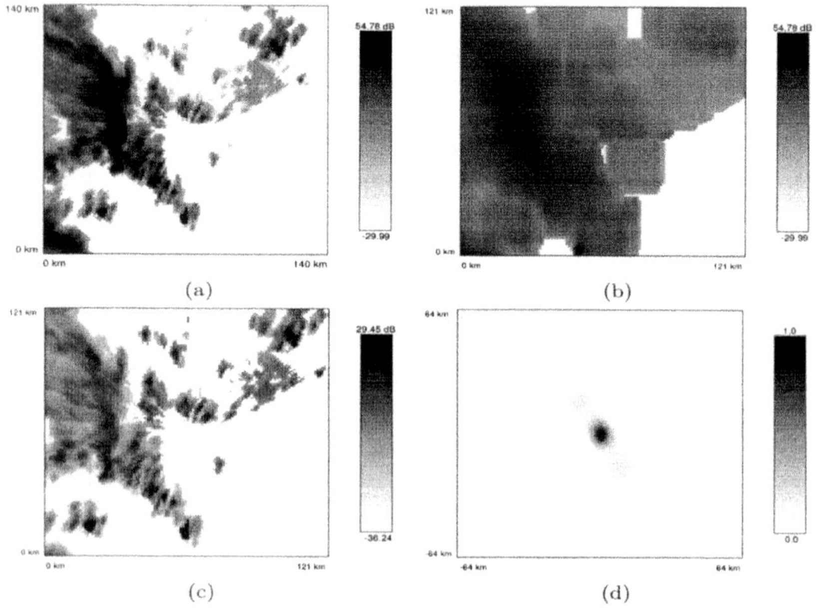

Figure 4. Kriging with an external drift: residual auto covariance modeling. (a) Radar image. (b) Smoothed radar image after applying a moving average with a window of 20 by 20 pixels. (c) Residual map. (d) Residual correlogram map.

1. Calculate a drift map of the radar data by smoothing it using moving averages.
2. Obtain the residual map subtracting the drift map calculated above from the radar data field.
3. Apply the automatic modeling algorithm to the residual map in order to calculate the residual correlogram map.
4. Extrapolate this residual correlogram map for rainfall.

In Figure 4 we can see the results of the methodology described above. After testing different sizes for the moving window averaging process, the selected size was 20 per 20 cells.

4. ESTIMATED FIELDS

The estimation codes of GSLIB (Deutsch and Journel 1997) were adapted in order to directly read correlogram tables. In the figure 5 we can see the rainfall estimated fields using the correlogram tables obtained. The artifacts that we can see in the fields estimated by ordinary kriging (5a and b) are caused by the limited number of raingage data that are available. The maps that use radar information (5c and d) clearly incorporate this information to produce more structured rainfall maps which are better tied to the patterns provided by the radar.

5. CONCLUSIONS

We have shown the use of an automatic (cross-)covariance modeling technique applied to a real hydrology data set. The motivation of this work was to find a methodology able to produce valid covariance and cross-covariances tables, in a reasonable time, in order to implement it in a real-time forecasting model.

The resulting correlogram maps shown in this paper, using the proposed algorithm, are obtained in a few seconds. These maps are licit for solving the kriging equations system and do not need any parametric model assumption. The currently available estimation programs have been easily adapted to use these maps directly.

The used technique has a particular interest for the cokriging case, in which we do not need to use a lineal model of coregionalization.

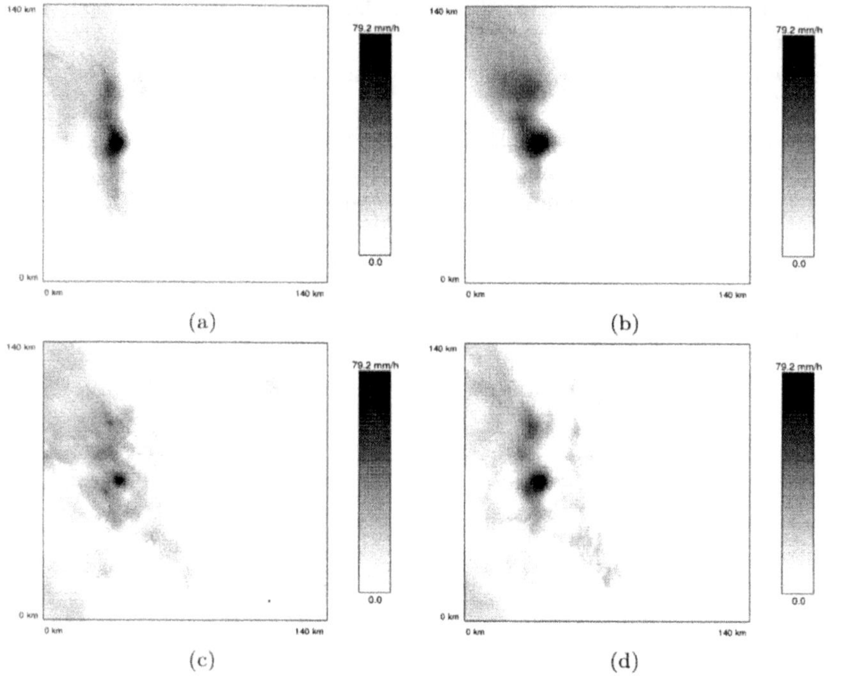

Figure 5. Rainfall estimated fields. (a) Estimated field by ordinary kriging using only the rainfall correlogram map. (b) Estimated field by ordinary kriging using the radar correlogram map. (c) Estimated field by ordinary cokriging. (d) Estimated field by kriging with an external drift.

ACKNOWLEDGEMENTS

We thank Prof. Daniel Sempere Torres from Polytechnic University of Catalunya who was so kind to provide us the data set to carry out the study. We also thank the support obtained from project *Desarrollo de técnicas hidrometeorológicas operativas para la previsión de inundaciones basadas en el radar meteorológico* financed by the Spanish Ministry of Science and Technology.

REFERENCES

1. Cassiraga, E. F., Gómez-Hernández, J. J. (1997). Improved Rainfall Estimation by Integration of Radar Data: A geostatistical approach. In A. Soares, editor, *geoENV I-Geostatistics for Environmental Applications*, Kluwer Academic Publishers, 363-374.
2. Creutin, J. D., G. Delrieu, and T. Lebel (1988). Rain measurement by raingage-radar combination: A geostatistical approach. *J.Atmos. Oceanic Technol.*, 5(1), 102-115.

3. Deutsch, C. and Journel, A. G. (1997). *GSLIB: Geostatistical Software Library and User's Guide*. Oxford University Press, New York.
4. Goovaerts, P. (2000). Geostatistical approaches for incorporating elevation into the spatial interpolation of rainfall. *Journal of Hydrology*, 228, 113-129.
5. Krajewski, W. F. (1987). Cokriging radar-rainfall and rain gage data. *Journal of Geophysical Research*, 92, (D8), 9571-9580.
6. Seo, D.-J., W. F. Krajewski, and D. S. Bowles (1990a).Stochastic interpolation of rainfall data from rain gages and radar using cokriging. 1. Design of experiments. *Water Resources Research*, 26 (3),469-477.
7. Seo, D.-J., W. F. Krajewski, A. Azimi-Zonooz and D.S.Bowles (1990b). Stochastic interpolation of rainfall data from rain gages and radar using cokriging. 2. Results. *Water Resources Research*, 26 (5),915-924.
8. Sun, X., Mein, R. G., Keenan, T. D. and Elliot, J. F. (2000). Flood estimation using radar and raingauge data. *Journal of Hydrology*, 239, 4-18.
9. Yao, T. (1998). *Automatic covariance modeling and conditional spectral simulation with Fast Fourier Transform*, PhD dissertation, Stanford University, 173 p.
10. Yao, T. and Journel, A. G. (1998). Automatic modeling of (cross) covariance tables using fast fourier transform, *Mathematical Geology*, 21(7), 715-739.

COMBINING RAINGAGES AND RADAR PRECIPITATION MEASUREMENTS USING A BAYESIAN APPROACH

C. Mazzetti and E. Todini
Dept. of Earth and Geo-Environmental Sciences, University of Bologna, Italy

Abstract: This paper examines a new technique, based upon the combination of block kriging and Kalman filter in order to optimally combine, in a Bayesian sense, spatial precipitation fields estimated from meteorological radar with the same fields estimated from point measurements of precipitation, such as the ones provided by a network of rain gauges. The Bayesian combination technique is tested by means of a numerical example, in order to demonstrate the potentiality of the proposed algorithm and to compare it with the methods developed in the past. The new method is shown to be superior, both in terms of bias and variance reduction, to the available ones, from Brandes' method (or similar) based on Barnes' objective analysis scheme, to the co-kriging approach.

Key words: Radar, rain gauges, Kalman filter, Bayesian combination

1. INTRODUCTION

At present, the most important and widely used systems for providing precipitation measurements, which can be used for real-time flood forecasting, are ground based tele-metering rain gauges and meteorological radar. Rain gauge data are typically considered to provide good point accuracy, since errors due to wind speed, which may reduce the funnel effective area can be corrected, but they offer little information on the spatial distribution of rain. On the other hand meteorological radar is capable of accurately delineating rainfall distribution but, because of various meteorological, equipment and methodological factors, its estimates of rainfall are burdened with errors that are very often quite significant, so in

X. Sanchez-Vila et al. (eds.), geoENV IV – Geostatistics for Environmental Applications, 401-412.

general radar can produce mainly biased estimates of rain. This has given rise to the interesting problem of using the point rain gauge measurements to improve the overall estimate of rainfall volume.

Although several techniques for merging rain gauges and radar data have been developed in the past, such combinations have generally produced good results in terms of bias reduction, while scant attention has been given to the reduction of variance.

The different nature of errors, which implies their independence, can be exploited to produce unbiased and more reliable precipitation estimates. Following this idea, Todini (2001, ref.5) recently proposed an original Bayesian combination technique, based on the use of block kriging and Kalman filter, that aims at eliminating the bias of meteorological radar precipitation estimates and at producing minimum variance precipitation estimates on pixels of variable size.

This paper extends the Todini (2001, ref.5) formulation to include the time evolution of the measurement error structure, which makes the technique suitable for real-time applications. Moreover, by means of a numerical example, the paper compares Todini's technique, both in its steady state and time variant formulation, to three techniques, available in literature, for adjusting radar data due to Brandes (1975), Koistinen and Puhakka (1981) and Krajewski (1987).

2. TECHNIQUES PROPOSED IN THE PAST

2.1 Brandes' technique

Following Brandes (1975), the radar field is calibrated with rain gauge observations by determining multiplicative calibration factors at each gauge site. Raw radar data from the cells containing the gauges are divided by the gauge amount to determine calibration factors G_k at each gauge site k; then Barnes (1964) objective analysis scheme is used to extrapolate corrections from the rain gauge site to all the other grid points representing the radar field. The weight WT_k each gauge calibration G_k receives at a grid point is:

$$WT_k = e^{-d^2/EP} \tag{1}$$

where d is the distance between the gauge site and the grid point and EP controls the degree of smoothing. Two steps through the objective analysis scheme are made to produce the final radar adjustment field. In the first step, a first guess grid point calibration F_1 is computed as:

$$F_1 = \frac{\sum_{k=1}^{N} WT_k \cdot G_k}{\sum_{k=1}^{N} WT_k} \tag{2}$$

where N is the number of gauges. In the second step, the final grid point adjustment factors are obtained as:

$$F_2 = F_1 + \frac{\sum_{k=1}^{N} WT'_k \cdot G_k}{\sum_{k=1}^{N} WT'_k} \tag{3}$$

where $D_k = G_k - F_1$ at each gauge location, and WT'_k is computed using (1) with $EP' = EP/2$. Multiplication of the adjustment field with the radar field produces the corrected (calibrated) radar precipitation field.

The radar precipitation field has been constrained to fit the gauge observations while retaining radar-observed precipitation variation between gauges. Moreover, the value of EP in equation (1) can negatively influence the posterior estimates and should be kept as small as possible to preserve details in the input observations.

2.2 Koistinen and Puhakka's technique

The method proposed by Koistinen and Puhakka (1981) is a modification of the Brandes (1975) method. It combines the uniform range dependent adjustment, by which the bias is removed from radar estimates, and the spatially varying adjusting method, by which radar measurements can be adjusted to fit individual gauge observations. In particular, in this paper we analyse the scheme used at ARPA-SMR, Meteorological Service of Emilia Romagna, Italy, which is a slightly modified version of Koistinen and Puhakka's algorithm.

In a first step, an adjustment factor A_k, i.e. the ratio between the gauge and the radar value in the same location, is computed. Then a regression analysis is performed, by which the range dependence of $log(A_k)$ is determined. As a result, a symmetrical range dependent adjustment factor field $A(r)$ is obtained:

$$A(r) = e^{\alpha + \beta \cdot r} \tag{4}$$

For each cell ij of the radar field, the adjustment factor field A^G can be determined as follows:

$$A_{ij}^G = \frac{\sum_{k=1}^{N} W_{ijk} \cdot A_k}{\sum_{k=1}^{N} W_{ijk}} \tag{5}$$

$$W_{ijk} = \exp\left(-\frac{r_{ijk}^2}{4\bar{r}_{ij}^2}\right) \tag{6}$$

where r_{ijk} is the distance between cell ij and gauge k and r_{ij} is the average distance between cells and gauges.

The final step computes the adjustment factor A^{ANA} by combining the range dependent function $A(r)$ and the adjustment factor A^G:

$$A_{ij}^{ANA} = A(r) + \exp\left(-\frac{\bar{r}_{ij}}{1.5 \cdot \rho}\right) * \left(A_{ij}^G - A(r)\right) \tag{7}$$

where ρ is the average density network. Multiplication of the adjustment factor field by the raw radar field produces the corrected (calibrated) radar field.

2.3 The co-kriging technique

In 1987 Krajewski proposed a new technique for merging rain gauges and radar data. It is based on an ordinary co-kriging procedure and, as opposed to the ones previously developed, it accounts explicitly for the different sampling characteristics of radar and rain gauge networks.

The first step in order to consistently combine the two fields is to interpolate rain gauge data onto the same grid blocks as those for which radar data are given, using block kriging technique. Then, the following model is proposed for merging the radar field R_{ij} and the field G_{ij}, obtained by block kriging rain gauge data:

$$V^* = \sum_{ij=1}^{Nl} \lambda_{Gij} \cdot G_{ij}(u_{ij}) + \sum_{ij=1}^{Nl} \lambda_{Rij} \cdot R_{ij}(u_{ij}) \tag{8}$$

where Nl is the number of radar cells and λ_{Gij} and λ_{Rij} are the coefficients (weights) that need to be estimated.

The weights λ_{Gij} and λ_{Rij} can be obtained minimizing the estimation variance under unbiased conditions. The problem can be solved using Lagrange multiplier technique, which leads to a set of simultaneous linear equations that can be written in matrix form as:

$$\begin{bmatrix} Cov_{RR} & Cov_{RG} & 1 & 0 \\ Cov_{GR} & Cov_{GG} & 0 & 1 \\ 1 & 0 & 0 & 0 \\ 0 & 1 & 0 & 0 \end{bmatrix} \cdot \begin{bmatrix} \lambda_R \\ \lambda_G \\ \mu_R \\ \mu_G \end{bmatrix} = \begin{bmatrix} Cov_{VR} \\ Cov_{VG} \\ 0 \\ 1 \end{bmatrix} \tag{9}$$

where Cov_{RR} and Cov_{GG} are the covariance of radar data and block kriged rain gauge data, respectively and Cov_{RG} and Cov_{GR} are the cross-covariance between gauges and radar data.

In order to ensure positive definiteness of the matrix of the co-kriging system, the matrices Cov_{RR}, Cov_{GG} and Cov_{RG} are modeled using exponential isotropic models, which are fitted using least squares technique. On the right hand side of the system, the vectors Cov_{VR} and Cov_{VG}, whose elements are covariances between radar and rain gauge data, respectively, and the true precipitation V, are approximated using:

$$Cov_{VR} = \beta_R \cdot Cov_{RR} \quad \text{and} \quad Cov_{VG} = \beta_G \cdot Cov_{GG} \quad \text{where} \quad \beta_R, \beta_G \in (0,1)$$

As opposed to the new proposed technique, the block kriging Bayesian combination, it is not possible to estimate from data the values of β_G and β_R, which are unknown scalar, reflecting the relative uncertainty of radar and gauges observations, and they have to be provided subjectively.

The co-kriging technique shows two major problems. The first relates to the fact that it is impossible to compute the right hand side of the co-kriging system (9), since the true precipitation values are not known and the approximation of the covariance matrices introduces additional uncertainty in the method. Moreover the solution of the co-kriging system implies the computation of the inverse of a matrix whose dimension is twice the number of cells of the lattice. For application on real basins the number of lattice cells is typically large and the computational burden becomes excessive.

3. THE BLOCK KRIGING BAYESIAN COMBINATION TECHNIQUE

As previously remembered, radar and rain gauges have measurement errors of different nature: rain gauge measurements tend to be more accurate in a point while their spatial significance decays with the distance and thus with the area; on the other hand radar provides better spatial (although biased) representations but a much poorer quantitative estimate. The different nature of errors, which implies their independence, has been exploited by Todini (2001, ref.5) to develop a Bayesian combination technique for merging rain gauges and radar data. The proposed technique aims at eliminating the bias of meteorological radar precipitation estimates and at producing minimum variance precipitation estimates on pixels of variable size.

3.1 Steady state formulation

Originally, Todini (2001, ref.5) developed the Bayesian combination technique on the basis of a steady state assumption, which means that the time evolution of the measurement error structure is not taken into account.

In order to consistently combine the two sets of data, Todini's method uses block kriging to regionalize the point rain gauge measurements on the pixels on which the radar data are given and to compute the block kriged variables error statistics on the pixels. Assuming that the rain gauge estimates are unbiased, once the estimation error statistics have been determined, a Kalman filter approach is taken to find the posterior estimates. In the Bayesian combination framework, the field y_t^R provided by the radar is taken as the a priori estimate, while the field y_t^G provided by the block kriging of the gauges is taken as the measurement vector z_t of a classical Kalman filter. The measurement equation of a classical Kalman filter is modified as follows to give a new measurement equation (10):

$$z_t = y_t^G = H_t \cdot y_t + \eta_t = y_t + \left(y_t^G - y_t\right) \tag{10}$$

where y_t is the true rainfall field at time t. At this point by taking:

$$y'_t = y_t^R - \mu_{\varepsilon_t^R} \quad \text{and} \quad P'_t = V_{\varepsilon_t^R} \tag{11}$$

as the a priori estimate of the state and the a priori estimate of its covariance matrix, it is possible to compute the innovation ν_t and the Kalman gain K_t, following the development of a classical Kalman filter.

In the end, the Kalman filter equations allow finding the posterior estimates by combining the a priori estimates and the measurements in a Bayesian framework to give:

$$y_t'' = y_t' + K_t \cdot v_t \tag{12}$$

$$P_t'' = P_t' - K_t \cdot H_t \cdot P_t' \tag{13}$$

where y_t'' is the posterior estimate of the rainfall field over the lattice and P_t'' its error of estimate covariance matrix.

3.2 Time variant formulation

The application of the Bayesian combination technique to real-time problems requires the development of real-time updates of the means and covariance matrices as a function of their evolution. For this reason means and covariance matrices are computed at each time step in order to take into account the time evolution of the measurement error structure and a new real-time estimator had to be developed. The gain equation is now modified to give:

$$K_t = P'_t \cdot (P'_t + V\eta_t)^{-1} = V_{\varepsilon_t^R} \cdot \left(V_{\varepsilon_t^R} + V_{\varepsilon_t^G}\right)^{-1} = V_{\varepsilon_t^R} \cdot V_{\varepsilon_t}^{-1} \tag{14}$$

To ensure positive definiteness, the matrix $V_{\varepsilon t}$, the covariance matrix of the deviations ε_t, used in the gain equation (14) needs to be modeled using a variogram. The estimation of the variogram parameters is performed using the Maximum Likelihood technique developed by Todini (2001, ref.6).

4. THE NUMERICAL EXAMPLE

The efficiency of the different methods, in terms of convergence of the posterior estimates toward the true, is demonstrated by means of a numerical example, for which the true rainfall field is perfectly known. This stochastic simulation is needed because in real world cases the actual precipitation field is not known.

In the proposed example, a 7x7 lattice with sides of 1 Km is considered, while 9 rain gauges are assumed to be set in the centers of the lattice cells as in Figure 1. The small dimension of the grid does not want to simulate operational conditions, its purpose is to create a numerical example in order to compare the statistical results of the different techniques. The high number of gauges and their symmetrical distribution were chosen to test the conservation of the symmetry in the percentage variance reduction.

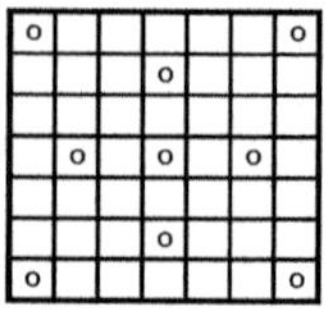

Figure 1. Distribution of the 9 rain gauges on the radar grid

A Gaussian random field y_t, taken as the true field, is generated 1000 times jointly on the lattice, representing the radar field, and on the measurement points, representing the rain gauges (Table 1).

Table 1. Random field parameters

Mean	Variance	Nugget (p)	Sill (w)	Range (a)
10	50	0	50	107

In addition, errors are generated both on the lattice and on the measurement points to simulate the different errors one could expect from the radar and from the gauges. In the example, radar measurements are considered biased and affected by noise. Therefore 1000 time realizations of a Gaussian random noise are generated on the lattice and are added to the true rainfall field y_t to give noise corrupted radar observations y_t^R. In practice, the error represents a large bias and a variance of the order of 30% of the signal (Table 2).

Table 2. Random noise field parameters

Mean	Variance	Nugget (p)	Sill (w)	Range (a)
5	15	0	15	10^6

Rain gauge observation errors are assumed to be random, uncorrelated in space and of the order of 10%. Errors are added to the value of the true rainfall field on the gauge points to give 1000 time realizations of 9 gauge like observations.

The data set used in this analysis consists in 1000 time realizations of radar estimates on the 49 cells together with 1000 time simultaneous realizations of 9 rain gauge measurements. Finally, 1000 time realizations of the true rainfall field on the lattice cells are also available and are used for the analysis of the convergence. After using the different approaches, the true rainfall field is subtracted from the posterior estimates and the error statistics, mean and variance, are computed for all the lattice cells.

In this numerical experiment, as opposed to what happens in real world cases, the knowledge of the true rainfall is used for assessing the performances of the different approaches.

4.1 Brandes' technique

The empirical exponential weighting method proposed by Brandes has been applied to the numerical data and the statistics have been computed for all the lattice cells. The results (Tab. 3) show that there is only a little bias over the lattice cells after the merging and that the posterior explained variance increases from 70% to 77%.

Table 3. Bias and explained variance improvements with Brandes' technique

	A priori	A posteriori
Bias	5.0550	0.3555
Ex. Variance	0.7033	0.7777

4.2 Koistinen and Puhakka's technique

Koistinen and Puhakka's method, which combines the uniform range dependent adjustment and the spatially varying adjusting method, has been applied to the numerical data. The results show (Tab. 4) an improvement in bias and explained variance toward Brandes' method: the value of the final bias (after the merging) is now smaller than the one obtained before and also the value of the final explained variance has increased.

Table 4. Bias and explained variance improvements with Koistinen's technique

	A priori	A posteriori
Bias	5.0550	-0.1390
Ex. Variance	0.7033	0.7847

4.3 The co-kriging technique

In order to compare the co-kriging technique with the steady state formulation of the block kriging and Kalman filter method, a steady state solution for the co-kriging system is proposed and two possibilities have been considered. The first one assumes that the right hand side of the co-kriging system can be computed using the true rainfall field. Although this is not possible in real world applications it was interesting to see if in the ideal case the approach performed well.

The results show that the bias has been completely removed while the explained variance has increased from 70% to 85%.

Table 5. Bias and ex. variance with steady state technique, computing the right hand side of the system

	A priori	A posteriori
Bias	5.0550	0.0229
Ex. Variance	0.7033	0.8515

Alternatively, the matrices Cov_{VR} and Cov_{VG} on the right hand side of the co-kriging system have been approximated using the values of the matrices Cov_{RR} and Cov_{GG} and two coefficients β_R and β_G as proposed by Krajewski (1987). Different couples of coefficients β_R and β_G have been used and the posterior radar estimates were compared with the true value of the rainfall field. The results show that the choice of the parameters strongly influences the quality of the posterior estimates and a bad choice of β_R and β_G can lead to poor quality results that do not converge toward the true value of the rainfall field. Table 6 shows the best results in terms of final bias and explained variance.

Table 6. Bias and ex. variance with steady state co-kriging technique, using the coefficients

	A priori	A posteriori
Bias	5.0529	0.0886
Ex. Variance	0.7035	0.9025

However in both cases the improvement in bias and explained variance shown by the co-kriging method is still smaller than the one provided by the Block Kriging Bayesian combination technique in the steady state formulation.

The time variant formulation of the co-kriging technique follows the algorithm proposed by Krajewski, which computes the co-kriging system for each time step, modeling the right hand side with the help of the coefficients β_R and β_G.

Again, the application of the method based upon co-kriging required the trial of several couples of coefficients β_R and β_G and Table 7 shows the results obtained using the couple of parameters that has shown to give the best results in terms of convergence of the posterior radar estimates toward the true value of the rainfall field.

Table 7. Bias and ex. variance improvements with time variant co-kriging technique

	A priori	A posteriori
Bias	5.0529	0.0634
Ex. Variance	0.7035	0.7859

The results show that the bias has been eliminated over the entire lattice and the posterior explained variance increases to 78%.

However, in real world there is no possibility to evaluate how good is a choice of parameters compared to another one. So there is no way of knowing which couple of parameters leads to the posterior radar estimate which is nearer to the true rainfall field.

4.4 The block kriging Bayesian combination technique

The Bayesian combination technique is tested for two different formulations of the Kalman filter: the steady state formulation, in which the time evolution of the measurement error structure is not taken into account, and the time variant formulation, in which real-time updates of the means and covariance matrices are considered as a function of their evolution.

For the steady state formulation of the Bayesian combination technique the results are impressive. Table 8 shows that the bias has been completely eliminated and the posterior explained variance reaches a very high value equal to 93%.

Table 8. Bias and ex. variance improvements with steady state Bayesian technique

	A priori	A posteriori
Bias	5.0529	0.0517
Ex. Variance	0.7035	0.9331

Some comparisons can be made with the results obtained with the steady state formulation of co-kriging: the bias has been eliminated in both cases, but the increase of the posterior explained variance is much higher for the Bayesian combination.

The block kriging Bayesian combination technique in the time variant formulation of the Kalman filter has been applied to the numerical data. The results are summarized in Table 9, which shows the posterior bias and explained variance, and they are quite similar to the values of Table 8.

Table 9. Bias and ex. variance improvements with time variant Bayesian technique

	A priori	A posteriori
Bias	5.0529	0.0942
Ex. Variance	0.7035	0.9103

The comparison between Table 3, Table 4, Table 7 and Table 8 shows that the best results in terms of posterior bias and explained variance are obtained using the Bayesian combination technique.

5. CONCLUSIONS

The test on numerical data has demonstrated the efficiency of the Bayesian combination technique and the improvements with respect to the most common and widely used method for merging rain gauges and radar data. This was proved both in the steady state and in the time variant formulation.

Extensive application of the technique is anticipated within the frame of the UE funded project MUSIC (Multi Sensor precipitation measurements Integration Calibration and flood forecasting).

ACKNOWLEDGMENTS

The study reported in this paper has been carried out within the frame of the UE funded project MUSIC (Multi Sensor precipitation measurements Integration Calibration and flood forecasting), contract number EVK1-CT-200-00058. The authors would also like to thank ARPA-SMR and Dr. Sandro Nanni for their supportive effort.

REFERENCES

1. Barnes S.L., A technique for maximizing details in numerical weather map analysis, J. Appl. Meteor., 1964; 3:396-409.
2. Brandes E.A., Optimizing rainfall estimates with the aid of radar, J. Appl. Meteor., 1975; 14:1339-1345.
3. Krajewski W.F., Cokriging radar-rainfall and rain gauge data, J. Geophysical Res., 1987; 92:9571-9580.
4. Koistinen J., Puhakka T., An improved spatial gauge-radar adjustment technique, 20th Conference on radar Meteorology, AMS Boston USA, 1981; 179-186.
5. Todini E., Bayesian conditioning of radar to rain gauges, Hydrol. Earth System Sci., 2001; 5:225-232.
6. Todini, E., Influence of parameter estimation uncertainty in Kriging. Part 1. Theoretical development, Hydrol. Earth System Sci., 2001, 5(2), 215-223.

SPATIO-TEMPORAL KRIGING OF SOIL SALINITY RESCALED FROM BULK SOIL ELECTRICAL CONDUCTIVITY

A. Douaik[1,2], M. Van Meirvenne[2] and T. Tóth[3]

[1]Department of Computer Science and Biometry, Institut National de la Recherche Agronomique, Avenue de la victoire, BP 415, Rabat, Morocco; [2]Department of Soil Management and Soil Care, Ghent University, Coupure Links 653, Gent, Belgium; [3]Research Institute for Soil Science and Agricultural Chemistry, Hungarian Academy of Sciences, Herman O. ùt 15, PO box 35, 1525 II Budapest, Hungary.

Abstract: Our spatial data consist of 413 measurements of the apparent electrical conductivity (ECa) obtained with electrical probes in the east of Hungary. Additionally, a limited subset of the locations (15 to 20) was sampled for laboratory analysis of soil electrical conductivity of 1:2.5 soil:water suspension (EC2.5), a simple proxy for the electrical conductivity of soil saturation extract (ECe). The latter formed our calibration data set. This procedure was repeated 17 times between November 1994 and December 2000 yielding a large spatio-temporal database. The first step was to rescale EC2.5 from ECa, based on the calibration data sets, using classical and spatial regression models. The residuals of the ordinary least squares model were tested for the absence of spatial dependence using the Moran's I test. This hypothesis was accepted, the EC2.5 was rescaled using the classical regression model. The next step was to identify the structure of the variability of the rescaled EC2.5 by computing and modeling the spatial, the temporal, and the spatio-temporal covariograms. Finally, soil salinity maps were produced for the study area and for any time instant using spatio-temporal kriging. The estimates were more precise compared to the ones obtained using only the spatial covariogram computed and modeled separately for each time instant.

1. INTRODUCTION

The effective control of soil salinity requires the knowledge of its magnitude and extent, and also its changes over time. Detecting trends,

X. Sanchez-Vila et al. (eds.), geoENV IV – Geostatistics for Environmental Applications, 413-424.

which occur in salinity conditions over time, is an important step to identify emerging problems, and to determine the progress of reclamation efforts.

Soil salinity assessment requires inventory and monitoring of soil salinity, since it is spatially variable and temporally dynamic in nature.

Soil salinity is conventionally determined by measurements of the electrical conductivity of the extract of a water-saturated soil-paste (ECe). This property can also be observed indirectly from measurements of the apparent electrical conductivity (ECa) of the bulk soil. The latter is measured in the field using electrical probes.

The conventional soil sampling and laboratory analysis procedure is very expensive. A cost-effective way is to use mobile techniques for rapidly measuring ECa as a function of the spatial position, to infer ECe from ECa, and to map ECe at any location in space and any instant in time.

Lesch et al. (1998) developed a statistical monitoring strategy. It requires the estimation of a conditional regression model to predict ECe from ECa, and the use of 2 statistical tests: one for detecting dynamic spatial variation in the salinity pattern and the other for detecting a change in the field median salinity level with time. The drawback of this approach is that we get salinity maps only for the observed time instants, and at the observed locations.

We propose in this work to use an alternative approach, based on geostatistical tools, which is capable of using the spatial and temporal dependencies as well as producing maps for any location in space and any time instant.

2. DATA SETS

The study area (of about 25 ha) is located in the Hortobagy National Park (47o 30" N and 21o 30" E), east Hungary. A lot of research on salinity/sodicity and its correlation to the vegetation has been done in this natural ecosystem (Toth et al., 1991; Van Meirvenne et al., 1995; Toth et al., 1998; and Toth et al., 2001).

We obtained measurements taken at 17 time instants over 7 years (from November 1994 to December 2000) with an approximate average temporal lag of 3 months, ranging from 2 to 9 months.

For each time instant, we have 2 data sets. The calibration data set for which we have the measures of the soil salinity in the laboratory (EC2.5 in dS.m-1) and the soil bulk electrical conductivity (ECa in dS.m-1). The measurements have been done in 15 to 20 locations depending on the time instant of sampling. The second data set involves only the measurements of ECa at 286 to 413 locations depending again on the time instant.

For the soil samples, we determined, in addition to EC2.5, the soil moisture content (%) and the soil pH. The soil samples were taken between 0 and 40 cm by 10 cm increments (bulked samples from 2 augerings).

The ECa was measured using electrical probes (with 4 probes), which were inserted down to 2 depths (8 and 13 cm) giving values corresponding to 0-20 and 0-40 cm soil depths, respectively. For the calibration sites, there were always 3 parallel measures. Fig. 1 illustrates the spatial location of the measurements and how the area was sampled.

3. ANALYSIS

The histograms of the ECa and also of EC2.5 showed skewed distributions and after a logarithmic transformation, the distributions became less asymmetric. All the analysis was based on the transformed data. The calibration data set was used to compute the calibration equations (one equation for each time instant) as a first step. We tried to relate EC2.5 to ECa using 2 approaches. The first method is the classical ordinary least squares regression (OLS):

$$\ln(EC2.5) = a + b\ln(ECa) + \xi$$

where a and b are the regression coefficients and x represents the independently gaussian errors.

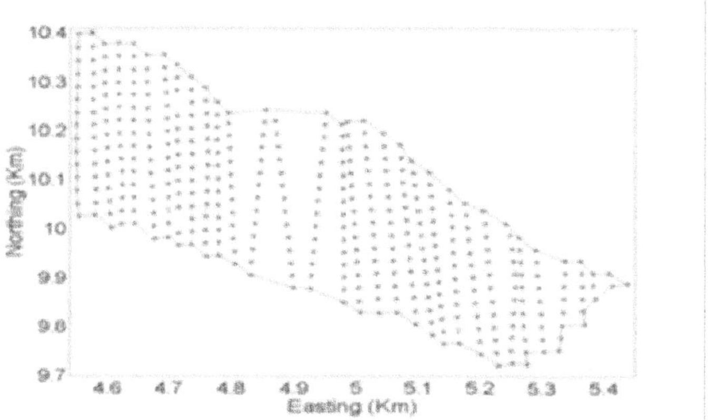

Figure 1. Spatial location of the samples where ECa were measured. The calibration data set is a sub-sample of these locations (relative position).

The other approach is the spatial regression (Anselin, 1988). For this method we checked 4 different models:

- the spatial autoregressive model (SAM):

$$\ln(\mathrm{EC2.5}) = \rho \mathrm{W} \ln(\mathrm{EC2.5}) + b \ln(\mathrm{ECa}) + \xi$$

- the spatial error model (SEM):

$$\ln(\mathrm{EC2.5}) = b \ln(\mathrm{ECa}) + \omega \qquad \text{with } \omega = \lambda \mathrm{W} \omega + \xi$$

- the spatial general model (SGM) which is the combination of the 2 above models:

$$\ln(\mathrm{EC2.5}) = \rho \mathrm{W} \ln(\mathrm{EC2.5}) + b \ln(\mathrm{ECa}) + \omega \qquad \text{with } \omega = \lambda \mathrm{W} \omega + \xi$$

In this equations ρ and λ are the spatial autocorrelation parameters, ω represents the errors with spatial dependence, and W is the matrix of the spatial weights build from the distance separating 2 observations using the Delaunay triangulation algorithm.

- The geographically weighted regression, GWR (Brundson et al., 1996):

$$\mathrm{W}_i^{1/2} \ln(\mathrm{EC2.5}) = \mathrm{W}_i^{1/2} b \ln(\mathrm{ECa}) + \xi$$

It uses distance-weighted sub-samples of the data to produce locally linear regression estimates for every point in space. Each estimated set of parameters is based on a distance-weighted sub-sample of neighboring observations based on distances separating the observations.

The residuals of the OLS regression were tested for the presence of spatial autocorrelation using the Moran's I test (Cliff and Ord, 1981). Further, we used maximum likelihood-based tests, on the results of the spatial regression, to check the significance of the spatial autocorrelation parameters (ρ and λ) and to choose the most adequate model.

At the end of this step, we got a data matrix of 17 columns (time instants) and 413 rows (locations) of EC2.5 values with some missing values corresponding to the locations for which ECa was not available.

This data matrix can be considered as a space time random field, STRF (Christakos, 1992):

$$Z(s,t),\ (s,t) \in D \times T \quad \text{with}$$
$$D \subset \Re^2 \text{ (real numbers set) and } T \subset \Re_+ \text{ (positive real numbers set)}$$

with s : 2-D spatial coordinates and t : temporal coordinate.

The second step in our analysis was to model the spatial, the temporal and the spatio-temporal dependencies of the salinity data matrix. There are mainly 3 conceptual approaches in modeling stochastically space-time data (Kyriakidis and Journel, 1999):

– Methods using a STRF (Christakos, 1992; Cressie, 1993);
– Methods based on vectors of independent spatial random fields (Goovaerts and Sonnet, 1993). The spatial variability is modeled either by a separate variogram for each time instant or by a single spatial variogram considering time instants as replicates as was done by Sterk and Stein (1997);
– Methods based on vectors of time series (Rouhani and Wackernagel, 1990).

The second approach is more suited in the case of rich data in the space domain and scarce data in the time domain but doesn't include the temporal dependence existing between observations and can predict only at the observed time instants.

The third approach is more adequate for data dense in time and scarce in space but it doesn't take into account the spatial dependence and it predicts only at the observed locations.

Only the first group of methods includes both the spatial and temporal dependencies so the interpolation is more precise and can be done for unsampled time instants at unsampled locations. This approach was used to analyze our salinity data set.

The procedure is as follows (Christakos et al., 2002):

– First the space-time mean trend is estimated. The smoothed spatial components (one for each location) were computed using an exponential spatial filter applied to the averaged measurements (for each location, over all the time instants). We computed also the smoothed temporal components (one for each time instant) using an exponential temporal filter applied to the averaged measurements (for each time instant, over all the locations);
– Then the above components of the space-time mean trend were interpolated to the data grid giving $m(s,t)$;
– The residuals were computed as the space-time mean trend subtracted from the original data matrix: $R(s,t) = Z(s,t) - m(s,t)$;
– The residual data matrix was used to compute the spatial $C(r,\tau=0)$, temporal $C(r=0,\tau)$ and spatio-temporal $C(r,\tau)$ covariograms :

$$C(r,\tau=0) = \mathrm{cov}[R(s+r),R(s)],$$

$$C(r=0,\tau) = \mathrm{cov}[R(t+\tau),R(t)],$$

$$C(r,\tau) = \mathrm{cov}[R(s+r,t+\tau),R(s,t)]$$

r and τ are the spatial and temporal lags, respectively, and cov is the covariance function.

– Finally we fitted theoretical models to the computed experimental covariograms.

The last step was the estimation at unobserved locations and time instants using space-time kriging (Chiles and Delfiner, 1999; Christakos, 1992) and the fitted covariograms.

As we used the residual covariograms, the resulting estimated data corresponded to the residual values. To get the values in the original scale, we interpolated the spatial and temporal components of the space-time mean trend to the kriging grid. These estimated values were added to the interpolated space-time mean trend values to obtain the kriged values in the original scale.

The classical regression was done using the SAS software (SAS Institute, 1990), the spatial regression was fitted using the Econometrics Toolbox (Lesage, 1999) running under Matlab software. The geostatistical computations were handled using the BMElib library (Christakos and al., 2002). The toolbox and the library are built on the Matlab software (Mathworks, 1999).

4. RESULTS

4.1 Calibration Equations

The OLS residuals showed no significant spatial dependence. This result was confirmed by the maximum likelihood-based tests of the non-appropriateness of an additional spatial parameter in the spatial regression models. However when we fitted a first autoregressive model to the ECa data (ECa regressed on its neighbors), we found a significant spatial dependence. The absence of spatial autocorrelation in the EC2.5-ECa relationship may be due to the fact that we have very few locations (15 to 20) which are far apart comparatively to the ECa data (286 to 413).

Consequently we adopted the classical OLS regression model to relate EC2.5 to ECa. This relation was very strong. Most of the correlation coefficients were higher than 0.85 (for 14 out of the 17 time frames) with a maximum value of 0.95. Different models were tried by adding other covariates than ECa, for example the coordinates and the vegetal coverage. The best model (having the highest adjusted coefficient of determination, the lowest mean square error and all its coefficients being significant) was the following:

$$\ln(EC2.5) = b_0 + b_1\ln(ECa) + b_2u + b_3u^2 + b_4v + b_{5j}cover_j$$

$u=(x - m_x)/ s_x$ with m_x and s_x: mean and standard deviation of the x coordinate

$v=(y - m_y)/ s_y$ with m_y and s_y: mean and standard deviation of the x coordinate

$cover_j$, j=1,…,4, represents the 4 categories of vegetal coverage

We fitted this model at each time instant separately, so finally we obtained 17 equations corresponding to the 17 time instants.

4.2 Descriptive statistics

The main statistic parameters of our data set are summarized in table1. The mean EC2.5 varied between 1.39 (November 1994) and 2.74 $dS.m^{-1}$ (September 1998). The minimum is enclosed between 0.06 (December 2000) and 0.68 (March 1996) and maximum varying between 2.59 (November 1994) and 9.41 (December 2000). The data are moderately to highly variable with coefficients of variation ranging between 0.28 (March 1997) and 0.64 (December 1997 and 2000).

Table 1. Statistic parameters of salinity data (EC2.5 in $dS.m^{-1}$). N: number of observations, cv: coefficient of variation, min: minimum, med: median, max: maximum.

EC2.5	N	mean	cv	min	med	max
Nov-94	411	1.39	0.37	0.45	1.38	2.59
Mar-95	411	2.03	0.32	0.60	1.94	4.78
Jun-95	412	1.74	0.39	0.48	1.62	4.66
Sep-95	410	1.77	0.33	0.54	1.70	3.66
Dec-95	413	1.65	0.38	0.53	1.53	3.93
Mar-96	392	1.96	0.29	0.68	1.86	3.30
Jun-96	411	1.54	0.42	0.31	1.41	4.52
Mar-97	310	1.48	0.28	0.42	1.43	2.91
Jun-97	286	1.69	0.59	0.24	1.48	8.33
Sep-97	411	1.50	0.58	0.13	1.32	5.83
Dec-97	412	1.44	0.64	0.13	1.18	6.99
Sep-98	411	2.74	0.55	0.31	2.47	8.35
Apr-99	409	1.43	0.63	0.17	1.19	6.90
Jul-99	409	1.96	0.50	0.16	1.81	5.85
Sep-99	411	1.93	0.63	0.11	1.58	6.39
Apr-00	312	1.78	0.58	0.09	1.57	7.07
Dec-00	411	2.10	0.64	0.06	1.76	9.41
Overall	**6640**	**1.78**	**0.54**	**0.06**	**1.59**	**9.41**

For all the time instants, the range in salinity values was large in comparison to the mean indicating that soil salinity is highly variable in space. Moreover, the differences in the statistic parameters (mean, median and range) between time instants are an indication of a temporal variation.

4.3 Covariography

The spatial, temporal and spatio-temporal dependencies in the salinity data were described and modeled using covariance functions. The spatial covariance was fitted with a nested exponential model as is illustrated in Fig. 2a. The small-scale range is about 70 m (with a sill equal to 70% of the total variance) and the large-scale range is beyond the dimensions of the study area (1500 m).

$$C(r,\tau=0) = c_{01}\exp(-3r/as_1)+c_{02}\exp(-3r/as_2)$$

with c_{01} and c_{02} the sills of the nested models and as_1 and as_2 their corresponding ranges.

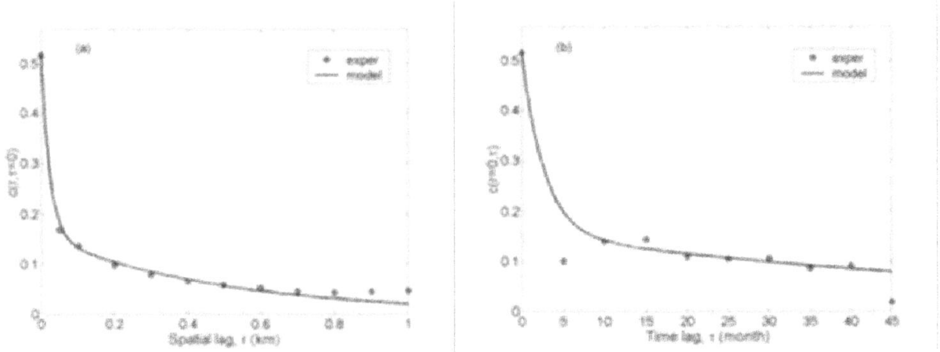

Figure 2. (a): Spatial covariogram; (b): temporal covariogram. Circles: experimental covariogram; curve: fitted model.

The same nested model was used to fit the temporal covariance (Fig. 2b) with a small-scale range of 8 months and a large-scale range far beyond the time period covered (200 months):

$$C(r=0,\tau) = c_{01}\exp(-3\tau/at_1)+c_{02}\exp(-3\tau/at_2)$$

with at_1 and at_2 the small-scale and large-scale ranges

The spatio-temporal covariance (Fig. 3) is a nested structure of two space/time separable covariance models:

$$C(r, \tau) = c_{01}\exp(-3r/as_1)\exp(-3\tau/at_1)+c_{02}\exp(-3r/as_2)\exp(-3\tau/at_2)$$

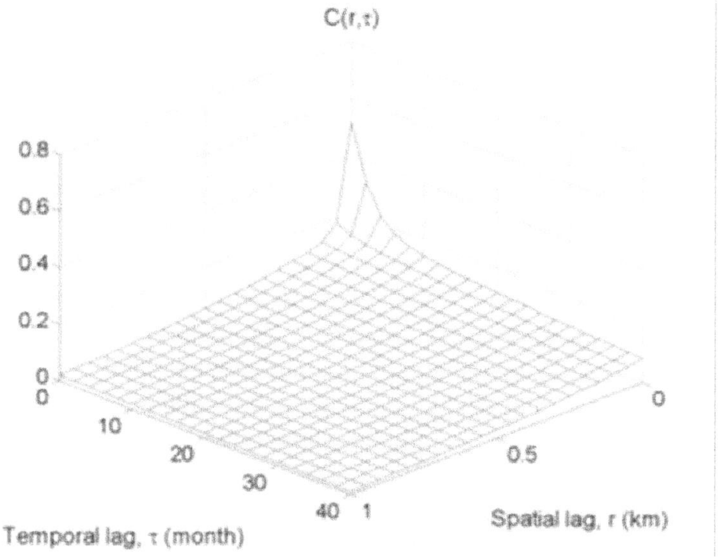

Figure 3. Spatio-temporal covariogram of the residual data R(s,t).

4.4 Spatio-Temporal Kriging

As the spatio-temporal dependence of EC2.5 was modeled using a nested separable space-time covariance function, it was used to estimate soil salinity at any location in space and any instant in time by defining a search neighborhood.

For illustration purposes we show only results for the most frequently observed month (September of the years 1995, 1997, and 1998). We estimated on a dense spatial grid including the 413 locations for which we have the observed EC2.5 values for September from 1995 to 1998 (Fig. 4).

First we note that for the non-observed time instant (September 1996), the smoothing effect is stronger than for the observed time instants (September 1995, 1997, and 1998). These are due mainly to the fact that for the latter ones, the neighbors come mostly from the simultaneous time instant but for the former one the neighbors are from different time instants. Also, there is a net general increase in soil salinity from September 1995 to 1998.

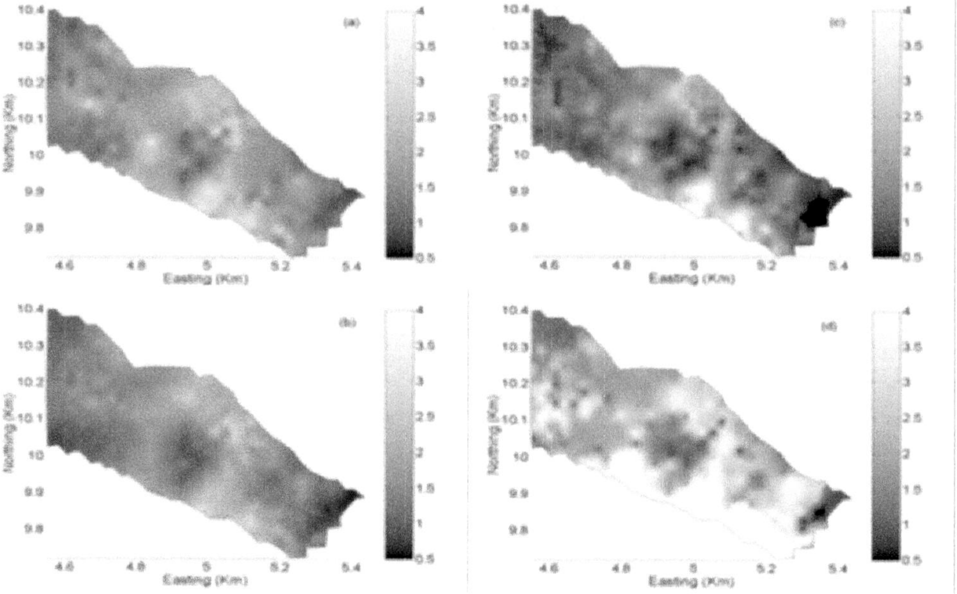

Figure 4. EC2.5 estimates ($dS.m^{-1}$) using the spatio-temporal covariogram models for each September between 1995 (a) and 1998 (d).

To check the contribution of the additional temporal dependence, we compared the spatio-temporal kriging to a single spatial kriging by modeling independently and separately the spatial dependence for each time instant. Sterk and Stein (1997) computed a single spatial variogram pooling the data of the 4 time instants that they had. This was required to circumvent the lack of sufficient observations (100 or more as reported by Webster and Oliver, 1992) to compute a reliable variogram. Ettema et al. (2000) adopted the same procedure. As we had sufficient observations at each time instant (at least 286), we computed the spatial variograms separately for each time instant. The results of the comparison are reported in Fig. 5 for the histograms of the estimated EC2.5 values and in Fig. 6 for those of their corresponding estimation errors. The estimated values are more or less the same but it is clear that the spatio-temporal estimates are more precise comparatively to the spatial estimates. Ettema et al. (2000) reached the same conclusions in their study of the spatio-temporal patchiness of nematode species.

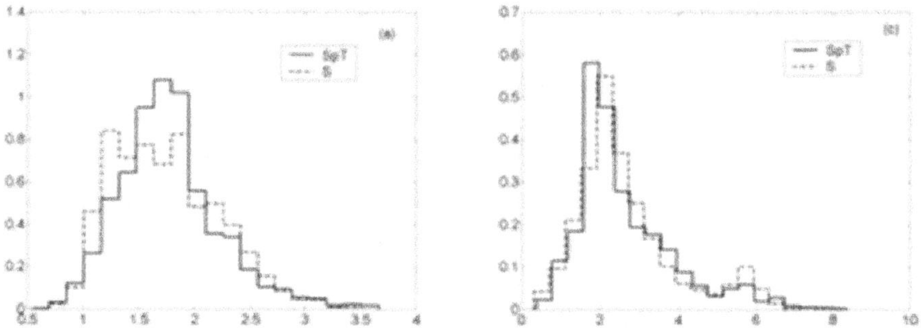

Figure 5. Spatial (S) and spatio-temporal (SpT) estimates of EC2.5 ($dS.m^{-1}$) for September 1995 (a), and 1998 (c).

5. CONCLUSIONS

The spatio-temporal kriging estimates were more precise than the estimates obtained using only the spatial component of the soil salinity dependence (the most frequent estimation error is bigger for the latter than for the former). Also the smoothing effect seems to be more pronounced in the case of the spatial kriging than in the spatio-temporal kriging (the extreme values are lesser for the former than for the latter). These conclusions were deduced from the graphic representation of the estimates and their estimation errors for the 2 approaches. For a more formal comparison, it would be better to use some quantitative criteria. So in this sense, it may be suitable to leave some locations for a validation data set that will be used in the computation of, for example, the mean error or the mean square error. Another possible improvement is to use the product-sum model of De Cesare et al. (2001).

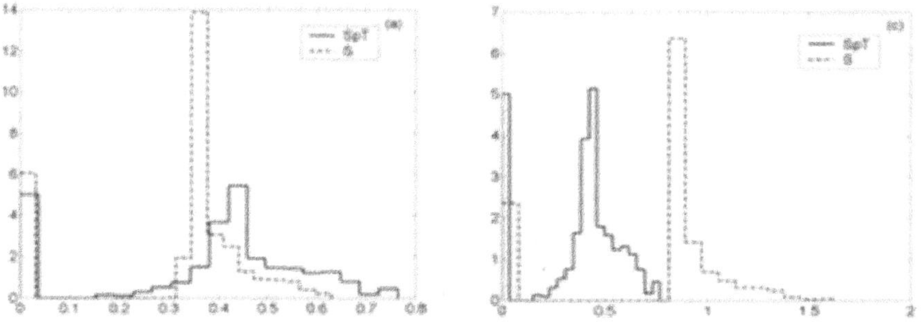

Figure 6. Spatial (S) and spatio-temporal (SpT) estimation errors of EC2.5 (dSm^{-1}) for September 1995 (a), and 1998 (c).

REFERENCES

1. Anselin L. (1988). Spatial econometrics: methods and models. Boston: Kluwer Academic.
2. Brundson C., Fotheringham AS and Charlton ME (1996). Geographically weighted regression: a method for exploring spatial non-stationarity. Geographical analysis, 28, 281-298.
3. Chiles JP and Delfiner P. (1999). Geostatistics: modeling spatial uncertainty. Wiley: New York.
4. Christakos G. (1992). Random field models in earth sciences. Academic Press: San Diego, California.
5. Christakos G., Bogaert P. and Serre ML (2002). Temporal GIS. Springer-Verlag: New York.
6. Cliff AD and Ord JK. (1981). Spatial processes: models and applications. Pion: London.
7. Cressie N. (1993). Statistics for spatial data. Wiley: New York
8. De Cesare L. Myers D and Posa D. (2001). Estimating and modeling space-time correlation structures. Statistics and Probability Letters, 51(1), 9-14.
9. Ettema, Rathbun and Coleman. (2000). On spatiotemporal patchiness and the existence of 5 species of Chronogaster (Nematoda Chronogasteridae) in a riparian wetland. Oecologia, 125, 444-452.
10. Goovaerts P. and Sonnet P. (1993). Study of spatial and temporal variations of hydrochemical variables using factorial kriging analysis. In Soares, ed., 'Geostatistics Troia'92, vol. 2, Kluwer: Dordrecht, The Netherlands, 745-756.
11. Kyriakidis and Journel. (1999). Geostatistical space-time models: a review. Mathematical geology, 31(6), 651-684.
12. Lesage JP. (1999). Applied econometrics using Matlab. The toolbox and the manual are available on the internet.
13. Lesch SM, Herrero J. and Rhoades JD. (1998). Monitoring for temporal changes in soil salinity using electromagnetic induction techniques. Soil Science Society of America Journal, 62(1), 232-242.
14. MathWorks. (1999). Using Matlab, version 5. The MathWorks Inc., Natick : MA.
15. Rouhani S. and Wackernagel H. (1990). Multivariate geostatistical approach to space-time data analysis. Water Resources Research, 26(4), 585-591.
16. SAS Institute. (1993). SAS/STAT user's guide, version 6, 2nd ed. SAS Institute, Cary: NC.
17. Sterk and Stein A. (1997). Mapping wind-blown mass transport by modeling variability in space and time. Soil Science Society of America Journal, 61, 232-239.
18. Toth T., Csillag F., Biehl LL and Micheli E. (1991). Characterization of semi-vegetated salt-affected soils by means of field remote sensing. Remote Sens. Environ., 37, 167-180.
19. Toth T., Kertsz M. and Pasztor L. (1998). New approaches in salinity/sodicity mapping in Hungary. Agrokemica es Talajtan, 47, 76-86.
20. Toth T., Kuti L., Kabos L. and Pasztor L. (2001). Use of digitalized hydrogeological maps for evaluation of salt-affected soils of large areas. Arid Land Research and Management, 15, 329-346.
21. Van Meirvenne M., De Groote P., Kertesz M., Toth T. and Hofman G. (1995). Multivariate geostatistical inventory of sodicity hazard in the Hungarian Puszta. In: Escadafal R., Mulders M. and Thiombiano L. (eds). Monitoring soils in the environment with remote sensing and GIS, ORSTOM éditions: Paris, 293-305.
22. Webster R. and Oliver M. (1992). Sample adequately to estimate variograms of soil properties. Journal of soil science, 43, 177-192.

CHARACTERIZATION OF ENVIRONMENTAL HAZARD MAPS OF METAL CONTAMINATION IN GUADIAMAR RIVER MARGINS

C. Franco[1], A. Soares[1] and J. Delgado–García[2]

[1]*Environmental Group of the Centre for Modelling Petroleum Reservoirs, CMRP/IST, Av. Rovisco Pais, 1049-001 Lisbon, Portugal, cfranco@ist.utl.pt;* [2]*Cartographical, Geodetical and Photogrammetric Engineering Department. University of Jaen, c/ Virgen de la Cabeza, 2 – 23071 Jaen, Spain*

Abstract: This paper presents a methodology to account for uncertainties in mapping of the probability of soil contamination by different heavy metals. The methodology is based on a co-simulation technique using direct sequential simulation of a multivariate set of variables, with each variable simulated based on hard data, the heavy-metal concentrations in the soil, and on soft data, consisting of a previously simulated map of one of the heavy metals. The suggest methodology is based on the influence of the spatial distribution of the heavy-metal concentrations; and on the influence of the soft data dependent on the global correlation with the hard data. With the 10 realizations of the simulated multivariate set, a "hazard" index was calculated for each pixel of the area, based on the simultaneous proportions (joint probabilities) of different levels of all metals. Finally, the intersection of the hazard map, based on the joint dispersion of all contaminants, with the environmental impact map for the different ecosystems, resulted in environmental hazard maps of the Guadiamar river margins. The performances of the multivariate set of co-simulated variables was compared considering two extreme alternatives: i) the soil is considered in need of treatment if all 5 heavy metals simultaneously exceed their concentration limit value at the same location; ii) the soil is considered in need of treatment if at least one heavy metal exceeds its concentration limit at the same location. The proposed simulation methodology improved the delineation of potential areas simultaneously contaminated with different pollutants.

X. Sanchez-Vila et al. (eds.), geoENV IV – Geostatistics for Environmental Applications, 425-436.

1. INTRODUCTION

On the morning of the 25th of April 1998, the waste damp basin of the Aznalcóllar mine, West of Seville (South-West of Spain), containing mud and acid waters, suffered a rupture spilling 4 Hm^3 of acid waters and 2 Hm^3 of acid mud directly into the Agrio river and consecutively into the Guadiamar river (C.M.A, 2000). This accidental spill spread to an area of around 49 km^2. This area, situated 60 km downstream the mine, has a great ecological importance because the Guadiamar river is the main hydrological resource of the National Park of Doñana (Biosphere Reserve, UNESCO 1994).

Nowadays the evaluation and management of environmental impacts due to residual soil contamination, is the main concern. Despite the direct remediation done to the whole area, primarily through mobilization and excavation of the acid mud, there still is a significant quantity of residual contamination, which can affect negatively all ecosystems. With this study, we intend to characterise the spatial dispersion of heavy metals – Cu, Pb, Zn, Cd and As on the Guadiamar river margins, to be able to elaborate and create environmental hazards maps as basic tools for important decision-making, such as the delineation of target areas for remediation or for additional sampling.

2. THE DATA SET

The study area is a 2 km^2 region located approximately 10 km in the South of the Aznalcóllar mine. The information available was obtained trough a soil sampling realized in August 1999, where 80 samples were collected. For the purpose of this study, only the samples located inside the study area, i.e. 40 samples, were used to characterize spatial dispersion of residual contamination with heavy metals. Soil samples were collected from the topsoil and analytical results of Cu, Pb, Zn, Cd and As recorded in terms of total concentration (ppm). Figure 1 shows the position of the soil sampling locations, as well as the heavy metal concentrations obtained from chemical analyses. The global descriptive statistics are presented in Figure 2 and the resulting descriptive statistics for each heavy metal in table 1.

Table 1. Descriptive statistics for each heavy metal

	Cd	Cu	Zn	Pb	As
Min	0.5	47.0	146.0	116.0	24.0
Max.	17.0	1074.0	4460.0	5150.0	2649.0
Mean	3.3	242.5	1042.5	686.4	336.6
SD	2.99	184.0	820.4	848.2	447.9

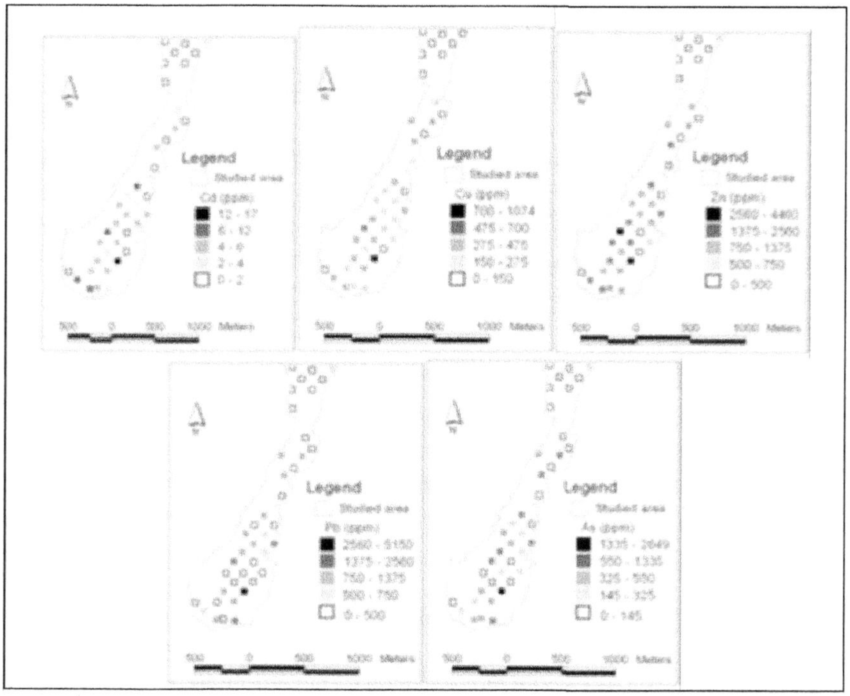

Figure 1. Heavy metal concentrations (ppm).

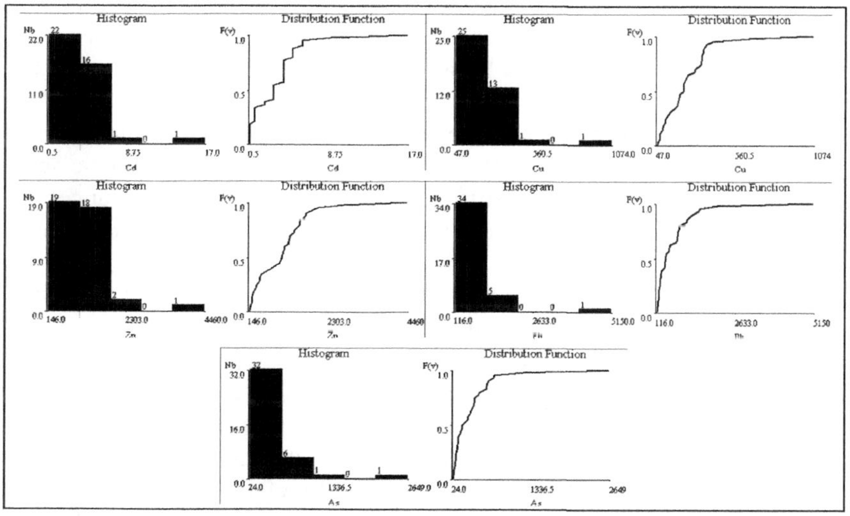

Figure 2. Histogram and distribution function of heavy metal concentrations (ppm).

3. METHODOLOGY

3.1. Co-simulation of the set of metals

The methodology applied relies on direct sequential simulation and co-simulation techniques (Soares A., 2001). The multivariate set of variables was co-simulated using the direct sequential co-simulation technique: each variable is simulated based on the hard data – experimental samples – and an image (secondary information) given by the previously simulated map of another metal.

Estimation of variograms of hard data

Correlation coefficients between the different metals are given in Table 2. The high correlation values between these metals clearly reflect the common origin of contamination. Variograms of hard data were calculated and fitted with exponential models (Figure 3a).

Simulation of the metals

Based on the correlations between the different variables and on the corresponding variograms (spatial continuity), it is possible to rank the variables to define the sequence of metals to be simulated.

For this study the first variable simulated was the Cd because it showed better spatial continuity (variogram ranges) as well as a good correlation with the other heavy metals (see Table 2). Cu was the next variable to be simulated using one simulated Cd map as *soft data*. This co-simulation process continued until the last heavy metal was simulated. The order used to simulate each heavy metal based on the previous variable was: Cd, Cu, Zn, Pb, As.

Given the high correlation coefficients between metals and the spatial continuity revealed by the variograms and after some tests, a set of 10 simulations was considered sufficient to represent the spatial uncertainty of those metals. Simulations and co-simulations were performed according to the following sequence:

1. First, a set of 10 realizations of the first element Cd was simulated with direct sequential simulation.
2. From the 10 previously simulated images, one is chosen to serve as *soft data* to simulate 10 images of the next element, Cu, using Direct Co-simulation. The "soft" image of Cd is chosen according to the better match of the basic statistics (standard deviation and mean). This step was repeated for the next 3 metals: Zn, Pb and As.

Table 3 and Figure 4 represent the global descriptive statistics and the histogram of the chosen images of each metal, and this can be compared with the correspondent statistics of *hard data* (Table 1 and Figure 2). The simulated maps show a similar concentration pattern for the different heavy metals (*Figure 5*), and reproduce quite well the variogram models.

Table 2. Heavy metal correlation

	Cu	Pb	Zn	Cd	As
Cu	1				
Pb.	0.91	1			
Zn	0.97	0.88	1		
Cd	0.97	0.92	0.98	1	
As	0.91	0.99	0.88	0.92	1

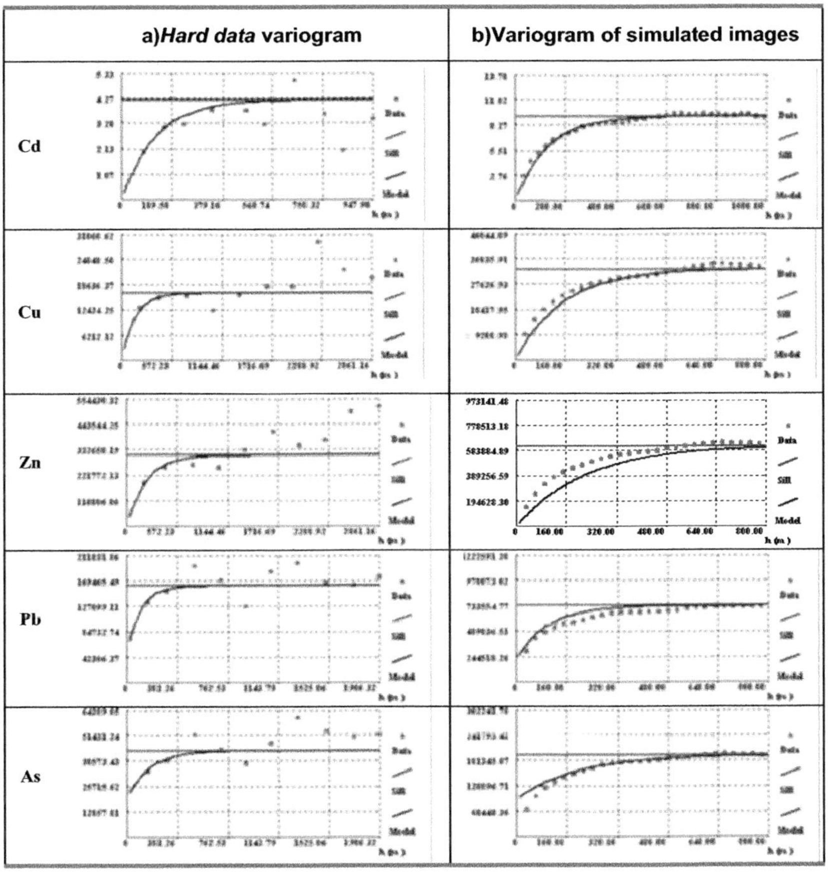

Figure 3. Spatial Variogram (omnidirectional) of a) *hard data*; b) one simulated image for each heavy metal

Table 3. Global descriptive statistics of the chosen simulated images

	Cd	Cu	Zn	Pb	As
Min	0.5	47	146	116	24.
Max.	17.0	1074.0	4460.0	5150.0	2649.0
Mean	3.3	241.01	1038.5	692.9	349.3
SD	3.05	181.99	787.73	856.57	440.01
Skewness	2.3	2.4	2.1	3.0	2.8
P95	9.1	621.7	2612.2	2092.4	1174.9

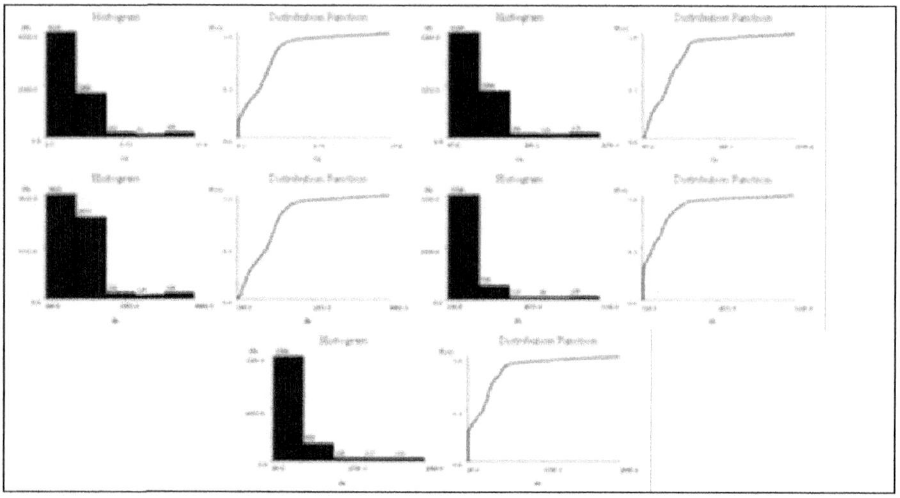

Figure 4. Histogram and distribution function of one concentration (ppm) simulation for each heavy metal (*soft data*)

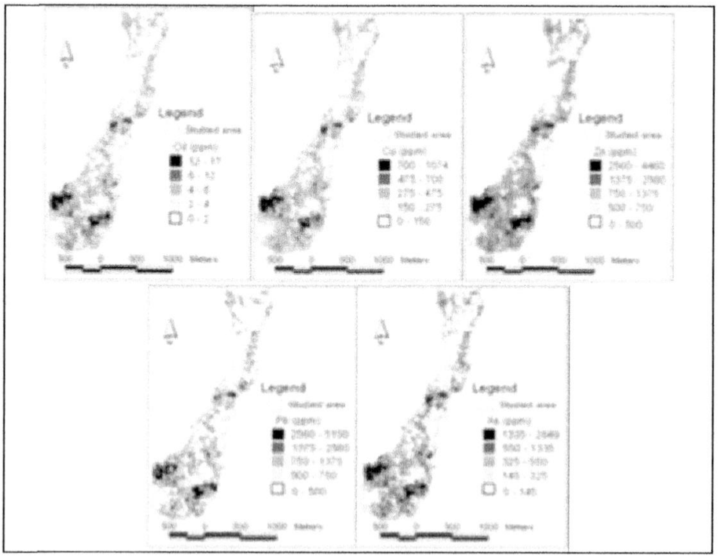

Figure 5. Simulated maps of the 5 metals.

3.2. HAZARD MAPS

The aim of this study was to delineate the areas that need future remediation based on intervention values. The *Consejeria de Medio Ambiente de la Junta de Andalucía* (Environmental Agency of the regional government of the South of Spain), C.M.A, (C.M.A., 2000) defined for each contaminant four different remediation levels: maximum level allowed, recommended investigation, compulsory investigation, compulsory treatment. Hence joint probabilities of different metals to be simulated simultaneously, above or under the remediation levels, can originate hazard maps of the region.

The study area mainly consists of agriculture soils with pH values lower than 7 (*Table 4*).

Table 4. Contamination levels proposal by C.M.A.

Heavy metal	Agriculture Soils: Maximum Level allowed		Recommended investigation		Compulsary investigation		Compulsory Treatment		PN	AI
	<7	>7	<7	>7	<7	>7	<7	>7	<7	>7
Cu	<50	<100	50-150	100-300	150-300	300-500	>300	>500	>500	>1000
Pb	<100	<200	100-250	200-400	250-350	400-500	>350	>500	>1000	>2000
Zn	<200	<300	200-300	300-500	300-600	500-1000	>600	>1000	>1000	>3000
Cd	<2	<3	2-3	3-5	3-7	5-10	>7	>10	>15	>30
As	<20		20-30		30-50		>50		>100	>300

Considering the thresholds: *z1* (maximum level allowed), *z2* (recommended investigation), *z3* (compulsory investigation) and *z4* (compulsory treatment), the following joint probabilities, at a given location x_0, can be identified with different hazard levels:

i) Prob $\{z_{Cd}(x_0) < z1_{Cd}, z_{Cu}(x_0) < z1_{Cu}, z_{Zn}(x_0) < z1_{Zn}, z_{Pb}(x_0) < z1_{Pb}, z_{As}(x_0) < z1_{As}\}$ corresponds to the most clean hazard scenario;

ii) Prob $\{z_{Cd}(x_0) =< z2_{Cd}, z_{Cu}(x_0) =< z2_{Cu}, z_{Zn}(x_0) =< z2_{Zn}, z_{Pb}(x_0) =< z2_{Pb}, z_{As}(x_0) =< z2_{As}\}$, corresponds to the intermediate clean hazard scenario, meaning that all metals at x_0 are lower or equal $z2$;

iii) Prob $\{z_{Cd}(x_0) >= z3_{Cd}, z_{Cu}(x_0) >= z3_{Cu}, z_{Zn}(x_0) >= z3_{Zn}, z_{Pb}(x_0) >= z3_{Pb}, z_{As}(x_0) >= z3_{As}\}$, corresponds to the intermediate contaminated hazard scenario, meaning that all metals at x_0 are greater or equal to z3;

iv) Prob $\{z_{Cd}(x_0) >= z4_{Cd}, z_{Cu}(x_0) >= z4_{Cu}, z_{Zn}(x_0) >= z4_{Zn}, z_{Pb}(x_0) >= z4_{Pb}, z_{As}(x_0) >= z4_{A}\}$, corresponds to the most contaminated hazard scenario.

If we define, for example for scenario i), the following marginal indicators:

$$I_{z1}{}^{Cd}(x0) = \begin{cases} 1 & \text{if } Z_{Cd}(x_0) < Z1_{Cd} \\ 0 & \text{otherwise} \end{cases} \qquad I_{z1}{}^{Cu}(x0) = \begin{cases} 1 & \text{if } Z_{Cu}(x_0) < Z1_{Cu} \\ 0 & \text{otherwise} \end{cases}$$

$$I_{z1}{}^{Zn}(x0) = \begin{cases} 1 & \text{if } Z_{Zn}(x_0) < Z1_{Zn} \\ 0 & \text{otherwise} \end{cases} \qquad I_{z1}{}^{Pb}(x0) = \begin{cases} 1 & \text{if } Z_{Pb}(x_0) < Z1_{Pb} \\ 0 & \text{otherwise} \end{cases}$$

$$I_{z1}^{As}(x0) = \begin{cases} 1 & if\ Z_{As}(x_0) < Z1_{As} \\ 0 & otherwise \end{cases}$$

A joint indicator can be computed by the following product:

$$I_{z1}(x_0) = I_{z1}^{Cd}(x_0) \cdot I_{z1}^{Cu}(x_0) \cdot I_{z1}^{Pb}(x_0) \cdot I_{z1}^{Zn}(x_0) \cdot I_{z1}^{As}(x_0)$$

The joint probability at x_0 – corresponding to scenario i) – can be estimated with the 10 simulated images:

$$prob_{z1}(x_0) = \frac{1}{10}\sum_{i=1}^{10} I_{z1}(x_0, i)$$ $I_{z1}(x_0, i)$ corresponds to $I_{z1}(x_0)$ of simulated image *i*.

Equivalent joint probabilities can be computed for the other scenarios: $prob_{z2}(x_0)$, $prob_{z3}(x_0)$ and $prob_{z4}(x_0)$.

Figure 6 shows the results of the proposed methodology to calculate hazard maps. The results showed that the first remediation level (scenario i), where all 5 heavy metals are jointly under the maximum level allowed, never occurs, which means that there is always at least one heavy metal that exceed the lowest threshold. Furthermore, scenario iii), where all metals are jointly above the remediation level, shows the highest probabilities of occurring.

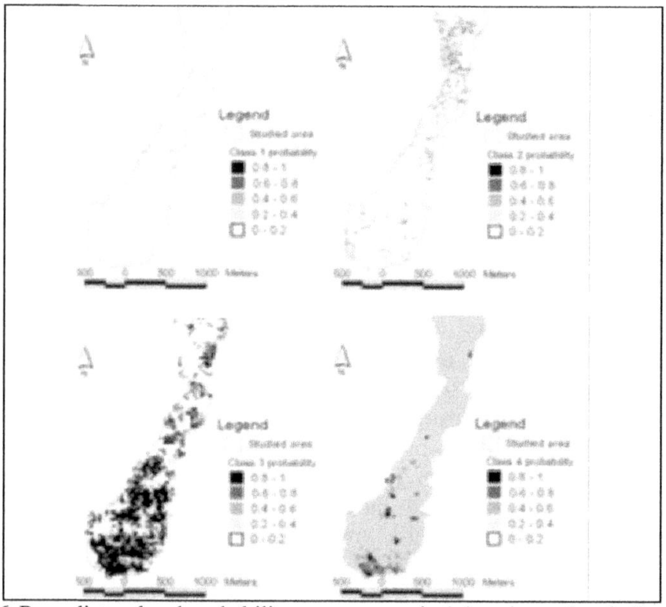

Figure 6. Remediaton-level probability maps: scenario 1 (maximum level allowed), scenario 2 (investigation recommend), scenario 3 (compulsory investigation) and scenario 4 (compulsory treatment)

Finally, a global hazard map was obtained by classifying each pixel in four defined scenarios (*Figure 7*). The results shows that approximately 44% of the study area need compulsory treatment and 40% of the study area need compulsory investigation (*Figure 8*).

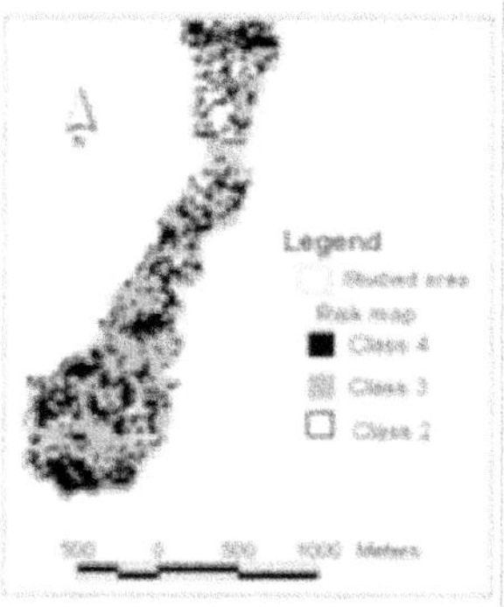

Figure 7. Global hazard map for scenario i.

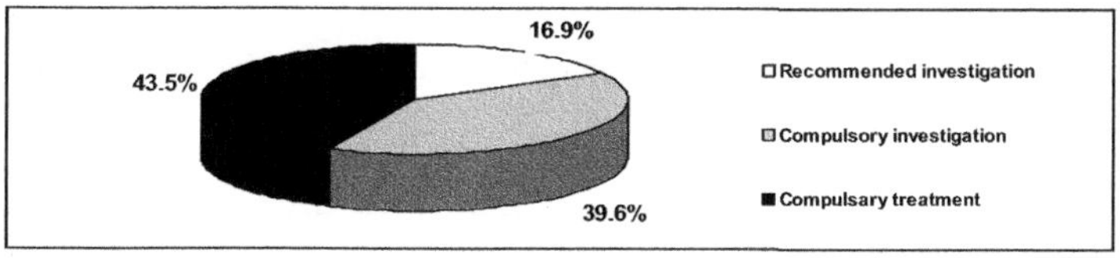

Figure 8. Graphic showing the % of area occupied by the 4 hazard levels.

An Alternative approach for Scenario iv – Compulsory Treatment

Since some European legislation impose treatment whenever one metal exceeds the highest threshold (compulsory treatment), an alternative for scenario *iv* was conducted: if at least one heavy metal exceeds the compulsory treatment threshold the pixel is considered to belong to scenario *iv*, i.e., treatment is imposed to that soil, with

Scenario iv) Prob $\{z_{Cd}(x_0) = z4_{Cd}$ or $z_{Cu}(x_0) = z4_{Cu}$ or $z_{Zn}(x_0) = z4_{Zn}$ or $z_{Pb}(x_0) = z4_{Pb}$ or $z_{As}(x_0) = z4_{A}\}$.

In this alternative, scenario *iv* has higher probabilities of occurring in comparison to the other remediation levels (*Figure 9*). The global hazard map of this scenario, *Figure 10*, shows that approximately 72% of the study area need compulsory treatment and 22 % of the study area need a compulsory investigation (*Figure 11*).

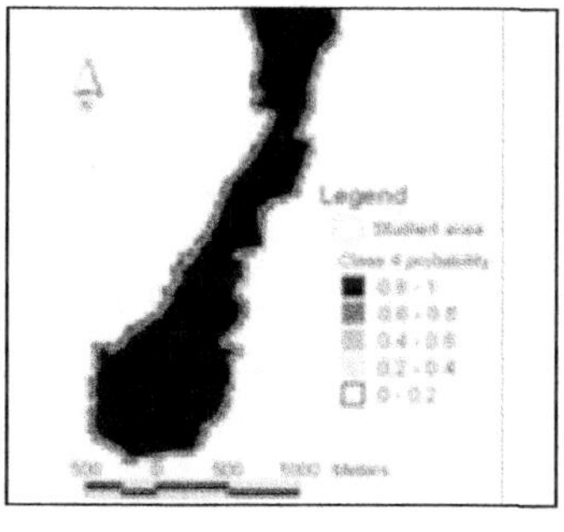

Figure 9. Remediaton-level probability map for the second alternative: class 4 (compulsory treatment)

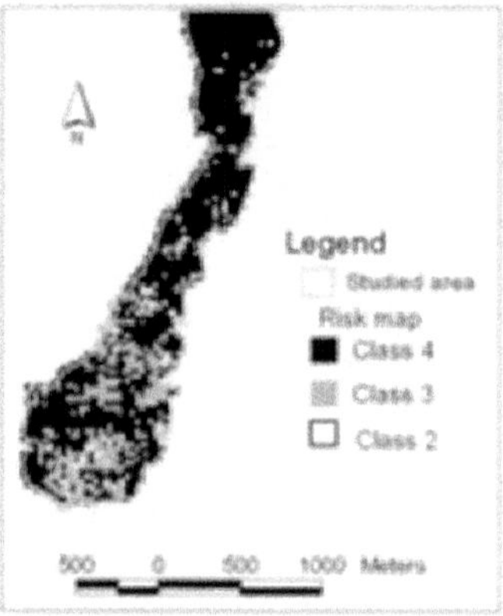

Figure 10. Global hazard map of the second alternative for scenario iv.

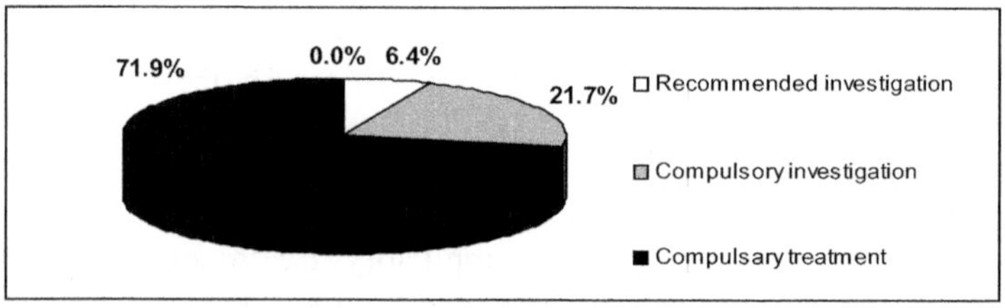

Figure 11. Graphic showing the percentages of area occupied by the 4 hazard levels.

3.3. ENVIRONMENTAL HAZARD MAPS

Finally, with the hazards maps obtained for the two alternatives it is possible to intersect them with an environmental impact map. An impact map on most surrounding sensitive eco-systems, *Figure 12*, was made by C.M.A. (C.M.A., 2000). It is mainly composed by 4 different impacts levels. The lowest impact level (impact 2) corresponds to an extensive culture occupation while the highest (impact 5) corresponds to the Guadiamar river margins. In *Figure 13* the percentage of area occupied by different impact levels is shown. The highest impact occupies 56% of the total study area while the lowest impact occupies 27% of the area.

Intersecting the impact map with the 2 hazards maps (resulting from the two alternatives for the scenario iv) 23% and 36% of the highest impact area need treatment for the first and second alternative, respectively (Figure 14 and Figure 15).

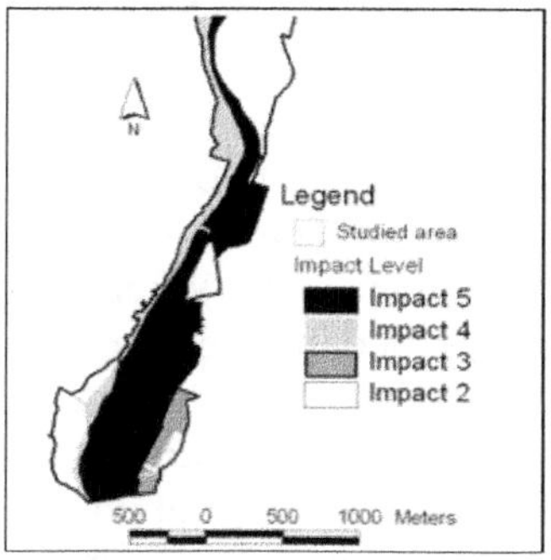

Figure 12. Impact map.

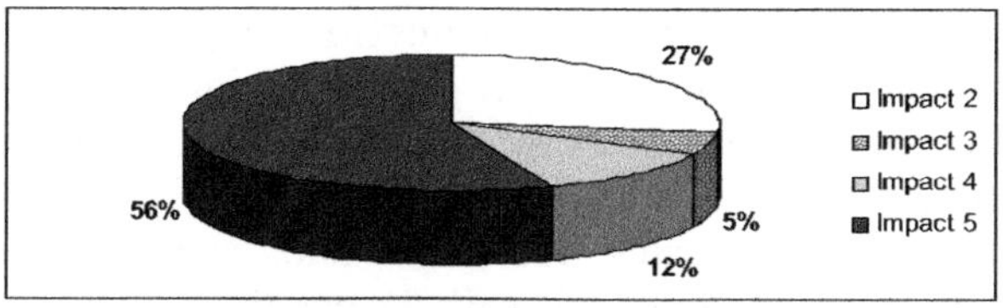

Figure 13. Area occupied by 4 impact levels.

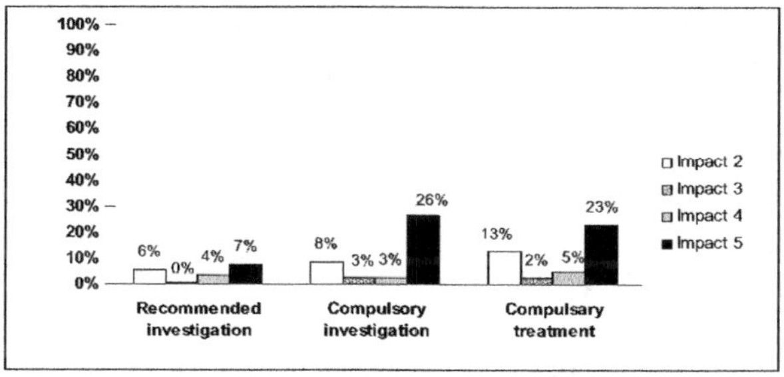

Figure 14. Area occupied by 3 remediation levels for each of the different impact levels, for alternative 1.

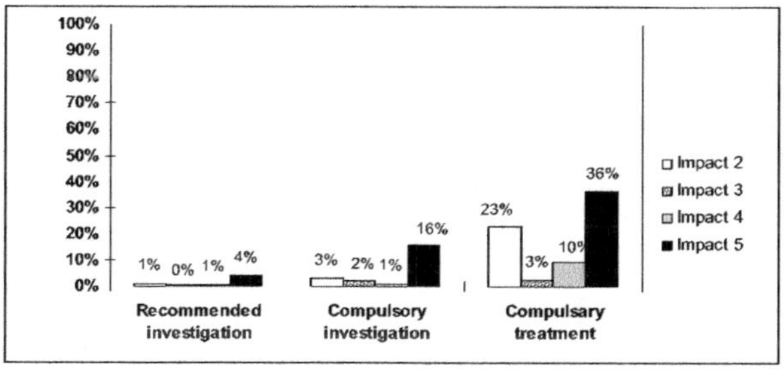

Figure 15. Area occupied by 3 remediation levels for each of the different impact levels, for alternative 2.

4. DISCUSSION AND CONCLUSIONS

When, in soil quality evaluations, more than one pollutant occur simultaneously, it is necessary to determine the total area affected by any of the pollutants. This leads to the problem of defining the area contaminated by the different pollutants simultaneously. When the concentration values of the pollutants exceed the established intervention values a future remediation will be considered.

Usually theses actions are extremely expensive and, for this reason, a good interpretation of the spatial dispersion of all pollutants will be reflected in the remediation costs. For this type of contaminations the delineation of the remediation/treatment areas should not be defined considering each pollutant separately. With the methodology presented in this paper it is possible to account for the uncertainties in mapping the probability that different pollutants are simultaneously contaminating the soil.

Application of this methodology was made considering two alternatives regarding the treatment level, extreme scenario: when at least one heavy metal critically contaminates the soil; and, when all pollutants are simultaneously contaminating the soil. Depending on the aim of the soil quality investigation and on the costs associated to the remediation actions, its possible to choose between these two alternatives. But, considering that European legislations imposes that the treatment actions should be carried out as long as one metal exceeds the highest threshold (compulsory treatment) in order to obtain a clean and safe area, the second alternative is certainly more appropriated.

ACKNOWLEDGMENT

The authors thank to the *Oficina Técnica del Corredor Verde del Guadiamar* for the data information and environmental expertise support.

REFERENCES

1. Soares A., 2001, "Direct Sequential Simulation and Co-Simulation", Mathematical Geology, Vol. 33, no.8, p. 911-926.
2. C.M.A., 2000, "Publicaciones da Consejeria de Medio Ambiente", www.cma.junta-andalucia.es.

DETECTING ZONES OF ABRUPT CHANGE: APPLICATION TO SOIL DATA

E. Gabriel[1], D. Allard[1] and J.N. Bacro[2]

[1]*Institut National de la Recherche Agronomique, Unité de Biométrie, Domaine St-Paul, site Agroparc, F-84914 Avignon Cedex 9, France.*
[2]*Institut National d'Agronomie, UMR INAPG/INRA 518, 16, rue C. Bernard, F-75231, Paris Cedex 05, France.*

Abstract: In the exploration of spatial data it is often of interest to locate boundaries between homogeneous zones. We propose a method for detecting zones of abrupt change, i.e. zones where the data present a discontinuity or a sharp variation in the mean. This method is based on the interpolation of the local gradient and relies mathematically on geometrical properties of χ^2 fields. We focus on the implementation issues raised by this method, illustrated on a soil data set in an agricultural field, in the context of precision agriculture.

1. INTRODUCTION

In many agricultural problems, it is of interest to map the zones where the variable under study changes abruptly. This is for example the case in precision agriculture, our motivating example. Precision agriculture aims at defining a location dependent management within a field, for nitrogen fertilization for example, instead of a unique management for the whole field. The main factor influencing the variability within a field is its soil. Hence, any location-dependent management first needs to delineate homogeneous zones. This can be done by spatial clustering techniques, but alternatively one can estimate the boundaries between the homogeneous zones which will be characterized by a sharp variation of the local average. We call these areas Zones of Abrupt Change (ZACs).

X. Sanchez-Vila et al. (eds.), geoENV IV – Geostatistics for Environmental Applications, 437-448.

In biology, the problem of detecting ZACs was first considered by Womble (1951). ZACs in gene frequencies could be linked to boundaries between populations. The ZACs were defined as the points where the gradient of a single variable computed on an interpolated grid varies the most, for example the upper 5% tail. Several authors improved this method: Barbujani et al. (1989) included the direction of the gradient in the definition of the ZACs and Bocquet-Apple and Bacro (1994) generalized it to the multivariate case. This method suffers from many flaws: a constant proportion of pixels is selected (e.g. 5%), the interpolation is not optimal, and the precision of the interpolation is not accounted for. Last, the significance of the ZACs can not be assessed since the only possible tests in this approach are permutation tests. But a permutation test corresponds to the null hypothesis of absence of any spatial structure, which is untenable in our applications, see Gleyze et al. (2001).

In this paper we propose to detect ZACs when the variable $Z(x)$ is a Gaussian random field in a domain D and the boundaries correspond to discontinuities of the expectation. The general method, presented in details in section 2 is in three stages. We first interpolate the gradient using geostatistics. Then, in the second stage, a local test for the existence of an abrupt change is built. The test statistic, denoted $T(x)$, depends on the observed data, their location and the covariance function of $Z(x)$, C_Z. The zones of abrupt change are defined as the set of points where the statistic is greater than a fixed level t_α, where t_α is the $(1-\alpha)$ quantile of the theoretical distribution of T. Under the null hypothesis of no discontinuity, the ZACs should be small and randomly scattered in the study area. On the contrary, if there is a discontinuity, we should expect large ZACs organized along the discontinuity. So, in the third stage, we test the significance of each ZAC using the theoretical distribution of its size under the null hypothesis. In section 3, several implementation issues are discussed. The method is then applied in section 4 to a soil data set in an agricultural field, showing that the field should be subdivided into two homogeneous zones.

2. THE METHOD

2.1 Variograms of dioxins for both lichens

Let $Z(x)$ be a centered, stationary, Gaussian random field defined on a fixed domain D of $\mathbf{R}^2$ and $Z=(Z(x_1),..., Z(x_n))^t$ be a sample of $Z(x)$ at $x_1,...,x_n$. The optimal predictor at an unsampled location x is the simple kriging,

$$\widehat{Z}(x) = C(x)^t \mathbf{C}^{-1} Z, \qquad (1)$$

with $C(x)=(C_Z(x-x_1),...,C_Z(x-x_n))^t$ and $\mathbf{C}=\mathbf{E}[ZZ^t]$ the matrix of covariance between the data. For sake of simplicity, we assume that the covariance function $C_Z(h)$ is continuous everywhere and that it is infinitely differentiable for all $|h|>0$. This is for example the case for the exponential covariance function, but it does exclude the spherical covariance function. Under this regularity assumption, the estimator of the gradient is the gradient of $\hat{Z}(x)$:

$$\widehat{W}(x) = \nabla\widehat{Z}(x) = (\nabla C(x))^t \mathbf{C}^{-1} Z, \qquad (2)$$

where $\nabla C(x)$ is the gradient of $C(x)$.

2.2 Definition of the zones of abrupt change

We consider that there is an abrupt change at a point x of D if the radient at x is "large". Thus, a test statistic for defining what "large" means is required. Since $\mathbf{E}[\widehat{W}(x)]=0$, the variance-covariance matrix of $\widehat{W}(x)$, denoted $\Sigma(x)$, is

$$\Sigma(x) = \mathbf{E}[\widehat{W}(x)\widehat{W}(x)^t] = \nabla C(x)^t \mathbf{C}^{-1} \nabla C(x) \qquad (3)$$

The test statistic is defined by

$$T(x) = \widehat{W}(x)^t \Sigma(x)^{-1} \widehat{W}(x). \qquad (4)$$

At the point x the statistic $T(x)$ compares a quadratic form of the estimated gradient to its variance and a standard result of statistics states that $T(x)$ has a marginal $\chi^2(2)$ distribution. We define a local test for deciding if a point x belongs to a zone of abrupt change by comparing the null hypothesis $H_0(x)$ "$\mathbf{E}[Z(x)]$ is continuous at x" versus $H_1(x)$ "$\mathbf{E}[Z(x)]$ shows a discontinuity at x". The null hypothesis is rejected if $T(x) \geq t_\alpha$ where t_α is the $(1-\alpha)$ quantile of the $\chi^2(2)$ distribution.

In practice, this procedure is applied at the nodes of a grid superimposed on the domain D. For a confidence level α, the set of the grid nodes whose statistic is above t_α defines the zones of abrupt change. If the field is stationary, we expect the ZACs to be randomly located, or non existent. On the contrary, if there is a discontinuity, ZACs are likely to be structured along the discontinuity.

2.3 Statistical significance of the ZACs

Once the zones of abrupt change have been estimated, we must test their statistical significance. Each connected component of the ZACs is tested in turn. We build a test to reject H_0 "the connected component is from a stationary random field" vs. H_1 "the connected component is from a random field showing a sharp variation". The test is based on the size of the connected component which is compared to the theoretical distribution of the size of a connected component belonging to a stationary random field.

ZACs are related to the theory of the excursion sets of χ^2 random fields (Adler 1981, Aronowich and Adler, 1988, Worsley 1994, Cao 1999). For $t_\alpha \to \infty$, i.e. for a confidence level $\alpha \to 0$, it can be shown that the size of a connected component, say C_0, of the excursion set is related to the local curvature of T at the maximum, say x_0, in this connected component. Allard et al. (2002) have shown that under some regularity conditions, the following convergence in law holds:

$$t_\alpha S_0(\alpha) \xrightarrow{L} \pi \mathbf{det}(\mathbf{\Lambda})^{-1/2} E, \tag{5}$$

as $t_\alpha \to \infty$, where $S_0(\alpha)$ is the area of C_0, ^ is the 2 X 2 matrix of the curvature of $T(x)$ at x_0 in C_0 and E is an exponential random variable with expectation 2, independant of T. ^ depends only on C_Z and the sampling pattern. A more detailed presentation of this result with explicit computation of ^ can be found in Allard et al. (2002). From equation (5) a p-value of each connected component can be computed:

$$p = \exp\left(-\frac{t_\alpha S_0(\alpha) \det(\mathbf{\Lambda})^{1/2}}{2\pi} \right). \tag{6}$$

The significance of each connected component is assessed by comparing this p-value to a confidence level, for example 0.05. If p is above this confidence level, it is considered as coming from a stationary random field (H_0 is not rejected) and the connected component is not significant. On the contrary, if p is below the confidence level, it is considered as significant

3. IMPLEMENTATION ISSUES

In practice, the method is run on a grid. On each grid node $[i,j]$, the gradient $\hat{W}[i,j]$, the matrix $\Sigma[i,j]$ and the field $T[i,j]$ are computed. Then for a confidence level α, the set of grid nodes whose statistic $T[i,j]$ is above the (1-α) quantile t_α of a $\chi^2(2)$ distribution define the ZACs. For each connected component of the ZAC, the p-value is then computed according to (6). When implementing this method, several parameters must be chosen: the mean for centering the variable, the covariance function, the discretization of the grid and the level α. Allard et al. (2002) carried out a simulation study to address these issues. A vector of 100 randomly located standard Gaussian random variables with an exponential covariance function was simulated on a unit square. A discontinuity was introduced along the line x_1=0.4 by adding a constant k to all samples with x_1<0.4, from k=0 (which corresponds to the null hypothesis: absence of discontinuity) to k=3.

The choice of the parameter α is the result of a trade-off between the accuracy of equation (5) and the power of the method. On the one hand, the convergence in law holds for $\alpha \to 0$, but on the other hand, a good detection of the existing discontinuities for low to moderate discontinuities is achieved if α is not too low. The simulation study showed that a good detection rate with a reasonable amount of false positives is achieved for $\alpha = 0.005$ for high k ($k \geq 2.5$). A slightly higher value of α (e.g., $\alpha = 0.01$) is preferable for low to moderate k ($k < 2.5$).

Equation (6) has been established on the continuous plane, not on a grid. The discretization has many effects: the local maximum is not correctly located and hence the curvature at the maximum is incorrect (it is usually underestimated), the size of the connected component is approximated and the connectivity of large clusters depends very much on the discretization (small clusters can be merged into a single one at a different resolution, or the contrary). The simulation study has shown that to lower values of α should correspond a higher discretization, which ensures that small connected components are detected.

In this method, the covariance function is assumed to be known. But in practice it is not the case, and it needs to be estimated. So, the robustness of the covariance estimation must be analyzed. For the simulation exercise described above, the method has been applied with three misspecified covariances: two exponential covariances with the parameter being divided or multiplied by two, and one spherical covariance with the same practical range. When the range is underestimated fewer ZACs near the discontinuity ("true" ZACs) and slightly more ZACs away from it ("false" ZACs) are detected. When the range is overestimated, more of both ``true" and ``false" ZACs are detected, specially in the case of "true" ZACs with intermediate values of k. On average, the ZACs are smaller (resp. larger) when the range is under- (resp. over-) estimated. Increasing the range of the covariance function is equivalent to increasing the regularity of the random function Z, leading ultimately to more rejection of H_0 in the presence of a discontinuity. With the spherical covariance function, ZACs are less often detected and the average size is similar to the baseline case.

4. APPLICATION TO SOIL DATA

4.1 The data

In order to better manage production in quality and in quantity, and minimize ground water pollution, precision agriculture is developing methods for applying "the right dose at the right place" of mineral nitrogen fertilizer. For this purpose, the spatial variability of the soil characteristics must be carefully assessed. In particular, it is of interest to delineate homogeneous zones and to estimate zones where these characteristics change abruptly. The data were collected in an agricultural field (10 ha) in Chambry, next to Laon in Northern France. The soil water content (QH) and the soil mineral nitrogen (QN = NH^{+}_{4} + NO^{-}_{3}) were measured on soil cores up to 150 cm. The sampling scheme (Figure 1a and 1b) included a regular grid with 75 nodes (distance between nodes 36 m) and two sampling crosses, with 17 and 19 nodes, used for the short distance estimation of the variogram. Similar data are presented in Mary et al. (2001), where a more detailed presentation, including a statistical analysis can be found. The histograms of the soil variables are shown in Figure 1c and 1d. A global variogram is computed on the whole field and exponential models were fitted. The estimated parameters are: range=110 m and sill=214 $(kg/ha)^2$ for QN and range=100 m and sill=3200 $(mm)^2$ for QH (Figure 2a and 2d). For visualization purpose, Figure 1a and 1b show the simple kriging interpolation of both variables (QN on the left and QH on the right). For both variables, it seems to be a transition for $y \approx 620$ m.

4.2 Running the method in practice

A first run

The field $T(x)$ is computed according to (4). Potential ZACs are defined as the set of pixels for which $T \geq t_\alpha$. An exploratory analysis has revealed that the standardized difference of means below and above $y = 620$ m is 1.2 and 2 for QN and QH respectively. Hence, considering the discussion in section 3, the level $\alpha = 0.01$ (i.e. $t_\alpha = 9.21$) is considered as appropriate to perform the test and the p-values are compared to the standard level of confidence 0.05. On Figure 2b and 2e the pixels $[i,j]$ (of a 31 X 49 grid with mesh size 10 m) where the statistic $T[i,j]$ is above this threshold are highlighted. Black clusters are significant ZACs, where as grey clusters are not significant. For QN, one large ZAC is detected at the top of the image ($p = 5.5\ 10^{-13}$) and a smaller one is visible where the lower sampling cross is located. This last one should not be considered as physically interesting because it mainly results from outlying data in the sampling cross. In addition, it does not really define a ZAC but rather a "hot spot" at a scale well under any precision agriculture management scale. For QH, there are two ZACs, one corresponding to the limit $y = 620$ m already mentioned ($p = 9.3\ 10^{-7}$) and one located on the lower sampling cross ($p = 3.2$

10^{-15}). For a better visualization of the picture, ZACs are also depicted at the higher level $\alpha = 0.05$ (i.e. $t_\alpha = 5.99$) on Figure 2c and 2f (but the test is still performed at the level $\alpha = 0.01$).

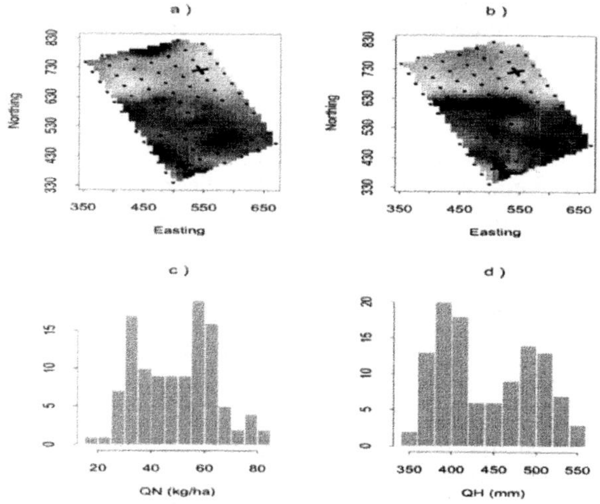

Figure 1. Simple kriging estimate and histogram of QN (left) and QH (right).

Improving the power

The analysis reported in the previous paragraph has shown that the hypothesis of no ZACs should be rejected. For QH an horizontal ZAC around $y = 620$ m was visible and for QN a non significant ZAC was visible. The method presented in Section 2 is exact under the null hypothesis i.e. in absence of discontinuities. In presence of a discontinuity however, we face two problems: the centering of the variable and the estimation of the variance. In the case of a bimodal histogram (see Figure 1c and 1d), we considered that the data should not be centered around the overall average, but rather around the principal mode. This does not alter the estimation of the variogram, but if the variable is not correctly centered, the simple kriging estimation of the variable and of the gradient is incorrect, leading to a non centered χ^2 distribution for $T(x)$. Note that this could be corrected with an ordinary kriging version of the method, yet to be developed. The second problem is more serious. The experimental variogram will pool all the data, resulting in a higher variance and ultimately in a loss of power. Indeed, $T(x)$ is proportional to σ^{-2}, where σ^2 is the sill of the variogram (the interpolated field $\hat{W}(x)$ does not depend on σ^2 and the covariance matrix $\sigma(x)$ is proportional to σ^2). Hence a higher variance leads to lower values of the field $T(x)$ and to smaller ZACs. To correct this over-estimation of the variance, the variogram was recomputed such that no pairs of points could intersect the line $y = 620$ m. The resulting variograms are depicted Figure 3b and 3d. The ranges are much shorter (30 m and 50 m for QN and QH) and the sills lower (110 $(kg/ha)^2$ and 1700 $(mm)^2$), illustrating the bias introduced by the existing ZAC in the estimation of the variogram. The new ZACs are depicted Figure 3b, 3c, 3e and 3f. Clearly, they are much more visible. This is particularly striking for QH, where a clear separation between the upper and the lower part of the field is visible. For QN a sharp variation is only detected on the left-hand side of the field. Looking back carefully at the image on Figure 1b, one can see that the variation is smooth in the right-hand side of the field and that there is no zone of abrupt change to be detected.

Robustness analysis

In the light of the general discussion of Section 3, the sensitivity of our method is explored, with respect to the discretization and the covariance estimation. Table 1 reports the number and the size of all connected components (significant and non significant) for various mesh size (from 5 X 5 m^2 for a 63 X 98 grid to 20 X 20 m^2 for a 15 X 24 grid). One can see that the discretization has only minor effects on the presence and total area of the significant ZACs. There is one exception however: for QH when changing from the coarser grid to the medium grid, one non significant ZAC is merged with a significant one, increasing its size from 9,200 m^2 to 11,100 m^2. As the grid gets finer, there are in general less non significant ZACs, in particular for the lower level $\alpha = 0.01$, and more significant ones. This is due to the combination of two effects: first, as the discretization effect increases, non significant connected components can become significant, because the maximum is more precisely located and hence the determinant of Λ is larger. Secondly, non significant connected components are merged to significant ones, thereby increasing their area. These two effects are clearly visible on Figure 4 where ZACs are depicted for QN at the level $\alpha =$ 0.01. One can notice that the 31 X 49 grid is a sufficiently fine to capture the main ZACs, whereas the coarser grid is not.

As already mentioned in Section 3, the method is mathematically correct if the covariance is known. But we only have an estimation of the covariance function. Therefore the method was run with different range parameters (± 30%) and with a spherical covariance with the same practical range. Table 2 reports the number of significant (and non significant) ZACs, along with their total area. As the range increases, the area of the ZACs increases, but the number and the area of non significant ones decrease. This is to be related to the results of the simulation study reported in Allard et al. (2002). Increasing the range amounts to increasing the alleged regularity of the variable. Thus $T(x)$ is increased (because $\Sigma(x)$ is decreased), resulting in larger clusters. Since the p-value of a cluster, as computed in equation (6), decreases when its size increases, more connected components are found to be significant as the range increases. The results with a spherical covariance function with the same practical range are very similar. We also checked graphically that the same ZACs were detected for all covariance functions. Only their size changed with the covariance function. Hence the method is found to be quite robust with respect to the covariance function.

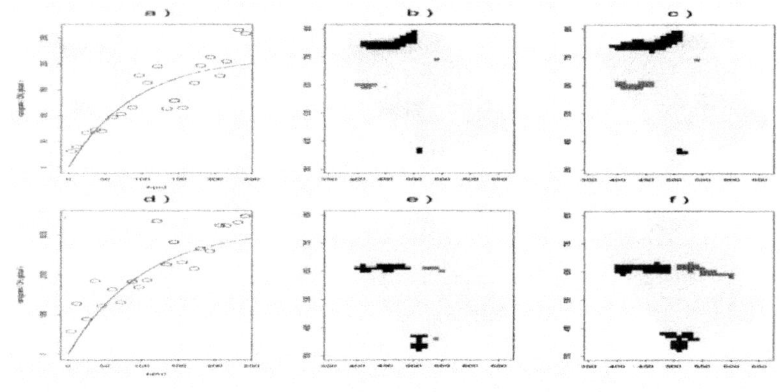

Figure 2. Global estimation of the ZACs for QN (upper row) and QH (lower row). Fitted variogram, detection of the ZACs for $\alpha = 0.01$ and representation of the same for $\alpha = 0.05$.

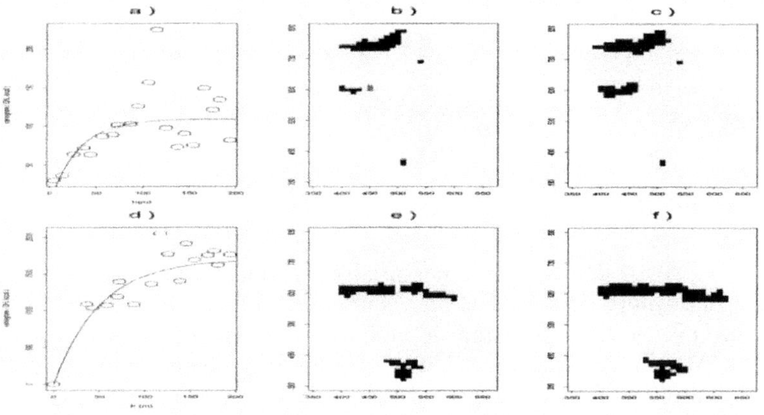

Figure 3. Estimation of the ZACs with the corrected variogram, as in Figure 2.

Table 1. Influence of the discretization and the level α on the number and total area (m^2) of the ZACs.

		QN				QH			
grid size		$\alpha = 0.05$		$\alpha = 0.01$		$\alpha = 0.05$		$\alpha = 0.01$	
15 × 24	sign. ZACs	2	*7200*	1	*3600*	2	*9200*	2	*6800*
	non sign. ZACs	2	*800*	1	*400*	3	*2000*	2	*800*
31 × 49	sign. ZACs	3	*7500*	4	*4600*	2	*11100*	3	*7600*
	non sign. ZACs	4	*700*	1	*200*	2	*200*	0	–
63 × 98	sign. ZACs	4	*6950*	5	*4575*	5	*11175*	5	*7375*
	non sign. ZACs	6	*575*	0	–	1	*100*	0	–

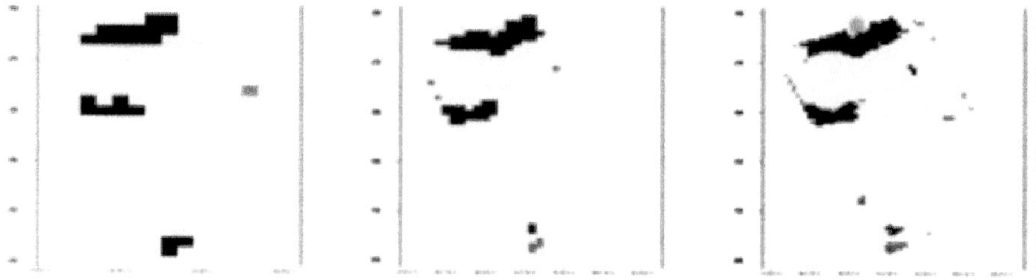

Figure 4. Non significant ZACs (in grey) and significant ZACs (in black) of QN at the level $\alpha = 0.01$ for three discretizations: left 15 X 24, middle 31 X 49 and right 63 X 98.

Table 2. Number of ZACs and their total area (m^2) for different covariance functions ($\alpha = 0.01$).

QN	range	exp(21)		exp(30)		exp(39)		sph(90)	
sign.	ZACs	3	*2300*	4	*4600*	4	*5800*	3	*3490*
non sign.	ZACs	4	*700*	1	*200*	0	—	0	—
QH	**range**	**exp(35)**		**exp(50)**		**exp(65)**		**sph(150)**	
sign.	ZACs	3	*5900*	3	*7600*	2	*8200*	2	*8000*
non sign.	ZACs	4	*1200*	0	—	0	—	0	—

4.3 Results

A first ZAC is detected for both variables. It runs horizontally for $y \approx 620$ m through the whole field for QH and is only visible on the left hand-side of the field for QN. On the right-hand side the transition is smooth, and hence cannot be detected as a significant ZAC. This ZAC is inagreement with the qualitative knowledge the soil scientists had of this field and is recognized as a boundary between two soil types.

A second ZAC appears at the bottom of the field for both variables. It is due to one of the sampling crosses. In this sampling cross there are a couple of very close samples with high differences, giving a large but localized gradient. For this reason, and because this ZAC defines an area at a scale below the precision agriculture management scale, it should not be considered as meaningful in precision agriculture, although being statistically significant.

There is a third ZAC existing for QN at the top of the field that was not related to a transition between soil types by the soil scientists prior to this analysis.

Only the first ZAC is really meaningful for precision agriculture. It is present for both variables and for all discretizations and range parameters that we could test. Hence we conclude that this field can be subdivided into two zones, approximately separated by the line $y = 620$ m.

5. CONCLUSION AND DISCUSSION

About the method

A new method has been proposed for estimating and testing zones of abrupt change (of the local mean) in the plane. It is based on an analysis of the interpolated gradient, and relies mathematically on properties of the geometry of $\chi^2(2)$ fields. Zones of Abrupt Change are defined as the points of the $\chi^2(2)$ field above a threshold t_α which is the $(1 - \alpha)$ quantile of a $\chi^2(2)$ distribution. This level α must be carefully chosen. On the one hand, the mathematical results require that $\alpha \to 0$, but due to the discretization, α should not be too high to be able to detect the clusters. It was found that for a reasonable discretization, $\alpha = 0.01$is a good trade-off.

Our method requires centered data and the knowledge of the variogram. We have found that it is quite robust with respect to the range parameter, but is sensitive to its sill. Under the null hypothesis, the procedure is correct, and there will be no problem. But under the alternative hypothesis of the presence of a discontinuity (or a sharp variation of the mean), we suggest an iterative procedure. ZACs are first detected with a global centering and a global variogram. If significant ZACs are detected, the variogram is re-estimated with the pairs of points that do not intersect them. This will result in a new variogram with a lower sill. ZACs are then re-estimated. This procedure is repeated until convergence. In our application, we reached convergence in one iteration. Note that this iterative procedure remains correct under the null hypothesis.

The method also relies on a Gaussian assumption of the variable $Z(x)$. It is well known in geostatistics that this assumption can not be tested and is usually inadequate. We did not explore its robustness with respect to the Gaussianity. On the soil data, the test performed quite well however.

About detecting ZACs

Detecting Zones of Abrupt Change is different than estimating (and testing) a global trend. In our method a global trend or a smooth variation will not be detected, as it is the case for QN on the right-hand side of the field. Only sharp local variations clustered in large zones can be detected.

There is a duality between the estimation of homogeneous zones and the estimation of their boundaries. Estimating homogeneous zones is a problem that can be addressed using clustering techniques, as in Allard and Monestiez (1999), and Allard and Guillot (2000). Clustering the data in groups always leads to boundaries between the groups. But in spatial clustering techniques, we do not have rules for selecting the number of groups. Spatial clusters do not always necessarily define ZACs along their boundaries if the transition is smooth. Conversely, ZACs do not always define spatially well separated groups. This is similar to a landscape. In a given domain, a cliff might be present, but not across the whole domain because on both ends of the cliff the landscape is smooth.

ACKNOWLEDGMENTS

We wish to thank Bruno Mary, Unité d'Agronomie de Laon, Martine Guérif, UnitéClimat, Sol, Environnement d'Avignon and Claude Bruchou, Unité de Biométrie, Avignon, INRA (France) for providing the soil data set and for enlightening discussions about its analysis.

REFERENCES

1. Adler, R. (1981). *The Geometry of Random Fields*. New-York: Wiley.
2. Allard, D., Gabriel, E. and Bacro, J.N. (2002). Estimating and testing zones of abrupt change for spatial data. *Technical report* (submitted).
3. Allard, D., Guillot, G. (2000). Clustering Geostatistical Data, in *Proceedings of the Sixth International Geostatistics Congress, Cape Town, South Africa, April 2000*, eds. Kleingeld, W.J. and Krige, D.G., pp.49-63.
4. Allard, D. and Monestiez, P. (1999). Geostatistical Segmentation of Rainfall Data, in *geoENV II: Geostatistics for Environmental Applications*, eds. Gomez-Hernandez J., Soares A. and Froidevaux R., Kluwer Academic Publishers, Dordrecht, pp. 139-150.
5. Aronowich, M. and Adler, R. (1988). Sample path behaviour of χ^2 surfaces at extrema. *Adv. Appl. Prob.*, **18**, 901-920.
6. Barbujani, G., Oden, N.L. and Sokal, R.R. (1989). Detecting regions of abrupt change of biological variables. *Systematic Zoology*, **38**, 376-389.
7. Bocquet-Appel, J.P. and Bacro, J.N. (1994). Generalized Wombling. *Systematic Biology*, **43**(3), 442-448.
8. Cao, J. (1999). The size of the connected components of excursion sets of χ^2, t and F fields. *Adv. Appl. Prob. (SGSA)*, **31**, 579-595.
9. Gleyze, J.F., Bacro, J.N. and Allard, D. (2001). Detecting regions of abrupt change: wombling procedure and statistical significance in *geoENV III: Geostatistics for environmental applications*, eds. Monestiez P., Allard D. and Froideveaux R., Kluwer Academic Publishers, Dordrecht, pp. 311-322.
10. Mary, B., Beaudoin, N., Machet, J.M., Bruchou, C. and Ariès F. (2001). Characterization and Analysis of soil variability within two agricultural fields: the case of water and mineral N profiles. In, *Proc. 3rd European Conference on Precision Agriculture*, eds. Grenier and Blackmore, pp. 431-436.
11. Womble, W.H. (1951). Differential systematics. *Science*, **114**, 315-322.
12. Worsley, K. (1994). Local maxima and the expected Euler characteristic of excursion sets of χ^2, F and t fields. *Adv. Appl. Prob.*, **26**, 13-42.

OPTIMAL REGIONAL SAMPLING NETWORK TO ANALYSE ENVIRONMENTAL POLLUTION BY HEAVY METALS USING INDIRECT METHODS. CASE STUDY: GALICIA (NW OF SPAIN)

C. Hervada-Sala[1] and E. Jarauta-Bragulat[2]

[1]Dept. Physics and Nuclear Engineering. EUETIT (UPC), Colom 1, 08222-Terrassa. Carme.Hervada@upc.es
[2]Dept. Applied Mathematics III. ETSECCPB (UPC). Jordi Girona 1-3, 08034-Barcelona. Eusebi.Jarauta@upc.es
Website: ftp://ftp-urgell.upc.es/Matematica/E.Jarauta/Geoenv2002

Abstract: Environmental pollution by heavy metals is a red-hot issue. It is being studied from many points of view, as it is not only an environmental problem but also a public health matter. The effect of pollution by heavy metals can be assessed directly, that is measuring heavy metal concentration in soils, or using indirect methods, that is measuring heavy metal contents on living beings of regional ecosystem, in particular on plants. One of the organisms that have proved to be the most faithfully and useful to do so are moss. So, heavy metal environmental pollution can be studied by taking moss samples and measuring their heavy metal contents. The aim of this work is to show the use of geostatistical tools in environmental pollution analysis applied to a case study of environmental pollution by heavy metals in Galicia (north west of Spain). To do so, two different information in that zone are available: on one hand, measures of heavy metal concentration in moss (Scleropodium purum), whose location points are known, also their level. On the other hand, situation of polluting sites (industrial areas and towns) and their classification taking into account their polluting capacity. This information allows assessing not only for the regional pollution, but also for its scattering. From this and using geostatistical tools, sampling network is being improved. Data set consists of 71 sample points where concentration of ten elements (Al, As, Co, Cr, Cu, Fe, Hg, Ni, Pb and Zn) is measured. For each of them classical statistical analysis is done. Furthermore, spatial variability is studied using a new methodology

X. Sanchez-Vila et al. (eds.), geoENV IV – Geostatistics for Environmental Applications, 449-460.

based on Fast Fourier Transform (FFT), which allows finding covariance matrix using all variables at the same time. FFT methodology improves the classical and tedious geostatistical methodology based on variogram and cross-variogram modelling to find data spatial variability. Finally contour maps of environmental pollution by heavy metals in Galicia are presented.

1. INTRODUCTION AND OBJECTIVES

Galicia is a region located at the northwest of Spain. It is about 29434 km^2 large. It ranges, approximately, between 7° and 9° western meridians and 40° and 42° northern parallels. The outline of this region is gently undulated, with hills and valleys; this smoothness defines its landscape with a series of high and low regions at several levels. So Galicia's landscape is full of high and low areas. The highest areas are in its east border.

Forests, cover the most part of Galicia. The most widespread trees are oak, chestnut, birch, cork and ilea. Since a few years ago there are also pine and eucalyptus. Another interesting aspect in Galicia is that their towns are small and scattered all over the whole country. Recently some of the towns have grown due to the enlargement of some industrial zones. The main industrial activities are cars (located at the north), woodwork, textiles, and craftsmanship.

As it is well known, the increasing of industrial activities implies a pollutant impact on the environment. Nevertheless, our society demands a quality of life compatible with technical progress, without renouncing to it. One of the most important contributors to environment pollution, caused by industry, is the presence of heavy metals in the air, which fall down when it rains and then are incorporated by living beings. In this article the presence of heavy metals from industrial origin are studied; in fact samples of them are measured on some moss: *Scleropodium purum* (Hedw.) Limpr. The metals that are measured are aluminium, cobalt, chromium, copper, iron, mercury, nickel, lead and zinc, and also the metalloid arsenic.

Accumulation of heavy metals over large areas and long periods causes damage to living organisms and it must be carefully controlled; it is also important to know the effects of these contaminants. To assess the pollution caused by metals there are two different methods: the direct one, which consists of measuring their concentration in the air or in soil, and the indirect one, which consists of studying their presence in some living beings. If previous monitoring is correctly done, indirect method can be useful in environmental assessment, because it is easiest and cheapest. This monitoring has been done in Galicia with moss (see reference 2).

The aim of this work is to build up contouring level maps of pollution by heavy metals using geostatistical methods; to do so, we take into account measures in several moss samples. The method used is kriging on a regular grid with correlogram tables obtained by applying the Fast Fourier Transform (FFT) methodology (see reference 5).

2. DATA SET: DESCRIPTION AND ANALYSIS

Full database can be found in our website. Professor J.A. Fernández, from Universidad de Santiago de Compostela, has supplied chemical analysis of 71 samples. Data consists on concentrations (ppm) of all ten (see above) elements. The number of sampling sites was the equivalent to a density of 2.6-samples/1000 km^2, higher than the density recommended for such studies at a regional scale. Sampling was carried out in 1995 (April-July) and covered almost all Galicia, with a higher density at most industrial areas. The concentration of Al, Co, Cr, Cu, Fe, Ni, Pb and Zn in moss extracts were determined using flame absorption spectrophotometry and Hg and As were determined using atomic fluorescence. Figure 1 shows a scatter plot of the location points; the co-ordinates are UTM scaled and in kilometres. Sampling points were taken at different levels, between 72 m and 1014.5 m high. In the light of sampling points and Galicia's dimension, a kriging grid of 9x9 nodes has been built. Distances between nodes are 21x23 km.

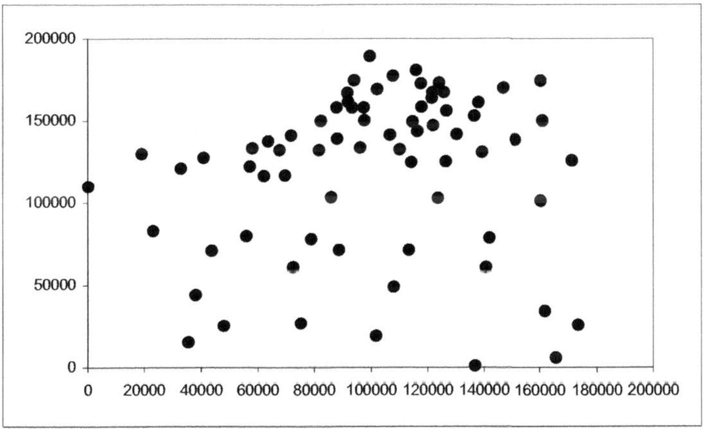

Figure 1. Scatterplot of sample locations.

To be able to carry out a bidimensional geostatistical study, we tried to find out a possible functional relation between data and altitude. The conclusion is that this dependence does not exist. Figure 2 shows, as an example of that, the scatterplots for Al and Cu concentration versus altitude (ALT); this figure shows also the regression line, which is quite horizontal.

The corresponding hypothesis test shows that it is not possible to reject the fact that correlation between data and altitude does not exist.

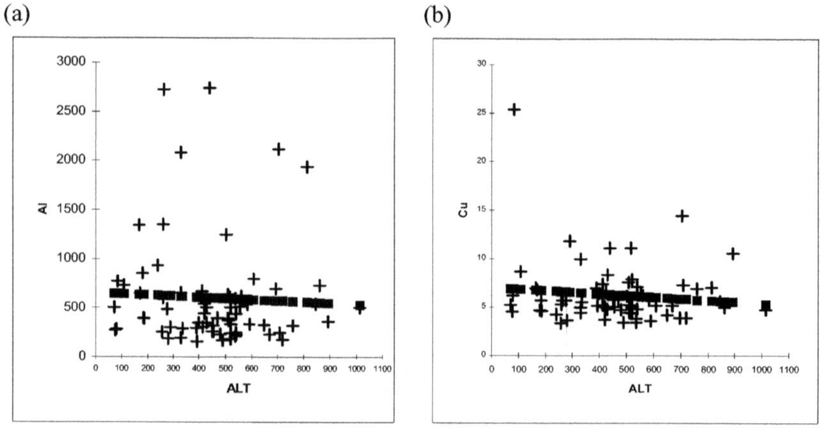

Figure 2. Scatterplots: Al (a) and Cu (b) versus altitude (ALT).

Table 1 shows the average concentration distributed at different levels, at altitude intervals of 100 m; the ALT values are the averages in the corresponding interval. In Table 2, there is a statistical descriptive analysis of data. In Table 3, correlation coefficients of the ten elements and altitude are shown. Finally, in Figure 3 there are shown the variable histograms (element concentration). For additional information about this data set, see reference 2.

Table 1. Concentration averages (ppm) at different altitude levels (m) for each element.

ALT	Al	As	Co	Cr	Cu	Fe	Hg	Ni	Pb	Zn
77.8	454.4	0.298	0.422	1.190	10.287	175.0	0.076	2.178	3.26	59.7
165.4	723.8	0.447	0.709	1.545	6.202	769.0	0.074	2.313	7.90	54.3
265.6	850.6	0.317	0.616	1.605	5.740	580.4	0.028	1.775	5.23	57.3
358.4	571.3	0.166	0.500	1.663	6.177	502.2	0.039	1.981	3.25	48.1
440.3	553.7	0.230	0.411	1.455	5.870	539.8	0.034	1.948	7.87	59.9
533.7	453.4	0.272	0.297	1.171	5.740	459.2	0.037	1.602	4.20	61.1
655.9	508.2	0.260	0.442	1.216	4.563	398.4	0.039	1.489	4.71	52.0
723.5	712.1	0.148	0.169	1.232	8.098	1218.6	0.033	1.610	14.54	69.7
885.2	810.6	0.432	0.298	1.133	6.547	435.4	0.045	1.629	2.74	64.1

Table 2. Descriptive statistics of sample data set (ppm).

	Al	As	Co	Cr	Cu	Fe	Hg	Ni	Pb	Zn
Mean	596.6	0.277	0.419	1.364	6.229	541.9	0.041	1.820	5.766	58.66
Median	436.5	0.204	0.307	1.105	5.553	388.4	0.034	1.722	3.658	57.27
Minimum	156.6	0.005	0.060	0.176	3.316	139.8	0.002	0.539	0.04	31.22
Maximum	2740.7	1.276	2.498	4.769	25.352	4028.9	0.203	5.018	60.42	117.67
Standard Dev.	548.3	0.263	0.413	0.833	3.122	580.9	0.030	0.845	8.637	18.14

Table 3. Correlation coefficients of elements concentration and altitude (ALT).

	ALT	Al	As	Co	Cr	Cu	Fe	Hg	Ni	Pb	Zn
ALT	1.000										
Al	-0.047	1.000									
As	-0.049	0.347	1.000								
Co	-0.290	0.696	0.283	1.000							
Cr	-0.174	0.634	0.185	0.605	1.000						
Cu	-0.116	0.337	0.036	0.254	0.200	1.000					
Fe	0.021	0.768	0.219	0.511	0.521	0.381	1.000				
Hg	-0.217	0.243	0.233	0.414	0.209	0.107	0.175	1.000			
Ni	-0.267	0.525	0.253	0.643	0.395	0.662	0.400	0.208	1.000		
Pb	-0.002	0.127	-0.087	0.035	0.135	0.213	0.361	-0.083	0.198	1.000	
Zn	0.139	0.062	0.063	-0.061	-0.226	0.264	-0.003	0.061	0.176	0.124	1.000

3. GEOSTATISTICAL ANALYSIS: POLLUTION MAPS

The first step in the geostatistical study, which is the most important goal of this work, is to calculate the data Normal Score Transform (NSCT), according to the GSLIB procedure (see reference 1). In this transformation, as Al and Fe do not have a Gaussian cumulative distribution function, some adjustments of their ties had to be done. Parameters used in this program are shown in Table 4.

The second step, which is the equivalent to calculate and model variograms in classical geostatistics, is the building of the initial correlogram matrix using Fast Fourier Transform (FFT) following the methodology established by Yao and Journel (1998) and Ma and Yao (2001); see references 5 and 4. This matrix consists of a 10×10-block matrix, which has in its diagonal the auto-correlations and the remaining the cross-correlations. So, we obtain 55 different correlation maps. The number of grid points in each map has to be of $1+2^n$; in this case n = 5, that is, we have a 33×33 element matrix. The correlations have been interpolated using a size 10 smooth window and then multismoothing using all correlations and variables with size 3 maximum half window has been carried out to have the final correlation. Some of those 55 maps are hanged at our website. Corresponding parameters are shown in tables 5, 6 and 7.

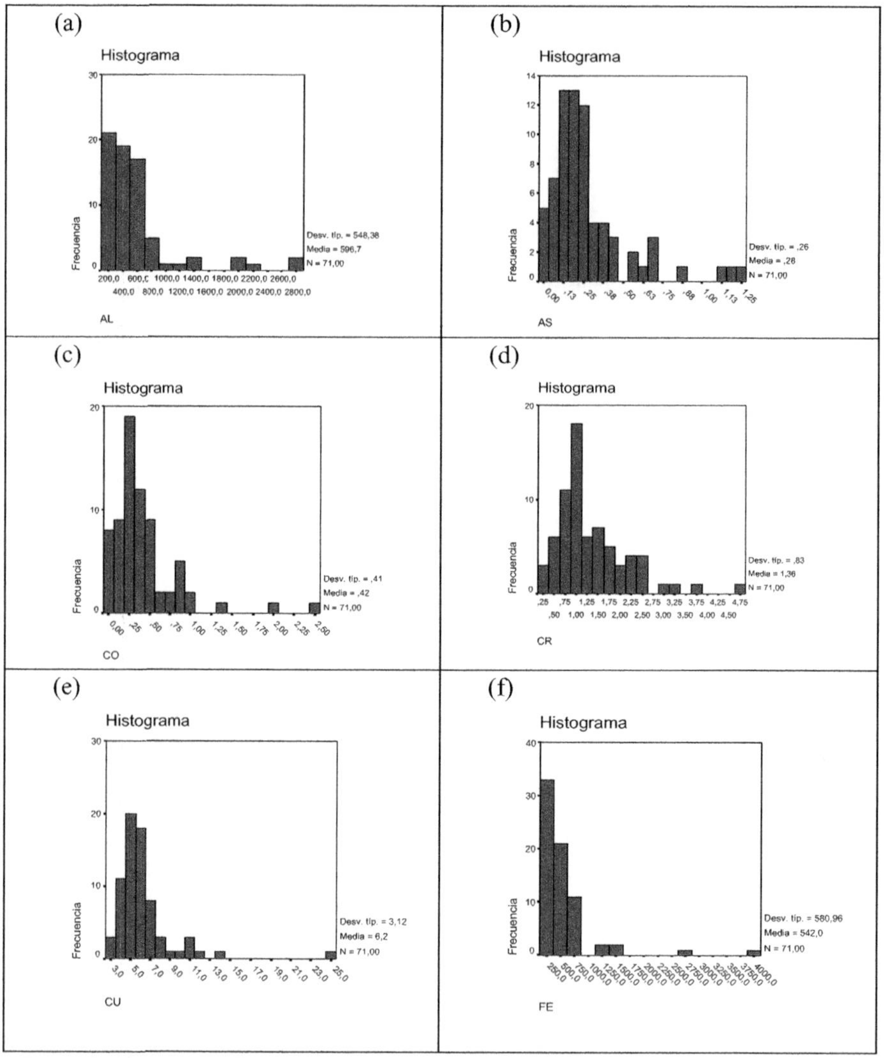

Figure 3. Histograms of elements' concentration (ppm): (a) Al, (b) As, (c) Co, (d) Cr, (e) Cu, (f) Fe, (g) Hg, (h) Ni, (i) Pb, (j) Zn.

The third step is to krige on a regular grid using those correlation maps. Kriging has been done using the program KB2D modified by Hervada-Sala and Jarauta-Bragulat (2001) see reference 3. After kriging, coordinates must be added taking in mind grid parameters; they are shown in tables 8 and 9. At last, back transformations of all results must be computed to recover original space and units. Parameters for that back transformation are shown in Table 10. Figure 4 shows the contour maps obtained from the kriging grid with the right back transformed values.

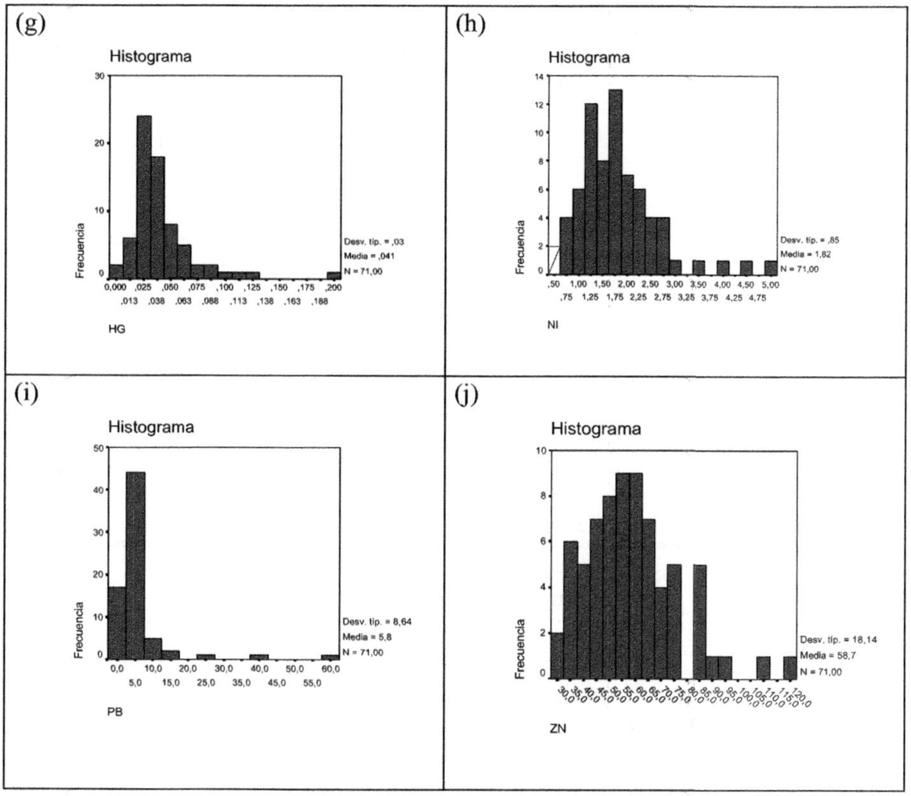

Figure 3. (Cont.).

Table 4. Parameters for NSCORE.

molsagal.dat	\file with data
13 0	\columns for variable and weight
-900 900	\trimming limits
0	\1=transform according to specified ref. dist.
hist1.out	\file with reference dist.
1 0	\columns for variable and weight
nsgal10.dat	\file for output
nsgal10.trn	\file for output transformation table

Table 5. Parameters for CORRMAP.

molsagal.dat	\file with data
10 4 5 6 7 8 9 10 11 12 13	\number of variables: column numbers
-999 999	\trimming limits
0	\1=regular grid, 0=scattered values
33 33	\if =1: nx, ny
1 1	\xsiz, ysiz
1 2	\if =0: columns for x,y coordinates
corgal	\file for correlogram output
16 16	\nxlagl, nylag
5250 5750	\dxlag, dylag
1 1	\xtol, ytol (in the grid unit)
1	\minimum number of pairs

Table 6. Parameters for INTPMAP.

10	-number of variables
33 33	-num. of nodes in x and y directions
corgal	-file with sample corr
intpgal	-output file with interpolated correlogram
indbg	-debug file
10	-smooth window
0.1 0.01	-ratio of the inner and outer radius of fan
4	

Table 7. Parameters for MULTSMTH.

10	-number of coregionalized variables
33 33	-number of nodes in x and y dir.
intpgal	-input file with original corr.map
mapagal	-output file of permissible corr.map
3	-maximum half smoothing window size
0	-minimum number of data for smooth.

Table 8. Parameters for KB2D.

molsagal.dat	\file with data
1 2 4	\columns for X, Y, and variable
-999 999	\trimming limits
3	\debugging level: 0,1,2,3
kb2d.dbg	\file for debugging output
krigal01.out	\file for kriged output
9 300 21000	\nx,xmn,xsiz
9 550 23000	\ny,ymn,ysiz
1 1	\x and y block discretization
1 8	\min and max data for kriging
2.11e4	\maximum search radius
1 2.302	\0=SK, 1=OK, (mean if SK)
mapagal.1	\cov file
31 31	

Table 9. Parameters for ADDCOORD.

krigal10.out	\file with data
krigal10.dat	\file for output
1	\realization number
9 300 21000	\nx,xmn,xsiz
9 550 23000	\ny,ymn,ysiz
1 1 0	\nz,...

Table 10. Parameters for BACKTRANS.

Krigal10.dat	\file with data
4	\column with Gaussian variable
-900 900	\trimming limits
bacgal10.out	\file for output
nsgal10.trn	\file with input transformation table
31.20 117.7	\minimum and maximum data value
1 0.05	\lower tail option and parameter
1 2	\upper tail option and parameter

4. CONCLUSIONS

The main conclusions of this work are the following:

1) It is possible to improve the statistical analysis of environmental pollution by heavy metals in Galicia done in 1, using a two-dimensional geostatistical analysis.

2) Sample density used in this study in not enough to reflect variability of environmental pollution, due to geography of Galicia; it is not possible to employ parameters fitted for a regional scale in that case.
3) Great problems arise with the use of classical geostatistical tools, based on variograms and cross variograms modeling. However, the use of modern FFT techniques allows for finding the full correlogram maps and so it is possible to krige adequately on a regular grid.
4) Finally, it has been possible to build contouring maps of all variables that reflect quite adequately the distribution and concentration of heavy metal pollution. This allows the design of a better sampling grid to control more accurately heavy metal pollution in that region.

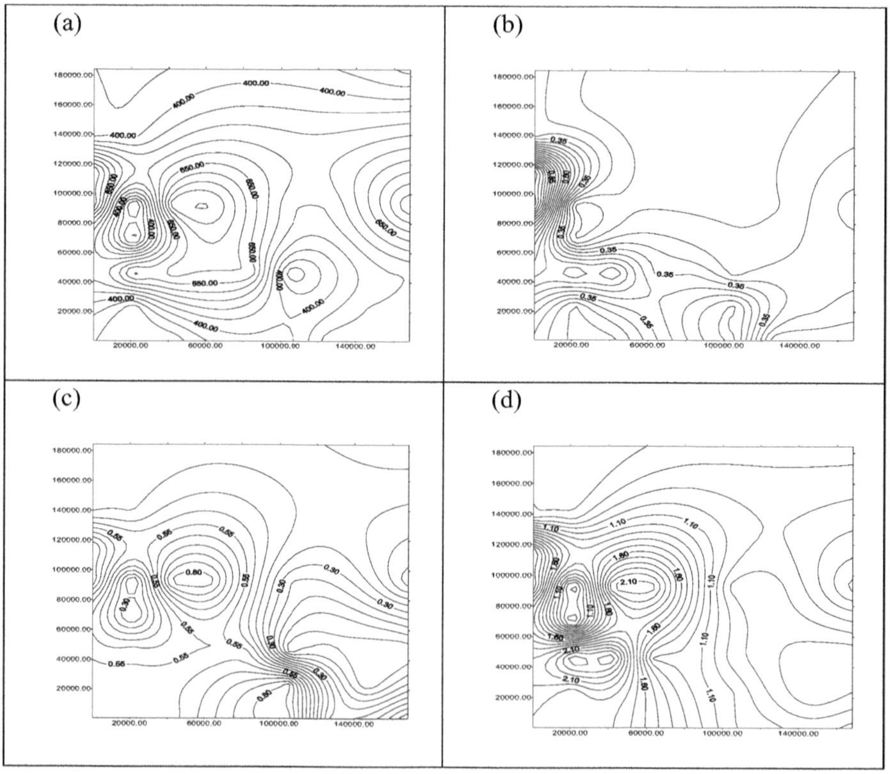

Figure 4. Contour maps of kriging results: (a) Al, (b) As, (c) Co, (d) Cr, (e) Cu, (f) Fe, (g) Hg, (h) Ni, (i) Pb, (j) Zn.

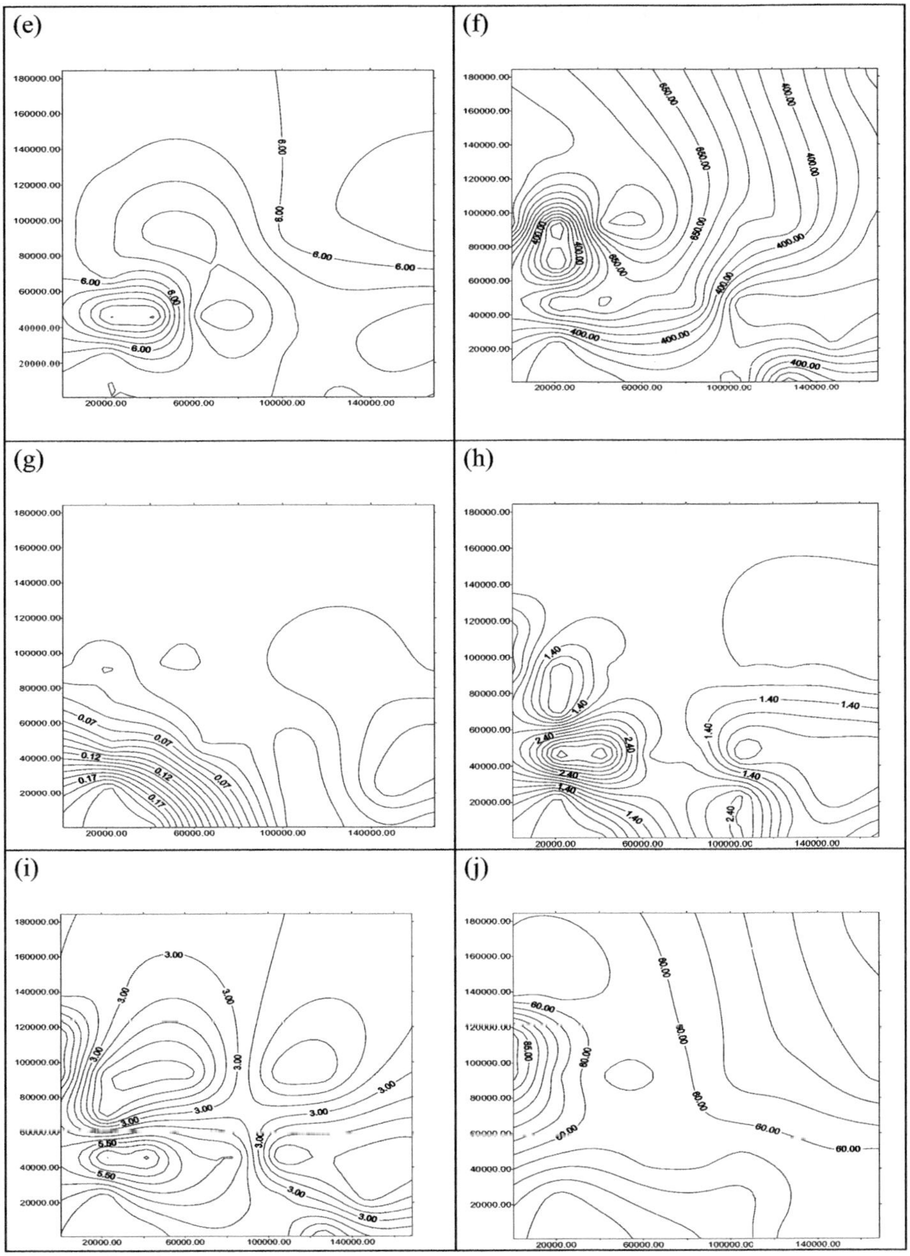

Figure 4. (Cont.).

ACKNOWLEDGEMENTS

Authors want to thank specially to professor J.A. Fernández, from the University of Santiago de Compostela (Spain), for furnishing the data set.

REFERENCES

1. Deutsch, C.V. and A.G. Journel (1998). *Geostatistical software library and user's guide – GSLIB*. Oxford University Press, 1 CD + 369 pp.
2. Fernández, J.A., A. Rey and A. Carballeira (2000). *An extended study of heavy metal deposition in Galicia (NW Spain) based on moss analysis.* The Science of Total Environment, vol. 254, pp 31-44.
3. Hervada-Sala, C. and E. Jarauta-Bragulat (2001). Modifications to kb2d program in GSLIB to allow use of tabulated covariances calculated with Fast Fourier Transform method. Computers & Geosciences, vol.27, num. 07, pp 887-889.
4. Ma, X and Yao,T. (2001). A program for 2D modeling (cross)correlogram tables using Fast Fourier Transform. Computers & Geosciences, vol.27, num. 07, pp 763-774.
5. Yao, T. and A.G. Journel (1998). Automatic modelling of (cross)covariance tables using fast Fourier transform. Mathematical Geology, 30(6), 589–615.

ESTIMATING THE GRADES OF POLLUTED INDUSTRIAL SITES: USE OF CATEGORICAL INFORMATION AND COMPARISON WITH THRESHOLD VALUES

N. Jeannee[1,2] and Ch. De Fouquet[2]

[1]Centre National de Recherche sur les Sites et Sols Pollués, 930 Bd Lahure, BP 537, 59505 Douai Cedex, France, currently at Geovariances, 49bis Av.Franklin Roosvelt,77212 Avon, Cedex, France. jeannee@geovariances.fr; [2]Centre de Géostatistique, Ecole des Mines de Paris, 35 rue Saint-Honoré, 77300 Fontainebleau Cedex, France. fouquet@cg.ensmp.fr

Abstract: Sampling of polluted sites often leads to inaccurate estimates, particularly because of the small number of samples, the importance of sampling errors and the high spatial variability at small distance. Auxiliary information like the history of the site or qualitative measurements are of interest to improve the quality of the grade estimates. The relationship between grades and soils information (presence of coal tar, smell, clay...) are examined on a former coking plant, polluted by PAH (Polycyclic Aromatic Hydrocarbons). A sensitivity analysis shows the utility of this auxiliary information known at additional points compared to the univariate kriging of the grades. Delineation of the zones to remediate is frequently carried out by selecting the areas where the estimated grades exceed the chosen remediation grade. If the estimation is subject to large uncertainties, this selection may generate bias. Estimation of the probability that the true grade is greater than the remediation value makes it possible to take into account the uncertainties associated to the estimated grades. Moreover, it is necessary to specify the support to be retained for this selection, which differs generally from the support of the samples. Neglecting this support effect leads to bias in the calculation of the soil volumes to remediate, as the proportion of the values exceeding a given grade varies with the support size (samples, blocks of various sizes). In this paper, conditional expectation and disjunctive kriging are compared for the estimation of the probability to exceed a threshold on blocks. The evaluation of polluted sites is then improved using a consistent methodology.

X. Sanchez-Vila et al. (eds.), geoENV IV – Geostatistics for Environmental Applications, 461-472.

1. INTRODUCTION

Nowadays, the stake of site investigation and remediation is high, due to the number of polluted sites, the sanitary risks they might represent and the high remediation costs. In France, site investigation is based on the estimation of the pollutants grades and the delineation of polluted areas. At this level, any error may lead to serious consequences, in sanitary and/or financial terms. However, despite the technical and financial investments, these estimations are often empirical. The knowledge of the pollution is usually derived from the historical information (often incomplete), which is occasionally verified by a few samples. Consequently, the number of soil samples and the resulting analyses of pollutants grades are usually scarce. In addition, a large grade variability is classically observed, even at small scale.

So, in order to improve the knowledge of the pollution without increasing too much the costs, there is an important need of additional information: historical information, organoleptical measures and soils information, or their combination. Besides, other pollutants easier to sample and analyze, or semi-quantitative in situ measures, might be of interest. The choice of a relevant auxiliary variable is of importance. Then, a modelling of the spatial relationship between the pollutant and the auxiliary variables is necessary.

Before site remediation, it is important to know which areas have to be treated depending on the chosen level of intervention. To achieve this task, a method consists in selecting the estimated values exceeding the intervention level by thresholding the grade estimate. In the case of an inaccurate estimate, it is well known that this kriged map excessively smoothes the always existing local variability that scarce data do not allow to reproduce. Therefore, using this kriged map to delineate polluted areas would potentially lead to an important bias. In order to take into account the lack of precision of kriging, it is useful to add to this estimate the probability that the true (unknown) grade exceed the intervention level. This probability will give access to the selection of areas where the pollutant grade exceed the intervention level, while knowing the risk to leave in place grades that are above the level (a risk which always exists).

Although frequently used, we will not discuss non parametric methods such as indicator kriging, due to the loss of information they imply, and the lack of consistency when considering successively several indicators. Indicator cokriging aims at minimizing the previous drawbacks by considering simultaneously the indicators at several cut-off values (Goovaerts, 1997, p.297). The larger the number of indicators, the smaller the loss of information, but also the heavier the modelling effort, as we need to model the covariance and cross-covariance functions of all the indicators (except is we assume an intrinsic correlation model, and turn to methods like Median Indicator Kriging).

Furthermore, the selection of polluted areas is usually performed on blocks consistent with the remediation management unit. The support effect implies that the proportion of blocks exceeding an intervention level varies with the size of the blocks. Moreover, the probability of a block grade to be above the threshold differs from the proportion of points within the block that exceed the threshold. Therefore, exceeding probabilities over blocks cannot be derived from punctual probabilities, and a consistent change of support model is necessary, if one wants to avoid the computation of simulations. To illustrate the impact of this support effect, the probability that a pollutant grade exceeds 10 ppm is computed by conditional expectation (CE) within a discrete gaussian model (see below for presentation of the case study, and theoretical details). This probability is estimated both punctually and over 5x5 m blocks. Figure 1 shows how the modelling of the support effect affects the probability estimates. The dispersion model tends to:

- increase the probabilities on blocks close to the ones containing large data values,
- decrease the probabilities on blocks close to blocks containing small values,

excepted for the blocks containing data. For the latter, whereas the punctual probability is equal to 0 if it coincides with a data inferior to the threshold 10 ppm, this probability is equal to 1 for a data larger than the threshold. Consequently, taking into account the support effect for these blocks sensibly changes the estimated probabilities, and therefore the delineation of polluted areas, and the remediation cost.

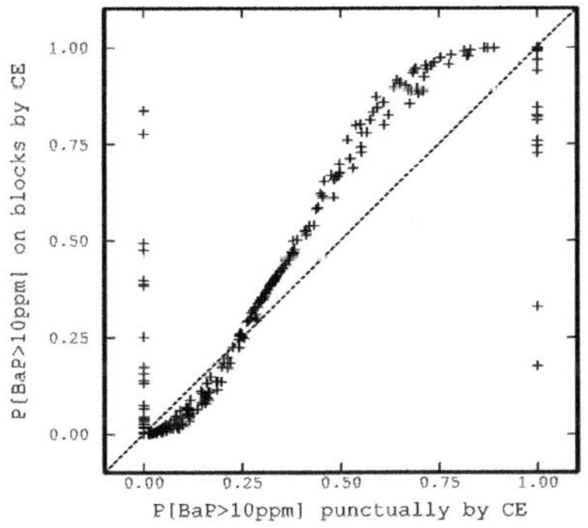

Figure 1. Scatter diagram between probability estimates to exceed 10 ppm of BaP computed by CE punctually (abscissa) and on 5 x 5 m blocks (ordinates).

In this context, the objective of the paper is firstly to discuss on a real case study the choice of an auxiliary variable and its interest. Then, when the goal is to obtain a probability map, we present and discuss from a practical point of view the pros and cons of conditional expectation and disjunctive kriging. These efficient estimation tools avoid the computation of conditional simulations, which could become prohibitive when dealing with large fields finely discretized.

2. CASE STUDY

We are interested in the soil pollution by Polycyclic Aromatic Hydrocarbon (PAH) compounds on a former coking plant. We focus in this paper on the benzo(a)pyren (BaP), a five cycles non volatile, non soluble and highly carcinogenic PAH. 52 points have been sampled on a main regular square grid of 10 x 10 m. The mean BaP grade equals 37.8 parts per million (ppm; S.I. units: mg kg-1) with a standard deviation of 96.2 ppm. Regarding the historical information, two pools of coal tar are located on the sampled area; they have been excavated and one of them, located in the south, has been filled in with non polluted material; backfill coming from the excavation of the north coal tar has been dumped in the north-west of the site (Figure 2).

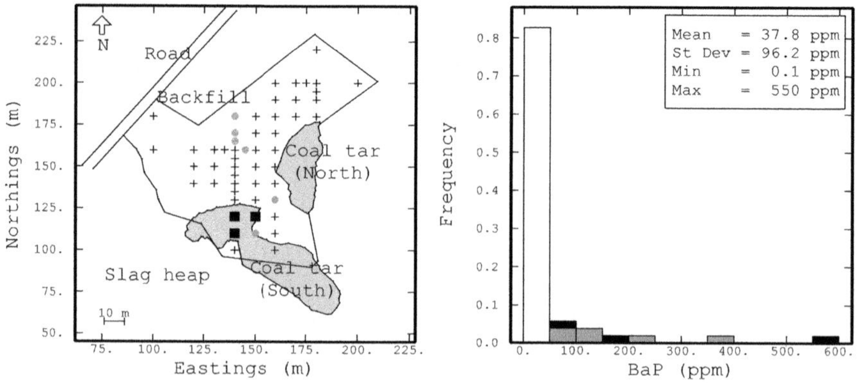

Figure 2. Site configuration, historical information. Histogram of BaP grades; indication of grades above 50 ppm, located on the previous pool of coal tar in south (black) and elsewhere (grey).

3. CONSTRUCTION OF A RELEVANT AUXILIARY VARIABLE

3.1 Which auxiliary variable?

3.1.1 Historical information

In the present case, historical information would lead to investigate mainly around the south pool of coal tar. Figure 2 shows that 3 out of 9 BaP grades larger than 50 ppm are located in this area. However, such a survey would miss the 6 other large grades, located in the north of the site, in the vicinity of the backfill heap, and between the two pools of coal tar.

Historical information therefore indicates areas of high concentrations, but is not sufficient to detect all of them. Consequently, using only this information to direct the sampling strategy is risky.

3.1.2 Soils information

Qualitative characteristics of samples have been observed on the sampled points: presence/absence of coal, coal tar, smell, limestone grains, stonework pieces, greenish colour of the sample, dross, etc.

Figure *3* illustrates the relationship between two qualitative information and the BaP grades. While the presence of stonework pieces is not preferentially associated to small of large BaP grades, the presence of coal tar systematically corresponds to large BaP grades. Consequently, with a reduced cost, this auxiliary variable brings some information about the pollution.

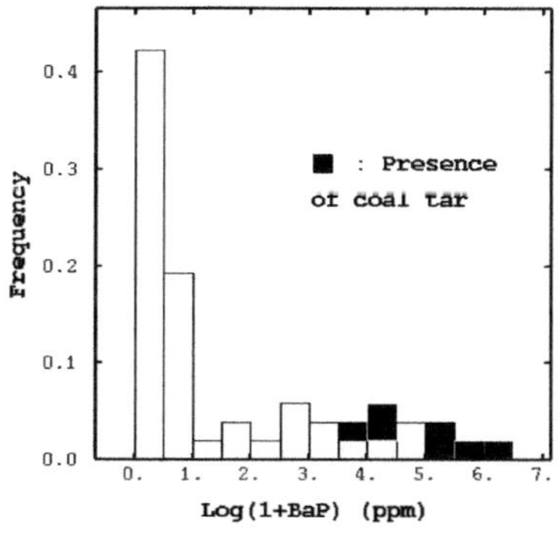

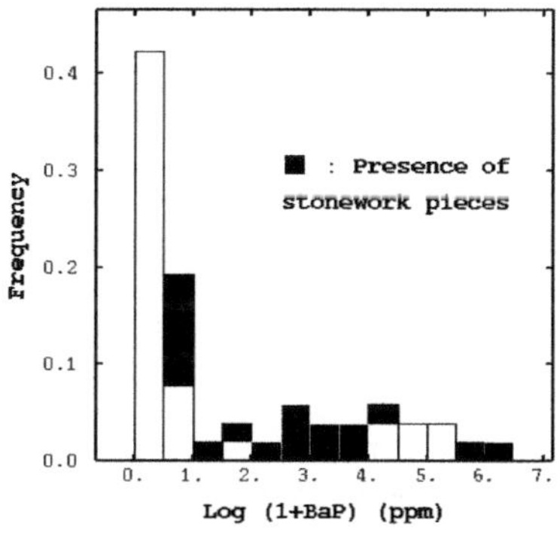

Figure 3. Histogram of BaP. Samples where coal tar (left) or stonework pieces (right) are observed are indicated in black.

The best empirical correlation between a numerical variable *Z* and a categorical variable with k categories is obtained by considering for each category the mean of the associated values of *Z* (Saporta, 1990, p.148). We apply this here to be consistent with BaP grades, even if the results won't be affected as our qualitative information only have two categories. The generalization to more than two categories is straightforward.

3.1.3 Combination of soils information

A correspondence analysis synthesizes the qualitative information. This factorial analysis technique reduces the high number of variables into a few number of non correlated factors containing the information about the data. Here, the first factors represent 33.6 % and 23.5 % of the total variance of the data (Figure 4-a). The greenish colour, limestone grains and presence of coal in soil mainly indicate that we are dealing with a soil in place, where as the other variables are more indicator of backfill; consequently, the first factor (called "auxiliary factor" hereafter) distinguish backfilled materials (high values) and soil in place. Figure 4-b shows that the small BaP grades mainly correspond to soil in place, whereas the medium and large grades correspond to backfilled materials (Jeannée, 2001, p.61). Compared to the presence/absence of coal tar, it has to be noted that the seven points where coal tar has been observed have an auxiliary factor value greater than 1. Therefore, most of the information brought by the auxiliary factor is already expressed by the presence of coal tar.

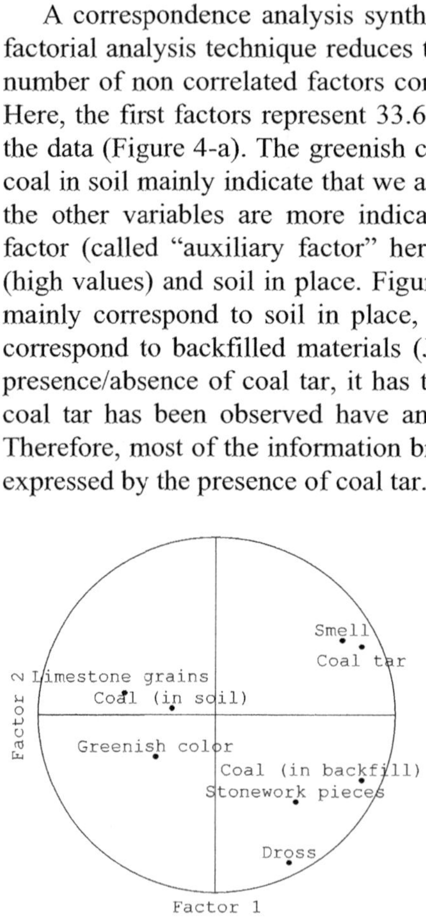

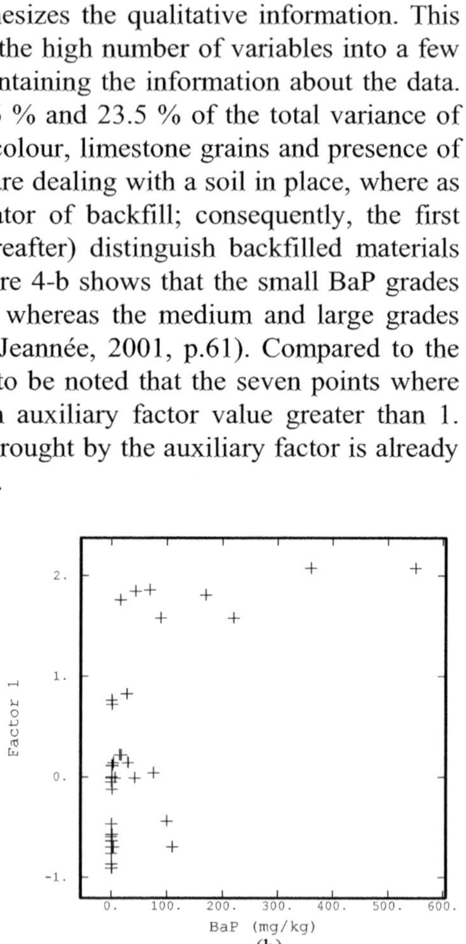

Figure 4. (a) Correspondence analysis on qualitative variables: projection of variables on the two first factors. (b) Scatter diagram between the first factor and the BaP grade.

3.2 Bivariate modelling

We aim at estimating BaP grades, and not the result of a transformation of these raw grades. For robustness purpose, as we are facing a variable with only 52 values, our interest is put on the modelling of the raw BaP grades, instead of considering any transformation attempting to reduce the skewness of the variable. Working in the framework of a linear model of coregionalization, bivariate variogram models are fitted on the BaP grade Z and (a) its mean for each category of coal tar (absence/presence) and (b) the auxiliary factor. An example is given in Figure 5 for the coal tar.

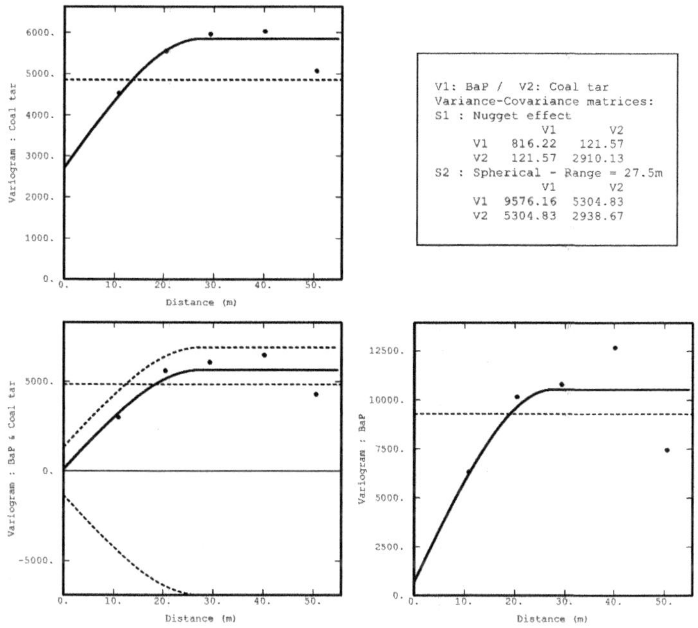

Figure 5. Bivariate variogram model between BaP grades and the mean BaP grade by category of coal tar.

4. CONTRIBUTION OF AUXILIARY VARIABLE TO THE ESTIMATION OF BaP GRADES

Ordinary kriging (OK) of BaP grades has been performed using the appropriate part of the model shown in Figure 5; the OK estimate will be used in the non linear section.

Cokriging BaP grades with isotopic (all variables known at the same locations) soft data does not improve sensibly the results. To assess the gain of precision obtained by using more densely sampled soft data, 50 % of randomly selected BaP grades are used as a validation set. OK of the other 50% BaP grades is performed at the location of the validation samples. This result is compared with those obtained by ordinary cokriging (OCK) with the presence of coal tar, assumed known on all samples, and OCK with the auxiliary factor. Figure 6 shows the improvement brought by the cokriging, particularly for the highest real value and the small grades.

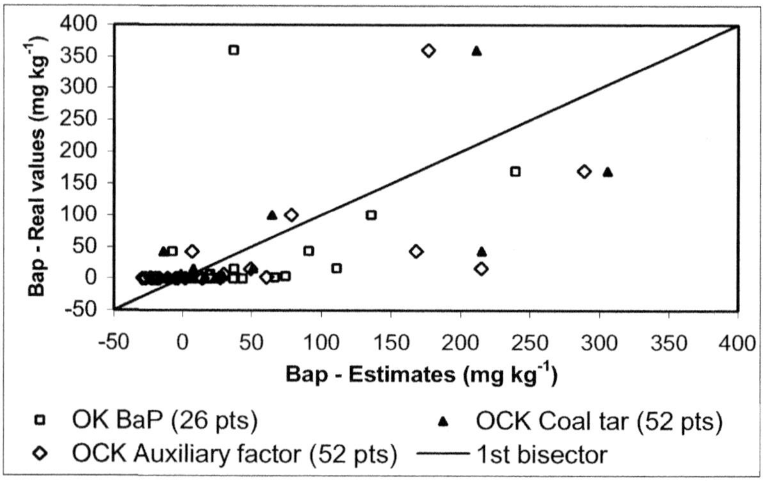

Figure 6. Scatter diagram on the validation set between BaP grades and their estimate by OK, OCK with the knowledge of all coal tar and all auxiliary factor information.

Mean error and mean square error between the OK estimates and the real BaP grades are computed on this validation set (Table 1). Both cokriging results lead to improved mean errors and mean square errors compared to the OK, in particular for the OCK with coal tar. Despite its influence, the removal of the maximum validation value (equal to 360 ppm) does not modify these conclusions.

Table 1. Computation of estimation mean error and mean square error on the validation set, for several estimations: OK, OCK with all the coal tar information, and finally with all the auxiliary factor information.

Estimation of BaP	Mean error (ppm)	Mean square error (ppm^2)
OK	15.28	6253
OCK with coal tar	5.93	3309
OCK with aux. factor	8.66	4980

5. NON LINEAR ESTIMATION

The usual remediation value for BaP is 10 ppm. Therefore, the mean concentration in BaP, equal to 37.8 ppm, already indicates that at least some areas will probably have to be cleaned up, which is confirmed by linear kriging techniques. We want to compare two parametric methods; firstly the disjunctive kriging (DK), which is equivalent to the cokriging of all the indicators, and requires only the modelling of the bivariate distribution (Rivoirard, 1994). Then, the conditional expectation (CE), which is the best estimator possible, but require stronger assumptions.

5.1 Theoretical background

Usually, the pollutant grades $Z(x)$ do not follow a gaussian distribution and it is useful to transform the raw histogram into a standard gaussian one. Therefore, we consider the stationary random function $Z(x)$ as a function $Z(x) = \Phi(Y(x))$ of the gaussian $Y(x)$. The "anamorphosis" function Φ is determined by the coefficients ϕ_i of its truncated development in Hermite polynomials $Z(x) = \sum_{i=0}^{n} \phi_i H_i(Y(x))$. Finally, we associate to each raw value z_i a gaussian transform value having the same cumulate frequency than z_i.

The discrete gaussian model allows the estimation by several methods of the probability to exceed a threshold value on blocks v of a given size. Every punctual value is considered as uniform in its block v; the block anamorphosis Φ_v is computed such that $Z(v) = \Phi_v(Y(v))$, where the block transformed values have a gaussian distribution. In this model, for any raw threshold value z_t, it is possible to compute the corresponding gaussian threshold $y_{V_t} = \Phi_v^{-1}(z_t)$, which varies with the block size. Consequently, we obtain $Z(v) \geq z_t \Leftrightarrow Y(v) \geq y_{V_t}$, which implies that $\mathrm{P}[Z(v) \geq z_t] = \mathrm{P}[Y(v) \geq y_{Vt}]$ or identically $\mathbf{1}_{Z(v)\geq z_t} = \mathbf{1}_{Y(v)\geq y_{Vt}}$.

Disjunctive kriging allows the estimation of any function of the variable of interest. In particular, the block disjunctive kriging of $\mathrm{P}[Y(v) \geq y_{Vt}]$ within a discrete gaussian model is given by

$$\left[\mathbf{1}_{Y(v)\geq y_{Vt}}\right]^{KD} = 1 - G(y_{V_t}) - \sum_{n\geq 1} \frac{1}{\sqrt{n}} H_{n-1}(y_{V_t}) g(y_{V_t}) H_n^{\ K}(Y(v)).$$

The conditional expectation is directly obtained from the gaussian cdf G and the block kriging of the gaussian transform

$$\left[\mathbf{1}_{Y(v)\geq y_{Vt}}\right]^{CE} = 1 - G\left(\frac{y_{V_t} - Y^K(v)}{\sigma^K(v)}\right).$$

The extension of the conditional expectation to the multivariate case is straightforward, the simple kriging of the gaussian transform being replaced by its simple cokriging with the transformed auxiliary variable.

For further details on the methods the reader should refer to Rivoirard (1994, p.78) or Chilès and Delfiner (1999, p.432).

5.2 Assumptions and preliminary comparisons

Both DK and CE require strict stationarity as soon as a change of support model is used. CE requires that the multivariate distribution of variables like $(Y(x), Y(x_1), ...)$ is multigaussian, i.e. any linear combination of these variables is normally distributed, whereas gaussian DK only necessitates that the variables $(Y(x), Y(x+h))$ are bivariate normal. Except in the case of an important systematic sampling, the validation of the multigaussian assumption is quite inextricable and is most of the time reduced to the validation of the bigaussian assumption. It is therefore difficult to assess how more constraining the multigaussian assumption is compared to the bigaussian assumption. Several tests exist to evaluate the bigaussian assumption: examination of h-scattergrams, computation of the ratio $\sqrt{\gamma(h)} / \gamma_1(h)$ between variograms of order 2 and 1, which has to be constant and equal to $\sqrt{\pi}$, validation of the relationship between raw and gaussian covariances (Lajaunie, 1993, p.40; Chilès & Delfiner, 1999, p.409).

CE is faster than DK, as it only requires the simple kriging of $Y(x)$, while DK necessitates the kriging of N Hermite polynomials.

As it is an indicator cokriging, DK does not ensure the consistency of the results and do not necessarily lie between 0 and 1. On the contrary, CE is by construction fully consistent.

DK allows easily the derivation of estimation variances, which might be useful. Indeed, if the probability to exceed a threshold remains approximately the same all over the field, this probability will probably be close to the a priori probability, and the kriging variances will be high. For example, if we consider the simple kriging case, then

$$\mathrm{P}[Z \geq z_t] = \mathrm{P}[Y \geq y_t] = \mathrm{P}[Y^{SK} + \sigma^{SK} X \geq y_t].$$

In poor estimation conditions (high nugget effect), Y^{SK} will be close to the mean 0, and σ^{SK} will be close to 1. Consequently, $P[Z \geq z_t]$ will be close to the a priori probability. It is therefore important to have in mind that in the case of poor estimation conditions, the probability estimate will also be of poor accuracy.

5.3 Computation of a probability may by DK and CE

This section aims at comparing in practice the estimates of the probability that the BaP grades exceed 10 ppm on 5 x 5 m blocks obtained by DK and CE within a discrete gaussian model. As the distribution of BaP grades is positively skewed, the raw grades have been transformed by anamorphosis into a gaussian variable. The analysis of the criteria previously mentioned lead us to accept the bigaussian assumption. Regarding the DK, 50 Hermite polynomials have been used. DK and CE results are consistent, even if CE leads to more contrasted estimates (Figure 7 and Figure 8).

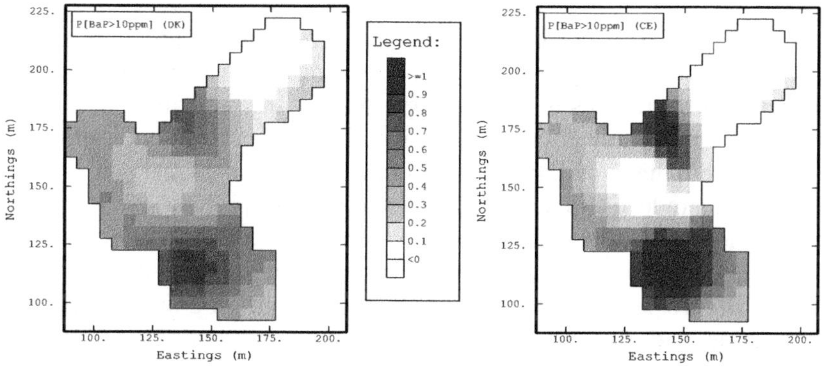

Figure 7. Probability to exceed 10 ppm of BaP estimated on 5 x 5 m blocks by DK (left) and CE (right) within a discrete gaussian model.

DK tends to smooth the large probabilities between the highly polluted areas. Indeed, the largest differences between CE and DK occur:

- in the areas where the grade estimates are small. In these areas, the probabilities estimates by DK are larger than the ones estimated by CE,
- close to high estimates, where the probabilities estimated by CE are larger.

These comparisons, reinforced by the easier and faster implementation of conditional expectation, lead us to prefer this method instead of the disjunctive kriging in our case. The use of a validation subset to assess the improvement due to the soft information - see section 4 - has been applied to the conditional expectation results; because of the absence of additional soft information, the results, being qualitatively comparable to what they were for the estimation, are not discussed here.

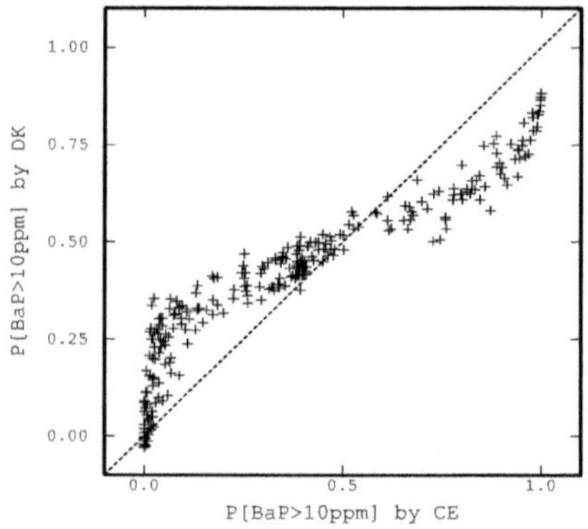

Figure 8. Scatter diagram between the probability to exceed 10 ppm of BaP estimated by DK and CE within a discrete gaussian model. The dotted line represents the first bisector.

6. CONCLUSIONS

Costs allotted to site investigation and remediation, although increasing, still limit the sampling effort. The interest of multivariate geostatistics, by improving the grade estimation using auxiliary soft information, is therefore intensified. From this point of view, we discussed how to choose in practice a relevant auxiliary variable. If site remediation is necessary, it is useful to add to the grade estimate the knowledge of the probability to exceed the remediation level, to evaluate the importance and the risk of selection errors. To achieve this goal, the paper discussed the underlying assumptions and the interest of two estimation tools, the disjunctive kriging and the conditional expectation, and compared their efficiency on a case study.

REFERENCES

1. Chilès J.P. and Delfiner P. 1999. Geostatistics: modelling spatial uncertainty, Wiley Series in Probability and Mathematical Statistics, 695p.
2. Goovaerts, P. 1997. Geostatistics for Natural Resources Evaluation. Oxford Univ. Press, New-York, 483p.
3. Jeannée, N. 2001. Caractérisation géostatistique de pollutions industrielles de sols. Cas des HAP sur d'anciens sites de cokeries. Thèse de Doctorat en Géostatistique, Ecole des Mines de Paris.
4. Lajaunie, C. 1993. L'estimation géostatistique non linéaire. Cours C-152, Centre de Géostatistique, Ecole des Mines de Paris.
5. Rivoirard, J. 1994. Introduction to disjunctive kriging and non-linear geostatistics. Oxford University Press, Oxford, 181p.
6. Saporta, G. 1990. Probabilités, analyses des données et statistique. Technip, Paris.

A CO-ESTIMATION METHODOLOGY FOR MAPPING DIOXINS MEASURED BY BIOMONITORS

M.J. Pereira[1,2], A. Soares[1], C. Branquinho[3,4], S. Augusto[4] and F. Catarino[5]

[1]*Environmental Group of the Centre for Modelling Petroleum Reservoirs, CMPR/IST, Av. Rovisco Pais, 1049-001 Lisbon, Portugal (e-mail: maria.pereira@ist.utl.pt).*
[2]*FE/UCP Faculdade de Engenharia da UCP, Estrada de Talaíde, Rio de Mouro 2635-631, Portugal*
[3]*Universidade Atlântica, Antiga Fábrica da Pólvora de Barcarena, 2745-615 Barcarena, Portugal*
[4]*CEBV - Centro de Ecologia e Biologia Vegetal, Campo Grande, Bloco C2, 4º Piso, 1749-016 Lisboa, Portugal*
[5]*Museu, Laboratório e Jardim Botânico, Rua da Escola Politécnica, 58, 1250-102 Lisboa Portugal*

Abstract: A biomonitoring survey was performed to measure PCDD/Fs deposition for mapping spatial dispersion of dioxins in the region of Setúbal, Portugal. Since, no single lichen species was found occurring in the whole study area, samples of two lichen species – *Ramalina canariensis* and *Xanthoria parietina* – were collected. These two species have different abilities to monitor the same pollutant concentration. As they are spread preferentially in two different areas, they should be viewed as two complementary indicators of dioxins concentration. The objective of this study was to build a geostatistical model that integrates, within a single coherent model, the two complementary visions of the same reality given by contaminant concentrations measured in the two sampled lichen species. For this purpose, a geostatistical model was built to integrate both lichen species' data to obtain a unique map of PCDD/Fs deposition. This model uses co-located cokriging with local spatial correlations in order to estimate the primary data. Some of the congeners of dioxins were also estimated with the same methodology.

X. Sanchez-Vila et al. (eds.), geoENV IV – Geostatistics for Environmental Applications, 473-484.

1. INTRODUCTION

The class of compounds made up of polychlorinated dibenzo-p-dioxins (PCDD) and polychlorinated dibenzofurans (PCDF) – usually called dioxins – is part of a wide group of persistent organic pollutants, which may cause adverse effects on human health through chronic exposure to as little as trace levels. The anthropogenic sources of PCDD/Fs are mainly combustion processes, manufacturing of chemicals, metallurgical processes and paper and pulp processing. Although there are 210 congeners of PCDD/Fs, only 17 are of concern, owing to their toxicity, stability and persistence in the environment (Buckley-Golder, 1999). Usually, a system of toxic equivalency factors is used to derive an equivalent concentration of the most toxic dioxin (2,3,7,8-TCDD), enabling the toxicity of complex mixtures to be expressed as a single number – the toxic equivalent or TEQ.

To map the spatial dispersion of PCDD/PCDFs in the region of Setúbal, a biomonitoring survey was performed to measure PCDD/F's deposition. Since, no single lichen species was found to be sufficiently representative for all of the area studied, samples of two lichen species – *Ramalina canariensis* and *Xanthoria parietina* – where collected. Sampling locations were selected according to climatic and orographic criteria. The measures of TEQ registered by the two species are different in terms of absolute values and territory cover. In fact, each species gives a distinct image of TEQ deposition for the area studied: *X. parietina* covers a larger area, but *R. canariensis* is more sensitive to TEQ concentration variability.

The idea of this study is to build a geostatistical model to integrate the different measures provided by the lichens – to obtain a unique map of PCDD/PCDF deposition. The *X. parietina* data, which is more widespread over the region, was assumed as primary data, and the *R. .canariensis* was used as secondary data.

2. BIOMONITORING CAMPAIGN WITH LICHENS

One of the main advantages of lichens used as biomonitors is the low cost of high-density sampling grids. Besides, biomonitoring measures the continuous and cumulative response of living organisms to (anthropogenic) environmental factors (Branquinho *et al.*, 2000).

In a first step, prior to the campaign, the species of lichens most appropriate to monitor atmospheric deposition were identified. Two basic criteria assisted this selection: biological and morphological aptitudes for performing as monitors of that specific pollutant and spatial representativeness, i.e., the species must be sufficiently robust to the air quality in order to be found in the whole area.

Two lichens species – *R. canariensis* and *X. parietina* – were chosen for covering complementary areas of the peninsula. Samples of both lichens were collected at 115 sampling locations (Figure 1). The sampling locations were selected according to climatic and orographic criteria, and the location of possible anthropogenic pollutant sources and conventional air quality monitoring devices. Note that *X. parietina* covers a larger area and was more frequently sampled than *R. canariensis*.

The samples were analysed regarding dioxins (the 17 most toxic congeners of PCDD/PCDF) with concentrations measured in TEQ (toxic equivalent), as well as other metals and gases – Pb, Cu, Ni, Cr, Co, S, Zn, Fe, Mn, Ca, N, K and Mg. For sake of simplicity, the sum of the 17 congeners of PCDD/PCDF concentration in TEQ's will henceforward be referred as dioxins concentration, for sake of simplicity. Histograms and basic statistics of dioxin concentrations at both lichens are presented in Figure 2. *X. parietina* samples present a positively skewed distribution, while *R. canariensis* samples show approximately a normal distribution. In general, *R. canariensis* samples present higher dioxins concentration and more variability than *X. parietina* samples. Data analysis showed that total concentration of PCDD/Fs in lichens was more similar to the concentrations reported for animals (top of the food chain) and soils (act as sinks) than those reported for plants. In general, the congeners and homologue profile observed in lichens (Figure 3) resemble that of the atmosphere more that of the soil showing that lichens are potential good biomonitors of PCDD/Fs (Branquinho et al., 2002).

3. GEOSTATISTICAL MODELLING

The two sampled lichens have different abilities to monitor the same pollutant concentration. As they are spread preferentially in two different areas (Figure 1), they should be viewed as two complementary views on dioxin concentration. The objective of this study is to build a geostatistical model that integrates, within a single coherent model, the two complementary visions of the same reality given by contaminant concentrations measured in the two lichen species *X. parietina* and *R. canariensis*. For this purpose, spatial co estimation was performed taking into account the spatial correlation between these two 'images' of reality.

Given the wide spatial cover of *X. parietina*, its sampling measurements are assumed as the primary variable to be interpolated for the entire region. *R. canariensis* dioxin concentrations are considered as secondary variable. Consider $Z_1(x)$ the primary variable – dioxin concentrations in *X. parietina* – and $Z_2(x)$ the secondary variable – dioxin concentrations in *R. canariensis* – known at spatial location x.

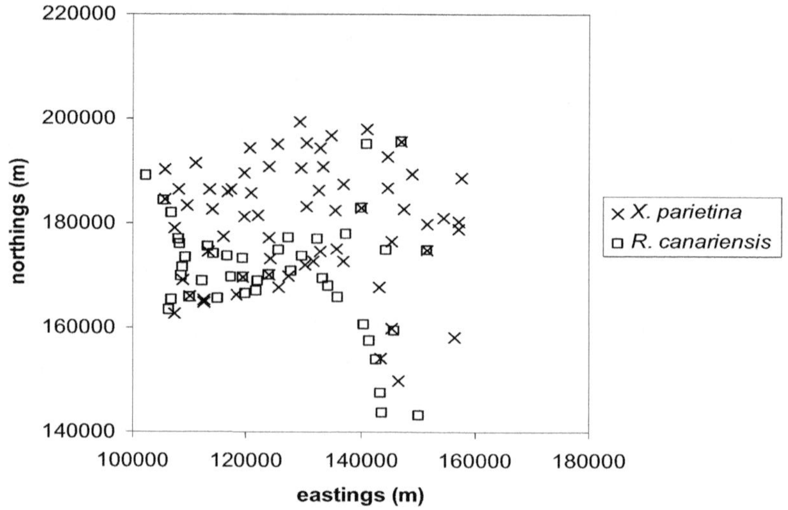

Figure 1. Sample locations for both lichens on the Setúbal peninsula (South of Lisbon).

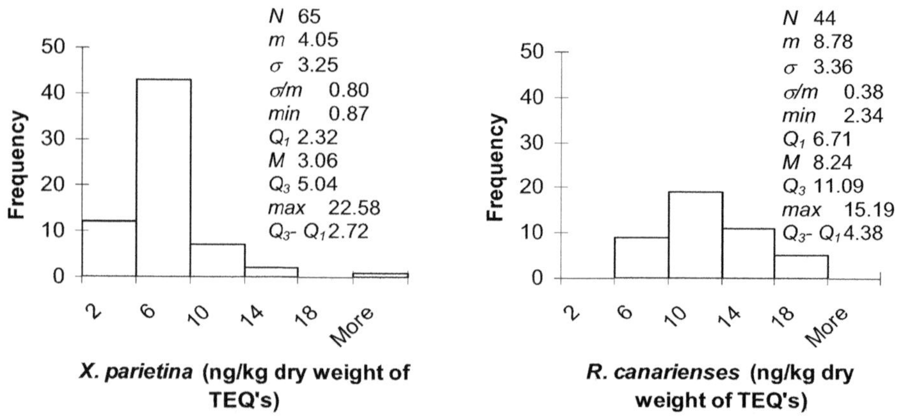

Figure 2. Histograms and univariate statistics of dioxin concentrations (in TEQ's) in *X. parietina* and *R. canariensis* samples.

The use of co-kriging was disregarded because the high density of samples of *R. canariensis*, the secondary variable $Z_2(x)$, mostly concentrated in the southern part of the peninsula, filtered out the influence of the scarce primary variable $Z_1(x)$ in that region. On the other hand, the "clustering" of $Z_2(x)$ values didn't allow for an estimation of reliable cross-covariances $C_{z1,z2}(h)$. Hence the idea of the co-estimation model proposed for this study, which is based on the following:

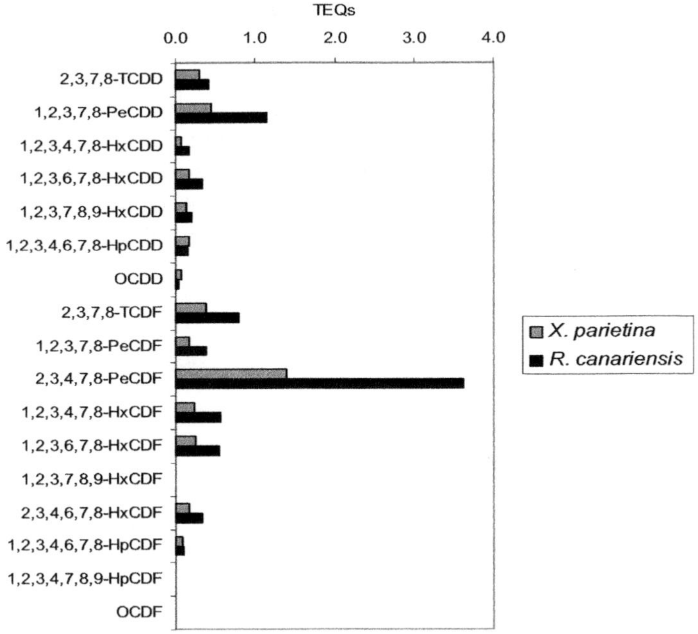

Figure 3. The congeners and homologue profile observed in *X. parietina* and *R. canariensis* samples at the studied area.

– $Z_2(x)$ is assumed to be a soft secondary image of dioxin concentrations after being estimated (by ordinary kriging) for the southern part of the study area:

$$\begin{cases} Z_2(x_0)^* = \dfrac{1}{N}\displaystyle\sum_{\alpha=1}^{N} \lambda_\alpha(x_0) Z_2(x_\alpha) \\ \displaystyle\sum_{\alpha=1}^{N} \lambda_\alpha(x_0) = 1 \end{cases}$$

– Co-located ordinary cokriging of $Z_1(x_0)$, with the estimated map of $Z_2(x_0)^*$ as secondary variable:

$$\begin{cases} Z_1(x_0)^* = \dfrac{1}{N}\displaystyle\sum_{\alpha=1}^{N} \lambda_\alpha(x_0) Z_1(x_\alpha) + \lambda\ (x_0) Z_2(x_0)^* \\ \displaystyle\sum_{\alpha=1}^{N} \lambda_\alpha(x_0) + \lambda\ (x_0) = 1 \end{cases}$$

– The use of local correlation coefficient maps to control the influence of the secondary variable $Z_2(x)$ on the dioxin concentrations estimates.

3.1 Variograms of dioxins for both lichens

Variograms of $Z_1(x)$ and $Z_2(x)$ – i.e. dioxin concentrations in *X. parietina* and *R. canariensis* samples – were computed. Isotropic spherical models were fitted to both lichens (Figure 4). *R. canariensis* presents more variability and a larger spatial continuity than *X. parietina*, since sill and range of the *R. canariensis* variogram are more or less two times those of the *X. parietina* model. Still, note that the spatial cover of *X. parietina* samples is larger than the *R. canariensis* samples (Figure 1).

(a) (b)

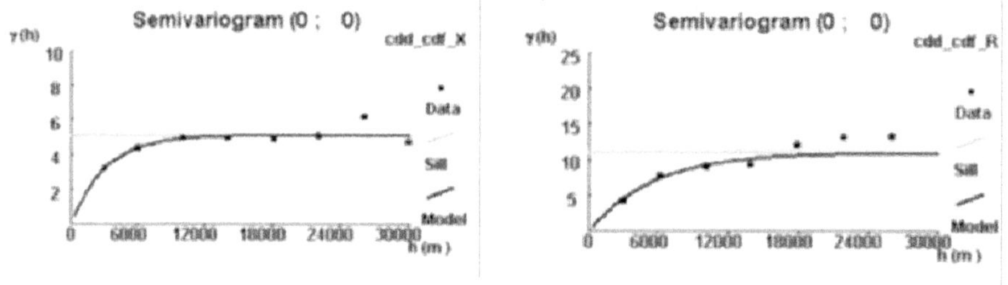

Figure 4. Variograms of dioxins in the two lichen species: (a) *X. parietina:* omnidirectional variogram, exponential model ($C_0 = 0.0$, $C_1 = 5.134$, range = 9000 m); (b) *R. canariensis:* omnidirectional variogram, exponential model ($C_0 = 0.0$, $C_1 = 10.943$, range = 16000 m).

3.2 Spatial estimation of dioxin concentrations

Dioxin concentrations in *R. canariensis*, $Z_2(x)$, were estimated (by ordinary kriging) for the southern part of the peninsula covered by this species' samples (Figure 5a). Dioxin concentrations in *X. parietina* , $Z_1(x)$, were estimated (by ordinary kriging) for the entire area (Figure 5b). Note that the dark spots of dioxine concentrations in *R. canariensis* could not be reproduced by *X. parietina* samples, because these were not found there.

3.3 Correlation between dioxin concentrations in both lichens

As mentioned, cross-variograms between Z_1 and Z_2 were hard to estimate, given the clustered location of Z_2 samples. However, it is known that there is good agreement between dioxin concentrations from both lichens. Hence, the Markov-Bayes approximation (Almeida and Journel, 1993) was adopted to perform the co-located cokriging of $Z_1(x)$. Under this approximation, the cross correlogram $\rho z_1 z_2(h)$ is linearly dependent on the univariate correlogram $\rho z_1(h)$ and the correlation coefficient $\rho z_1 z_2(0)$: $\rho z_1 z_2(h) = \rho z_1(h)\, \rho z_1 z_2(0)$.

a)

b)

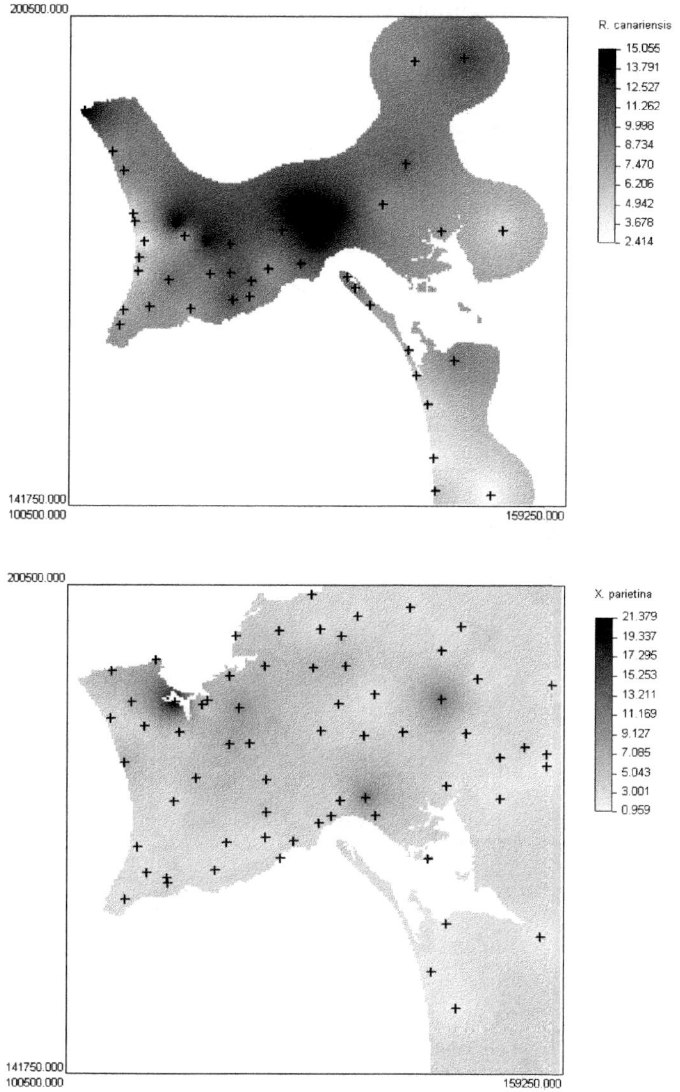

Figure 5. Dioxin concentration estimates for *R. canariensis* (a) and *X. parietina* (b); + lichen samples for each species, respectively.

The correlation coefficient between $Z_1(x)$ and $Z_2(x)$ was calculated (r = 0.76) based on the few common sampling points (Figure 6). The linear relationship between lichen dioxin concentrations was considered valid only for the area covered by the soft image (i.e. the representative area where samples of both lichens coincide, Figure 1), with estimations for the complementary area solely influenced by the primary data. For this purpose the map of local correlations shown in Figure 7 was used. This map was built considering a maximum correlation between the two lichen species of 0.76 at sample locations of *R. canariensis,* and decreasing correlation with increasing distance from these locations.

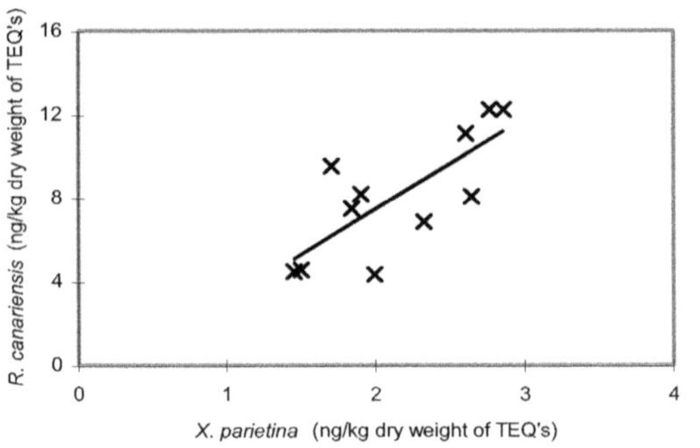

Figure 6. Linear regression between the dioxin concentrations in the two lichen species: × samples; — regression line (*X. parietina = -1.25 + 4.37 R. canariensis*).

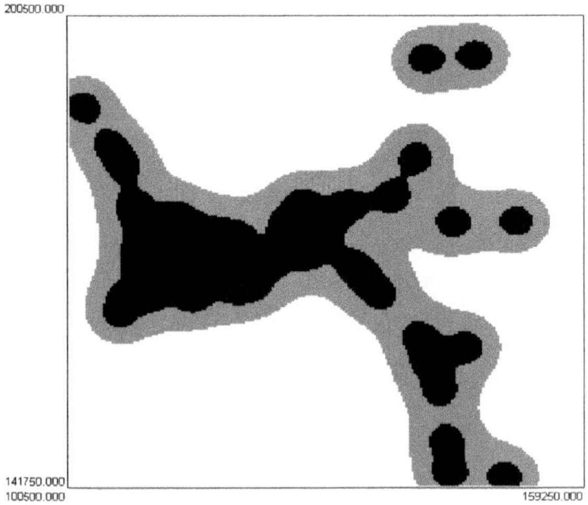

Figure 7 .Map of local correlations between dioxin concentrations of *R. canariensis* and *X. parietina* samples: ■ linear correlation > 0.4; ■ linear correlation between 0.4 and 0.2; □ linear correlation ≤ 0.2.

3.4 Spatial co-estimation of dioxin concentrations

The final dioxin concentration map (Figure 8) for the entire area was achieved by colocated co-kriging of $Z_I(x)$, taking into account the hard data of *X. parietina* samples, the soft estimated dioxin concentration image for *R. canariensis* (Figure 4a) and the local correlations map (Figure 7).

In comparison to the dioxin map estimated only with *X. parietina* values (Figure 5b), it is worth noting that the southern part of the study area (where the influence of *R. canariensis* prevails) shows more clearly the high and low values of dioxin concentrations. This can also be seen in Figure 9 where the map of differences between the values of dioxin concentrations estimated by co-located cokriging (Figure 8) and by ordinary kriging (Figure 5b) are shown.

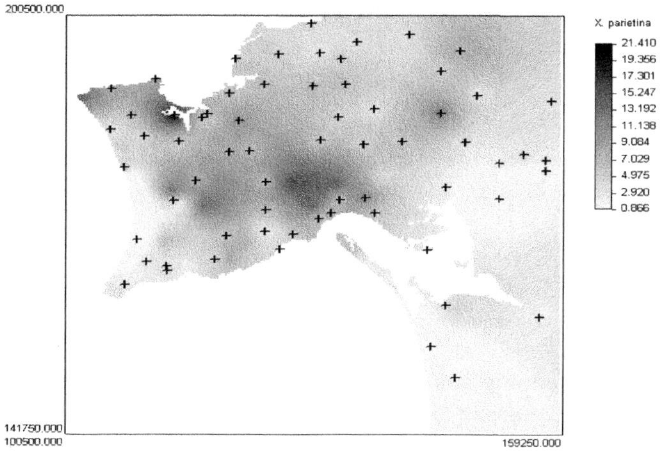

Figure 8. Final estimated map of dioxin concentrations (TEQ's); + *X. parietina* samples location.

3.5 Spatial estimation of congeners

Among the 210 congeners of PCDD/F, only 17 are of concern, owing to their toxicity, stability and persistence in the environment. A profile of these 17 congener may serve as a signature of the types of PCDDs and PCDFs associated with particular environmental sources of these compounds (Cleverly, *et al.*, 1997). Thus, the estimation of concentration maps of these congeners may be very useful in explaining source contributions to environmental measurements.

3.6 Spatial estimation of congeners

In this paper and just for illustration purposes, three congeners were selected: 2,3,7,8 TCDD which is the most toxic congener, essentially emitted from oil combustion sources; 2,3,7,8 TCDF which the dominant congener in combustion processes occurring in cement kilns not burning hazardous waste; and 2,3,4,7,8 PeCDF which is the most abundant congener in the Setúbal Peninsula.

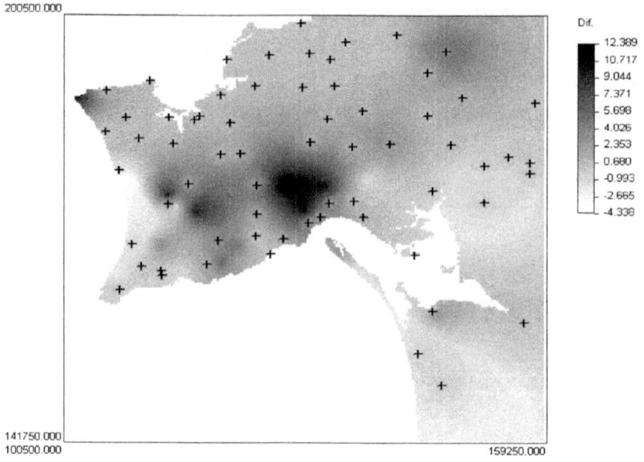

Figure 9. Map of differences between the colocated cokriging and ordinary kriging estimated values of dioxin concentrations (TEQ's); + *X. parietina* samples location.

The same methodology as described above was used for the estimation of these three dioxin congener concentrations: colocated cokriging of *X. parietina* measurements with the estimated image of *R. canariensis* as secondary information (Figure 10,11 and 12).

4. FINAL REMARKS

Biomonitoring with lichens is a very promising and consistent way of sampling airborne pollutants as it measures pollutant effects on living organisms and can cover, with relatively low costs, a large study area.

Measurements from classical physical monitoring stations represent mainly the time component of air quality, given that the, usually, few monitoring stations available are spatially unrepresentative of the phenomenon. Hence, while measurements from physical monitoring stations can give a detailed image of time series of contaminants, for a few points in space, they are most of the times useless when, for instance, a regional image of a pollutant's impact is required.

Biomonitoring with lichens allows for another view on air-pollutant dispersion: a cumulative effect in time, with good representativeness in space. In situations where the evaluation of pollutant impact in a given region resulting from several and different sources is the main issue, biomonitoring the air quality with lichens is often the most appropriate sampling technique.

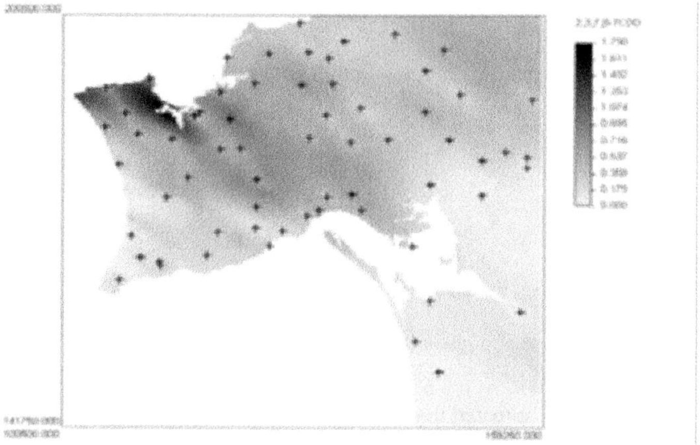

Figure 10. Final estimated map of 2,3,7,8 TCDD; +*X. parietina* samples location.

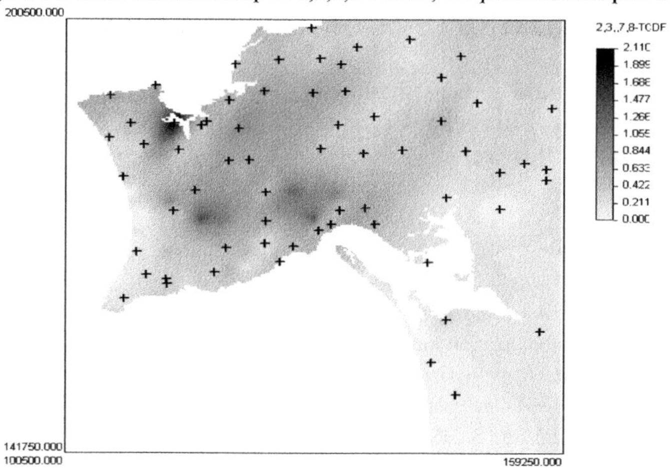

Figure 11. Final estimated map of 2,3,7,8 TCDF; + *X. parietina* samples location.

Anothcr issuc addrcsscd in this study rcgards thc cnvironmcntal impact, for eco-toxicological and epidemiological purposes, of a class of compounds made up of polychlorinated dibenzo-p-dioxins (PCDD) and polychlorinated dibenzofurans (PCDF) – usually called dioxins. This study pioneered the use of lichens to measure PCDD/Fs atmospheric deposition.

Finally, a geostatistical co-estimation model is proposed to integrate two types of biomonitoring samples with complementary spatial cover of the studied area. Co-located cokriging, with a previously estimated map as secondary information, has shown to be a valid and coherent way of approaching such situations.

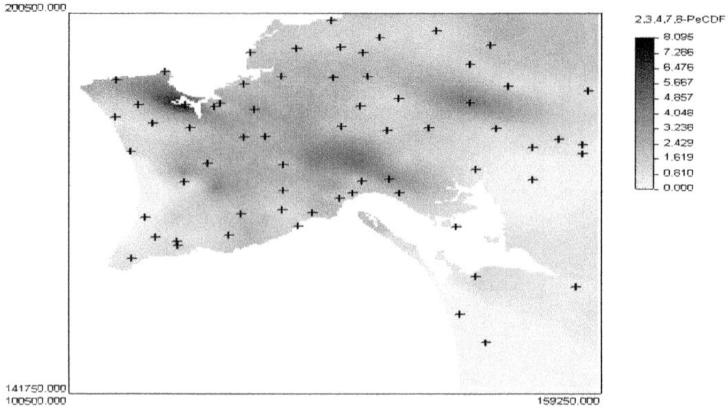

Figure 12. Final estimated map of 2,3,4,7,8 PeCDF; + *X. parietina* samples location.

ACKNOWLEDGMENTS

The authors would like to thank Secil-Companhia Geral de Cimento e Cal, S.A., and the European Commission (LIFE 98 ENV/P/000556) for financially support this work.

REFERENCES

1. Almeida, A.S. and A. G. Journel. 1994. Joint simulation of multiple variables with a Markov-type coregionalization model, *Mathematical Geology*, **26**(5): 565-588.
2. Bio, A., Carvalho, J., Rosário, L., 2002. Improving satellite image forest cover classification with field data using co-located cokriging. GeoENV2002. Barcelona.
3. Branquinho, C., Augusto, S., Pereira, M.J., Soares, A., Catarino, F., 2002, Lichens as biomonitors of PCDD's and PCF's at urban environment, *Klumpp A. eds*., EuroBionet2002, Conference on Urban Air Pollution, Bioindication and Environmental Awareness, University of Hohenheim, Stuttgart, pag. 11.
4. Branquinho, C., Catarino F., Brown, D.H., Pereira, M.J , Soares A. 1999. Improving the use of lichens as biomonitors of atmospheric metal pollution. *Science of Total Environment.*232 67-77.
5. Buckley-Golder, 1999, *Compilation of EU Dioxin Exposure and Health Data*, Report produced for European Commission DG Environment and UK Department of the Environment Transport and the Regions (DETR), Oxfordshire, UK.
6. Cleverly, D., Schaum, J., Schweer, G., Becker, D., Winters, D., 1997, *The congener profiles of anthropogenic sources of chlorinated dibenzofurans in the United States.* Presentation at Dioxin'97, the 17th International Symposium on Chlorinated Dioxins and Related Compounds, held August 25-29 in Indianapolis, IN, USA. Short paper in, Organohalogen Compounds, Volume 32:430-435.
7. Lillesand, T.M. and R. W. Kiefer. 1994. *Remote Sensing and Interpretation*, John Wiley and Sons, New York, 750p.

SPATIAL VARIABILITY OF SOIL PROPERTIES AT HILLSLOPE LEVEL

M. Ulloa[1] and J. Dafonte[2]

[1]*Inst. de Geología. Univ. de A Coruña, Spain. monseed@mail2.udc.es*
[2]*Dept. Ingeniería Agroforestal. Univ. Santiago de Compostela, Spain. jdafonte@lugo.usc.es*

Abstract: The aim of this paper is to analyze the spatial structure of several soil properties at 0-30 cm soil depth: pH in water and in KCl, contents of organic matter, sand, silt and clay, at a 2.1 ha hillslope in northwest Spain. The semivariograms and correlations between these soil properties and several variables derived from DEM (Digital Elevation Model) data, such as slope and elevation, were calculated. A medium correlation was found between pH and elevation, and the results obtained using kriging with external drift were similar to those obtained with ordinary kriging. Estimations of sand, silt and clay contents were used to calculate texture maps using ordinary kriging and Gaussian conditional simulation. Small differences were observed between maps obtained with these two methods.

1. INTRODUCTION

Traditionally, agricultural fields have been managed as uniform units. However, for many years it has been recognized that properties of the soil and crop yields vary within the field (Frogbrook et al., 2002). The spatial variability is governed by the processes of soil formation, which are in turn interactively conditioned by lithology, climate, biology, and relief through geologic time (Wilding et al., 1994).

X. Sanchez-Vila et al. (eds.), geoENV IV – Geostatistics for Environmental Applications, 485-496.

To study the spatial variability of soil attributes, geostatistical methods can be applied to collect such information. Geostatistics provides a set of statistical tools for modeling spatial patterns and allows making predictions at unsampled locations and assessment of the uncertainty attached to these predictions for the different attributes/soil properties of the soil (Goovaerts, 2000a).

If the values of the different variables are known at unsampled locations, it is possible to improve the recommendations given for the application of fertilizers, pesticides or liming to protect the environment, this being one of the objectives of precision agriculture. There has been growing interest in the management of within-field soil variability, and by nature, involves the collection of high-resolution secondary information (Bishop and McBratney, 2001). An objective of this paper is to analyze the possibility of using secondary information derived from DEM data to improve variable estimation using techniques like kriging with external drift, and the analysis of two different geostatistical tools, ordinary kriging and Gaussian conditional simulation, for the construction of texture maps.

2. STUDY AREA

The study site is a 2.1 ha hillslope (UTM 0559100; 4788100) with permanent grassland, located in Mabegondo (A Coruña), northwest Spain (Fig. 1).

For this study, 44 soil samples (Fig. 2) were taken at 0-30 cm depth. For each sample, pH in water and in KCl, and the contents of organic matter (OM), sand, silt and clay were measured.

The mean slope of the study site is 8.6%. The topographic map (Fig. 2) was constructed using digital elevation model (DEM) data with 4 m grid-size cell.

The study area belongs to the geological formation called Ordenes complex described in Martínez et al. (1982), and the geological material is Ordenes schist (Parga Pondal, 1956). The type of soil found in this hillslope is Cambic Umbrisols according to the FAO classification (ISSS-FAO-ISRIC, 1994).

The study area belongs to the geological formation called Ordenes complex described in Martínez et al. (1982), and the geological material is Ordenes schist (Parga Pondal, 1956). The type of soil found in this hillslope is Cambic Umbrisols according to the FAO classification (ISSS-FAO-ISRIC, 1994).

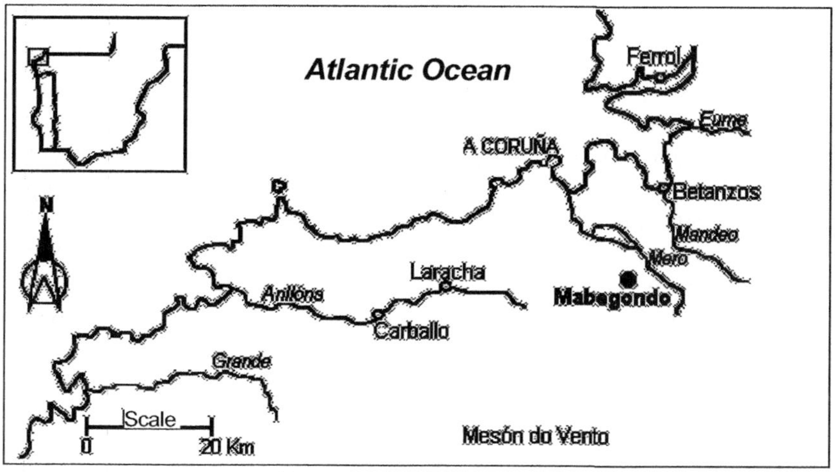

Figure 1. Situation of the study area.

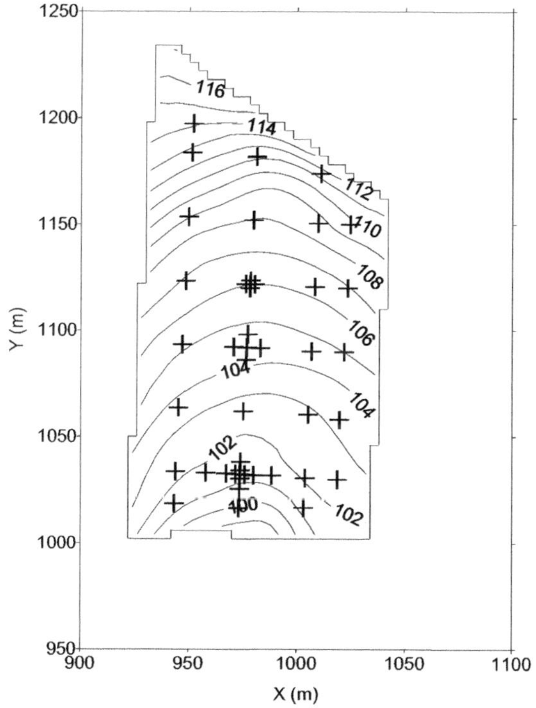

Figure 2. Topographic map and location of the samples (elevation values are in meters above sea level)

3. ANALYTICAL METHODS

Prior to analysis, all samples were air-dried and sieved (2 mm mesh). Soil pH was measured in two solutions, water and 0.1 M KCl (ratio 1:2.5 soil:solution), using a pH meter (Guitián and Carballas, 1976; MAPA, 1994). To calculate organic matter content, first the amount of organic carbon was determined in the soil samples, using a multiplication factor (1.724) to convert it to organic matter (Guitián and Carballas, 1976; MAPA, 1994). Soil particle size distribution (sand, silt, clay) was determined by standard methods (MAPA, 1994, official methods of Spanish government).

4. GEOSTATISTICAL ANALYSIS

4.1. Semivariogram analysis

All semivariograms were estimated using the program Variowin (Panattier, 1996). The method used to initially choose the parameters (nugget effect, range, sill) of the different semivariogram models was the weighted least-squares method, where the weights were the number of datapairs that contributed the information needed to calculate the experimental semivariograms. Then, the cross-validation method, a powerful model validation technique to check the performance of the model for kriging (Chilés and Delfiner, 1999), was used to check model performance. The criteria applied to evaluate this performance were the standardized error (Eq. 1) close to one and the correlation coefficient between the measured and estimated values.

$$SE = \frac{1}{m}\sqrt{\sum_{i=1}^{m}\frac{(\hat{z}(x_i) - z(x_i))^2}{\sigma_{ki}^2}} \tag{1}$$

where m is the number of points measured, $\hat{z}(x_i)$ and $z(x_i)$ are the estimated and measured values, respectively, of variable Z at location x_i, and σ_{ki}^2 is the value of the variance of kriging at location x_i.

4.2. Kriging and simulation

The geostatistical methods used in this paper are:

– Ordinary kriging (OK)
– Kriging with external drift (KED)
– Gaussian conditional simulation (GS)

These methods are considered standard geostatistical techniques (Goovaerts, 1997; Samper and Carrera, 1990; Deustch and Journel, 1998). OK is not described here, while KED and GS are described only briefly.

Estimation values obtained with kriging minimize local criteria such as local error variance, whereas stochastic simulation aims to reproduce global statistics such as the histogram or semivariogram (Goovaerts, 2000a).

4.2.1. Kriging with external drift

This method incorporates secondary information into the kriging system when the main and secondary variable are correlated. It is necessary to know the value of the secondary variable at all points where the primary variable is going to be estimated (Goovaerts, 1997). The secondary information is used to find the local means of the primary variable and performs simple kriging on the residuals (Goovaerts, 2000b). To use this geostatistical method, it is necessary to estimate the residual semivariogram using datapairs that are unaffected or slightly affected by the trend, e.g. perpendicular to the trend (Goovaerts, 1997).

4.2.2. Gaussian conditional simulation

The method used for the simulations was the Gaussian conditional simulation through LU (lower-upper) decomposition of the variance matrix. This method is the preferred Gaussian-based algorithm when the total number of conditioning data plus the number of nodes to be simulated is small and many realizations are requested (Goovaerts, 1997; Deutsch and Journel, 1998).

5. RESULTS AND DISCUSSION

5.1. Statistical analysis

The statistical moments of the variables measured are shown in Table 1. The distribution of these variables appears normal, using the analysis of the skewness coefficient, the kurtosis and the histograms of these parameters.

The coefficients of variation can suggest what type of variability is present, using the schelle of Gomes (1984). The CV of OM content is very high, of pH is medium, and of soil particle sizes is between low and high (5-21%). The pH value in water is strongly acidic, while that in KCl is even lower, because it includes the acidity present in the interchange cationic complex of the soil. The organic matter content is low for the region and silt is the predominant size of the soil particles.

Table 1.- Summary statistics of soil properties measured. N=Number of Samples; Mean=Arithmetic Mean; Var.=Variance; CV=Coefficient of Variation (%); Skew.=Skewness; Kurt.=Kurtosis; Min.=Minimum; Max.=Maximum.

Attribute	N	Mean	Var.	CV	Skew.	Kurt.	Min.	Max.
pH(H_2O)	45	4.94	0.206	9.20	0.20	2.14	4.17	5.89
pH(KCl)	45	4.39	0.227	10.85	0.47	2.25	3.60	5.53
OM (%)	45	2.52	3.312	72.14	0.70	3.84	0.00	8.52
Sand (%)	45	21.14	20.080	21.19	-0.38	2.26	11.31	28.34
Silt (%)	45	56.70	9.685	5.49	0.40	3.85	50.66	66.86
Clay (%)	45	22.16	11.670	15.42	0.40	2.31	16.48	29.81

The coefficients of correlation were calculated among the different variables and the variables were derived from DEM data (elevation and slope). A high correlation value was not observed among the variables, except between pH(H_2O) and pH(KCl), as seen in Table 2. With respect to the correlation between the variables and elevation and slope, there is medium correlation between pH and elevation. The correlation with other variables derived from DEM values, such as area that drains each cell, the perpendicular curvature and parallel curvature, was investigated but correlation coefficient values were low.

Table 2.- Correlation matrix.

	pH (H_2O)	pH (KCl)	OM (%)	Sand (%)	Silt (%)	Clay (%)	Elev. (m)	Slope
pH(H_2O)	1	0.89	0.20	0.15	0.02	-0.21	0.45	0.37
pH(KCl)		1	0.24	0.25	0.00	-0.33	0.59	0.35
OM (%)			1	-0.27	0.22	0.15	0.26	0.24
Sand(%)				1	-0.65	-0.72	0.07	-0.27
Silt(%)					1	-0.06	0.17	0.33
Clay(%)						1	-0.25	0.05
Elev. (m)							1	0.53
Slope								1

5.2. Semivariogram estimation and fitting

Figure 3 shows the semivariograms for all variables; the organic matter content has a nugget effect, and spatial dependence is not observed at the study scale. The pHs measured in H_2O and KCl show similar spatial structure. The pH values are correlated with elevation, and there is an elevation gradient in north-south direction, hence, the directional semivariogram in east-west direction was used as estimator of semivariogram of residuals for kriging with external drift. All the different texture sizes have well-defined spatial structure, with a low nugget structure value.

The theoretical semivariogram models chosen are shown in Table 3.

Table 3. Parameters of the theoretical semivariograms fitted to the experimental semivariograms (C_0=nugget effect; C_0+C_1=sill; a=range).

Attribute	Model	C_0	C_1	C_0+C_1	A
pH(H_2O)	Exponential	0.08	0.12	0.20	30**
pH(KCl)	Exponential	0.05	0.3	0.35	30**
pH(H_2O)*	Spherical	0.11	0.06	0.17	50
pH(KCl)*	Exponential	0.05	0.10	0.15	10**
OM (%)	Nugget effect				
Sand (%)	Spherical	0	24.096	24.096	45
Silt (%)	Exponential	0	11.138	11.138	16.66**
Clay (%)	Spherical	0	14.004	14.004	50

* Directional semivariogram in E-W direction

** Theoretical range

5.3. pH

The pH in H_2O and KCl was estimated using OK and KED. Fig. 3 shows the omnidirectional and residual semivariograms for the pH measured in water and in KCl.

The results of the parameters of cross-validation of KED with elevation such as external drift were similar than those using OK (table 4). The map obtained with OK is compared with that obtained with KED, it is possible to observe the likeness. The maps in the figure 4 show a variable estimation values distribution very similar in the two methods.

Table 4. Cross validation parameters for pH (SE= standardized error; r= correlation coefficient between measured and estimated values)

Attribute	Kriging	SE	r
pH(H_2O)	OK	0.9978	0.6469
	KED	0.9215	0.6412
pH(KCl)	OK	0.9921	0.7002
	KED	0.997	0.6991

5.4. Size and texture of soil particles

The maps of sand, silt and clay contents obtained using OK appear in Fig. 5. The clay values estimated by OK are maximal at the outlet of the hillslope and the sand content estimates minimal in the same area. The

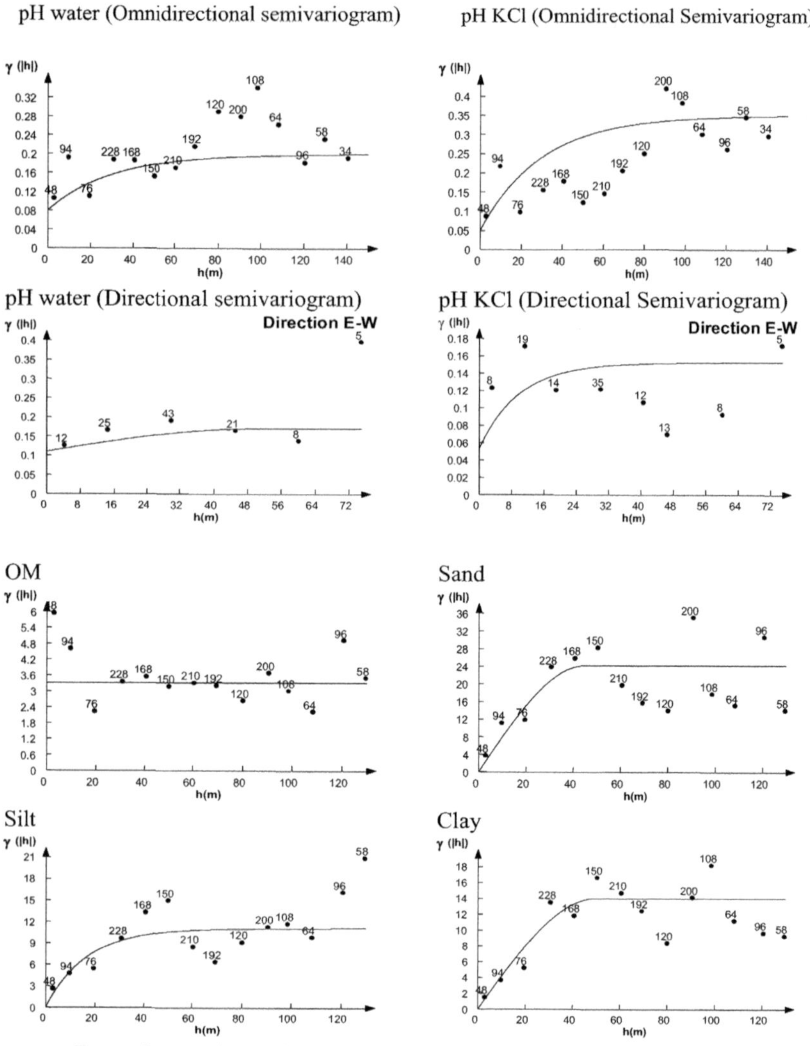

Figure 3. Experimental and fitted model semivariograms for all variables.

silt values are high in all areas, but even higher at the central drainage area. The sum of the sand, silt and clay contents is 100% at all estimation points.

The OK estimate maps of sand and clay contents were used to calculate the texture map (Fig. 6), using the USDA triangle and script file of a raster GIS: PCRaster software (PCRaster Environmental Software, 1997). This map shows that the main textural class is silt loam, but some areas present silt texture. The GS method was used to simulate the

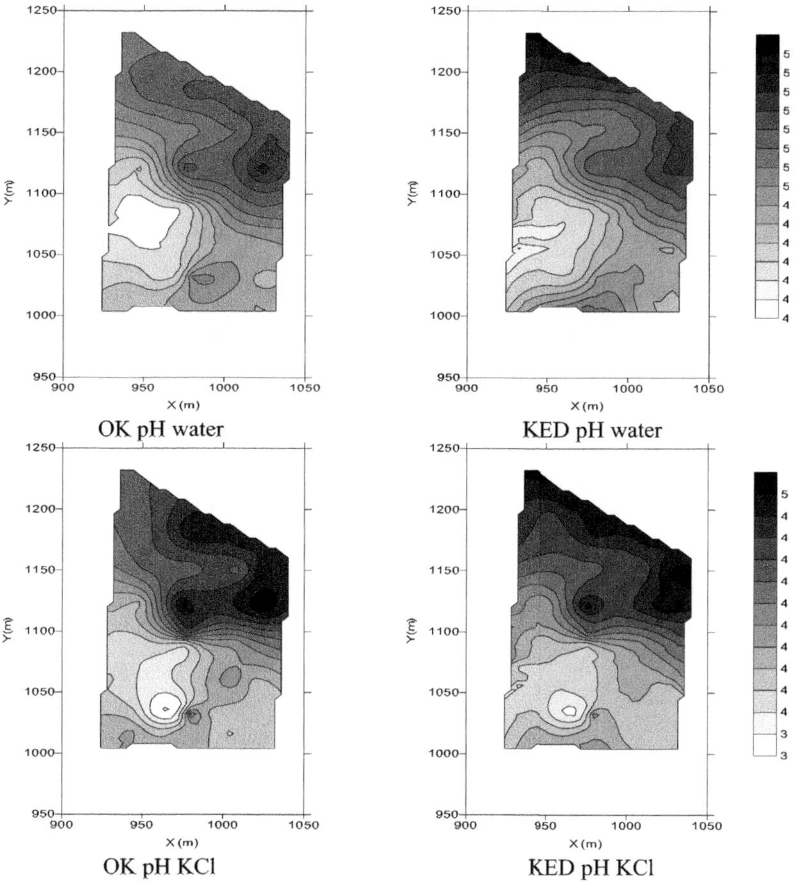

Figure 4. Maps for pH obtained using OK and KED

contents of sand, silt and clay, and then texture maps were difficult to see on the maps. The different kinds of surface textures are shown in Table 5. If the mean of 100 simulations of sand, silt and clay contents is calculated, the texture map obtained is similar to that produced with OK and the areas are very similar. However, the distribution areas of the two textural classes are slightly different. The use of simulations seems to be convenient in cases where it is important to know the existence of different textural classes, and not only the main textural class.

5. CONCLUSIONS

In this study the spatial variability of six variables, measured at 0-30 cm depth, was studied. All the variables had spatial structure at the study scale,

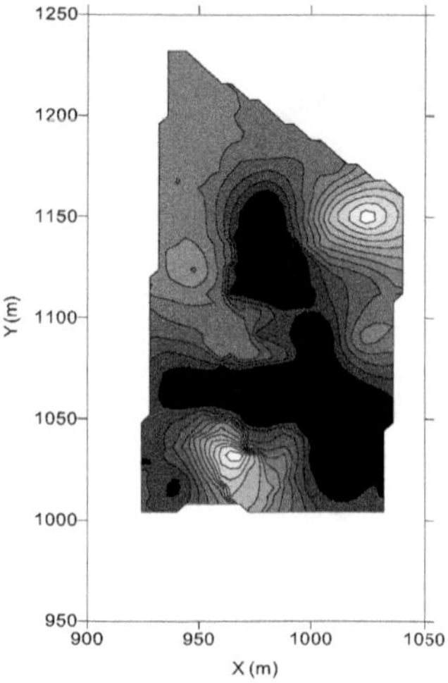

OK Sand content

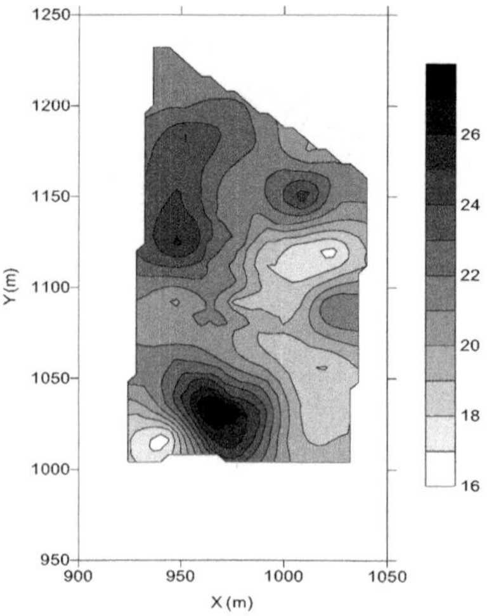

OK Clay content

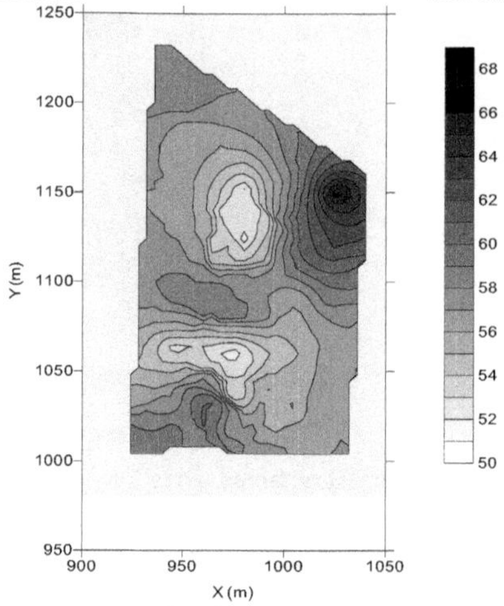

OK Silt content

Figure 5. Maps of sand, silt and clay contents obtained with OK.

Table 5.- Area occupied by every texture classes using different geostatistical methods.

Texture Class	OK (%)	GS mean (%)	GS No. 1 (%)	GS No. 100 (%)
Si	14.70	10.72	11.22	7.24
SiL	85.30	89.28	87.11	91.96
L	-	-	1.16	0.51
SiCL	-	-	0.29	0.29
CL	-	-	0.07	0.22
C	-	-	0.07	-

Si- Silt; SiL- Silt Loam; L- Loam; SiCL- Silt Clay Loam; CL- Clay Loam; C- Clay

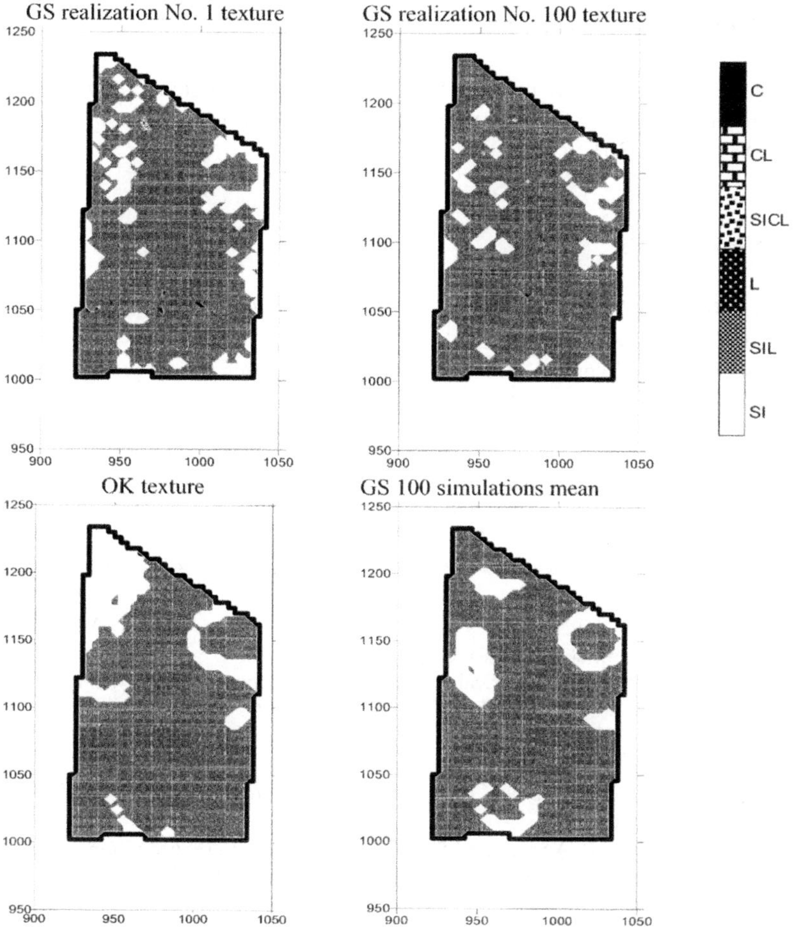

Figure 6. Texture maps obtained by ordinary kriging, Gaussian simulations No. 1 and 100 of sand, silt and clay contents, and arithmetic mean of 100 simulations of sand, silt and clay contents.

except for organic matter content. The values of pH showed medium correlation with elevation. For the possible application of geostatistical techniques in precision agriculture, the correlation with variables derived from DEM data was investigated, but the results were not good. The use of KED for pH estimation does not improve the results obtained with OK.

Lastly, the texture maps obtained from OK and those of 100 GS are slightly different and show the same texture classes, but the individual simulations show new texture classes that do not appear in the maps obtained with OK. It can be interesting using simulation if it is necessary to obtain all the texture classes present in the field and not only the main texture classes.

ACKNOWLEDGMENTS

Financial support was provided by the Ministry of Science and Technology with project REN2000-0445-C02-02 HID and by Xunta de Galicia with project PGDIDT99MA20301.

REFERENCES

1. Chilès, J-P. and Delfiner, P. (1999). *Geostatistics Modeling Spatial Uncertainty*. New York: John Wiley & Sons.
2. Deutsch, C.V. and Journel, A.G. (1997). *GSLIB, Geostatistical Software Library and User's Guide*. New York: Oxford University Press.
3. Frogbrook, Z.L.; Oliver, M.A.; Salahi, M.; Ellis, R.H. (2002). Exploring the spatial relations between cereal yield and soil chemical properties and the implications for sampling. Soil Use and Management. 18:1-9
4. Gomes, F.P. (1984). *A estatistica Moderna na Pesquisa Agropecuaria*. Piraçicaba: Associaçao Brasileira para Pesquisa da Potassa e do Fosfato.
5. Goovaerts, P. (1997). *Geostatistics for Natural Resources Evaluation*. New York: Oxford University Press.
6. Goovaerts, P. (2000a). Estimation or simulation of soil properties? An optimization problem with conflicting criteria. Geoderma, 97: 165-186
7. Goovaerts, P. (2000b). Geostatistical approaches for incorporating elevation into the spatial interpolation of rainfall. J. of Hydrology; 228: 113-129
8. Guitián, F. & Carballas, T. (1976). *Técnicas de Análisis de Suelos*. Santiago de Compostela: Ed. Pico Sacro.
9. ISSS-FAO-ISRIC. (1994). *World Reference Base for Soil Resources*. Wageningen/Rome: ISSS-ISRIC-FAO.
10. MAPA (Ministerio de Agricultura, Pesca y Alimentación) (1986). *Métodos Oficiales de Análisis de Suelos, Aguas y Plantas. Tomo III.*, Madrid: Ministerio de Agricultura, Pesca y Alimentación. Servicio de Publicaciones.
11. Martínez, J.R.; Klein, E.; de Pablo, J.G. & González, F. (1984). El Complejo de Órdenes: subdivisión, descripción y discusión sobre su origen. Cadernos Lab. Xeololóxico de Laxe; 7:139-210.
12. Pannatier, Y. (1996). *VARIOWIN: Software for Spatial Data Analysis in 2D*. New York: Springer-Verlag.
13. Parga Pondal, I. (1956). Nota explicativa de mapa geológico de la parte NO de la provincia de La Coruña. Leidse Geol. Med.; 21:468-484.
14. Pebesma, E.J. (2001). *Gstat User's Manual*. Utrecht: Dept. of Physical Geography, Utrecht University.
15. PCRaster Environmental Software. (1997). *PCRaster Version 2*. Utretch: Dept. of Physical Geography, Utrecht University.
16. Samper, J. and Carrera, J. (1990). *Geoestadística. Aplicaciones a la Hidrología Subterránea*. Barcelona: CIMNE.
17. Wilding, L.P.; Bouma, J.; Boss, D.W. (1994). "Impact of spatial variability of interpretive modeling" In *Quantitative Modeling of Soil Forming Processes SSSA Special Publ., No 39,* R.B. Bryant and R.W. Arnold (Ed.). Madison: SSSA.

POSTERS

The simple but meaningful contribution of Geostatistics: three case studies

by Bonduà S.[1], Bruno R.[2], Guêze R.[2], Morosetti M.[2], Ricciardi O.[2]

[1] DICASM – Univ. Bologna - Viale Risorgimento, 2 - 40136 Bologna, Italy, m@il: stefano.bondua@mail.ing.unibo.it

[2] DICMA – Univ. Bologna- Viale Risorgimento, 2 - 40136 Bologna, Italy, m@il: roberto.bruno@mail.ing.unibo.it

Abstract

Many of the basic geostatistical tools still can "make the difference" in any environmental study. In fact, simple geostatistical concepts give the possibility of solving the existing problems in a way that looks like "advanced" for many sector operators. The following three case studies show it, by different approaches.

CASE STUDY A: SOIL CONTAMINATION

Problem
Correct selection of the contaminated area where it's necessary to proceed through the appropriate decontamination methods for restoring natural environmental status.

Location
The contaminated area belongs to a dismissed industrial site of about 0.5 square kilometres.

In the underground there is the presence of 6 heavy metals (Pb, Cu, Zn, Cd, As, Hg).

Below the soil surface, at a deep from 10 to 15 metres, there is a confined phreatic zone.

Sample data are available from 50 monitoring – wells.

Methodology
For each pollutant, estimated values map by punctual Kriging was built.

The aquifer limit and geometry was reconstructed from available well-data.

According with Italian law, indicator variables was built by choosing Zcut for each pollutant.

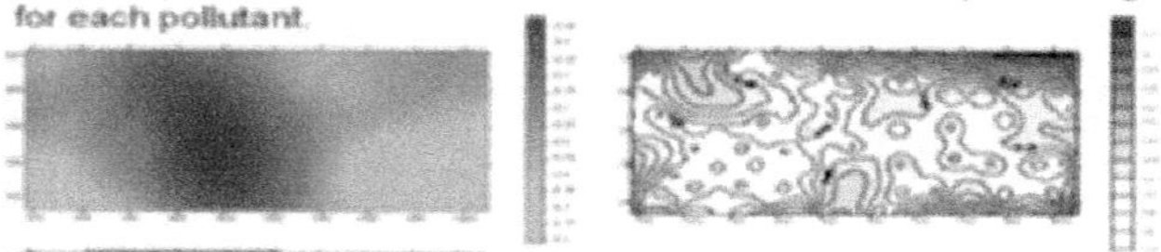

Pb Zcut= 1000 mg/kg

The Italian law about contaminated soil indicates a risk probability not upper to 10%, for this reason we calculated the 90% probability maps ($Z_{90\%}i = Zi^* + 1.28\ \sigma$), moreover in this section we used a block Kriging of 20 m x 20 m dimension.

CASE STUDY B: WATER POLLUTION

Problem
Flow simulation study of an aquifer to predict water pollution of heavy metal by percolation of residual waste from industrial area. Geostatistical estimation and simulation was indispensable to give input data to flow simulation program.

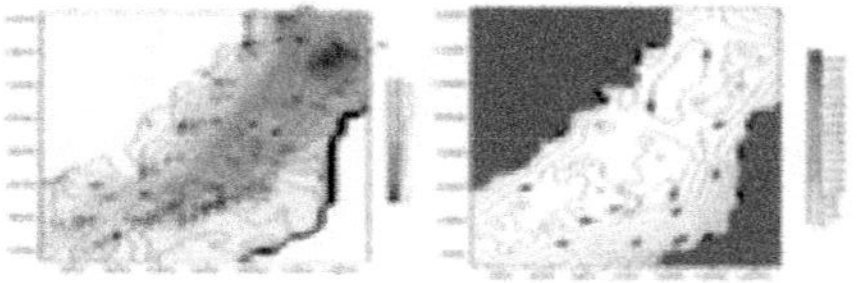

Fig. B.1. B2 bottom aquifer and variance estimation map.

Methodology
To represent the bottom and the piezometry of the aquifer, geostatistical estimation by FAI-k Kriging was used.

To obtain the spatial variability model, the few available sample data were integrated with official cartography information, opportunely elaborated.

Integration between available sample and information from literature was necessary for getting the input to flow simulation model different hypothesis of leaching and regime flow was applied to the flow simulation model for studying several possible propagation schemes of pollutant in the aquifer.

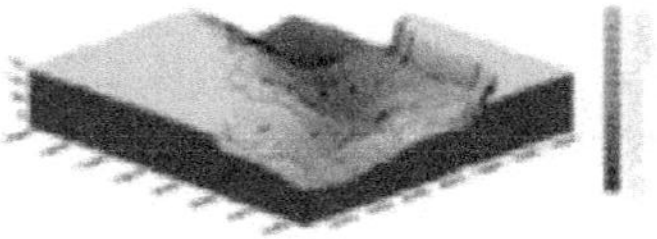

Fig. B.3 3D representation of bottom aquifer

The overlay of the 6 risk maps, lets to identify the mostly contaminated areas. Then it is possible to identify the most contaminated areas and the volume of the soil to be treated.

Reconstruction of the clay thickness for the individuation of the barrier to aquifer contamination (layer with thickness > 1m.).

Reconstruction of the clay surface level for the individuation of the flowing plane of the infiltration water.

To test the sensibility of the aquifer simulation model about pollutants dispersion, it was considered also the weight of different parameters used. The example below shows the propagation of pollutant using constant permeability.

Fig. B.4 constant permeability

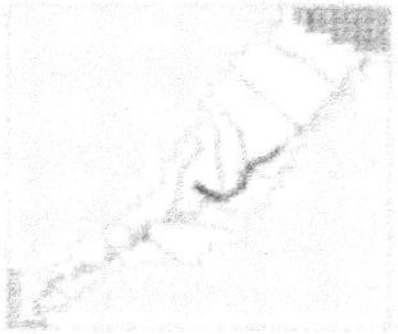

Fig. B.5 permeability distribution

In a second step the flow model was calculated with a different stochastic distribution. Due the properties in porous media, permeability distribution has to respect as far as possible the original variability distribution.
It is possible to assert that in this kind of environmental application the permeability distribution is on of the most important parameters of the aquifer flow model.

CASE STUDY C: AIR POLLUTION

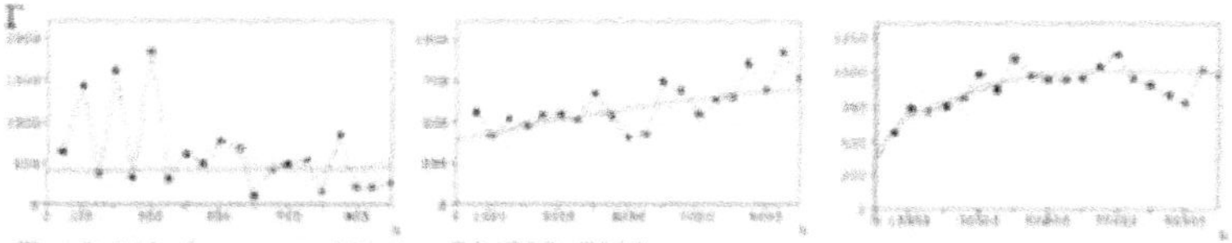

Fig. C.1 Variograms with step 50, 500, 5000 m.

Problem
Understanding the origins of indoors air radon pollution in Emilia Romagna Region. The observation data were collected for a study of Environmental Regional Agency in the school buildings of the region. More of 500 sample data were available.

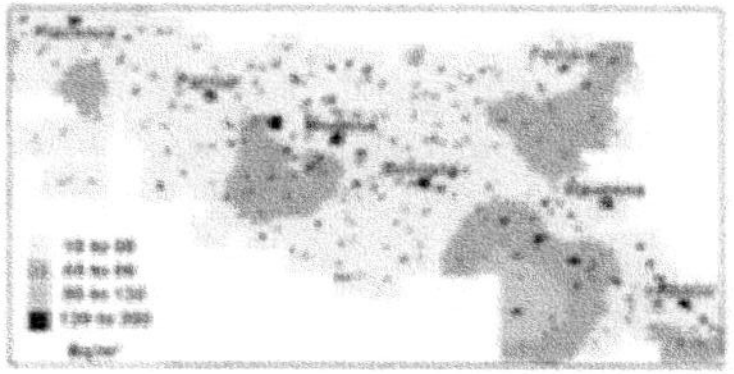

Fig. C.2 Radon estimation map in the Emilia Romagna Region

Methodology
The spatial correlation observed at any scale (50m, 500m, 5000 m) of variogram from one side confirms the hypothesis of the radon link with geology, but it gives also very useful insights for understanding the origin, the transport mechanism and the accumulation conditions.
Results
The link with radon and geology is clear in the mountain areas where the nugget effect is very low and the experimental model has a nested structure with two spherical models. In the plain area the sedimentary rocks work as film barrier and this reduces the spatial correlation and introduces a nugget effect.

Evolution, Neoteny, and Semi-Variogram Model Fitting Search

by Bonduà S.[1], Ramos V.[2],

[1] DICASM – Univ.Bologna - Viale Risorgimento, 2 - 40136 Bologna, ITALY, stefano.bondua@mail.ing.unibo.it
[2] CVRM - IST Geo-Systems Center, Avenida Rovisco Pais, 1049-001,Lisboa, PORTUGAL vitorino.ramos@alfa.ist.utl.pt

Abstract

The basic tool, and probably one of the most used in geostatistics, the variogram, is used to quantify spatial correlations between observations. In fact, this method has become one of the most important - although not necessary - in any natural resource modelling and planning study. For instance, once a mathematical function (or functions) has been fitted to the experimental variogram, this model can be used to estimate values at unsampled points, by an estimation procedure called kriging (Krige, Matheron, 1970's), which has been found to be an exact interpolator.

The most old and common method for variogram fitting, is "by eye", where an operator visually interacts finding the bests parameters (number of structures, ranges, angles of anisotropy, etc) for the best model fitting. In the present work, automatic modelling of variograms is however required because there is a need for the estimation of variogram parameters for a high number of digital images. Indeed we are looking to avoid subjective interpretation of the data by the user. In order to overcome this problem, a Genetic Algorithm was, for the first time, fully applied, and a novel bio-inspired strategy developed in recent works – entitled, Neoteny - was used for better convergence.

Variogram model selection and common fitting methods

Current practice in selecting a variogram model is often rather subjective. Sometimes, a priori knowledge about the underlying stochastic process can be helpful. However, when such information is not available or not relevant enough, one has to rely on some empirical guidelines, often related to the range, sill, and nugget effect of the variogram. These rules are often lacking because the shape of some variogram models are very similar: for instance, the exponential, gaussian, and spherical model.

The most common method is fitting "by eye", that is, using a graphical computer program to fit a model that the operator judges satisfactory, in a trial-and-error process. Another, more quantitative, method is weighted least squares. Weighted least squares take into account the uncertainty of each semi-variogram lag, but it assumes that the different semi-variogram lag estimates are uncorrelated. The manner in which the uncertainty of the semi-variogram lag is evaluated used to be a crude approximation, for example the number of pairs used to compute the semi-variogram lag estimate. Generalised Least-Squares (GLS) accounts for both the uncertainty of the semi-variogram lag estimates and the correlations among them. More recently and in order to fit any theoretical model, Pardo and Dowd (2001) have used non-linear GLS and estimation of uncertainty of the semivariogram model parameters. As mentioned before, the usual approach is to perform this fitting "by eye" in a trial and error process that continues until the fitting is considered by the operator satisfactory. This procedure allows the user more freedom to include his/her own knowledge about the geology, geophysics, etc., of the spatial variable. Nevertheless, a program of automatic fitting may be of considerable help for a number of different reasons: (a) as a preliminary estimation of the parameters, (b) when a high number of variograms must be modelled on a routine basis, (c) in order to make a more objective modelling (so that different practitioners will obtain the same results) and (d) in order to perform a more statistically sound fitting (i.e., each experimental variogram point should be weighted according to its own statistical uncertainty).

Evolving solutions using Genetic Algorithms and Neoteny

In the present work however, automatic modelling is required because there is a need for the estimation of variogram parameters for a high number of digital images. Indeed we are looking to avoid subjective interpretation of the data by the user. The data available are in the form of grey tone images of slabs tiles of Serizzo Antigorio and from some types of Portuguese Granites. The models of variogram are required for two kinds of problems, namely, classification and simulation. In this particular case, the data have homogeneous spatial distribution, and the dimension of each image is 650 x 650 pixels.

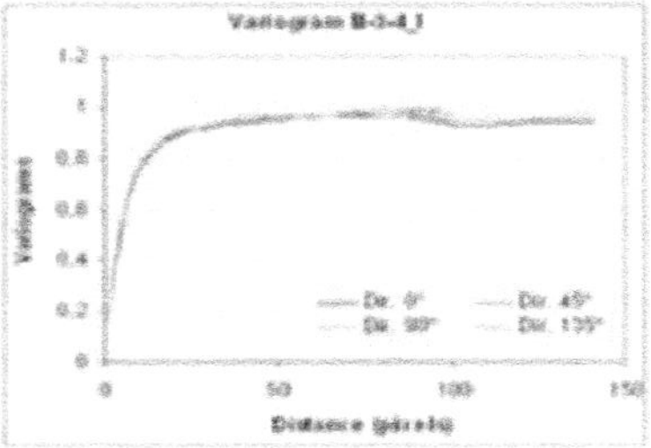

Typical variogram model

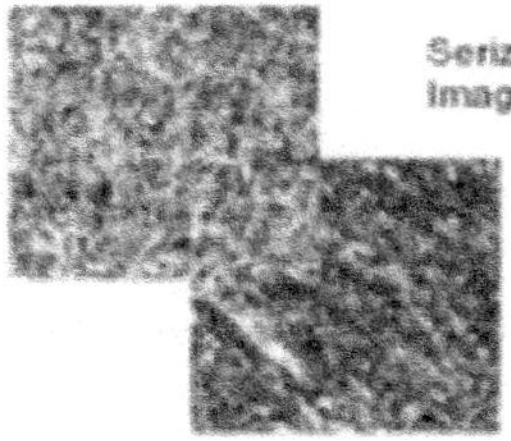

Serizzo Antigorio slab images
Images was token with linear cam

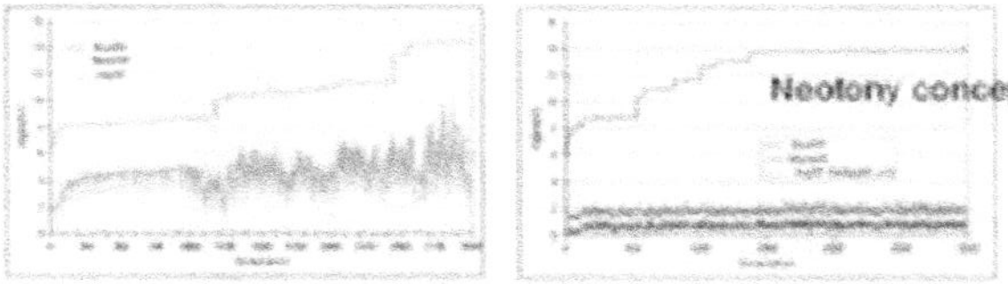

Real Fitness with (a) and without Neotony method (b)

The estimation of the experimental variogram is performed with a program in FORTRAN language. Fast performance is granted due the homogeneous spatial data configuration. No distance tolerance and angular distance was taken in account. Finally, the Genetic algorithm (GA) is used and combined with weight least squares methods. The weight function chosen for this kind of data is based only on the inverse of distance, which gives satisfactory results. Genetic algorithms are search procedures based on the mechanics of natural selection and natural genetics.

The basic operation of a GA is conceptually simple: maintain a population of solutions to a problem, to select the better solutions for recombination with each other, and to use their offspring to replace poorer solutions. In order to speed-up the GA convergence, and the quality of solutions, artificial Neoteny was used (i.e., the retention of juvenile characteristics in adulthood) to maintain diversity in GA's. This strategy was incorporated in simple non-Memetic genetic algorithms, by simply preserving some individuals in the earlier generations (using elitism), and by randomly re-injecting this genetic material into the later generations, allowing for substantial increases in diversity, and (as it seems) for an appropriate balance between exploration and exploitation of the search space.

Results on several test variogram models (and on their parameters), point to square mean errors less than 1%. GA optimise computing time and solution confidence giving best fit model trough large variety of structures.

Kriging of hydraulic head field for a confined aquifer

Youri Brochu, Denis Marcotte and Robert P. Chapuis; *École Polytechnique de Montréal, Canada*

Abstract: The estimation and mapping of a realistic hydraulic head field is one of the major goals of hydrogeological studies. A flexible and simple approach is the direct kriging of the head field using the measurements obtained at observation wells. We modify kriging to incorporate available information such as: precision of head measurements, presence of impervious boundaries, pumping or injecting wells, and known local head gradients. Using the dual kriging formalism enables simultaneous estimation of the head field and the aquifer mean hydrogeological parameters from head observations either in steady or transient state conditions. The approach is used to analyze a simple finite element synthetic model. The improvement brought to the kriged head field by each modification to the usual ordinary kriging system is assessed. The approach is applied to a real unconfined aquifer. The head fields obtained prior to and after 4 months of intensive pumping appear very realistic and help determine the influence zone of the well and the effect of pumping on regional flow.

Introduction

- Need for realistic 2D head field: Transport direction, zone of influence of wells, regional gradient, estimate of boundary conditions for numerical modeling.
- Scarcity of data: few observation wells; few T determination; poorly known boundary conditions.

Objectives

a) Obtain a realistic head field directly from kriging with available data (i.e. without using inverse analysis and flow simulation)

b) Obtain mean aquifer parameter estimates of transmissivity and regional gradient.

Materials and methods

We modify kriging to include (see Brochu and Marcotte, 2003):

a) The effect of a regional gradient → *add linear drift constraints;*

b) Precision of head measurement → *add nugget effect (can vary for each observation);*

c) Effect of a well → *add a drift term in log(r/t);* r : distance to the well; t: pumping time

d) Effect of hydrogeological boundaries → *include pairs of dummy points to represent boundary conditions* (Chilès and Delfiner, 1999; Delhomme 1979);

e) Different lithological units → *Split the unbiasedness constraints in distinct blocks, one block per lithology.*

Results (synthetic model)

1) Reality: finite elements model

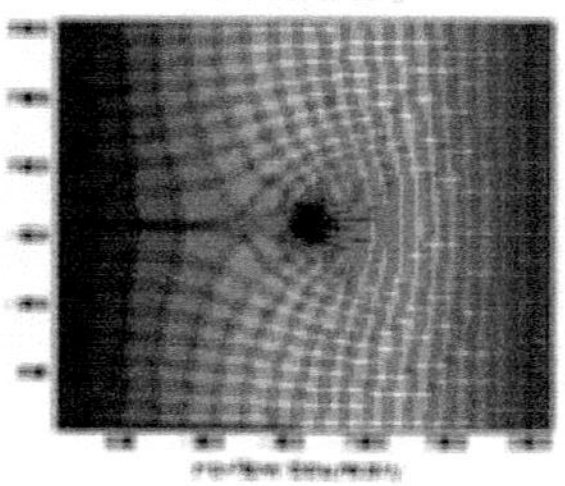

2) Ordinary Kriging (5 obs. wells)

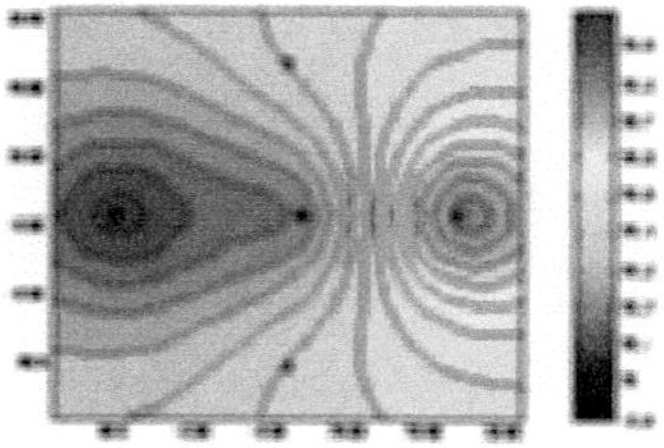

3) Kriging: linear drift

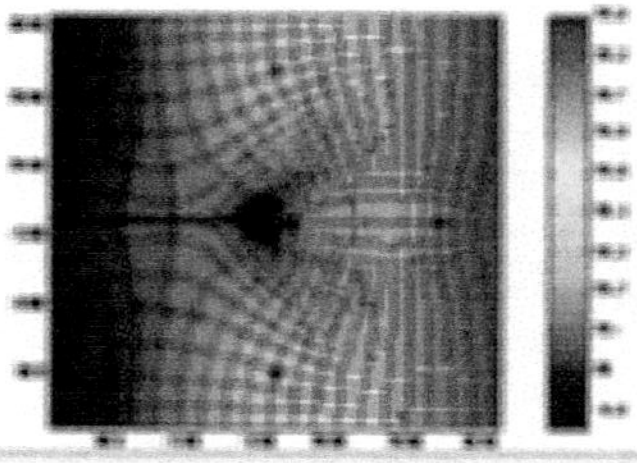

4) Kriging: linear drift + well drift

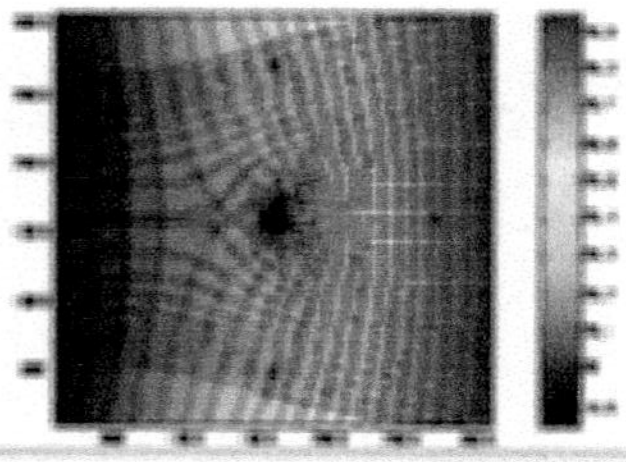

5) Kriging: linear drift + well drift + boundary conditions

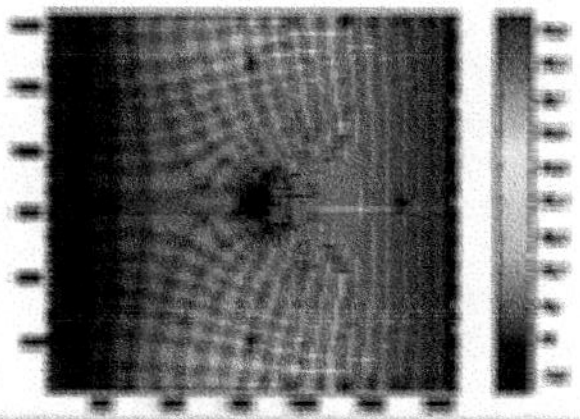

6) As case 5 but with 16 observation wells instead of 5

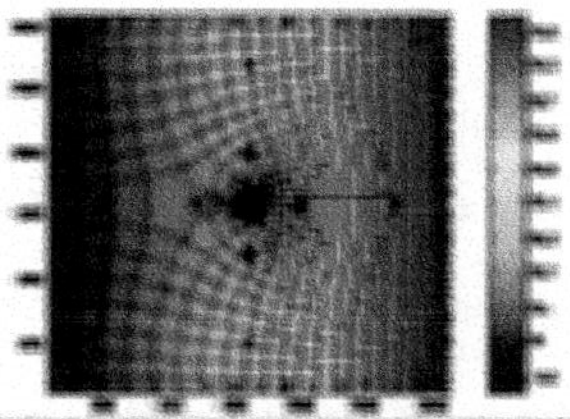

Results (case study)

Bois de Sepey aquifer, Switzerland (Parriaux and Jolépin, 2001)

1) Natural state

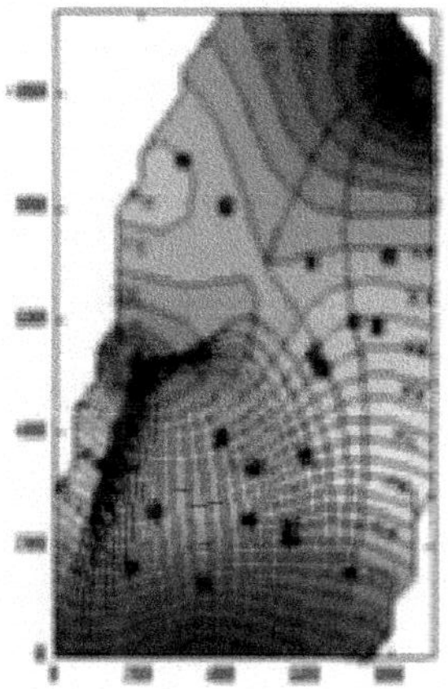

2) After 4 months of intensive pumping

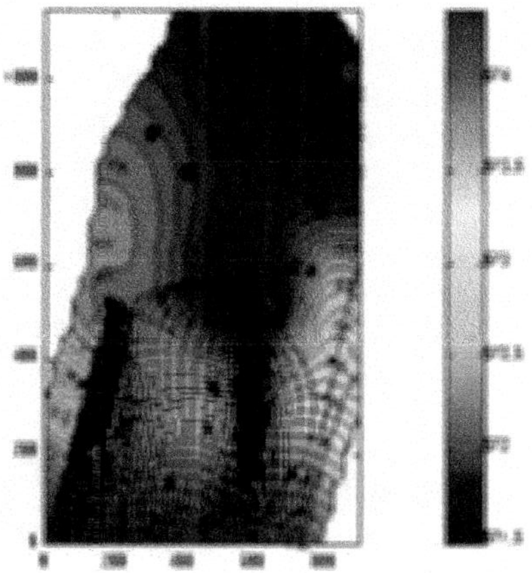

Discussion and conclusion

- Transmissivity, regional gradient and storage coefficient are estimated precisely using dual formalism of kriging. For example, T estimate for case 6 is 1.13 and the true value used in the flow simulation is 1.10.

- The extension is immediate to account for « n » wells: drift terms of the different wells can be merged, or not, depending on the presumed homogeneity of the field. Wells within a single lithological unit can be merged to provide a unique T estimate.

- Image wells can reproduce no-flow or constant head rectilinear boundaries. A more flexible alternative is the use of double-points.

- Kriging retrieves the exact analytical solution for the steady state homogeneous infinite confined aquifer. Kriging is almost exact for the homogeneous unconfined aquifer. Cooper-Jacob approximation allows precise estimation in the unsteady state. Kriging provides a good approximation to finite elements numerical solutions in a variety of hydrogeological contexts (Brochu and Marcotte (2003)).

- Boundary conditions do not need to be known everywhere; very few head observations are required and no transmissivity data is required.

- The kriged solution provides realistic boundary conditions for flow simulations.

- Case study: The pumping modifies substantially the natural flow regime. Both head maps appear realistic.

References

Brochu, Y., and Marcotte, D., 2003, A simple approach to account for radial flow and boundary conditions when kriging hydraulic head fields for confined aquifers. *Math. Geol.*, 35, 2, 111-139.

Chilès, J.P. and Delfiner, P., 1999, *Geostatistics, Modeling spatial uncertainty*, Wiley, New York.

Delhomme, J.P., 1979, Kriging under boundary conditions. Presented at the *American Geophysical Union Fall Meeting*, San Francisco, December 1979.

Parriaux, A., and Jolépin, P., 2001, Commune de Gry : Prospection d'eau au Bois de Sepey. Étude # 9821, rapport du 10.07.01, GEOLEP- Laboratoire de géologie, École Polytechnique Fédérale de Lausanne, 12 p.

This research was financed by a *National Sciences and Engineering Research Council of Canada* (NSERC) research grant and an NSERC student scholarship.

Bathimetric Morphological Classification using a Geostatistical Approach: an application to the Alboran Sea (S of Spain)

Delgado-García, Jorge[1], Sánchez-Gómez, Mario[2], Román-Alpiste, Manuel J.[3], Gracia-Mont, Eulàlia[4] and the HITS Cruise Scientific Party

[1] Dpto. Ingeniería Cartográfica, Geodésica y Fotogrametría. Universidad de Jaén. c/ Virgen de la Cabeza, 2 - 23071 Jaén (Spain). e-mail: jdelgado@ujaen.es
[2] Dpto. Geología. Universidad de Jaén. Campus Las Lagunillas Edif. B-3 - 23071 Jaén (Spain). e-mail: msgomez@ujaen.es
[3] Instituto Andaluz de Geología Mediterránea. Universidad de Granada-CSIC. Avda. Fuentenueva s/n - 18071 Granada (Spain). e-mail: mjroman@ugr.es
[4] Instituto de Ciencias de la Tierra "Jaume Almera". CSIC. c/ Lluís Solé Sabarís, s/n - 08028 Barcelona (Spain). e-mail: Eulàlia Gracia@ict.csic.es

I. INTRODUCTION

We present a bathimetric data geostatistical application. The main objective of this treatment is to obtain a morphological classification of the sea-floor based on the parameters derived from the structural analysis. The use of Geostatistics methodology to the bathimetric data analysis must be approach differently that the classic (mining) applications perspective. In this case, the available information volume are usually very high (frequently, several millions of data) so it is not necessary to complete the information but it is necessary to provide tools that allows the data interpretation from a topoprobabilistic approach.

The study area is located in the Alboran Sea (Western Mediterranean) (fig.1) and constitutes an atypical continental margin, where geological processes of opposite sense occur to configure a unique physiography. Since early Miocene a non-conventional episodic continental rifting is superimposed to the collision of Africa and Iberia. The submarine topography is a direct result of the modifications of the last tectonic events over sediments with present active tectonics and volcanic builds. The submerged structures are basic for the knowledge of the geology of the coastal surroundings areas. In this context, analysis of sea floor morphology constitutes a powerful tool in the multidisciplinary geophysical and geological studies of complex inaccessible areas that could implies an hazard. Faults on near-shore submarine areas cannot be followed by the geodetic and paleoseismic standard studies and new approach are required to establish the risk. The Alborán sea is an unique place to characterize unusual tectonic activity and establish its intrinsic hazard.

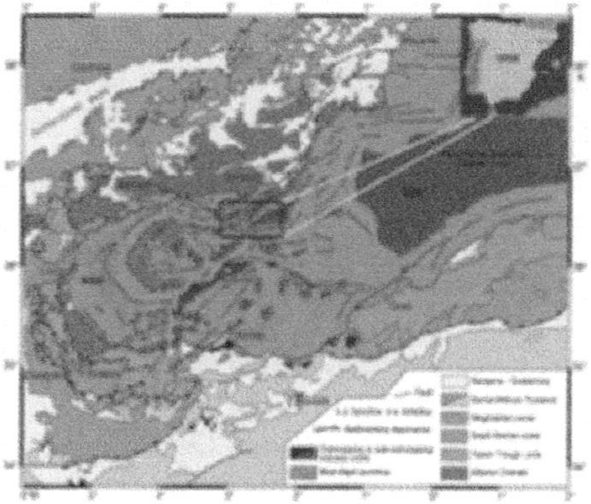

Fig 1. Betic and Rif chains and Alboran sea geologic map (from Comas et al., 1999)

II. METHODOLOGY & RESULTS

A total of 4.936.200 XYZ bathymetric data captured with a SIMRAD EM-12 multibeam sounder has been used. This survey mapped an area between 36.65ºN and 36.30ºN, 3.1ºW and 2.2ºW. The total area coverage, with a few gaps (produced by the coverage loss with the depth decrease) corresponds to 3.300km² (100km x 33.3 km), with a vertical accuracy around 60cm (0.25% depth). Ship navigation and individual depth measurements were cleaned and processed on board, and gridded at 0.0002º (around 20m resolution). Water depths extend from 80m at the upper portion of talus to 1900m in the lower end. The talus slopes down from N to S.

Local variograms following a moving windows schema has been calculated. The used overlapping windows have a 0.01ºx0.01º size (50x50 data) and 0.001ºx0.001º spacing (a total of 304.172 windows). The windows with more than 1500 data (190.749 windows) have been selected for the variogram calculation. The local variograms in the 4 main directions (N-S, E-W, NE-SW, NW-SE) have been calculated and 58 parameters have been extracted. The parameters are: mean; std. dev.; coef. var. of the windows; and, experimental, maximum, minimum and ratio between directional variograms for several lags.

The parameters have been included in a ISODATA unsupervised classification procedure obtaining the final map (Fig.3). In this map, an important morphological structure that seems to be an offshore extension of Carboneras fault (A) and high variability zones near of Chella (B) and Sabinal (C) banks have been detected.

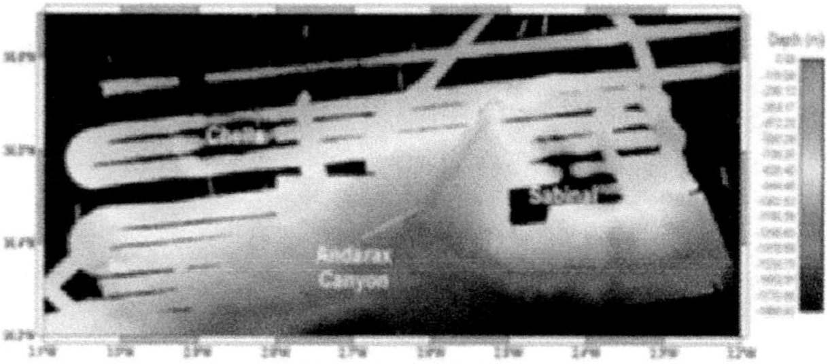

Fig 2. Representation of the DEM (Black: No Data)

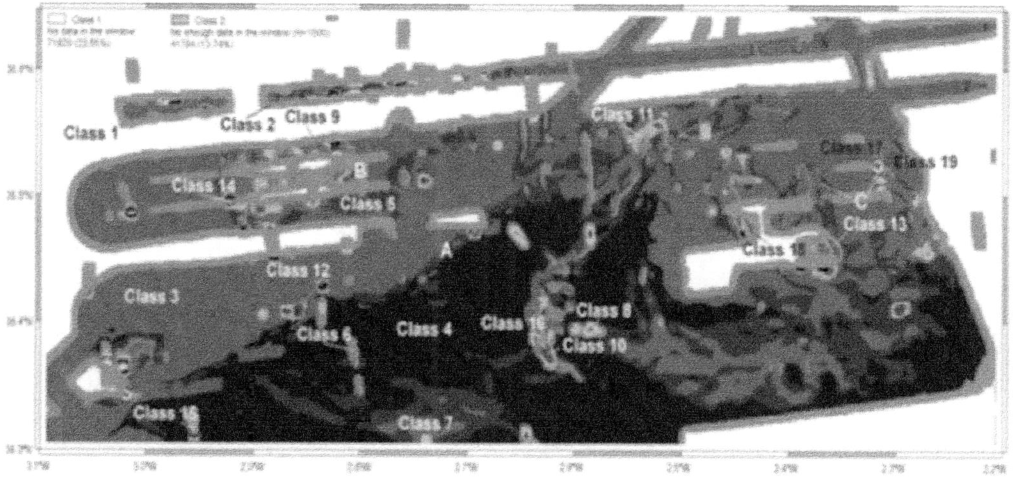

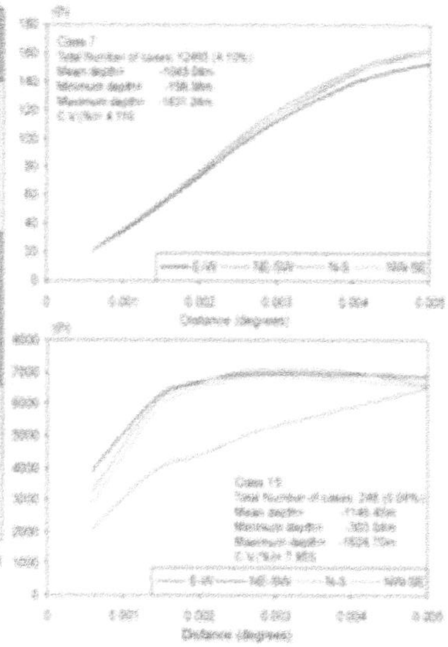

Fig.2 Morphological classification map and examples of the directional mean variograms of several classes.

III. CONCLUSIONS

The main conclusion is the capability of the geostatistical methodology for the bathimetric data analysis. The used methodology has been specially designed for treat a large volume of data and has allowed to obtain a morphological classification of the area. The classification has been based in 58 parameters obtained from the local structural analysis. Actually, the results are being revised using additional information (geophysical, geological and sea-floor high-resolution reflective images).

The principal advantage of this approach is the possibility of obtain an objective classification of the data based in the geostatistical methodology. The classification incorporates anisotropies and different spatial variability scales, both important aspects in the bathimetric data analysis.

ACKNOWLEDGMENTS

Bathymetric data and SIMRAD data processing were done during HITS Cruise (2001), funded by EC Project EASSS HPRI-ZT-1999-00047 and Spanish Project REN2000-2130-E. The geostatistical treatment has been funding by the Spanish Project REN2001-3868-CO3-01/MAR. We are gratefully thanks to Menchu Comas (IACT-CSIC) for the encouraging and geological background about the area.

A strategy for groundwater protection from nitrate leaching using spatial and geostatistical analyses

Sarah Evers, Steve Fletcher, Rob Ward and Bob Harris, National Groundwater and Contaminated Land Centre, Environment Agency, UK. Margaret Oliver, Reading University, UK. AndrewLovett, Iain Lake and KevinHiscock, University of East Anglia, UK.

This poster illustrates an approach for defining land zones to protect groundwater resources from the effects of agriculturally derived diffuse nitrate pollution. The approach operates in two distinct stages. The first stage collates and statistically examines groundwater nitrate data from all available monitoring points. At the same time a number of groundwater vulnerability scenarios are modelled to determine the theoretical risks to groundwater based on land use, climate, soil and (Solid and Drift) geological characteristics. The maps derived from this latter exercise are compared with the monitoring data to determine which of the scenarios best predicts the vulnerability of groundwater to nitrate pollution. In the second stage, the spatial distribution of nitrate concentrations is estimated bygeostatistical analysis of the monitoring data and is used in conjunction with the selected vulnerability map to define land use zones.

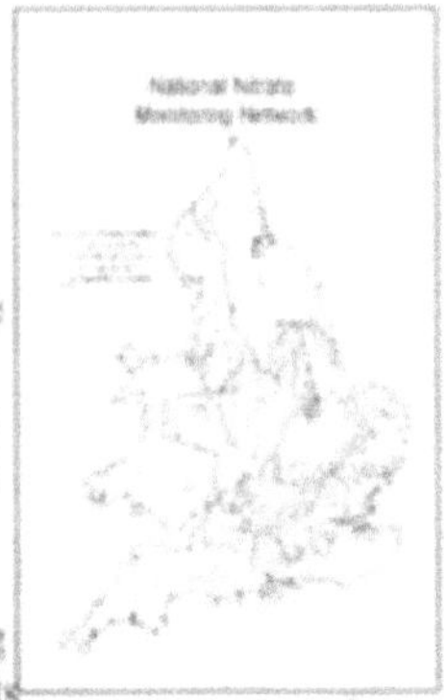

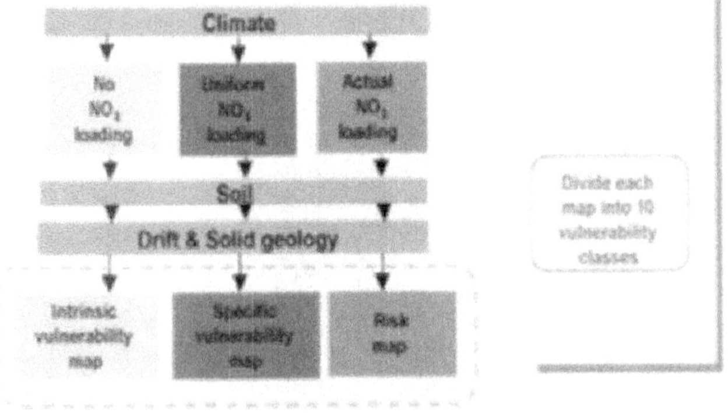

Schematic of GIS vulnerability and risk models

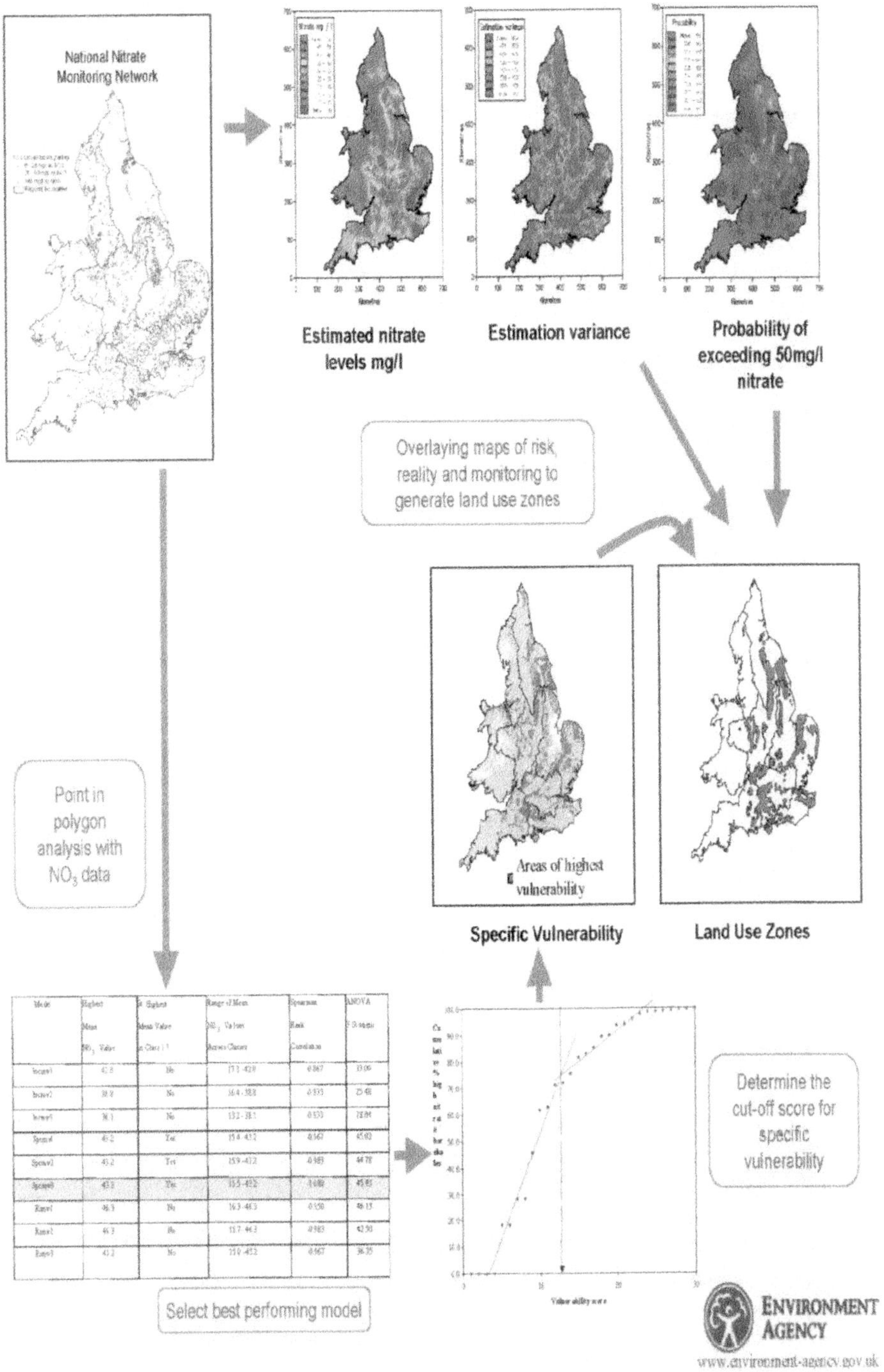
National Nitrate Monitoring Network
Estimated nitrate levels mg/l
Estimation variance
Probability of exceeding 50mg/l nitrate
Overlaying maps of risk, reality and monitoring to generate land use zones
Areas of highest vulnerability
Specific Vulnerability
Land Use Zones
Point in polygon analysis with NO3 data
Determine the cut-off score for specific vulnerability
Select best performing model
Environment Agency
www.environment-agency.gov.uk

Visualizing Hake Recruitment A Non-Stationary Process

Ernesto Jardim (ernesto@ipimar.pt)
IPIMAR - Instituto de Investigação das Pescas e do Mar
Av. Brasilia, 1400-006 Lisboa, Portugal.

ABSTRACT
Hake is an important fisheries resource for Portugal and EU. The spatial distribution of recruits (0-old individuals) is particularly important to define candidate zones to Marine Protected Areas. In this work the method of Sampson and Guttorp (1992) is used to overcome the problem of non-stationarity of a process, trough a smoothing function of spatial covariance. The results are interesting although is not clear how to back-transform the results and how smoothed they are.

Figure 1: Study Zone

INTRODUCTION
Hake (*Merluccius merluccius*) is an important fisheries resource in Europe and is distributed from the Barents Sea to the Mauritanean Coast. The population is splited in two different stocks (Northern and Southern) for management purposes. Like other fish species Hake recruits (age 0) ocurre in particular zones, where the environmental conditions are the best. This grounds are important to study because they are natural candidates to Marine Protected Areas.
This poster presents an atempt to visualize the most important Hake recruitment area on the Portuguese Coast.

DATA & METHODS
Data were collected by IPIMAR's Bottom Trawl Surveys, between 1989 and 2001. The study variable $\{z_{s,t}; s \in R^2, t=1,\dots,T\}$, is the number per hour of recruits caught in Southeast area of the Portuguese Continental Coast between 20m and 500m (Figure 1). Data was log-transformed and a polynomial (2nd degree) regression was applied to model the Spatial trend.
The spatial covariance was modelled using the method described by Sampson & Guttorp (1992). This method implements a non-parametric approach to model spatial dispersion ($d_{ij}^2 = Var(Z_{it}-Z_{jt})$ where i and j are locations) as a smoothed function of the geographic coordinates. It doesn't assume stationarity or isotropy. With Multidimensional Scaling (MDS) it defines a new configuration of points $\{u_i\}$ in R^2, where stations are arranged so that the distance between them is a function of the spatial dispersion. Then it models $d_{ij}^2 = ghat(|u_i-u_j|)$, where ghat is a function similar to the variogram (conditionally nonpositive definite). Finally maps the G-Plane in D-plane using thin plate splines (Figure 2). A regular grid covering all area was build and mapped in the D-plane for estimation purposes.

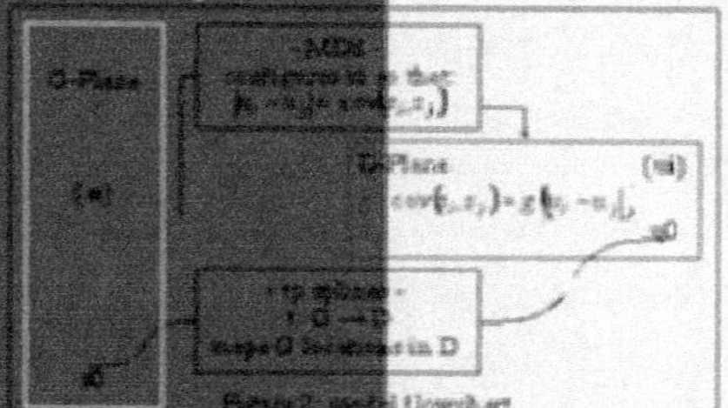

Figure 2: model flowchart

RESULTS

Figure 3 shows the mean and variance of r. Figure 4 shows G-Plane (top) and D-Plane (bottom) differences. The results of MDS (left), ghat (middle), tps by mapping the prediction grid (right). The adjustment parameters of MDS and thin plate splines are shown on table 1. It's obvious that the ghat in D-Plane shows a visible structure while in G-Plane it doesn't. The mapping of the prediction grid in the D-plane is concentrated in the middle of the Plane. Figure 5 shows the resulting maps of the spatial distribution of Hake recruits (log-transformed) and the kriging variance.

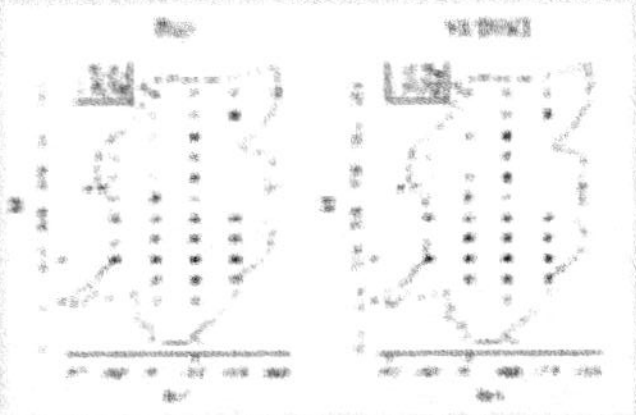

Figure 3: Observed values of Hake recruits

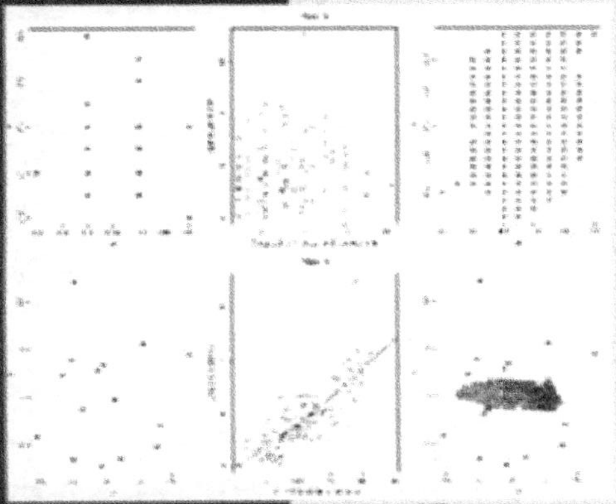

Figure 4: G-Plane and D-Plane differences

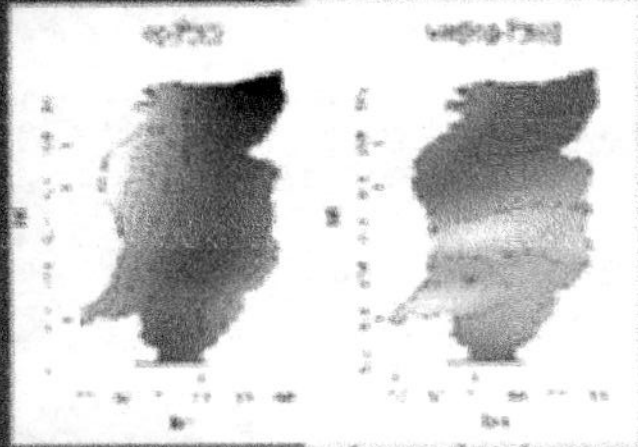

Figure 5: Estimated values of Hake recruits

DISCUSSION/CONCLUSIONS

- The method allows the variable visualization whithout explicity violate any method's requests, has would have happenned if a traditional geostatistical analysis was performed.
- The spatial distribution shows some differences from SESITS (1999). In particular there is a high value in the northeast that doesn't apear in SESITS (1999).
- It's interesting to notice that the variance distribution doesn't follow the variable distribution. Some locations with high values of the variable have very low variance and vice-versa.
- The mapping function may be responsible for smoothing, by placing all locations concentrated on the D-Plane.
- The several transformations performed on the data may cause problems of back-transformation and interpretation.

MDS	tps		ghat	
Stress Crit.	R2	Smooth	Alpha	Beta
17.7	[illegible]	[illegible]	1.71	0.24
[illegible]	[illegible]	[illegible]	[illegible]	

Table 1 - adjustment and ghat parameters

BIBLIOGRAPHY

Sampson, P.D. And Guttorp, P. (1992) Nonparametric Estimation of Nonstationary Spatial Covariance Structure. Journal of the American Statistical Association, 87(417): 108-119.

SESITS (1999) Evaluation of Demersal Resources of Southwestern Europe from Standardised Groundfish Surveys - Study Contract 96/029. Final report, DG XIV /EC.

Monitoring in Two Markov
Chain Markov Field Models
Ph. D. Eric Järpe
Halmstad University
Sweden
eric.jarpe@ide.hh.se
MARKOV CHAIN MARKOV FIELD (GUYON 1995)
SPATIO-TEMPORAL MONITORING
DYNAMIC MODEL
PERFECT SIMULATIONS
MODEL (STIM)
PERFECT SIMULATIONS

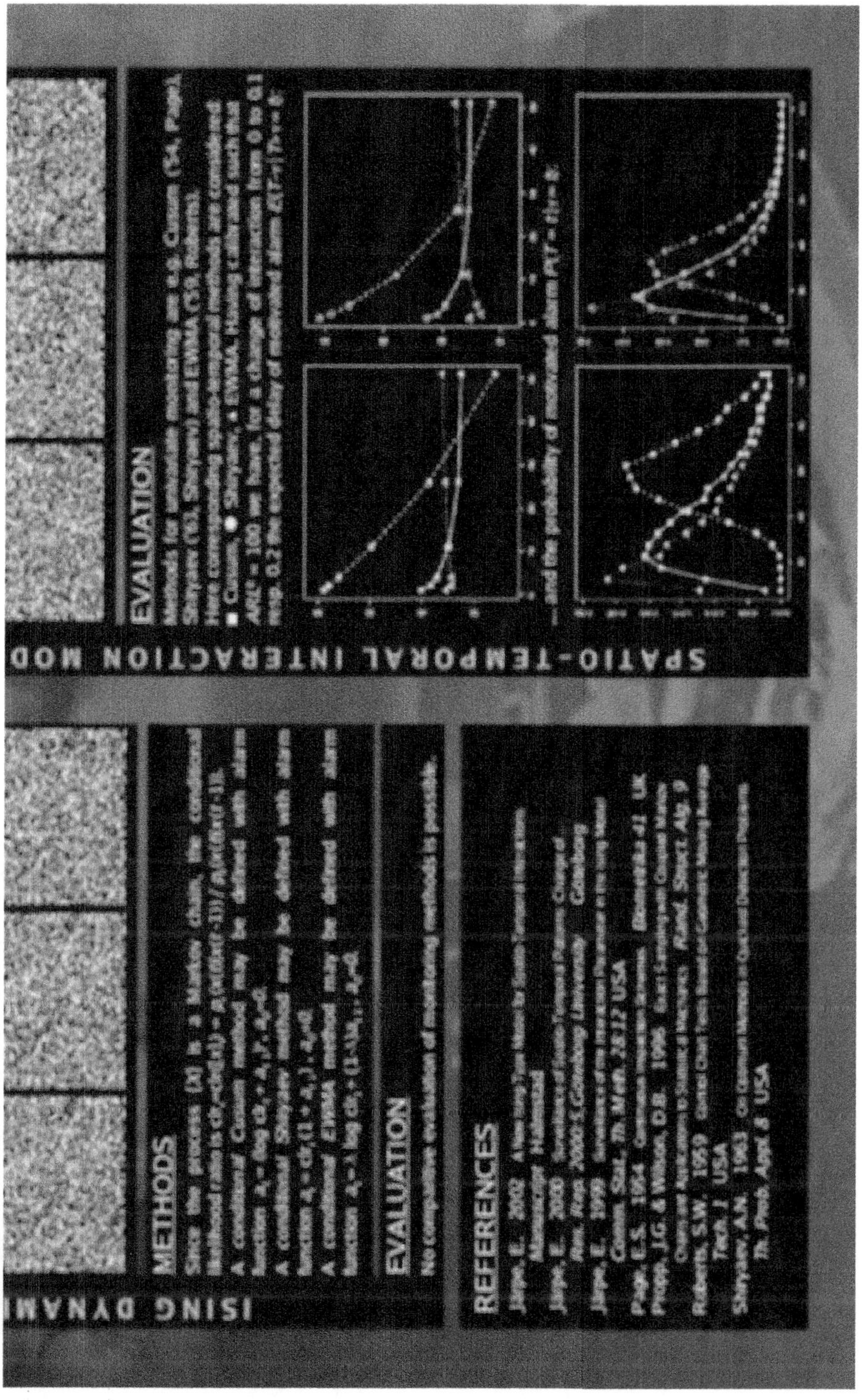
ISING DYNAM
METHODS
Since the process {X} is a Markov chain, the conditional likelihood ratio is
A conditional Cusum method may be defined with alarm function
A conditional Shiryaev method may be defined with alarm function
A conditional EWMA method may be defined with alarm function
EVALUATION
No comparative evaluation of monitoring methods is possible.
REFERENCES
Järpe, E. 2002
Manuscript Halmstad
Järpe, E. 2000
Res. Rep. 2000:5,Göteborg University Göteborg
Järpe, E. 1999
Comm. Stat., Th. Meth. 28:12 USA
Page, E.S. 1954 Biometrika 41 UK
Propp, J.G. & Wilson, D.B. 1996
Rand. Struct. Alg. 9
Roberts, S.W. 1959
Tech. J USA
Shiryaev, A.N. 1963
Th. Prob. Appl. 8 USA
SPATIO-TEMPORAL INTERACTION MOD
EVALUATION
Methods for univariate monitoring are e.g. Cusum ('54, Page), Shiryaev ('63, Shiryaev) and EWMA ('59, Roberts).
Here corresponding spatio-temporal methods are considered:
Cusum, Shiryaev, EWMA. Having calibrated such that
... and the probability of motivated alarm

Geostatistical inversion of flow and transport data. Application to the CRR project

Jódar[1] J., Meier[1], P., Medina[2] A., Carrera[1] J.
(1) Dep. Geotech.Eng.&Geosci.; School of civil Eng.;UPC
(2) Dep. Applied Mathematics; School of civil Eng.;UPC

INTRODUCTION

Cross-hole pumping tests and tracer test were performed in the framework of the migration experiment (MI) within a subvertical shear zone in the granitic host rock of the Grimsel Test Site (GTS) (see Smith et al. 2001 and Meier et al. 2001). Hydraulic tests were interpreted using a geostatistical inverse approach. Flow and transport parameters were estimated leading to a successful hydraulic and tracer data fitting.

New hydraulic and tracer test have been performed in the framework of the Colloid and Radionuclide Retardation (CRR) project in the same shear zone. The observed well responses during these new tests have been predicted (blind prediction) using the heterogeneous transmissivity field and the spatially homogeneous transport parameters derived from the calibration of the existing MI model. In order to obtain a good agreement between the new observed and computed data a recalibration of the transmissivity field and of the spatially homogeneous transport parameters has been performed.

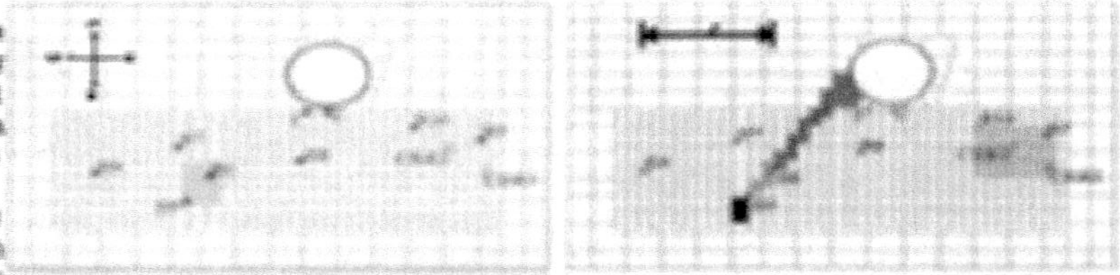

Figure 1: Left: Detail (most discretized zone) of the MI model domain; Right: Detail (most discretized zone) of the CRR model domain. In both cases is presented the intersection of Boreholes (B4, B5, B6, B7, B8, B9, B10, I, C, CRR1, CRR2) and the access gallery with the MI Shear-Zone. Gray filled zones are used to define the overcored EP borehole in the plane of the fracture. The three gray different patterns are referred to the three different filling material within the overcored borehole.

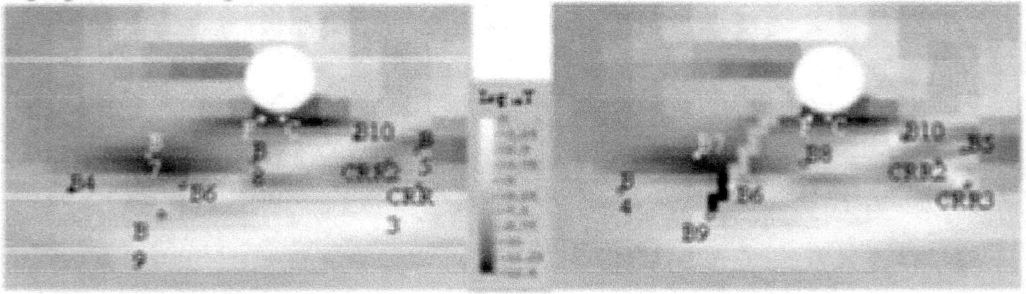

Figure 2: Most discretized zone of the transmissivity (Log10T) fields calibrated in the framework of the MI project (left side), and also in the framework of the CRR project (right side).

MATERIALS AND METHODS

Transmissivity is treated as a regionalized variable because of its strong heterogeneity as revealed from the MI experiment results (Meier, 1999). The model domain is divided into 922 T zones which are shown in the upper part of Fig.1. The more discretized zone from the former MI model, located in the MI test zone (between boreholes B6 and B9) make no sense in this new model, in which the area of interest in the shear zone is located between boreholes B10 and CRR3 (CRR test zone). A homogeneous storativity for the model domain is assumed.

The hydraulic properties of the MI test zone have changed. A new borehole was drilled parallel to the shear zone, following the direction of the line which links boreholes B9 and B6 with the access gallery (ED borehole). The borehole was filled with sand, resin, and a mixture of resin and sand depending on the borehole depth.
Two covariance structures have been considered. One corresponding to the general fracture zone using the same MI variogram, and a second one to take into account the excavation of the ED borehole in the MI test zone:

- General fracture zone: Spherical anisotropic variogram $\sigma^2(\mathrm{LnT})=100$, $R_x=9m$, $R_y=3.5m$
- ED borehole fracture zone: Spherical isotropic variogram $\sigma^2(\mathrm{LnT})=200$, $R_x=R_y=0.5m$

RESULTS AND DISCUSSION

A recalibration of the transmissivity field and of the spatially homogeneous transport parameters has been performed (Jódar et al., 2002), leading to a successful agreement between observed and computed drawdowns and breakthrough curves. Neither cubic nor the porous filling models are optimal representations of the relationship between transmissivity and porosity within this shear zone. Still, the observed behaviour falls consistently between the predictions of these two models, which confirms the findings of Meier (1999).

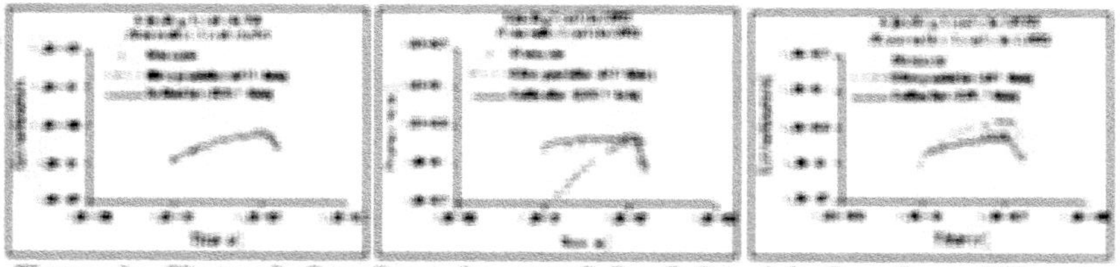

Figure 3: Observed drawdown in several boreholes, injecting from different boreholes, blind prediction using the MI T-field (Meier 1999), prediction using the calibrated CRR T-field (Jódar et al 2002).

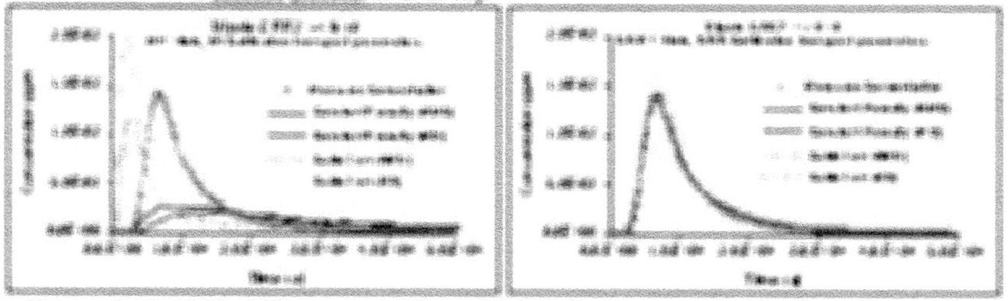

Figure 4: Breakthrough curve obtained during the tracer test (dipole configuration) performed between boreholes CRR7 (injection) and B10 (pumping). Left side: Blind prediction breakthrough curves obtained using the MI T-field and the MI transport parameters (Meier 1999). Right side: Predicted breakthrough curves obtained using the CRR T-field and the CRR transport parameters (Jódar et al 2002). Tracer tests were interpreted using two different assumptions about the relationship between transmissivity and porosity (constant porosity and cubic law respectively). Matrix diffusion (MD) and no matrix diffusion (NMD) processes are considered.

REFERENCES

- Meier, P., M. 1999 Estimation of representative groundwater flow and solute transport parameters in heterogeneous formations. Ph.D. Dissertation. School of Civil Engineering, Barcelona.
- Jódar, J., A. Medina, J. Carrera 2002. Actualization of geostatistical inverse model with new crosshole pumping tests and non-sorbing tracer test evaluation. Nagra Technical Report (in preparation).

ACKNOWLEDGEMENTS

This work has been funded by ANDRA and IDRESA within the framework of the CRR project.

Statistical Learning

Kanevski M.[(1)], Pozdnukhov A.[(1)], Mc

[(1)] IBRAE Institute; Russian Academy of Sciences, [(2)] Sandia National

Abstract

A current problem across many different fields is how to handle, to understand and to model data if there are too many or too few of them. Traditionally, geostatistics is one of the well-established approaches for working with spatially distributed data. It is a model-dependent approach based on the exploratory analysis and modelling of spatial correlation structures. On the other hand, recent explosive growth in the development of adaptive methods for learning from data have resulted in data-driven and model-free contemporary approaches, particularly based on Statistical Learning Theory -SLT (Vapnik-Chervonenkis theory) [1].

In the present report the models of Statistical Learning Theory – Support Vector Machines (SVMs), are adapted for spatial data binary classification tasks, (dichotomies), for multiclass classification problems (classification of hydrogeological units, classification of soil types, etc.), for robust decision-oriented mapping of environmental and pollution data (e.g. soil pollution by radionuclides).

SLT and Learning Algorithms

In Machine Learning one's aim is to find ("learn") an algorithm (modeling/mapping function) that describes training data and has good generalization abilities – allows for accurate predictions at the unknown points. SLT is devoted to such problems of extracting knowledge from finite empirical data.

The following bounds of the generalization error were derived in SLT:

$$R(\lambda) \le R_{emp}(\lambda) + R_{conf}(\lambda)$$

where $R_{emp}(\lambda)$ is an empirical risk on the training data (training error), and $R_{conf}(\lambda)$ is a confidence term which depends on the "complexity" of the modeling function (or their hyper-parameters λ).

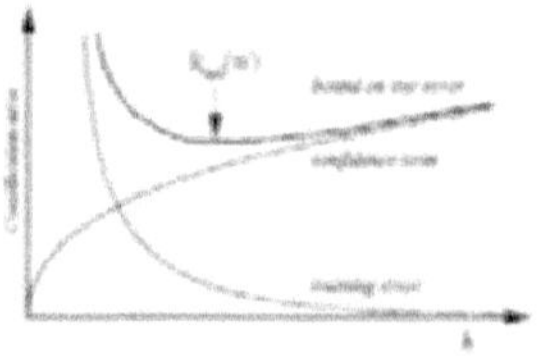

"Complexity" is characterized with a parameter h – VC-dimension of the modeling functions. Hence the relevant strategy for constructing a learning machine is to minimize the training error maintaining h small (see figure above). This idea is realized for the specific learning tasks and results in a family of Support Vector algorithms.

SVM provides non-linear and robust solutions by mapping the input space into a higher-dimensional space using kernel functions and building a linear model in this feature space. Using different kernels we can obtain learning machines analogous to well-known architectures such as Radial Basis Function neural networks and multilayer perceptrons as well as linear and polynomial algorithms. SVMs seek for regression decision function in the form

$$f(x) = \sum_{i=1}^{N} \alpha_i K(x, x_i) + b$$

where K(.,.) is a symmetrical positive definite function – kernel. For classification task one uses $sign(f(x))$ as the decision function. The weights α are obtained from the solution of the convex QP optimization problem. Gaussian Radial Basis Functions were found to be well suited for environmental applications.

Binary Classification: ^{137}Cs Activity.

The task is to classify the spatial areas of "high" and "low" contaminant activity based on some pre-defined level. This classification can establish the basis for site remediation activity and other decisions.

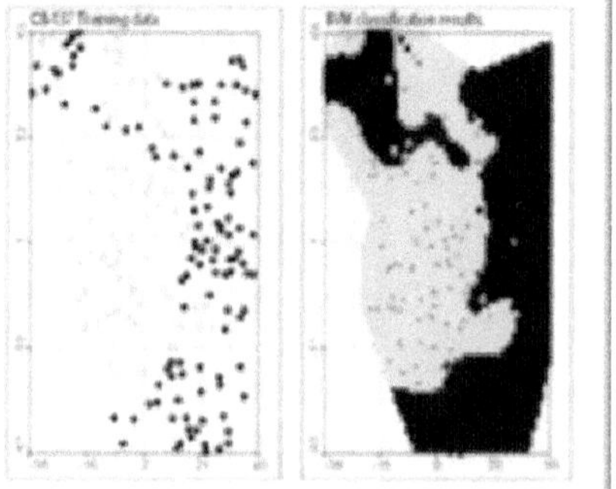

Figures illustrate training data and resulting classification with validation data shown with colored circles marked by crosses. Validation error 9%.

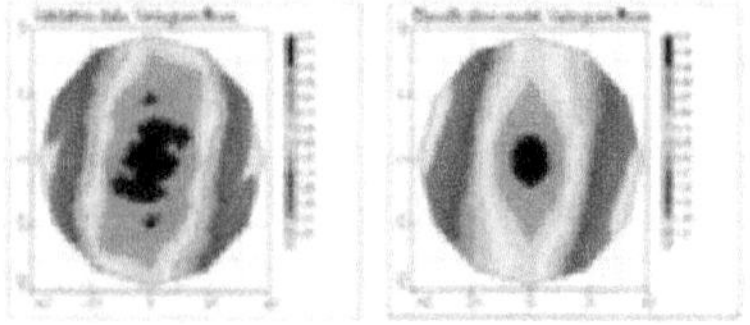

Variogram roses show that spatial structure of the data is reproduced, in spite of the fact that SVM avoids direct modeling of the spatial structure.

Theory for Spatial Data

Kenna S. [2], Murray Ch. [3], Maignan M. [4]

Lab., USA, [3] Pacific Northwest Lab., USA, [4] University of Lausanne, CH

Multi-class Classification

The straightforward way to solve a multi-class classification task is to train and combine several binary classifiers. It was found that the simple "one-against-rest" combination scheme gives good results in environmental classification tasks.

1. **Hydrogeological units.** The task is to classify the inner sub-structure of the hydrogeological unit based on 220 data of structure type. The task is of great importance for accurate ground water flow model predictions.

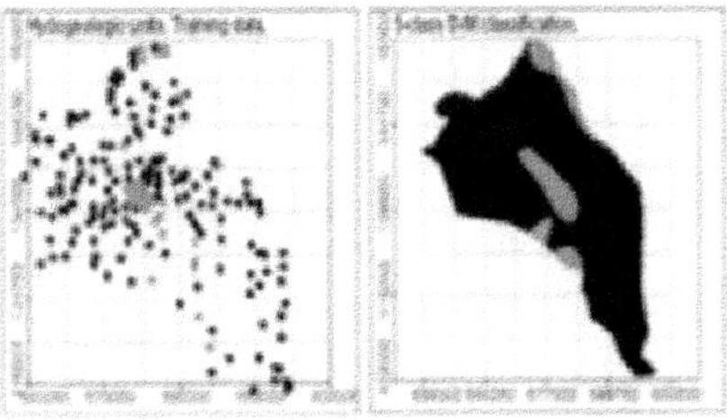

2. **Soil types classification.** Migration of radionuclides in soils are strongly dependent on the soil types. At the same time the data on the soil types often accompany the data on radionuclide activity etc.

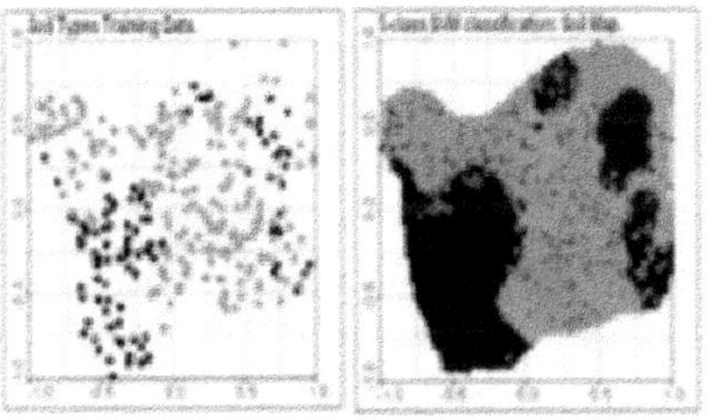

The postplot of the prediction mapping is accompanied with the validation data, shown by crosses. Geostatistical classification method - indicator kriging - fails since there are too few data in some classes to model the correlation structure adequately. The results of comparison with other methods: probabilistic neural network (PNN) and nearest neighbor classifier (NN) is presented in the table below.

Method	Training error	Validation error
NN	0	17.8%
PNN	1.3%	18.2%
SVM	0.65%	12.8%

Spatial Regression

The algorithm for regression estimation based on SLT is a Support Vector Regression (SVR). The flexibility of the SVR allows the model to obtain a wide variety of solutions, including spatial trend models.

Prediction of ^{241}Am activity. The data consist of 193 measurements, the task is to predict the activity in 917 points.

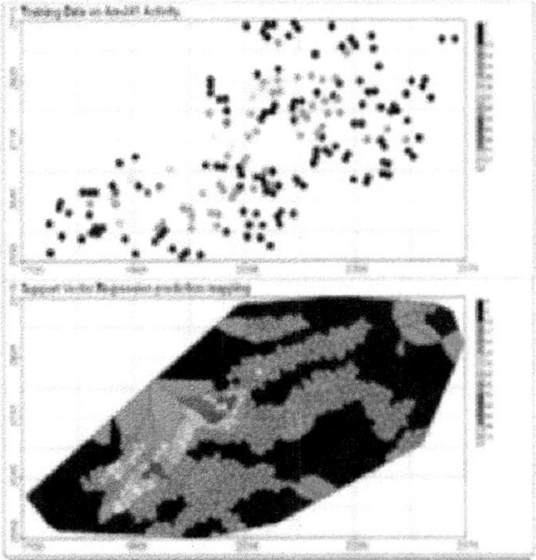

Method	RMSE	MAE	Corr. Coeff.
Ordinary Kriging	9.92	5.94	0.669
Transductive Regr.	9.73	5.98	0.674
SV Regression	9.73	6.56	0.665

Conclusions.

Machine Learning algorithms gives promising results for a number of tasks such as binary/multi-class classification and spatial regression. These approaches are data-driven and model-free. They also allows to model the data when geostatistical methods fail.

Acknowledgements

This work was supported in part by INTAS 99-00099, CRDF grant RG2-2236 and by the collaboration under Memorandum of Understanding between the U.S. Department of Energy and the Russian Academy of Sciences.

References

[1] Vapnik V. Statistical Learning Theory. New York: John Wiley & Sons, 1998.

[2] Kanevski M., Pozdnukhov A., Canu S., Maignan M. Advanced Spatial Data Analysis and Modeling with Support Vector Machines. Int. J. of Fuzzy Systems, Vol. 4, No. 1, pp. 606-616, March 2002.

GEOSTATISTICAL ANALYSIS OF TRACE ELEMENTS AT SMALL CATCHMENT IN FINISTERRE (SPAIN)

López Candia, A.[1] & Paz González, A.[2]
Facultad de Ciencias. Universidad de A Coruña. A Zapateira s/n. 15.071. A Coruña. Spain
albertlc@mail2.udc.es[1], tucho@udc.es[2]

INTRODUCTION

Semivariograms have been widely used in geochemistry for mapping metal concentrations at a range of scales and with different sizes of sampling grids. Geostatistical analysis of metal concentration data sets has also contributed to the discussion on general subjects such as optimal spacing of a regular grid for kriging interpolation.

The objective of this study was to examine the spatial variation of six geochemical variables (V, Cr, Co, Cu, Zn and Pb) at small scale, on a site located near Finisterre, in Coruña Province, Spain.

MATERIALS AND METHODS

The chemical composition of the weathered material under the soil is used to explore sampling issues and to analyse the spatial distribution. Data were recorded at 323 points using a sampling grid scheme (150 m x 50 m). The data set was collected from the subsoil level.

The concentrations of six elements were measured at each location, using standard analytical techniques. The prediction set was used to identify the magnitude and scale of spatial variation of six selected elements. Experimental semivariograms of individual variables were computed and modeled by a nugget component and a structure.

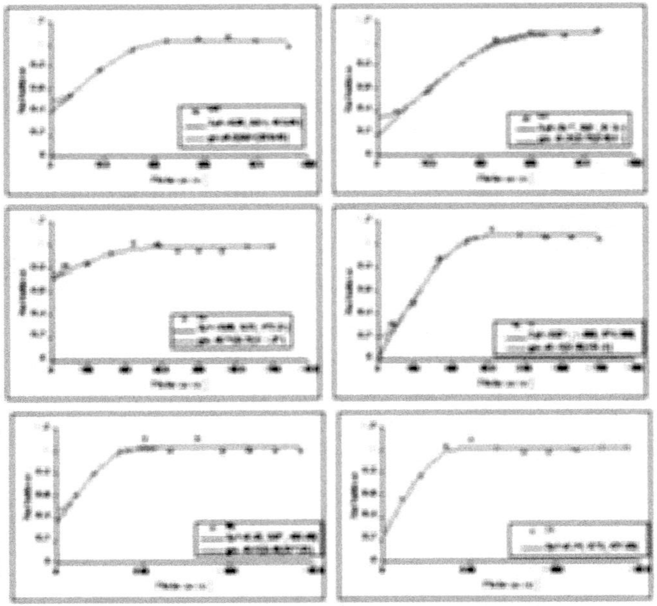

Figure 1. Experimental semivariograms and models fitted to them.

RESULTS AND DISCUSSION

According to the structural analysis, the best fitting models were spherical for the six elements. However, fitting gaussian models showed small differences in the validation parameters and estimation variance for five out of six elements. The autocorrelation ranges for most of the chemical elements studied varied between 163 and 477 m. Nugget effect varied between 0 and 0.75 (Figure 1).

The spatial dependence was used to obtain kriging maps, allowing the identification of regions with different metal concentrations and the corresponding kriging estimation variances.

The prediction quality of the method used was determined based on the root mean square error (RMSE) and the mean error (ME) of the validation set. Block kriging was used to optimally obtain interpolated values for the places not sampled. In addition, correlation coefficient of the scatter plots of the actual element contents versus the kriged predicted values were also calculated.

Mean error and correlation coefficient values indicated that the studied variables were generally well predicted. Measured data of the validation set were significantly correlated ($p<0.001$) with predicted values. The poorest prediction was for Pb, followed by Cu.

The results of the estimates are presented in map form, showing the distribution of each element in the area studied (Figure 2). From maps of metal concentrations, a real distribution of main mineralizations was assessed.

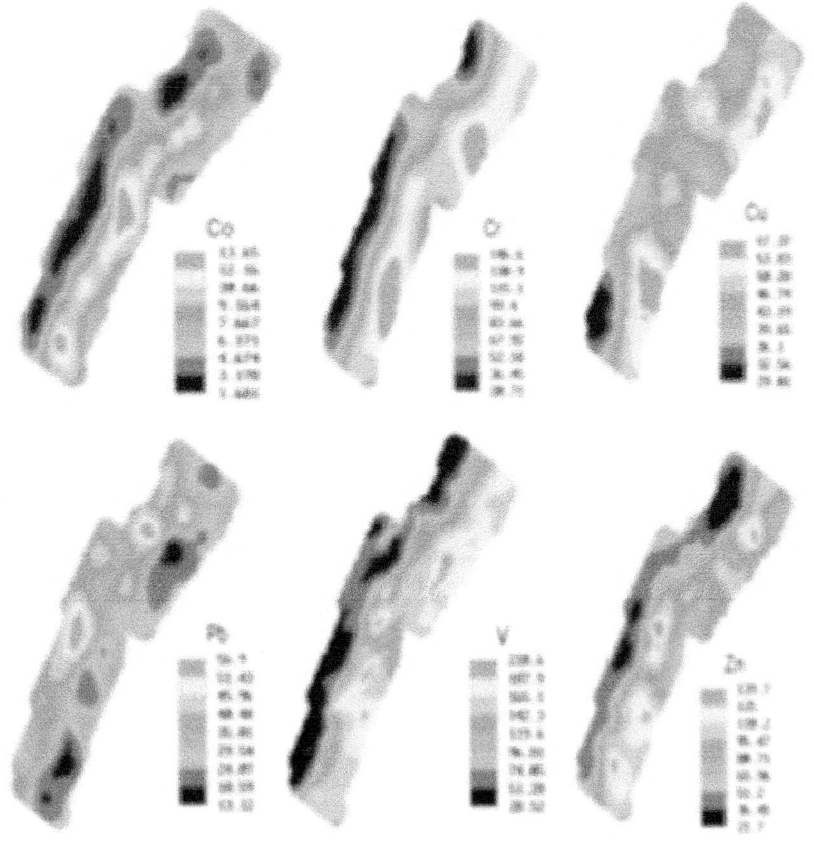

Figure 2. Kriging maps for different elements.

Comparison between Kriging, MLP and RBF in a slate mine

J. M. Matías[1], A. Vaamonde[1], J. Taboada[2], W. González-Manteiga[3]
(1) jmmatias@uvigo.es, vaamonde@uvigo.es, Dpt. Statistics and (2) jtaboada@uvigo.es; Dpt. Mining, Univ. Vigo; (3) wences@zmat.usc.es Dpt. Statistics, Univ. Santiago de Compostela, Spain.

Abstract

In order to asses the exploitability (Y=1 or Y=0) of a new unexploited area of a slate mine, we apply the following estimation techniques to a sample of drilling data: Kriging, Regularization Networks, (RN), Multilayer Perceptron networks (MLP) and Radial Basis Function Networks (RBF).

Regularization Networks (RN)

$Y = Y(\mathbf{x}) = E(Y/\mathbf{x}) + \varepsilon$ with ε random noise

$$\min_{f \in H} L(f) = \min_{f \in H} \left\{ \|\mathbf{y} - \mathbf{f}\|^2 + \lambda \|f\|_H^2 \right\}$$

with H (Semi) Reproducing Kernel Hilbert Space,

$$\text{with } \langle f, g \rangle = \int_{\mathbb{R}^d} \frac{\hat{f}(\boldsymbol{\omega}) \hat{g}^*(\boldsymbol{\omega})}{\hat{k}(\boldsymbol{\omega})} d\boldsymbol{\omega}$$

k is semidefinite positive of order m and the solution is:

$$f(\mathbf{x}) = \sum_{i=1}^{n} c_i k(\mathbf{x}, \mathbf{x}_i) + \sum_{j=1}^{m} d_j q_j(\mathbf{x})$$

$$\text{with } q_j \in \Pi_m^d \text{ and } \begin{cases} (K + \lambda I)\mathbf{c} + Q\mathbf{d} = \mathbf{y} \\ Q^t \mathbf{c} = 0 \end{cases}$$

Examples are Splines and Gaussian RN

A Comparison between Kriging and RN

- Universal Kriging and RN aproaches provide the same solution when the generalized covariance function coincides with the semi-reproducing kernel and the regularizer is the variance of the process.

- RN solution with m=1 is equivalent to Ordinary Kriging, and with m=0 to Simple Kriging.

- The non-consideration of noise in kriging is equivalent to a null regularising parameter.

Radial Basis Function Networks (RBF)

They arose in the context of interpolation techniques and inspired by smoothing techniques:

$$f(\mathbf{x}) = \sum_{i=1}^{r} c_i k_i(\|\mathbf{x} - \mathbf{t}_i\|) + b,$$

with $r << n$ and $\mathbf{c} = (K^t K)^{-1} K^t \mathbf{y}$

Variables selection by Orthogonal Least Squares (Chen et al. 1991) and model selection by analytical methods (CV, etc.).

Multilayer Perceptron Networks (MLP)

One hidden layer with tanh and one linear output:

$$f(\mathbf{x}) = \psi\left(\sum_{k=1}^{h} c_k \phi_k(\mathbf{w}_k^t \mathbf{x} + w_{k0})\right) + b,$$

Bayesian training method implemented by Foresee and Hagan'97 following MacKay'92.

Aplication to a Slate Mine

Sample : $\{(\mathbf{x}_i, y_i)\}_{i=1}^{N}$ with $N = 1,932$,

$n_1 = 1,000$ for training, $n_2 = 932$ for test.

$y_i \in \{0,1\}$ (1: exploitable, 0: non exploitable).

Decision :

$\hat{Y} = 1$ if $\hat{f}(\mathbf{x}) = \hat{P}(Y = 1/\mathbf{x}) > 1/2$ and zero otherwise.

Estimated Covariogram :

$k(h) = 0.23 - 0.104(1 - e^{-[illegible]}) - 0.126(1 - e^{-[illegible]})$

RESULTS				
Model	**Error**	**NBF**	**ENP**	**Int.**
Ordinary Kriging	**0.11695**	**1000**	**1000**	**Yes**
Gaussian RN	**0.12124**	**1000**	**311**	**No**
Covariogram RN	**0.11370**	**1000**	**143**	**Yes**
Gaussian RBF	**0.12661**	**163**	**164**	**No**
Covariogram RBF (SK)	**0.11910**	**1000**	**1000**	**Yes**
MLP	**0.11803**	**14**	**65**	**No**

Error: Test Error (%), **NBF**: Number of basic functions.
ENP: Equivalent number of parameters. **Interpolator**: (Yes/No).

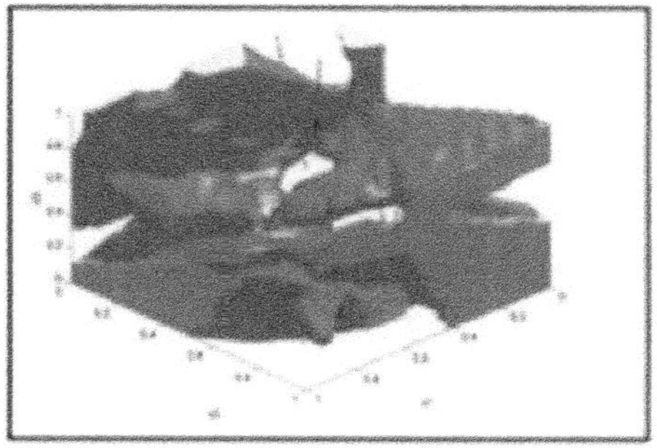

MLP

Drilling Data

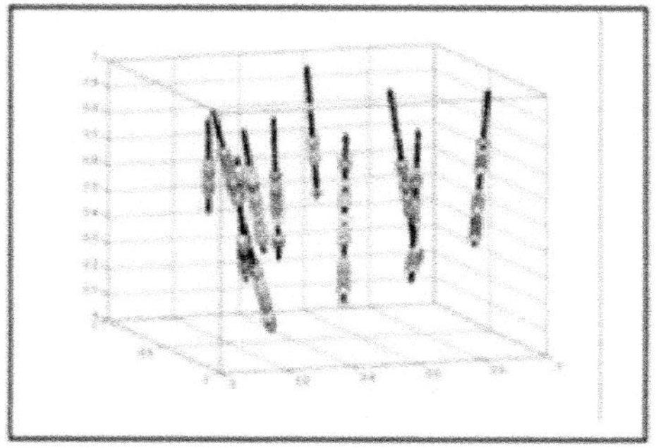

RBF

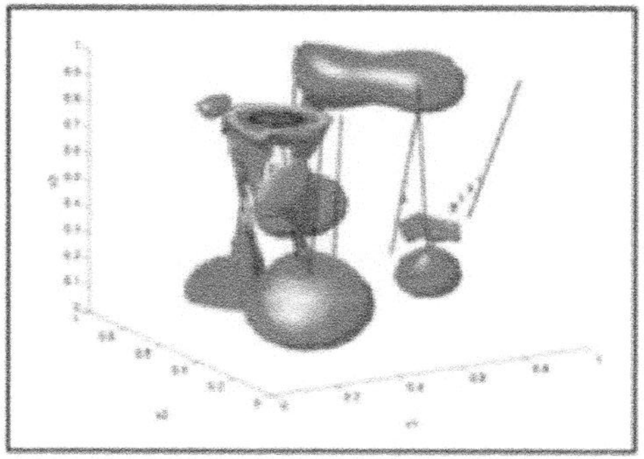

References

(1) Chen, S., Cowan, C. F. N., Grant, P. M., 1991. Orthogonal Least Squares Learning Algorithm for Radial Basis Function Networks. IEEE Transactions on Neural Networks, v. 2, no. 2, p. 302-309. **(2)** Foresee, F. D., Hagan, T., 1997. Gauss-Newton Approximation to Bayesian Regularization. Proceedings of the 1997 International Joint Conference on Neural Networks, p. 1930-1935. **(3)** Girosi, F., Jones, M., Poggio, T., 1995. Regularization Theory and Neural Networks Architectures. Neural Computation, no. 7, p. 219-269. **(4)** MacKay, D. J. C., 1992. A Practical Bayesian Framework for Backpropagation Networks. Neural Computation, v. 4, p. 448-472.

ESTIMATING SPATIAL VARIABILITY OF TEMPERATURE

Mirás Avalos[1], J.M. & Paz González, A[2].
Facultad de Ciencias. Universidad de A Coruña.
A Zapateira s/n. 15071. A Coruña (Spain) e-mail: [1]josemanu@mail2.udc.es, [2]tucho@udc.es

INTRODUCTION

Any attempt to analyse complex systems such as the weather involves a certain amount of complication due to the fact that any of its variables are difficult to predict and, as such, we assume that it is a random or chaotic system.

After precipitation, temperature is the most widely measured weather variables and one of the most important in order to characterise climate at a regional level. Temperature is an important factor for economic activity and agricultural production. It is a variable displaying large spatio-temporal variation. The aim of this study is to obtain an efficient mapping of the temperature at the regional level. Different aspects of temperature mapping are of interest from a geostatistical perspective, in particular the potential use of kriging algorithms that account for secondary information in the prediction process.

MATERIALS AND METHODS

Monthly and annual temperature data sets analyzed were obtained from 61 weather stations located in Galicia, north west of Spain. Data have been provided by the National Institute of Meteorology. In addition exhaustive data of terrain altitude derived from a digital elevation model and discretized into grids of 500 m x 500 m were used.

The spatial variability was assessed by means of scaled semivariograms, using the sample variance as a scaling factor. The best fitting models and the ideal number of neighbours for kriging estimation were selected using least square approximation and cross validation.

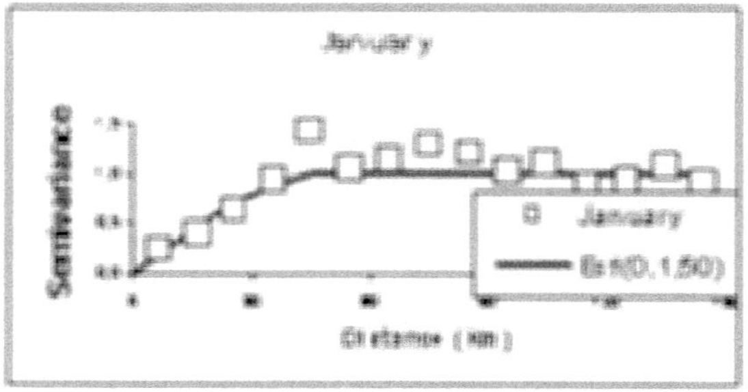

Figure 1. Example of experimental data and the model fitted to them.

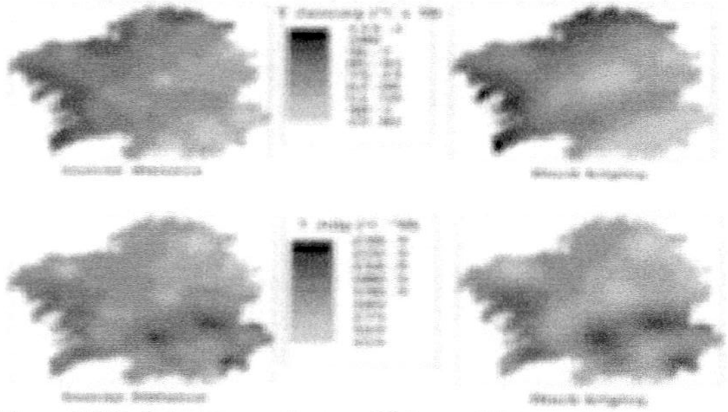

Figure 2. Block kriging vs. Inverse distance of temperature estimates (°C x10) of January and July in Galicia.

RESULTS AND DISCUSSION

Spherical and gaussian functions were fitted to experimental semivariograms (Figure 1). These provided a clear description of the spatial structure of the temperature, with a range of spatial dependence between approximately 16 and 60 km, depending on the month.

Kriging maps were compared with results obtained by traditional methods that do not account for spatial structure, such as inverse squared-distance (Figure 2). Semivariogram analysis and kriging maps illustrate possible environmental processes determining temperature distribution and allowed inferences to be made about factors controlling its spatial and temporal variation. Kriging maps provide additional evidence that topography and distance to the coast are the principal factors controlling the pattern of temperature distribution in the studied region.

In a second step, temperature data were also combined with topographical information to improve estimation at the regional level. In this case, the used geostatistical algorithms were ordinary cokriging and collocated cokriging.

Estimation accuracy of the different methods was evaluated using cross validation. Generally, cokriging and collocated cokriging were more accurate and performed better than ordinary kriging. The basic difference was in the size of areas in which there are high values for the estimation variance, which was somewhat larger for the map obtained with ordinary kriging. However, maps obtained by different geostatistical interpolation procedures, basically present the same results, even if there was a little more detail on the maps resulting from cokriging and collocated cokriging, owing to the contribution of the altitude as a secondary variable. Thus, the gain for mapping purposes by incorporating altitudes as secondary information was limited, so that not very different results were obtained by the application of the different geostatistical approaches.

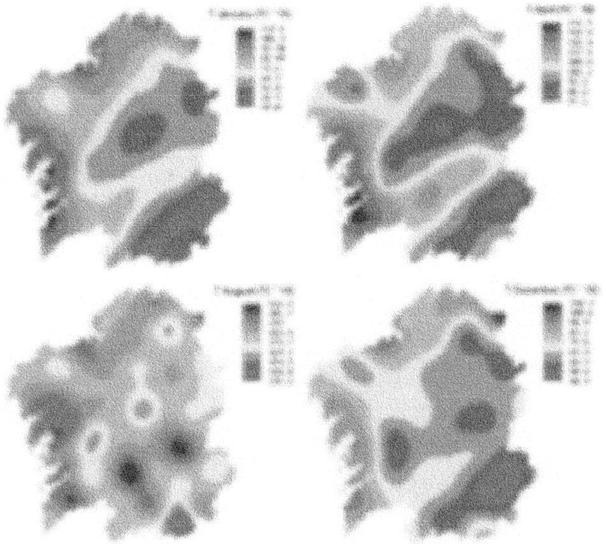

Figure 3. Temperature estimates (°C x10) using ordinary kriging

LITERATURE CITED

Carrera López, J.L. 1999. *Análisis geoestadístico y fractal de la precipitación en Galicia durante 1993 y 1994*. Tesis de Licenciatura. Universidade da Coruña. Facultad de Ciencias. 171 pp.

Goovaerts, P. 1997. *Geostatistics for natural resources evaluation*. Applied Geostatistics Series. New York. 483 pp.

Ministerio de Medio Ambiente y Energía. *Resúmenes Climatológicos Mensuales de Galicia*. Centro Meteorológico Territorial de Galicia. Sección de Climatología.

Pebesma, E.J. 2000. *Gstat User's Manual*. Dept. of Physical Geography. Utrecht University. 100 pp.

SPATIAL VARIABILITY OF SOIL pH AND Eh BEFORE AND AFTER FLOODING A RICE FIELD

MORALES[1], L. A. PAZ GONZALEZ[2], A.

[1]Secretaría General de Ciencia y Técnica - Facultad de Ciencias Agrarias UNNE Argentina

[2]Facultad de Ciencias. Campus A Zapateira s/n 15071. Coruña. Spain.

INTRODUCTION

Soil properties may vary at a range of spatial resolutions, comprising variation over short distances of a few centimetres and longer distances of ten of meters and the challenge is to detect the scale of interest. Spatial variability of natural soils is inherent to how the soil formation factors interact within the landscape; moreover, variability can occur also as a result of cultivation and land use.

There is now increasing evidence that spatial variation of soil physical and chemical properties can give rise to patterns at all levels of resolution and these patterns are more likely subject to temporal changes. Thus, in assessing soil properties variability not only spatial issues but also temporal aspects, should be taken into account. The interaction between spatial and temporal variability of soil nutrient status may be very complex and valuable information is only gained by taking into account both aspects simultaneously. Rice cultivation in wetlands or paddy soils is economically important in Latin America, for which soil inundation is currently utilised. Saturation by water strongly influences physical, chemical and biological soil properties; as a consequence of anaerobiosis, dramatic changes are induced during the rice growing season. Lowland soils in Corrientes (Argentina) have been also increasingly devoted to rice production. Most of these soils are extremely acid in natural conditions (pH in H_2O < 4.0) so that adequate liming and soil management practices to improve soil properties and increase crop yield are now a common practice.

The objective of this study was to examine the changes of spatial variation in soil pH and Eh on an acid soil of the Corrientes area, Argentine.

MATERIALS AND METHODS

A regular sampling grid scheme (11.9 m x 51.9 m) with 96 sites was used, in order to identify magnitude and scale of spatial variation of the selected physico-chemical soil properties in a 5.1 ha field. Data sets were collected within this field for three different times, during a growing season, first before flooding and two times more after flooding. Soil pH and Eh were determined by standard routine methods.

RESULTS AND DISCUSSION

Statistically the experimental field was more or less heterogeneous. The smallest Cvs were for pH (0.88 to 4.20%) whereas the Eh Cv before flooding was 1.99% and after flooding heterogeneity increased dramatically, so that Cvs between 20.6 and 60.8% were obtained. Rice cultivation modifies heterogeneity pattern of both studied variables. The pattern of spatial variability for pH and Eh was found to change along the time, as the soil reduction processes were more important. For pH, the nugget effect was about 23% in the first sampling data, and steadily increased until about 50% of the total variance in the third sampling date. Also Eh nugget variance increased in successive sampling dates from about 34% to 43%. Thus, the effect of agricultural soil use was to diminish the spatial continuity of the studied variables at close distances. The autocorrelation ranges Eh varied between 46 and 119 m, whereas for pH were between 57 and 90 m. Semivariograms provided a description of the spatial structure of the studied physico-chemical parameters and some insight into possible processes affecting their distribution. Kriging maps allowed the identification of small regions with distinct pH and Eh values and illustrate its heterogeneity in the studied field for different sampling dates. The prediction performance of the used geostatistical techniques was evaluated.

Table 1. Summary statistics for the physicochemical data in wetland rice

	pH1	pH2	pH3	Eh1	Eh2	Eh3
Mean	4.310	5.851	6.711	542.64	-24.69	-194.44
Std Dev	0.1108	0.2456	0.0592	10.78	15.02	38.97
C Variation	2.5696	4.1981	0.8820	1.987	-60.823	-20.042
Variance	0.0123	0.06033	0.0035	116.26	225.46	1518.61
Skewness	-0.2137	-0.0044	-0.0474	0.0108	-0.1809	-0.2145
Kurtosis	-0.5465	-0.6485	-0.1747	-1.0239	-0.3254	-0.8406
Pr < W**	0.2435	0.3877	0.6049	0.0315	0.3239	0.0622

Figure 2. Experimental pH semivariogram for the first (a), second (b) and third (c) of soil sampling data.

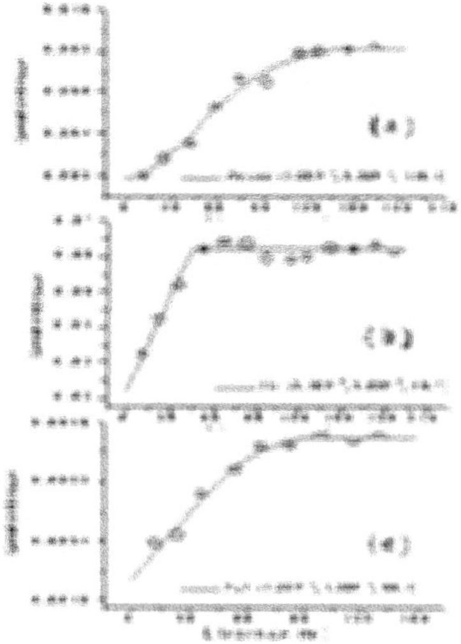

Figure 3. Kriging maps of pH for the first (1), second (2) and third (3) of soil sampling data.

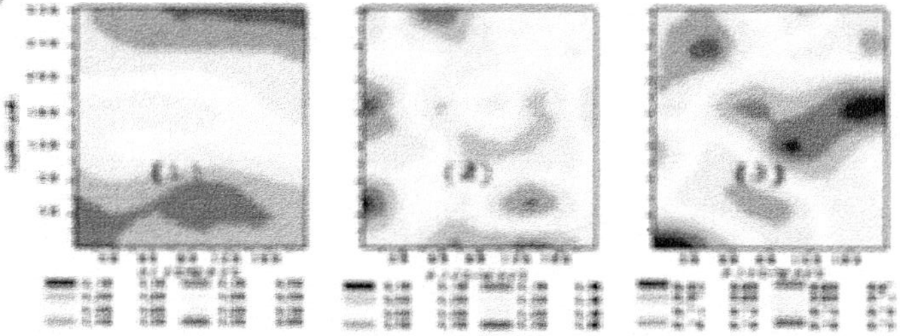

Figure 4. Kriging maps of Eh for the first (1), second (2) and third (3) of soil sampling data.

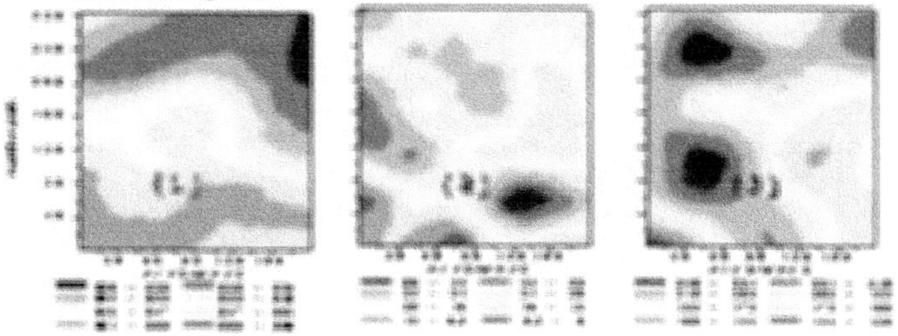

Figure 5. Experimental Eh semivariogram for the first (d), second (e) and third (f) of soil sampling data.

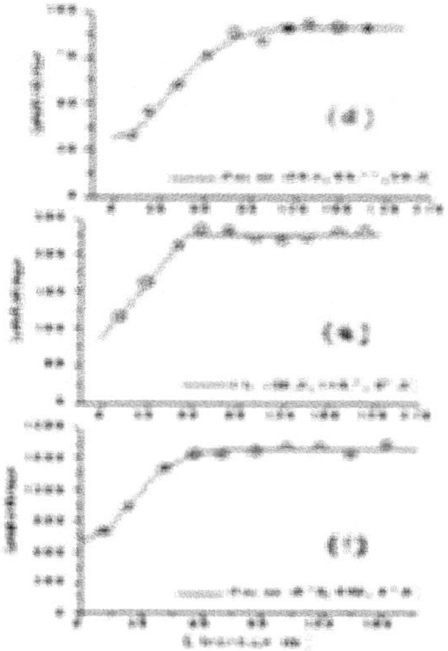

Total Catch and Effort in the Shark Bay King Prawn Fishery.

U.A. Mueller[1], L.M. Bloom[1], M.I.Kangas[2], J.M. Cross[1] and A.M. Denham[1].
[1] Edith Cowan University, Australia [2] Department of Fisheries, Australia

Abstract

This poster details the variography and the Ordinary Kriging estimation maps for catch rates (kg/h) of western king prawns (*Penaeus latisulcatus*) in the Shark Bay Managed Prawn Fishery.

Introduction

The Shark Bay Prawn Fishery is located within a World Heritage area in the northwest of Western Australia. It is a multi-species fishery and its management is a cooperative arrangement between industry and the WA Department of Fisheries. Commercial prawn fishing vessels record daily logbook data noting exact location, catch (by species and size) and effort (minutes trawled) for each trawl shot. The results presented here are part of a pilot study to assess the feasibility of using intrinsic geostatistics for measuring the spatial variability of such data.

Methods

Data were aggregated into total catch (kg) and catch rate (kg/h) at average locations. The fishery is closed for 3 to 7 days closest to the full moon so lunar weeks were used as the basic unit of time for each modelling period. Traditional intrinsic variography and Ordinary Kriging estimation were carried out for both total catch and catch rate.

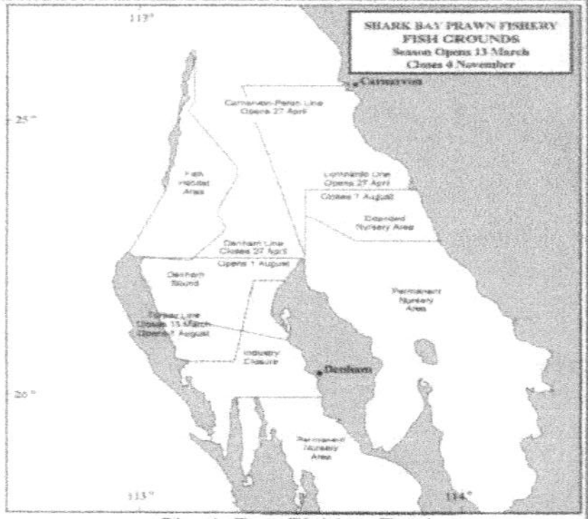

Shark Bay Fishing Region.

Shark Bay Prawn Trawler.

Shark Bay, Western Australia.

Results

Semivariogram surfaces, experimental semivariograms, semivariogram models and subsequent Ordinary Kriging estimates for the catch rates of king prawns caught in the April/May 2000 lunar month are shown. The semivariograms typically exhibit nested structures consisting of a nugget together with a short and a long range component. This prevails for the other lunar months as well. The semivariograms are isotropic with the sole exception of those for Week 1 and Week 3 of April/May 2000. On occasion it was necessary to use the pairwise relative semivariogram to infer the range. In April/May 2000, the estimates of the catch rates in Week 2 and Week 3 are higher than those for the other two weeks and are, in fact, the highest for the entire time period investigated.

Conclusions

Intrinsic geostatistical methodology appears to lend itself well to the analysis of prawn total catch and catch rate data at average locations. The next step is to investigate these variables at individual locations and for various prawn sizes. We also consider the relative abundance of king prawns with respect to other target species such as tiger prawns (*Penaeus esculentus*) and saucer scallops (*Amusium balloti*).

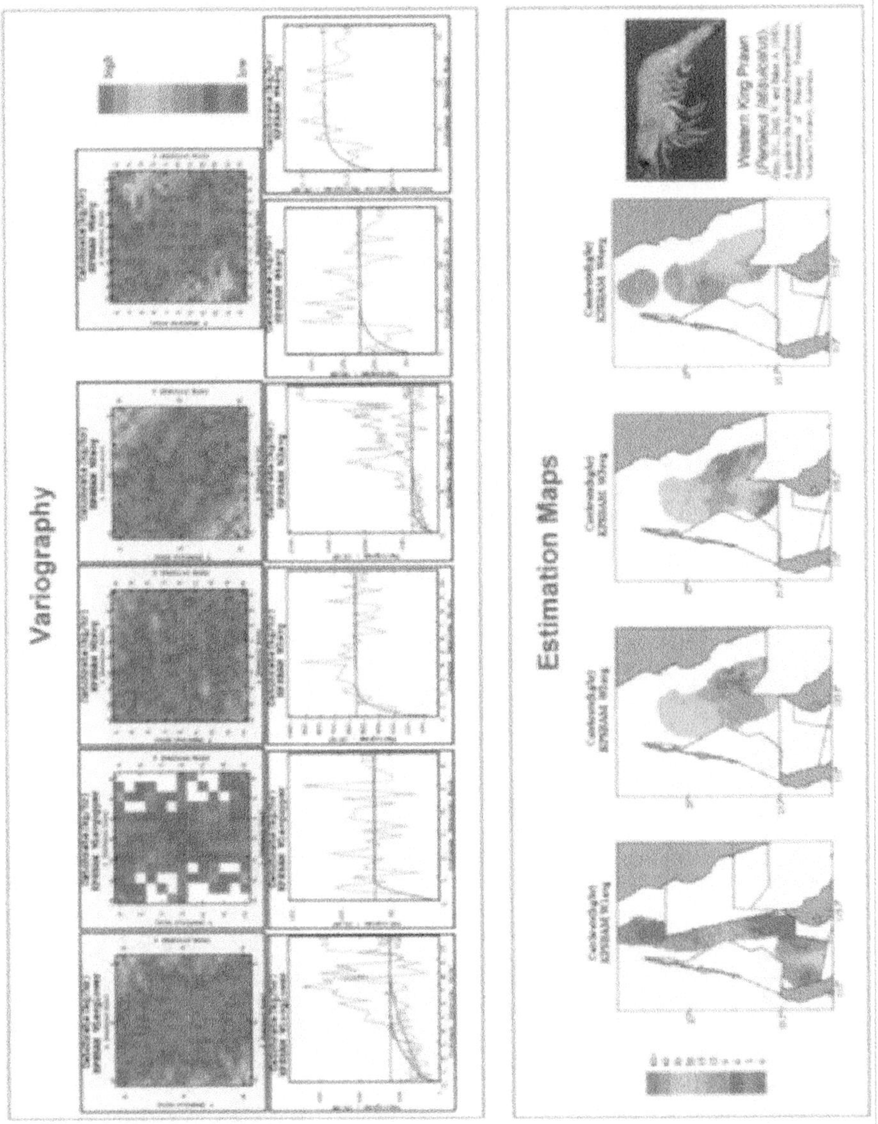
Variography
Estimation Maps
Western King Prawn

USING GEOSTATISTICS TO ASSESS THE AREA OF SPATIAL REPRESENTATIVITY OF AIR QUALITY MONITORING STATIONS

PERDRIX Esperanza (perdrix@ensm-douai.fr), FOURCHÉ Benoît and PLAISANCE Hervé
Ecole des Mines de Douai, Dépt. Chimie et Environnement, 941 rue Charles Bourseul, BP 838, 59508 Douai cedex, France

ABSTRACT: In the field of air quality, common pollutants are continuously monitored by way of a limited number of fixed stations. In the case of nitrogen dioxide (NO_2), the automatic monitoring stations are generally spread across metropolitan areas, giving priority to densely populated zones. The knowledge of the area of spatial representativity of the stations may be useful to position the analysers in an optimal manner, avoiding redundant sites and unnecessary expenditures. Taking profit of temporary, consecutive measurement campaigns by NO_2 passive sampling (carried out with more than a hundred sampling sites), we applied kriging in order to perform NO_2 pollution maps. Then estimated concentrations were screened in order to get the locations where estimations are always (whatever the campaign) equal to the concentrations measured by a given fixed station. The selected locations allow to draw up the area of spatial representativity of a fixed station.

INTRODUCTION

In the field of ambient air quality, common pollutants are continuously monitored by way of a limited number of fixed stations. The question is how to locate the fixed stations in order to have the measurements which are representative of pollution from the maximum area with the minimum of analysers. One way to solve this problem is to assess the area of representativity of an analyser.

This can be done by using the results of temporary measurement campaigns, run in addition to the continuous automatic monitoring. During a campaign a large number of diffusive samplers are spread all over the area of interest (**Fig. 1**). Usually campaigns are successively conducted in order to get various atmospheric conditions. The datasets are large enough to be treated by geostatistical techniques in order to perform pollution maps. According to a desired precision, a selection interval can be applied to the kriged values of each campaign, in order to reveal the area where the pollutant concentration can be considered as not different to the value given by a fixed station.

The area of spatial representativity of the fixed station is then determined as the intersection of the areas selected for all the campaigns.

Figure 1: Location of the fixed stations and of the temporary samplers in the city of Lille

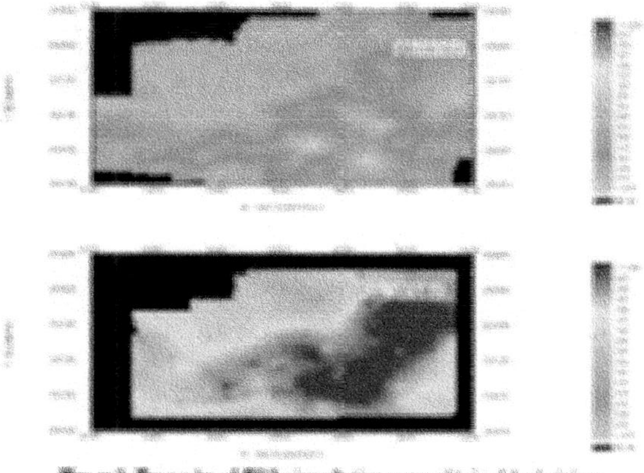

Figure 2. Examples of NO2 air pollution maps obtained by kriging

RESULTS AND DISCUSSION

[illegible]

CONCLUSION

[illegible]

MATERIALS AND METHODS

[illegible]

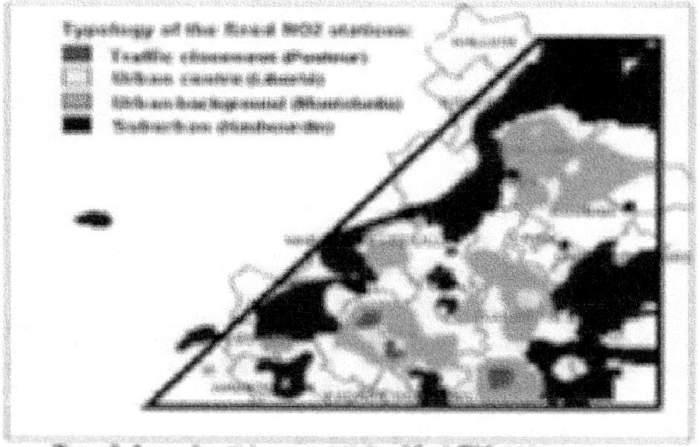

Figure 3. Area of spatial representativity of fixed NO2 stations depending on their typology

[illegible]

ARCHITECTURE OF FAULT ZONES IN A GRANITIC MASSIF DETERMINED FROM OUTCROP, CORES, 3-D SEISMIC TOMOGRAPHY AND GEOSTATISTICAL MODELING

A. Pérez Estaún, D. Martí, R. Carbonell, M.J. Jurado. Instituto Ciencias de la Tierra Jaume Almera, CSIC Barcelona. **J. Escuder Viruete** Universidad Complutense de Madrid

1. INTRODUCTION. Brittle fault zones are lithologically heterogeneous and structurally anisotropic discontinuities in the Earth upper crust. Evaluating the mesoscopic structure of fault zones at scales between one meter and hundreds of meters is important for a variety of applications, including hydrogeology, hydrocarbon migration, toxic and non-toxic waste isolation, ore deposits, earthquake nucleation and propagation, and the rheological/mecanical behavior of faults. For example, in granitic rocks where the rock-matrix hydraulic conductivity is commonly very low, the main transport of water and other crustal fluids is through fault zones. However, in fluid flow models through fault zones results critical to know the 3-D architecture of fault zones.

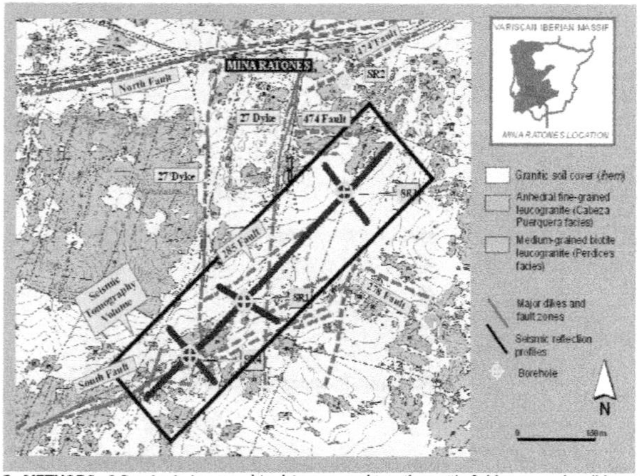

Fig 1. Geological-Structural Map of the Mina Ratones area showing the location of the wells (SR-1 to SR-4) and grid limits of the seismic tomographic experiment.

2. MATERIALS. Following field observations and conceptual models of fault zone architecture, faults can be divided into two distinct components: the fault core, where most of the displacement is accommodated, and the damaged zone, that is mechanically related to the growth of the fault zone. The fault core and the damage zones are surrounded by relatively undeformed rock. From a lithologic, structural, rheological and hydrogeological point of view, fault core, damage zone and undeformed rock are distinct units that reflect the material properties and deformation conditions within a fault zone. This architecture controls the physical properties of fault zones as porosity, permeability, fluid flow, their mechanical behavior and geophysical/geochemical signatures.

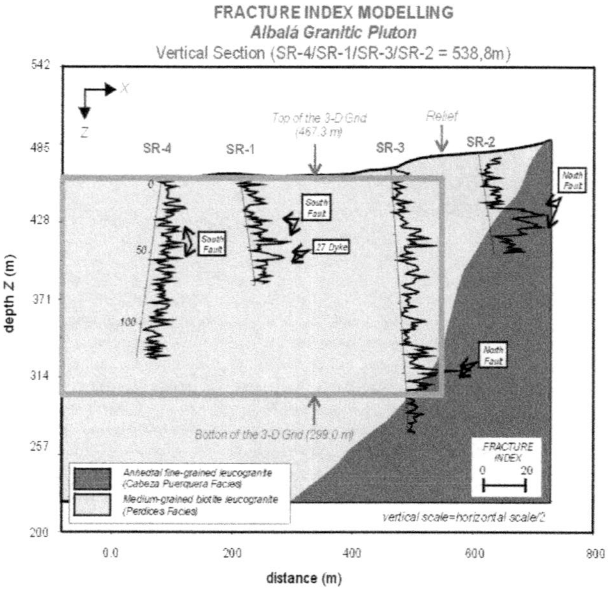

Fig. 2. Well-log data of the Fracture Index (FI)

3. METHODS. 3-D seismic tomographic data are used together with field, core and well-log structural information to determine the detailed 3-D architecture of fault zones in a granitic massif of volume 500575168 m, at Mina Ratones area in the Albalá Granitic Pluton (Cáceres, SW Spain). To facilitate the integration of different data, geostatistical simulation algorithms are applied to interpolate the relatively sparse structural (hard) control data conditioned to abundant but indirect 3-D (soft) seismic tomographic data. To effectively integrate geologic and tomographic data, 3-D migration of the velocity model from the time domain into the depth domain was essential. The resulting 3-D model constitutes an image of the fault zone architecture within the granitic massif that honours hard and soft data and provides an evaluation of the spatial variability of structural heterogeneities based on computation of 3-D experimental variograms of Fracture Index (fault intensity) data. This probabilistic quantitative 3-D model of spatially heterogeneous fault zones is suitable for subsequent fluid flow simulations.

4. RESULTS AND DISCUSSION. The modeled image of the 3-D fault distribution is consistent with the fault architecture in the Mina Ratones area, which basically consists of two families of subvertical structures with NNE-SSW and ENE-WSW trends, that displace the surfaces of low-angle faults (North Fault) and follows their seismically detected staircase geometry. These brittle structures cut two subvertical dykes (27 and 27' Dykes) with a NNE-SSW to N-S trend. The faults present high FI (FI>12) adjacent bands of irregular geometry in detail that intersects in space delimiting rombohedral blocks of relatively less fractured granite (FI<6). Both structural domains likely correspond with the protolith and the damage zone/fault core in the widely accepted model for fault zone architecture. Therefore, the construction of 3-D grids of the FI in granitic areas affected by brittle tectonics permits the quantitative structural characterization of the rock massif.

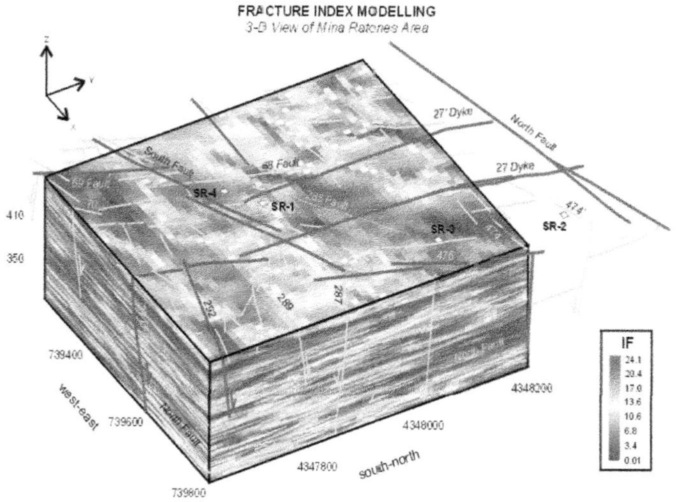

Fig. 4. Fig. 13. Interpreted relationships between fault architecture and the 3-D grid of the FI in the Mina Ratones area obtained from geostatistical conditional simulation (average grid from 10 realizations).

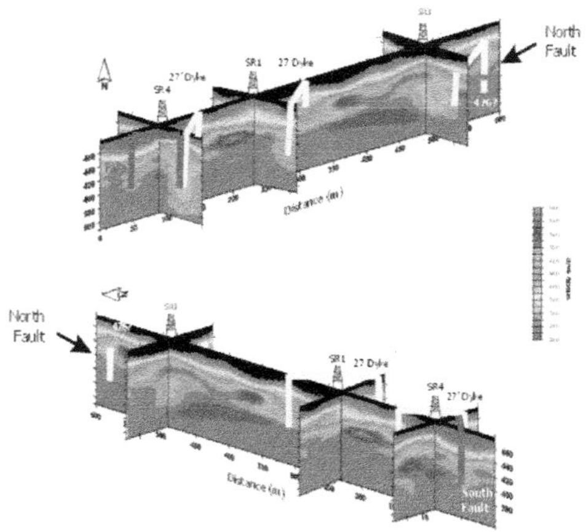

Fig. 3. Tomographic images illustrating the velocity distribution within the block. The low velocity values area is interpreted as altered or fractured zones within the granitic massif. The mined 27 and 27' Dykes, South fault and other high-angle structures can be identified as subvertical low velocity anomalies.

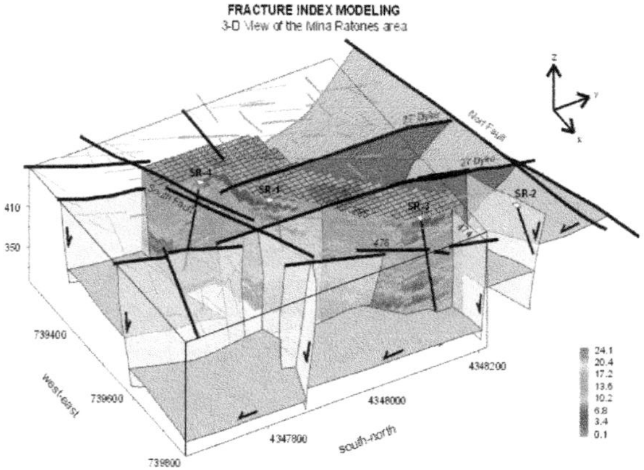

Fig. 5. Interpreted relationships between fault architecture and the 3-D grid (slices) of the FI in the Mina Ratones area. Two structural domains in base to the Fracture Index values are defined: variably irregular fault zones (FI>12) and rhomboidal blocks of relatively undeformed granite located between them (FI<6).

Creation of a Digital Elevation Model of the Water Table in a Sandstone Aquifer

Paulette Posen
School of Environmental Sciences, University of East Anglia, Norwich NR4 7TJ, UK
(E-mail: p.posen@uea.ac.uk)

1. Interpolating water level data

The thickness of the unsaturated zone of aquifers is an important factor in the assessment of groundwater vulnerability. One approach to estimating this thickness is to create a digital elevation model (DEM) of the water table, which can be subtracted from surface topography within a GIS, to create a 'depth to water table' data set. Water level data from a subset of observation boreholes in the Midlands region of the UK have been used to construct DEMs of the water table. The suitability of three different methods of interpolation between data points is investigated (Fig. 1).

Due to the global nature of the spline interpolation method, erratic values or warping of the surface may occur where data are sparse (e.g. in the area around point A, Fig. 1(i)) and artefacts of this distortion appear as 'edge effects' (seen on the southern and eastern edges of the map). Inverse distance weighting (IDW) is an exact local interpolation method, which produces a surface whose value changes smoothly between the data points to which it is tied (Fig. 1(ii)). The extent of smoothing is dependent on the chosen value for the decay parameter. Kriging (Fig. 1(iii)) also interpolates locally, but uses the underlying spatial dependence of the data to calculate the most appropriate value for the decay parameter.

Cross-validation is a method which removes each data point in turn, and interpolates from the remaining points to estimate a value at the corresponding location. A cross-validation of the kriged interpolation (Fig. 3) shows that the estimated water table surface follows the actual surface very closely. The regression analysis indicates that the kriged surface describes 92% of the variability in the actual water table values.

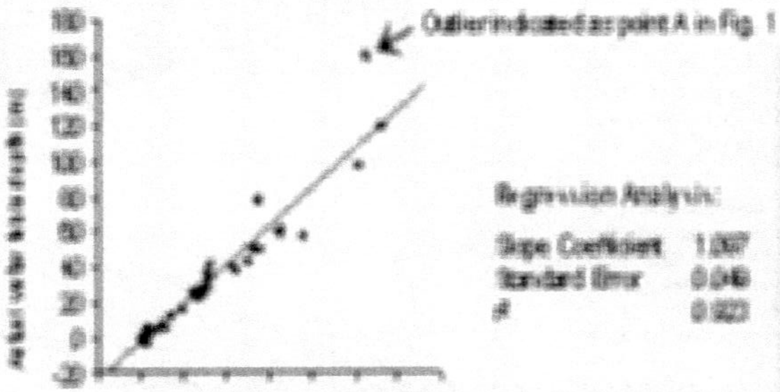

Fig. 3 *Cross-validation of the kriged surface.*

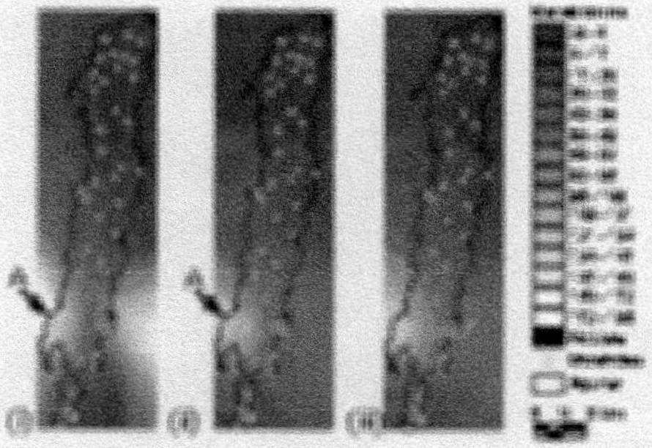

Fig. 1 *Water table surfaces created using (i) spline, (ii) IDW and (iii) kriging interpolation methods. Elevation values are relative to sea level. Methods (i) and (ii) were performed using ArcView GIS¹ and method (iii) was performed using the GS+ program².*

2. Evaluating the surfaces

A simple test for evaluating the effectiveness of an interpolation method is to re-calculate the surface after the removal of one or more significant data points. Fig. 2 shows that IDW provides a much truer interpretation of the surface than the spline method, by following the change in data distribution closely.

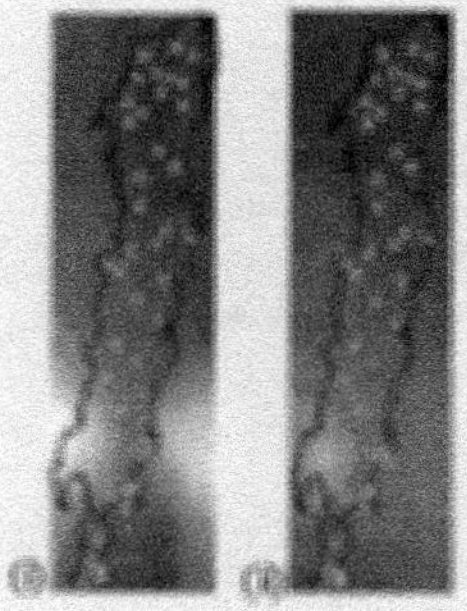

Fig. 2 *Revised water table surface interpolations using (i) spline and (ii) IDW methods after removing point A (Fig. 1) from the data set. It can be seen from (i) that removal of a single peak value has little effect on the spline interpolation. However, the revised IDW interpolation (ii) shows a significant local change in the shape of the surface.*

References: ¹ArcView GIS Version 3.2, http://www.esri.com. ²GS+ Version 5, http://www.gammadesign.com.

This project is part of a Natural Environment Research Council PhD studentship, with CASE partner support from the UK Environment Agency National Groundwater & Contaminated Land Centre, who kindly supplied data used in this study.

3. Interpolation for hydrogeological purposes

The potential for edge effects and erratic values when using spline interpolation makes this method unsuitable for modelling surfaces with non-uniform distribution of data points, such as borehole observation sites. The exact nature of IDW leads to very accurate representation of the water table in areas where data are abundant, but less certainty where data are sparse. The ability of kriging to accurately estimate and take account of the underlying data trend leads to a closer approximation of the true surface in areas where data are lacking. Kriging is therefore thought to be the most suitable interpolation method for hydrogeological purposes.

4. Application of the resultant model

The kriged surface has been used to produce a map of 'depth to water table' (Fig. 4(c)), which clearly shows channels indicative of groundwater in contact with the topographic surface. The channels correspond closely with a digital overlay of the major river network.

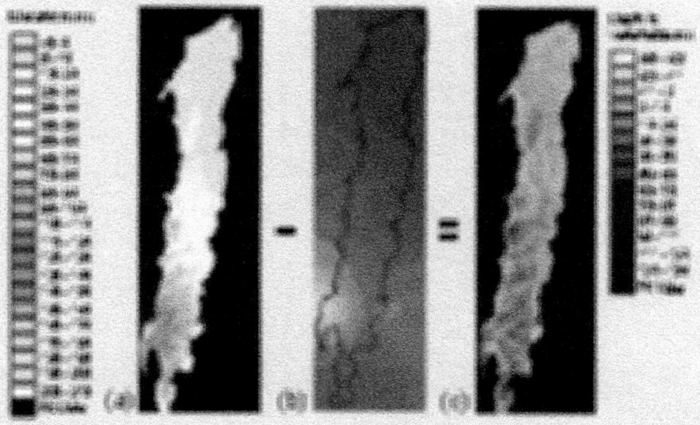

Fig. 4 *The kriged surface (b) was subtracted from the surface topography in the aquifer study area (a) to produce a 'depth to water table' map (c). Negative values indicate surface flow.*

A geostatistically interpolated Digital Elevation Model of Galicia (Northwest Spain)

Ivo Thonon & Antonio Paz
Laboratorio de Edafoloxía, Facultade de Ciencias, Universidade da Coruña, Campus da Zapateira s/n, 15 071, A Coruña, Galicia, España. E-mail: tucho@udc.es

Introduction
Using block kriging and a GIS, we constructed a Digital Elevation Model of Galicia (29,730 km²) on the basis of 118,492 data points.

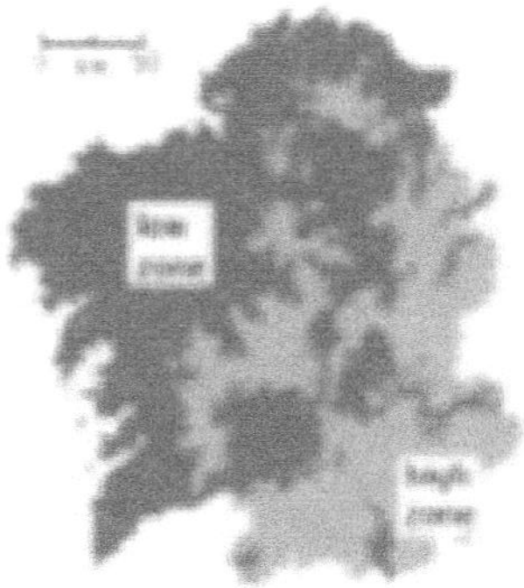

Experimental variograms and variogram models
With the freely available program gstat, we constructed experimental variograms on two different levels, one pair on a regional scale and another on a more local scale (see right). For the lower areas, we detrended the variogram by subtracting a first order trend surface (see lower right). The low zone was fitted by an exponential model (a' = 16734), the high zone variogram was fitted by a spherical + linear model, with ranges of 5000 and 15000 m, respectively.

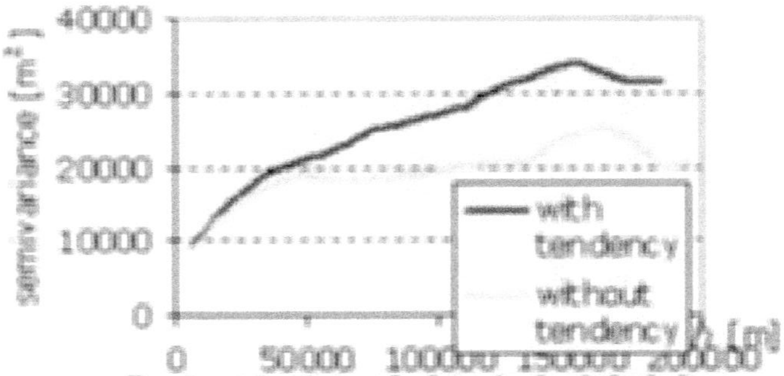

Experimental variograms for the regional scale for the lower zone, both with and without an underlying first order trend.

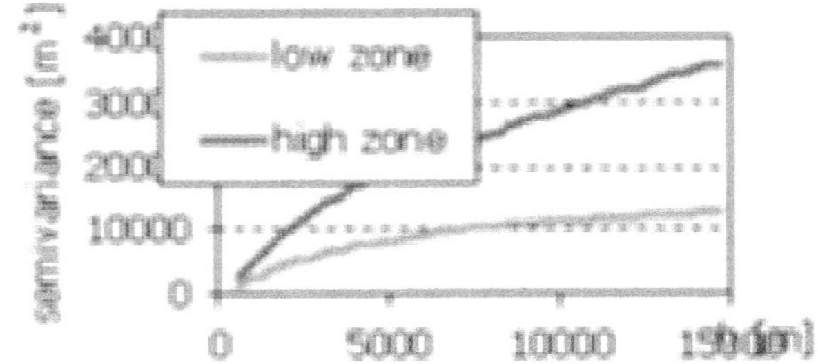

Experimental variograms for the higher and lower zone

Pre-analysis
We checked the data set on the presence of a normal distribution (see right). Next, the study area was split up in two areas, i.e. above and below 550 m a.s.l. (see above) for the construction of experimental variograms. In both subareas, a random sample of 25 % was taken for the construction of the experimental variograms.

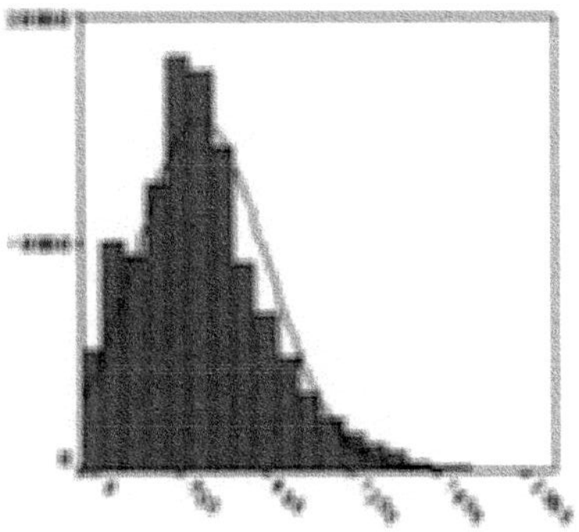

Histogram of the original data set with a Gaussian distribution added for comparison

Conclusions & prospects
The tight coupling of geostatistics (gstat) and GIS (PCRaster) has proven to be very helpful in making a DEM, which can subsequently be used in a wide range of analyses, such as that of the climate and hydrology of Galicia.

Acknowledgements
Funding was kindly provided by the *Xunta de Galicia, Red Temática*, contract number PGIDT-PR426A-02/2.

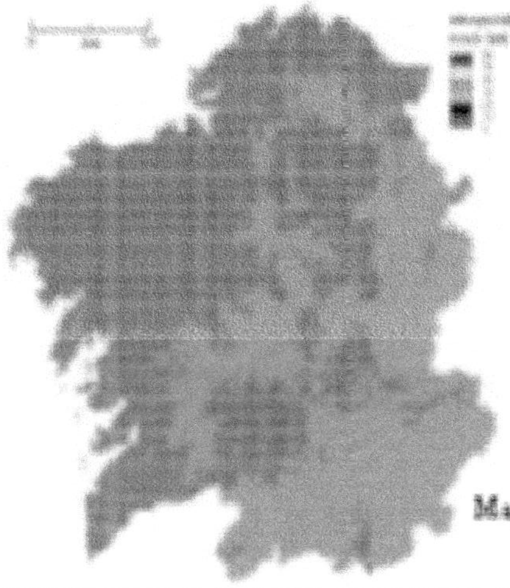

Map of kriging errors

Results
Using the fitted models, we applied stratified block kriging with blocks of 100 by 100 m. The final map was made with PCRaster (see on the right). The maximum interpolation error was 54 m, but in general the errors were much lower (see left map).

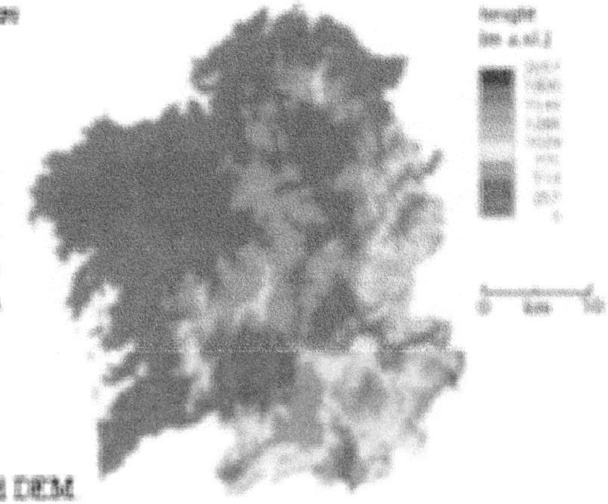

Final interpolated DEM

THE SPATIAL DEPENDENCE OF SOIL SURFACE MICRORELIEF

VIDAL VAZQUEZ, E. & TABOADA CASTRO, M. M.

Facultad de Ciencias. Universidade da Coruña. A Zapateira 15071. A Coruña. Spain

ABSTRACT

Surface microtopography of tilled soils is subject to spatial and temporal changes. The objective of this study was evaluate the spatial dependence of point elevation data on two cultivated soils located at Northwest Spain. Sixty-six microrelief data sets were obtained on bare plots. Each data set consisted of 4624 point elevations on a 20 mm grid. The spatial structure of the surfaces was very continuous and could be modeled by spherical and exponential semivariograms without nugget effect.

INTRODUCTION

Surface micro-topography is an indicator of soil structure and is also a factor in preventing water erosion by providing micro-catchments for rain. On agricultural land soil surface roughness is influenced by several factors such as tillage, vegetation, soil type and previous amount and intensity of rainfall. A greater surface roughness also increases the infiltration capacity. Both processes reduce surface runoff and erosive soil loss.

The properties of the soil surface and plough layer are subject to rapid spatial and temporal changes. Tillage operations result in abrupt changes in roughness, depending on the type of tillage. It is recognised that right after tillage a large roughness is often associated with a high infiltration capacity, so that no runoff is generated. The main objective of this study was to analyse the spatial dependence of soil microrelief at the small scale (1-2 m^2) from point elevation data measured on a 20 mm x 20 mm sampling grid.

MATERIAL AND METHODS

Surface roughness data were obtained from two different experimental fields located at Liñares (Culleredo) and Mabegondo (Abegondo) in A Coruña, Spain. The study period was from December 1998 to September 1999 during which microrelief of 66 plots was determined. Elevation measurements were performed by means of a pin board which allow to take point data along a profile. The profiles were registered by means of photographs using a digital camera. Image analysis was applied to obtain point heights. The Profile Meter Program developed by the USDA-ARS Wind Erosion Research Unit of Kansas State University was utilised for this purpose.

Measurements were performed in different dates, the first one just after tillage and before rain and subsequently with increasing quantities of natural rain. The microrelief of each plot was characterised by 68 profiles separated by 20 mm. Each profile consists of 68 height measurements spaced 20 mm from each other. The total number of sampled points was 4624 inside each plot and the plot area covered was ca. 1.80 m^2.

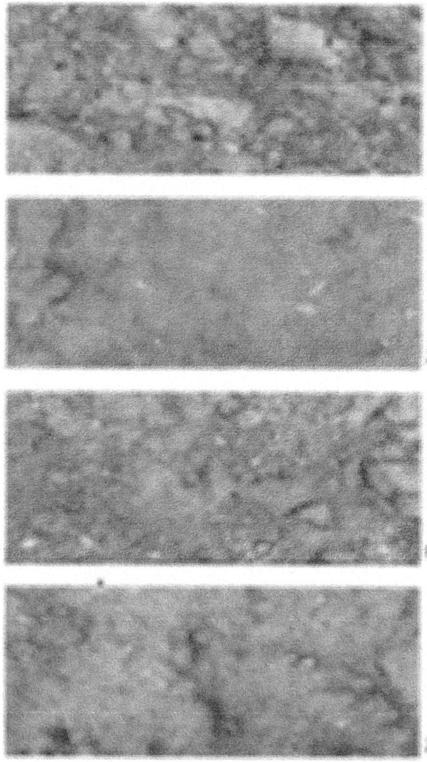

Figure 1.- Initial soil surfaces just after tillage versus surfaces degraded by rainfall.

RESULTS AND DISCUSSION

All the studied surfaces showed a very distinct trend. After a non-deterministic trend removal for withdrawal of slope and tillage mark effects, residual surfaces were obtained, representing the structural units of the soil microrelief as determined by soil properties and management practices.

Semivariogram analysis evidenced a strong spatial dependence for the studied sampling grid. 35 exponential and 11 spherical semivariogram models were fitted to the experimental data of the 46 data sets. All semivariogram models were fitted without nugget effect, indicating a high continuity at small distances for the imposed sampling grid. The autocorrelation range varied between 8 and 70 cm. Because literature on microrelief at the plot or field scale (0.1 ha to 1 km^2) extensively reports the use of Gaussian variogram models, possible explanations for the inability of fitting a Gaussian variogram model by the microrelief data at the studied scale were discussed. The first hypothesis is the large heterogeneity of the soil microrelief, rendering a spatial structure that favours the fitting of exponential and spherical semivariogram models. The second hypothesis is the absence of data points with smaller lags than 200 mm, causing uncertainty of the semivariograms in the nugget region and the possible blurring of a Gaussian spatial structure.

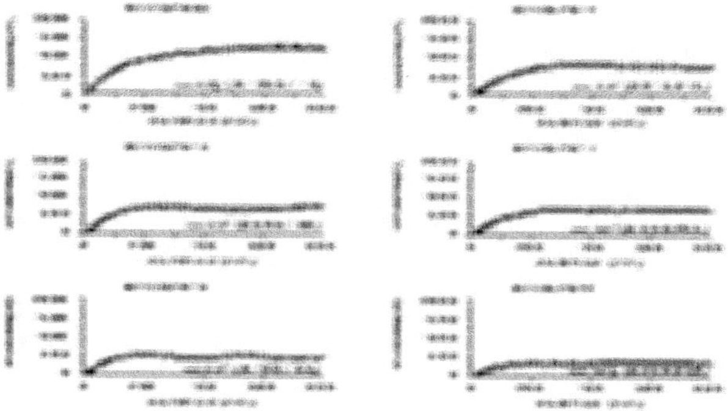

Figure 2.- Soil microrelief variography for different dates and evolution states.

Author Index

Quantitative Geology and Geostatistics

1. F.M. Gradstein, F.P. Agterberg, J.C. Brower and W.S. Schwarzacher: *Quantitative Stratigraphy*. 1985 ISBN 90-277-2116-5
2. G. Matheron and M. Armstrong (eds.): *Geostatistical Case Studies*. 1987 ISBN 1-55608-019-0
3. *Cancelled*
4. M. Armstrong (ed.): *Geostatistics*. Proceedings of the 3rd International Geostatistics Congress, held in Avignon, France (1988), 2 volumes. 1989 Set ISBN 0-7923-0204-4
5. A. Soares (ed.): *Geostatistics Tróia* '92, 2 volumes. 1993 Set ISBN 0-7923-2157-X
6. R. Dimitrakopoulos (ed.): *Geostatistics for the Next Century*. 1994 ISBN 0-7923-2650-4
7. M. Armstrong and P.A. Dowd (eds.): *Geostatistical Simulations*. 1994 ISBN 0-7923-2732-2
8. E.Y. Baafi and N.A. Schofield (eds.): *Geostatistics Wollongong* '96, 2 volumes. 1997 Set ISBN 0-7923-4496-0
9. A. Soares, J. Gómez-Hernandez and R. Froidevaux (eds.): *geoENV I - Geostatistics for Environmental Applications*. 1997 ISBN 0-7923-4590-8
10. J. Gómez-Hernandez, A. Soares and R. Froidevaux (eds.): *geoENV II - Geostatistics for Environmental Applications*. 1999 ISBN 0-7923-5783-3
11. P. Monestiez, D. Allard and R. Froidevaux (eds.): *geoENV III - Geostatistics for Environmental Applications*. 2001 ISBN 0-7923-7106-2; Pb 0-7923-7107-0
12. M. Armstrong, C. Bettini, N. Champigny, A. Galli and A. Remacre (eds.): *Geostatistics Rio 2000*. 2002 ISBN 1-4020-0470-2
13. X. Sanchez-Vila, J. Carrera and J.J. Gómez-Hernández (eds.): *geoENV IV - Geostatistics for Environmental Applications*. 2004 ISBN 1-4020-2007-4; Pb 1-4020-2114-3

KLUWER ACADEMIC PUBLISHERS – DORDRECHT / BOSTON / LONDON

GPSR Compliance

The European Union's (EU) General Product Safety Regulation (GPSR) is a set of rules that requires consumer products to be safe and our obligations to ensure this.

If you have any concerns about our products, you can contact us on ProductSafety@springernature.com

In case Publisher is established outside the EU, the EU authorized representative is:

Springer Nature Customer Service Center GmbH
Europaplatz 3
69115 Heidelberg, Germany

Batch number: 10755248

Printed by Printforce, the Netherlands